Annotated

Contemporary Business Mathematics
for Colleges

14e

Annotated Instructor's Edition

Contemporary Business Mathematics
for Colleges

James E. Deitz, Ed.D.
Past President of Heald Colleges

James L. Southam, Ph.D.
San Francisco State University

THOMSON
SOUTH-WESTERN

Australia · Brazil · Canada · Mexico · Singapore · Spain · United Kingdom · United States

Contemporary Business Mathematics for Colleges, Fourteenth Edition
Annotated Instructor's Edition
James E. Deitz, Ed.D. and James L. Southam, Ph.D.

VP/Editorial Director:
Jack W. Calhoun

VP/Editor-in-Chief:
Alex von Rosenberg

Senior Acquisitions Editor:
Charles McCormick

Senior Developmental Editor:
Alice Denny

Marketing Manager:
Larry Qualls

Production Project Manager:
Magaret M. Bril

Manager of Technology, Editorial:
Vicky True

Technology Project Editor:
Chris Wittmer

Web Coordinator:
Scott Cook

Manufacturing Coordinator:
Diane Lohman

Production House:
Pre-Press Company

Printer:
Quebecor World
Versailles, Kentucky

Art Director:
Stacy Jenkins Shirley

Internal Designer:
Grannan Graphic Design, Ltd.

Cover Designer:
Grannan Graphic Design, Ltd.

Cover Photo Images:
© Getty Images

Photography Manager:
John Hill

Photo Researcher:
Rose Alcorn

COPYRIGHT © 2006
Thomson South-Western, a part of The Thomson Corporation. Thomson, the Star logo, and South-Western are trademarks used herein under license.

Printed in the United States of America
1 2 3 4 5 08 07 06 05

Annotated Instructor's Edition: ISBN 0-324-31810-3

Annotated Instructor's Edition with CD: ISBN 0-324-31809-X

ALL RIGHTS RESERVED.
No part of this work covered by the copyright hereon may be reproduced or used in any form or by any means—graphic, electronic, or mechanical, including photocopying, recording, taping, Web distribution or information storage and retrieval systems, or in any other manner—without the written permission of the publisher.

For permission to use material from this text or product, submit a request online at
http://www.thomsonrights.com.

Library of Congress Control Number: 2005923758

For more information about our products, contact us at:
Thomson Learning Academic Resource Center
1-800-423-0563

Thomson Higher Education
5191 Natorp Boulevard
Mason, OH 45040
USA

To the Instructor

Contemporary Business Mathematics for Colleges presents an arithmetic-based, basic approach to business mathematics. It emphasizes a practical, skill-building approach to prepare students for future careers in business through step-by-step development of concepts, numerous practice exercises, and a focus on real-world application of techniques. The text progresses from the most basic to more complex business mathematics topics.

During its previous editions, *Contemporary Business Mathematics for Colleges* has sold more copies than any other business mathematics textbook. The goal of this new fourteenth edition is to make a successful book even better. This edition is shorter and more focused, yet still maintains its coverage of practical, real-world, business math problems and offers step-by-step solutions to help your students solve these problems. The new edition content is focused entirely on business mathematics with an eye toward the needs of today's business students as well as the requirements of shorter regular and online courses. *Contemporary Business Mathematics for Colleges* presents the basic principles of mathematics and immediately applies them in a series of practical business problems. This new edition is designed to provide a balance among conceptual understanding, skill development, and business applications.

In the business world, everyone (employees and managers alike) needs knowledge of and skill in business mathematics. While computers and calculators are used for many calculations, it is important to understand the concepts behind mechanical computations. The purpose of the business mathematics course is to increase your students' mathematics knowledge and skill as it applies to many aspects of business, and to help make them more valuable players in the business arena.

NEW TO THIS EDITION

The material in *Contemporary Business Mathematics for Colleges* has been refined throughout and reorganized when necessary to enhance learning. The new edition is significantly shorter and more focused.

Organization: There are now 24 chapters arranged in six parts that are each focused on a particular use of business math. For example, all of the chapters on percentages are together in Part 2, and all chapters concerning interest are together in Part 4. The Brief Course version of the text includes the 16 chapters in the first four parts.

iLrn Business Math Homework: iLrn Business Math Homework facilitates classroom management, finally allowing you to test the way you teach. iLrn Business Math Homework assesses students through homework, as well as on quizzes and exams, in the process of doing real data analysis on the Web. Student responses get automatically graded and entered into the iLrn grade book, making it easy for you to assign and collect homework over the Web.

Updated and New Examples and Exercises: Much new material appears throughout the text.

New Coverage of the Use of Scientific and/or Business Calculators: Appears in the coverage of compound interest in Chapter 16 and annuities in Chapter 23.

Microsoft® Excel Coverage: Where relevant, problems using templates provided on the Student CD-ROM illustrate how Excel functions can be applied to the chapter topic. The Excel problems are located at the end of chapters so instructors who don't wish to use them may skip them.

KEY FEATURES

Before you begin to teach with the new edition *of Contemporary Business Mathematics for Colleges,* use the visual preface to take a guided tour through the special features.

Integrated Learning Objectives: These icons call out the locations throughout the chapter where each Learning Objective is addressed, helping students to assimilate key topics from the very beginning, throughout each chapter.

Concept Checks: Following each major chapter section, concept checks provide students with the opportunity to immediately assess their understanding and their ability to apply the material they've just learned.

Highly Successful Step-by-Step Problem-Solving Approach: Short, concise text sections are followed by examples with step-by-step solutions. Students learn mathematical concepts by immediately applying practical solutions to common business problems and gain confidence in their own problem-solving skills by studying the way example problems are worked out.

Real-World Examples and Problems: Abundant practical business problems and business examples from a variety of real companies help students relate to the material better as they see how it is applied to everyday life.

Bottom Line: These end-of-chapter features tie each learning objective to self-test problems (with answers). Students have the opportunity to check whether they have mastered the chapter's key skills before moving on to the assignments.

Self-Check Review Problems: Located at the end of each chapter, they provide yet another opportunity for students to test themselves before completing the end-of-chapter assignments. Answers are provided at the end of the text.

Exceptional Method for Solving Word Problems: This proven methodology appears in Chapter 4, and algebra is also introduced at this point early on in the text to give students exposure to this key concept for mastering business math.

Video Icons: Video icons are placed where appropriate throughout the text to direct students to the video clips available on the included Student CD-ROM.

ANCILLARIES FOR TEACHING AND LEARNING

Thomson/South-Western is committed to providing you, our educational partners, with the finest educational resources available. *Contemporary Business Mathematics for Colleges* comes with an integrated teaching package.

Annotated Instructor's Edition (032431809X): The AIE contains the student textbook with marginal teaching notes throughout the chapters as well as answers to all of the end-of-chapter assignments.

Instructor's Resource CD (0324318154): The IRCD contains Microsoft® PowerPoint® Presentation slides of the solutions and teaching transparencies and the Test Bank in Word files.

ExamView (0324318146): A computerized version of the test bank allows the instructor to quickly and efficiently produce professional-quality tests.

Solutions and Teaching Transparencies (032431812X): Solutions to all end-of-chapter assignments and teaching transparencies are an aid to class lectures. The transparencies are available in PowerPoint format on the IRCD.

Topic Review Video (0324318162): The video covers 12 major mathematical concepts and applies them to a series of practical business problems. They can be used in class or for individual review by your students. A digital version of the video segments is included on the Student CD-ROM for easier access.

Printed Test Bank (0324318111): Written by the text authors, the Test Bank provides additional problems and solutions for each chapter. Test Bank questions can be used either as testing material or as homework assignments. Please note that the IRCD has the Test Bank in Word files.

Microsoft® Excel Templates: Spreadsheet templates give students practice with both mathematics and spreadsheet software. The Excel templates were prepared by text authors Deitz and Southam as well as by Adele Stock of Normandale Community College, and are available on the Student CD-ROM.

Student Resource CD-ROM: The Student CD-ROM is packaged with every new text, and includes the Excel templates, digitized Topic Review Video, and the Math in Employment Tests supplementary material for use in class or for review by the individual student.

Product Web Site: The text Web site at http://deitz.swlearning.com provides online quizzes, Internet links for the text, teaching resources, and more. The online quizzes can be assigned as homework and submitted to the instructor for credit or grading, or used as practice before assignments or exams. Some teaching resources can be downloaded directly from this site.

WebTutor™ Advantage on Blackboard® (0324318081) and WebTutor™ Advantage on WebCT™ (0324318073): With WebTutor Advantage's text-specific, preformatted content and total flexibility, you can easily create and manage your own custom course Web site. WebTutor Advantage's course management tool gives you the ability to provide virtual office hours, post syllabi, set up threaded discussions, track student progress with the quizzing material, and much more. Instructors can also access resources for the use of lectures and class preparation. WebTutor Advantage also provides robust communication tools, such as a course calendar, asynchronous discussion, real-time chat, a whiteboard, and an integrated e-mail system. There is also an instructor's resources page with help in using WebCT™ and Blackboard™. For students, WebTutor Advantage offers real-time access to a full array of study tools, including tutorial videos, chapter outlines, summaries, learning objectives, glossary flashcards (with audio), practice quizzes, and Web links. WebTutor Advantage now comes with a daily news feed from NewsEdge, an authoritative source for late-breaking news of interest to you and your students.

ACKNOWLEDGEMENTS

We would like to acknowledge the work of reviewers who provided suggestions about this edition's reorganization and comments about other ways to continue to improve our text.

Joseph Amico, Utica School of Commerce
William Barkemeyer, American Commercial College
Karen Bean, Blinn College

Yvonne Block, College of Lake County
Sharon Brown, Randolph Community College
Randy Burns, Cochise College
Veronica Cook, Austin Community College
Sandra Copa, Anoka-Ramsey Community College
Brian Fink, Danville Area Community College
Amanda Hardin, Mississippi Delta Community College
Steve Hixenbaugh, Mendocino College
Thomas Howlin, Germanna Community College
Jeffrey Kroll, Brazosport College
Anne Leonard, San Jose Valley Community College
Paul Martin, Aims Community College
Rodney Murray, Compton Community College
Cheri Nelson, Northeast Iowa Community College
S. Owens, Blinn College
John Palafox, Ventura College
Charles Shatzer, Solano Community College
Dawn Stevens, Northwest Mississippi Community College–Desoto Center
Philip Walsh, Berks Technical Institute
Theresa Wickstrom, Minneapolis Community and Technical College
Michael Wissen, Northcentral Technical College

We continue to owe a debt to many colleagues for their helpful comments and suggestions in the development of earlier editions of our text. Among them are:

Jimmy Anderson, College of the Albemarle
Ann Aron, Aims Community College
Dale Dean, Athens Technical College
Nellie Edmundson, Miami Dade College
Zona Elkins, Blue Ridge Community College
John Falls, North Central State College
Kay Finlay, Indiana Vocational Technical College
William Foster, Fontbonne University
William Harrison, DeVry Institute of Technology
Dianne Hendrickson, Becker College
Linda Johnson, Northern Illinois University
Elizabeth King, Heald Business College
Estelle Kochis, Suffolk County Community College
Kenneth Larson, LDS Business College
Cheryl Macon, Butler County Community College
Fran March, Chattanooga State Technical Community College
Alan Moggio, Illinois Central College
Russ Nail, Pasco-Hernando Community College
John Northrup, Bismarck State College
Karen O'Rourke, Lane Community College
Carol Perry, Marshall Community and Technical College
Allan Sheets, Indiana Business College
Steven Teeter, Utah Valley State College
Charles Trester, Northeast Wisconsin Technical College
LaVerne Vertrees, St. Louis Community College
Queen Young, DeKalb Technical Institute

We want to recognize Professor Pam Perry of Hinds Community College, who provided great insight into the needs of instructors who teach business math as an online distance learning course, and Jean Hunting, Heald College, Hayward, California, who provided practical insight and current classroom experience in developing text to meet the needs of contemporary students. Our appreciation also goes to verifier Sheila Feeney Viel, who is a CPA and a Lecturer at the University of Wisconsin, Milwaukee. She checked the text, test bank, and materials for the WebTutor product. We also thank the staff at Thomson/South-Western who worked to make this new edition the best business mathematics text possible: production editor Marge Bril, senior acquisitions editor Charles McCormick, Jr., senior marketing manager Larry Qualls, and senior developmental editor Alice Denny.

James E. Deitz
James L. Southam

Dear Colleague:

In the midst of today's progress and breakthroughs in business technology, success in business math still continuously comes back to a basic understanding of the principles. That's why the latest edition of *Contemporary Business Mathematics for Colleges* helps you take students back to the best of the basics with a proven presentation and complete support.

Time and time again, dedicated instructors like you have turned to the reliability of our text for a blend of contemporary issues in business and the principles most important for business math success. In fact, throughout 13 previous editions, this text has trained more students for business math success than any other book.

Now, this streamlined edition brings only the best from years of success. You'll find a shorter text, more focused on what's most important for your students with both proven and new strengths.

- **New** *iLrn Business Math Homework* provides a complete software system for creating homework and other assessments, efficiently grading text assignments, and posting grades using an electronic gradebook.
- **Step-by-step presentation** – a hallmark that has defined this text's success – guides students through building the skills for business math proficiency with immediate opportunities to apply the knowledge they're gaining.
- **Learning features** like *Concept Checks*, *Learning Objectives*, *Bottom Line* and numerous examples and exercises direct and verify student progress.
- **New Excel coverage**, new organization, video support, and a complete support package ensure reliability and the resources for progress.

Choose from either a full or brief version for the *Contemporary Business Mathematics for Colleges, 14e* option that best meets the needs of your course. We look forward to bringing you and your students the best of the basics for a solid understanding and foundation for future business math success.

Sincerely,

James E. Deitz

James E. Deitz

James L. Southam

James L. Southam

Bringing You the Best of the Basics

Clear Focus on the Concepts Behind the Computations

Prepare today's students with the practical, contemporary math skills they need to build their future career success in business. *Contemporary Business Mathematics for Colleges, 14e* provides a proven, arithmetic-based basic approach to business math. The text's step-by-step development of concepts is reinforced with numerous practical exercises and a focus on the real-world application of techniques.

Now Streamlined for Success

Take a look for yourself at the business advantages and solid foundation the streamlined edition of this proven text offers you and your students in the pages that follow. This edition focuses on what's most important for a clear, more inviting presentation that progresses from the most basic to more complex business math topics. Learning features and resources that have worked well in the past combine with new Excel coverage, new *iLrn Business Math Homework* software, instructional videos, and other exciting improvements to guide learning.

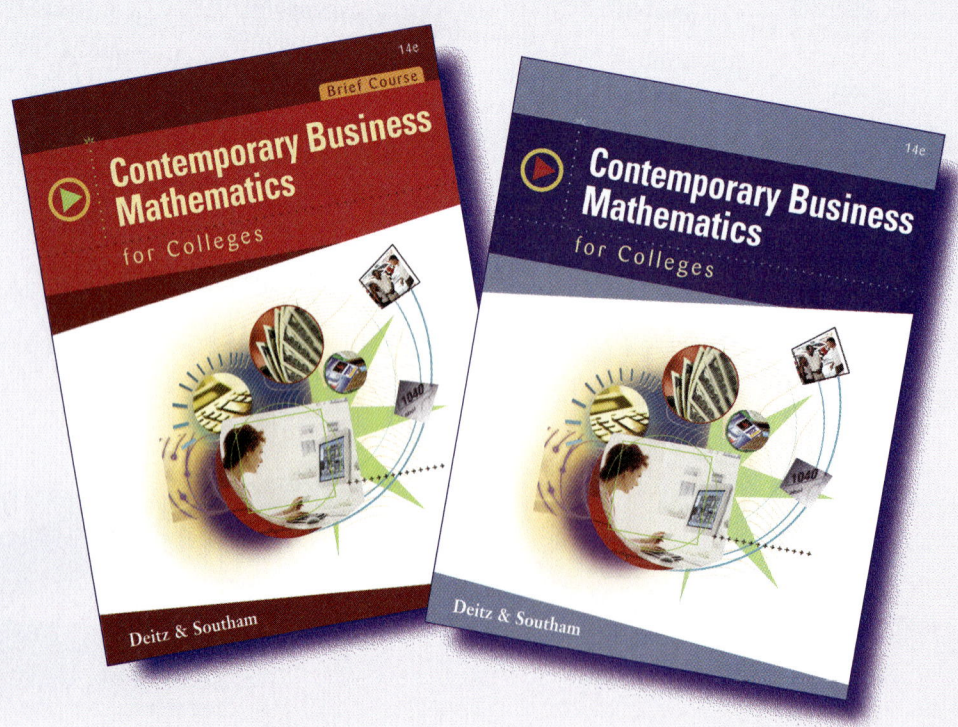

Back to the Basics

Discover the tried and true approach that's brought more students to business math success throughout the decades than any other business math text.

HIGHLY SUCCESSFUL, STEP-BY-STEP, PROBLEM-SOLVING APPROACH –
Short, concise instructional sections are followed by examples with step-by-step solutions.

- Students learn mathematical concepts by immediately applying practical solutions to common business problems.
- Worked-out, clear examples help build student confidence in their own understanding and their problem-solving skills.

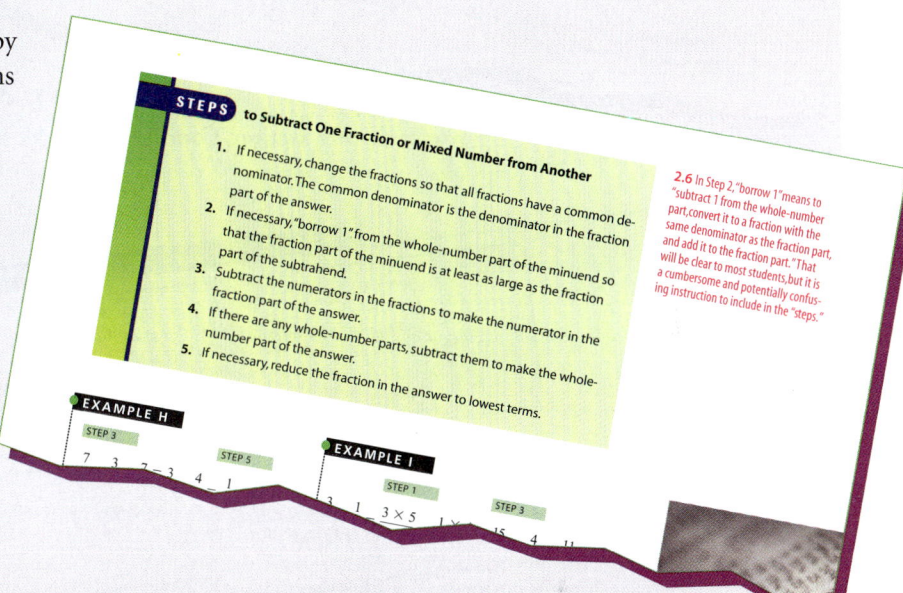

FOCUSED, INTEGRATED LEARNING OBJECTIVES –
Clear *Learning Objectives* are now more focused with *icons throughout the chapter* that indicate where *Learning Objectives* are discussed within the text to help students recognize and assimilate key topics.

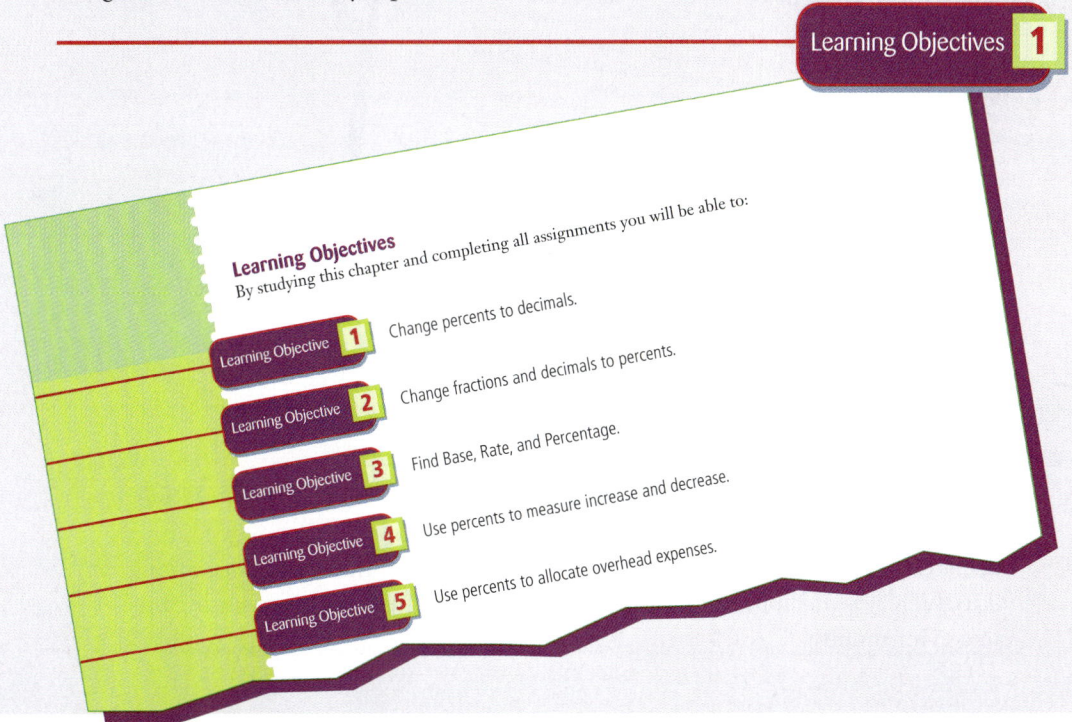

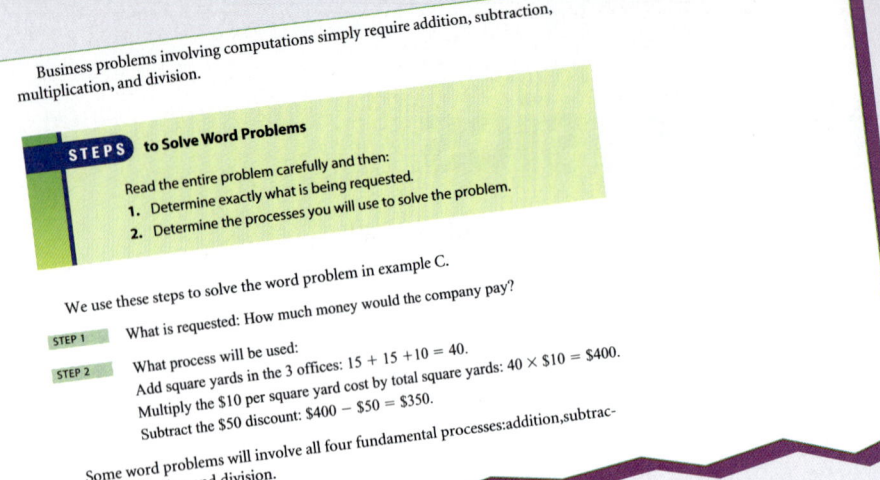

Business problems involving computations simply require addition, subtraction, multiplication, and division.

STEPS to Solve Word Problems

Read the entire problem carefully and then:
1. Determine exactly what is being requested.
2. Determine the processes you will use to solve the problem.

We use these steps to solve the word problem in example C.

STEP 1 What is requested: How much money would the company pay?

STEP 2 What process will be used:
Add square yards in the 3 offices: $15 + 15 + 10 = 40$.
Multiply the $10 per square yard cost by total square yards: $40 \times \$10 = \$400$.
Subtract the $50 discount: $\$400 - \$50 = \$350$.

Some word problems will involve all four fundamental processes: addition, subtraction, multiplication, and division.

EXCEPTIONAL METHOD FOR SOLVING WORD PROBLEMS –
This text's proven methodology introduces word problems and the use of algebra early in the text, giving students the ability to master this key concept in business math.

CONCEPT CHECKS –
Following each major chapter section, *Concept Checks* let students immediately assess their understanding and ability to apply the material they've just learned.

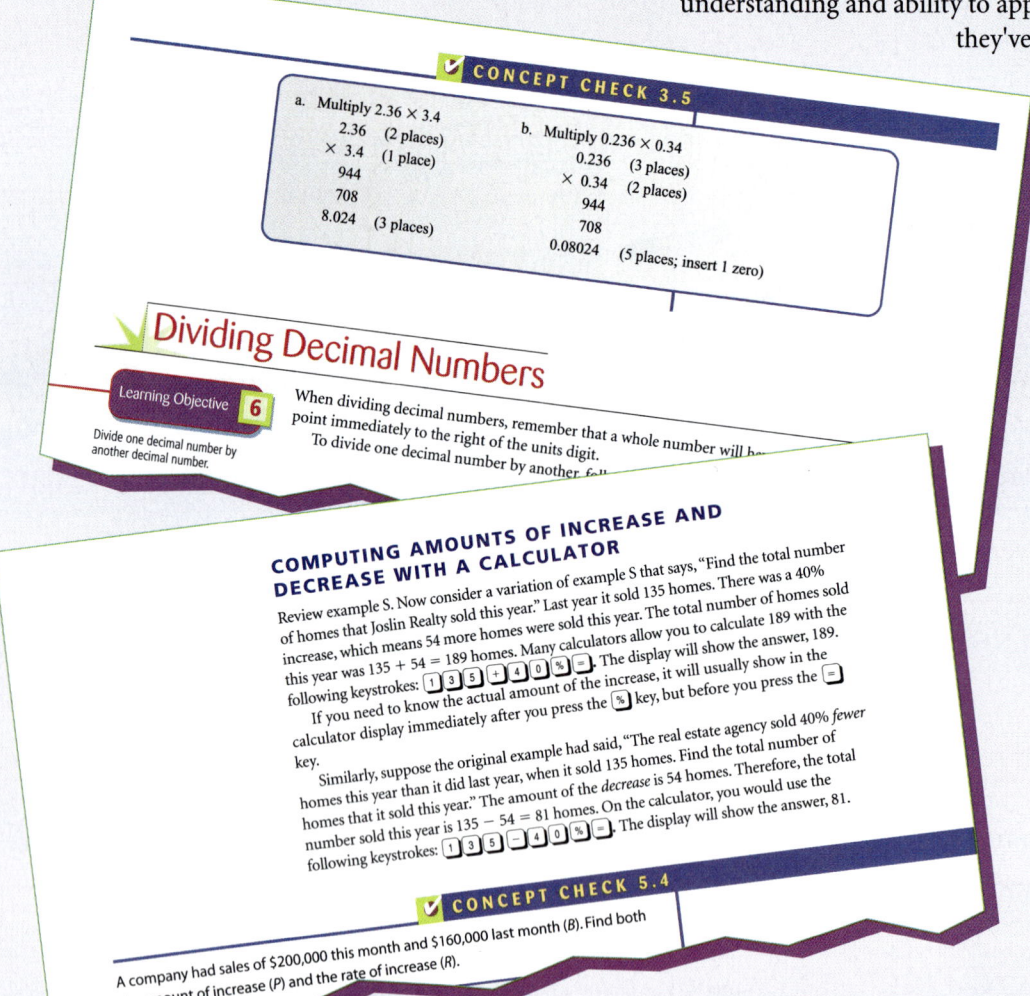

CONCEPT CHECK 3.5

a. Multiply 2.36×3.4
$$\begin{array}{r} 2.36 \text{ (2 places)} \\ \times\ 3.4 \text{ (1 place)} \\ \hline 944 \\ 708 \\ \hline 8.024 \text{ (3 places)} \end{array}$$

b. Multiply 0.236×0.34
$$\begin{array}{r} 0.236 \text{ (3 places)} \\ \times\ 0.34 \text{ (2 places)} \\ \hline 944 \\ 708 \\ \hline 0.08024 \text{ (5 places; insert 1 zero)} \end{array}$$

Dividing Decimal Numbers

Learning Objective 6 — Divide one decimal number by another decimal number.

When dividing decimal numbers, remember that a whole number will have a decimal point immediately to the right of the units digit. To divide one decimal number by another, fo...

COMPUTING AMOUNTS OF INCREASE AND DECREASE WITH A CALCULATOR

Review example S. Now consider a variation of example S that says, "Find the total number of homes that Joslin Realty sold this year." Last year it sold 135 homes. There was a 40% increase, which means 54 more homes were sold this year. The total number of homes sold this year was $135 + 54 = 189$ homes. Many calculators allow you to calculate 189 with the following keystrokes: [1][3][5][+][4][0][%][=]. The display will show the answer, 189.

If you need to know the actual amount of the increase, it will usually show in the calculator display immediately after you press the [%] key, but before you press the [=] key.

Similarly, suppose the original example had said, "The real estate agency sold 40% *fewer* homes this year than it did last year, when it sold 135 homes. Find the total number of homes that it sold this year." The amount of the *decrease* is 54 homes. Therefore, the total number sold this year is $135 - 54 = 81$ homes. On the calculator, you would use the following keystrokes: [1][3][5][-][4][0][%][=]. The display will show the answer, 81.

CONCEPT CHECK 5.4

A company had sales of $200,000 this month and $160,000 last month (B). Find both the amount of increase (P) and the rate of increase (R).

NEW FOCUS ON HOW TO MOST EFFECTIVELY USE CALCULATORS –
Wherever applicable, special tips covering how to use calculators walk students, step-by-step, through specific formulas and calculations for useful real world applications such as compounding interest or computing the value of an annuity.

THE BOTTOM LINE – These end-of-chapter summaries – enhanced in this edition – connect *Learning Objectives* to self-test problems, allowing students to check their mastery of the chapter's key skills before completing the assignments.

SELF-CHECK REVIEW PROBLEMS – These valuable self-check problems at the end of each chapter provide additional opportunities for students to test themselves before completing end-of-chapter assignments. Answers at the end of the text allow students to easily check their own progress before moving on.

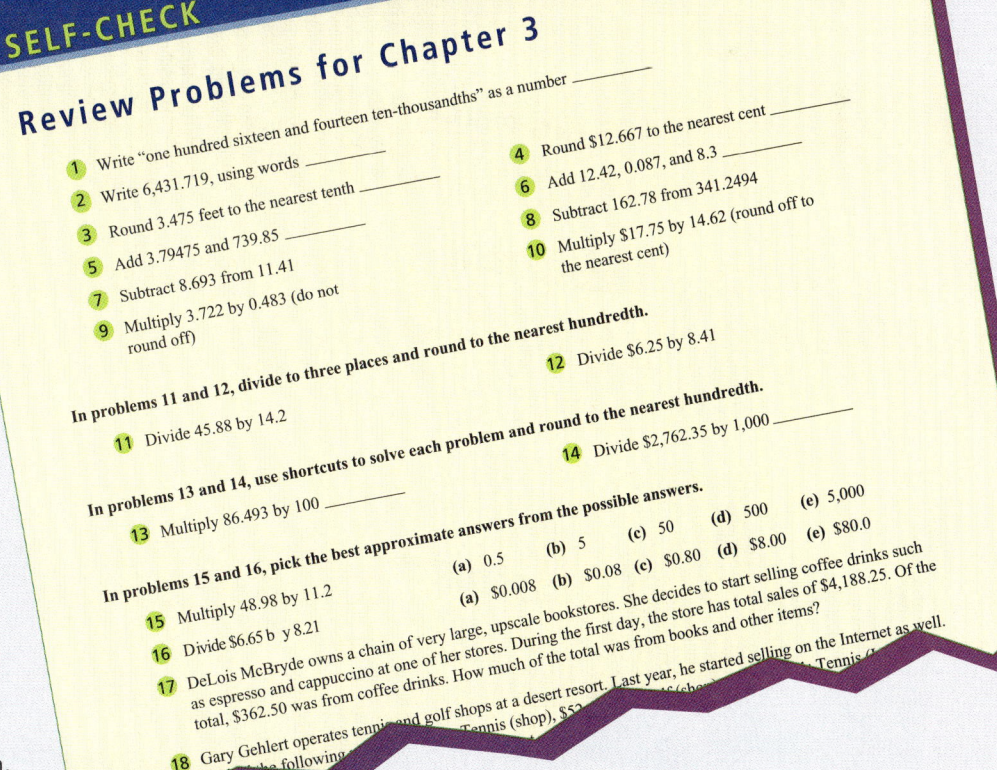

THE BOTTOM LINE

Summary of chapter learning objectives:

Learning Objective	Example
3.1 Read decimal numbers	1. Write 8.427, using words. 2. Write forty-one and eleven ten-thousandths, using digits.
3.2 Round decimal numbers	3. Round 0.506489 to the nearest thousandth (that is, to three decimal places). 4. Round up 13.26012 to the next hundredth (that is, to two decimal places).
3.3 Add two or more decimal numbers	5. Add 82.9, 14.872, and 2.09.
3.4 Subtract one decimal number from another	6. Subtract 14.5977 from 19.34.
3.5 Multiply two decimal numbers	7. Multiply: 4.68 × 3.5

8. Di...

SELF-CHECK

Review Problems for Chapter 3

1. Write "one hundred sixteen and fourteen ten-thousandths" as a number _____
2. Write 6,431.719, using words _____
3. Round 3.475 feet to the nearest tenth _____
4. Round $12.667 to the nearest cent _____
5. Add 3.79475 and 739.85 _____
6. Add 12.42, 0.087, and 8.3 _____
7. Subtract 8.693 from 11.41 _____
8. Subtract 162.78 from 341.2494 _____
9. Multiply 3.722 by 0.483 (do not round off) _____
10. Multiply $17.75 by 14.62 (round off to the nearest cent) _____

In problems 11 and 12, divide to three places and round to the nearest hundredth.

11. Divide 45.88 by 14.2
12. Divide $6.25 by 8.41

In problems 13 and 14, use shortcuts to solve each problem and round to the nearest hundredth.

13. Multiply 86.493 by 100 _____
14. Divide $2,762.35 by 1,000 _____

In problems 15 and 16, pick the best approximate answers from the possible answers.

(a) 0.5 (b) 5 (c) 50 (d) 500 (e) 5,000
(a) $0.008 (b) $0.08 (c) $0.80 (d) $8.00 (e) $80.0

15. Multiply 48.98 by 11.2
16. Divide $6.65 b y 8.21
17. DeLois McBryde owns a chain of very large, upscale bookstores. During the first day, the store has total sales of $4,188.25. Of the total, $362.50 was from coffee drinks. She decides to start selling coffee drinks such as espresso and cappuccino at one of her stores. How much of the total was from books and other items?
18. Gary Gehlert operates tennis and golf shops at a desert resort. Last year, he started selling on the Internet as well.

VIDEO ICONS – Icons throughout the text direct students to related instructional video segments on the accompanying CD-ROM that further clarify concepts.

VIDEO
Markup Based on Cost/Selling Price

REAL WORLD EXAMPLES AND PROBLEMS – Abundant practical business problems and business examples profile a variety of real experiences to help students relate to the material and see the value of math principles applied to everyday life.

USING PERCENTS IN BUSINESS

Percent problems occur frequently in business. Examples Q and R are typical fundamental applications, in which we solve for the Base (*B*) amount and the Rate (*R*), respectively.

EXAMPLE Q

Lena Hoover is a financial analyst. In December, she received a $600 bonus, which equaled 15% of her monthly salary. What was her monthly salary?

P = amount of bonus = $600
R = rate of bonus = 15%
B = monthly salary = ?
As $P \div R = B$,
$P \div R = \$600 \div 15\% = \$600 \div 0.15 = \$4{,}000$ monthly salary

EXAMPLE R

Last year Bayside Coffee Shop had total expenses of $300,000. Of that total, $210,000 was the expense for employee salaries. At Bayside, employee salary expense is what percent of total expenses?

P = employee salaries = $210,000
R = ?
B = total expenses = $300,000

Moving Forward with a Focus on the Future

Dynamic changes and a streamlined focus throughout this new edition clarify instruction for students and provide the contemporary connections that prepare them for business career success.

New *iLrn Business Math Homework*

This new online tool allows you to easily assign and automatically grade all assignments from the text. An online gradebook saves you time and allows you to assess class progress and adjust your presentations accordingly. See page ix for more details.

NEW MICROSOFT® EXCEL COVERAGE – Wherever relevant, end-of-chapter problems show how Excel functions apply to chapter content and give students important hands-on practice.

- Excel templates included on the Student CD-ROM save valuable time.
- Answers to all Excel problems are available on the Student CD-ROM so students can check their understanding.
- For maximum flexibility, all Excel material is located at the end of the chapter so you have the option of incorporating it into your course or not – depending on your needs.

NEW EXAMPLES AND EXERCISES – Throughout the text, fresh content and examples reflect the most recent development in business today.

MATH IN EMPLOYMENT TESTS – This supplement, available on the Student CD-ROM, may be used within your course or independently by the student when needed to prepare for future employment.

NEW ORGANIZATION – This new edition is significantly shorter with a refined, more focused approach to enhance student understanding.

Table of Contents

BRIEF CONTENTS

Part 1: Fundamental Review
1. Fundamental Processes
2. Fractions
3. Decimals
4. Word Problems and Equations

Part 2: Percentage Applications
5. Percents
6. Commissions
7. Discounts
8. Markup

Part 3: Accounting Applications
9. Banking
10. Payroll
11. Taxes
12. Insurance

Part 4: Interest Applications
13. Simple Interest
14. Installment Purchases
15. Promissory Notes and Discounting
16. Compound Interest

Part 5: Business Applications
17. Inventory and Turnover
18. Depreciation
19. Financial Statements
20. International Business

Part 6: Corporate and Special Applications
21. Corporate Stocks
22. Corporate and Government Bonds
23. Annuities
24. Business Statistics

Brief Edition includes Chapters 1-16.

The reorganized, shortened text includes 24 chapters arranged in six parts, each focused on a particular use of business math.

Chapters covering related topics are grouped together within each part to streamline understanding.

Students review fundamentals early to establish a strong understanding of the basics. A proven strategy for mastering Word Problems is introduced early in the text (Ch. 4).

Optional Excel problems at the end of chapters gives you the choice of incorporating this important business application into your course.

Dynamic content, streamlined and tightened throughout, focuses on mastery of the basic math principles most critical to student success without the distraction of extraneous materials.

New content on how to use calculators in today's business world most effectively prepares students to maximize this critical mathematical tool.

Once part of the main text, the "*Math in Employment Tests*" supplement on the Student CD provides students with key questions for use in class or on their own.

Technology and Tools for Today's Times

Comprehensive tools help you prepare and students learn.

This edition's comprehensive support package stands alone with traditional and technology-driven tools designed to enhance class lectures, streamline your preparation time, and further clarify student understanding to make your job easier and more rewarding.

iLrn BUSINESS MATH HOMEWORK (ISBN 0-324-31817-0)

Managing the growing size of Business Math courses in the face of fewer budgetary resources is a major challenge facing Business Math instructors. So the better your tools, the easier your job becomes. Use this online tool to automate your homework process. This program includes most of the textbook exercises and helps you efficiently grade all text problems and easily record grades using a simple electronic gradebook.

- The online gradebook not only saves you time, but also allows you to assess class progress and adjust your presentations accordingly.
- You can automate and manage your class roster in a few easy steps.
- You have complete control over the exercises to assign. Create homework and other assessments in four simple steps using the Wizard.
- Students' work is graded automatically. You have the option of giving immediate feedback, delayed feedback, or no feedback.
- Class exercises records are available at a glance. You can view any exercise for any student and see both the student's answer and the correct answer. You may also change the score on any one question or the score for an entire exercise.

WEBTUTOR™ ADVANTAGE ON WEBCT® OR BLACKBOARD®

With WebTutor Advantage's text-specific, pre-formatted content and total flexibility, you can easily create and manage your own custom course Web site! WebTutor Advantage's course management tool gives you the ability to provide virtual office hours, post syllabi, set up threaded discussions, track student progress with the quizzing material, and much more.

- For students, WebTutor Advantage offers real-time access to a full array of study tools, including videos, that bring the book's topics to life with chapter outlines, summaries, learning objectives, glossary flashcards (with audio), practice quizzes, Weblinks, and more.
- Instructors can access password-protected resources to create dynamic lectures and minimize class preparation. WebTutor Advantage also provides communication tools, such as a course calendar, asynchronous discussion, real-time chat, a whiteboard, and an integrated e-mail system.

WebTutor™ Advantage for WebCT® (ISBN 0-324-31807-3)

WebTutor™Advantage for Blackboard® (ISBN 0-324-31808-1)

CONTEMPORARY BUSINESS MATHEMATICS FOR COLLEGES, 14E WEB SITE
http://deitz.swlearning.com

This companion Web site provides comprehensive tutorial tools and business links for students.

- Students find Course Resources including Learning Objectives from the book, interactive quizzes, flash cards, and other dynamic learning links that make business math come alive.
- An instructors-only, password-protected site offers downloadable instructor's resources.

STUDENT CD-ROM (ISBN 0-324-31804-9)

This invaluable CD-ROM is available FREE with each new text and can also be purchased separately. The CD includes Topic Review Videos that reinforce 12 key concepts from the text for the visual learner, Excel templates to save time, and the *Math in Employment Tests* supplement for students to use in class or on their own.

ANNOTATED INSTRUCTOR'S EDITION (ISBN 0-324-31809-X)

Find everything you need to create a dynamic learning environment with minimal preparation. Student pages are surrounded by margin notes that provide comprehensive teaching notes. You'll also find answers to all end-of-chapter assignments.

INSTRUCTOR'S RESOURCE CD (ISBN 0-324-31815-4)

This comprehensive all-in-one electronic resource places all key instructor resources at your fingertips, including Solutions and Teaching Slides in Microsoft® PowerPoint® format, Topic Review Videos in digital format, Test Bank questions in Microsoft® Word files, and ExamView® Computerized Test Bank.

TRANSPARENCIES (ISBN 0-324-31812X)

Solutions and Teaching slides in acetate format.

PRINTED TEST BANK (ISBN 0-324-31811-1)

Written by the text authors, this test bank provides problems and solutions for each chapter, with questions that are ideal for tests or homework assignments.

EXAMVIEW® ELECTRONIC TEST BANK (ISBN 0-324-31814-6)

This easy-to-use software allows you to select problems at random or by learning objective. You can customize or add test questions and create multiple versions of the same test.

TOPIC REVIEW VIDEO (ISBN 0-324-31816-2)

The video covers 12 major mathematics concepts and applies them to a series of practical business problems. Video segments are available on the Instructor's Resource CD, the Student CD-ROM and within *WebTutor™ Advantage*.

About the Authors

JAMES E. DEITZ
PAST PRESIDENT OF HEALD COLLEGES

Author **James E. Deitz** brings both a thorough understanding of effective education today and a practical business knowledge to the latest edition of this leading text. Dr. Deitz earned his bachelor's degree in accounting from Memphis State University and doctorate of education from UCLA. Dr. Deitz has been an educator for more than 35 years, including professorships with UCLA and Los Angeles State College and a long-standing position as President of Heald Colleges. An active member of the business community, Dr. Deitz is a recognized international speaker and has served on regional educational accrediting commissions. He has authored several texts in addition to this best-selling *Contemporary Business Mathematics for Colleges*.

JAMES L. SOUTHAM
SAN FRANCISCO STATE UNIVERSITY

Author **James L. Southam** provides a wealth of first-hand knowledge about business throughout the world as well as a strong background in mathematics. With a diversity of business and teaching experience, Dr. Southam holds bachelor's and master's degrees in mathematics education from Southern Oregon College, a doctorate in mathematics from Oregon State University, a master's of business administration in finance from University of California, Berkeley, and a law degree from University of California College of Law. Dr. Southam's 40 years of teaching experience include Southern Oregon College, California State University, Stanislaus, and San Francisco State University. Dr. Southam has led several international business ventures and has served as an international business consultant as well as a successful author. He is also a member of the San Francisco State University Athletics Hall of Fame.

Resource Integration Guide

When you start with a new text or even a new edition of a familiar text, sometimes the amount of change and supplemental material can seem overwhelming. It can be daunting to lay out an entire course, and piece together the ancillaries that fit your particular needs.

We have created this resource guide to help you and your students extract the full value from *Contemporary Business Mathematics for Colleges,* 14e and its supplements. This guide organizes the book's resources to help you plan and conduct your class and evaluate your students' mastery.

PART 1: Fundamental Review
CHAPTER 1: Fundamental Processes

Class Preparation / Lecture Tools	Testing Tools / Course Management	Student Mastery / Homework and Tutorials
Instructor's Annotated Edition with Teaching Notes and Solutions, pages 3–28	**Test Bank,** also available in Word files on the IRCD	**Chapter Learning Objectives,** page 3
PowerPoint® Assignments with Solutions Slides Includes the chapter's five assignments with solutions	**ExamView** ® computerized version of the Test Bank	**The Bottom Line Summary of Learning Objectives,** page 16
Assignments with Solutions Acetates Includes the chapter's five assignments with solutions	**iLrn Business Math Homework** Includes the chapter's five assignments; grades are recorded in instructor's online grade book	**Self-Check Review Problems,** page 18
Topic Review Video • Estimating Answers	**WebTUTOR Advantage** Lecture notes, discussion threads	**Key Terms as Crossword Puzzles** http://deitz.swlearning.com
		Internet Quizzes http://deitz.swlearning.com
		Topic Review Video • Estimating Answers
		WebTUTOR Advantage Lecture notes, discussion threads, flash cards and quizzes

CHAPTER 2: Fractions

Class Preparation / Lecture Tools	Testing Tools / Course Management	Student Mastery / Homework and Tutorials
Instructor's Annotated Edition with Teaching Notes and Solutions, pages 29–46 This material was included in Chapter 5 of the 13e. **PowerPoint® Teaching Slides** Slides 2-1 through 2-4 illustrate using fractions **PowerPoint® Assignments with Solutions Slides** Includes the chapter's two assignments with solutions **Teaching Acetates** Acetates 2-1 through 2-4 illustrate using fractions **Assignments with Solutions Acetates** Includes the chapter's two assignments with solutions **Topic Review Video** • Reducing Fractions to Lowest Terms and Raising Fractions to Higher Terms • Adding and Subtracting Fractions • Multiplying Mixed Numbers with Fractions	**Test Bank,** also available in Word files on the IRCD **ExamView®** computerized version of the Test Bank **iLrn Business Math Homework** Includes the chapter's two assignments; grades are recorded in instructor's online grade book **WebTUTOR Advantage** Lecture notes, discussion threads	**Chapter Learning Objectives,** page 29 **The Bottom Line Summary of Learning Objectives,** page 40 **Self-Check Review Problems,** page 41 **Key Terms as Crossword Puzzles** http://deitz.swlearning.com **Internet Quizzes** http://deitz.swlearning.com **Topic Review Video** • Reducing Fractions to Lowest Terms and Raising Fractions to Higher Terms • Adding and Subtracting Fractions • Multiplying Mixed Numbers with Fractions **WebTUTOR Advantage** Lecture notes, discussion threads, flash cards and quizzes

CHAPTER 3: Decimals

Class Preparation / Lecture Tools	Testing Tools / Course Management	Student Mastery / Homework and Tutorials
Instructor's Annotated Edition with Teaching Notes and Solutions, pages 47–68 This material was included in Chapter 4 of the 13e. **PowerPoint® Teaching Slides** Slide 3-1 illustrates multiplication and division of decimals **PowerPoint® Assignments** Includes the chapter's three assignments with solutions **Teaching Acetates** Acetate 3.1 illustrates multiplication and division of decimals **Assignments with Solutions Acetates** Includes the chapter's three assignments with solutions	**Test Bank,** also available in Word files on the IRCD **ExamView** ® computerized version of the Test Bank **iLrn Business Math Homework** Includes the chapter's three assignments; grades are recorded in instructor's online grade book **WebTUTOR Advantage** Lecture notes, discussion threads	**Chapter Learning Objectives,** page 47 **The Bottom Line Summary of Learning Objectives,** page 61 **Self-Check Review Problems,** page 62 **Key Terms as Crossword Puzzles** http://deitz.swlearning.com **Internet Quizzes** http://deitz.swlearning.com **WebTUTOR Advantage** Lecture notes, discussion threads, flash cards and quizzes

CHAPTER 4: Word Problems and Equations

Class Preparation / Lecture Tools	Testing Tools / Course Management	Student Mastery / Homework and Tutorials
Instructor's Annotated Edition with Teaching Notes and Solutions, pages 69–85 This material was included in Chapter 2 of the 13e. **PowerPoint® Assignments with Solutions Slides** Includes the chapter's two assignments with solutions **Assignments with Solutions Acetates** Includes the chapter's two assignments with solutions **Topic Review Video** • Word Problems	**Test Bank,** also available in Word files on the IRCD **ExamView** ® computerized version of the Test Bank **iLrn Business Math Homework** Includes the chapter's two assignments; grades are recorded in instructor's online grade book **WebTUTOR Advantage** Lecture notes, discussion threads	**Chapter Learning Objectives,** page 69 **The Bottom Line Summary of Learning Objectives,** page 78 **Self-Check Review Problems,** page 80 **Key Terms as Crossword Puzzles** http://deitz.swlearning.com **Internet Quizzes** http://deitz.swlearning.com **Topic Review Video** • Word Problems **WebTUTOR Advantage** Lecture notes, discussion threads, flash cards and quizzes

PART 2: Percentage Applications
CHAPTER 5: Percents

Class Preparation / Lecture Tools	Testing Tools / Course Management	Student Mastery / Homework and Tutorials
Instructor's Annotated Edition with Teaching Notes and Solutions, pages 87–106 This material was included in Chapter 6 of the 13e. **PowerPoint® Teaching Slides** Slides 5-1 and 5-2 cover decimals and percents and an example illustrating distribution of overhead costs **PowerPoint® Assignments with Solutions Slides** Includes the chapter's four assignments with solutions **Teaching Acetates** Acetates 5-1 and 5-2 cover decimals and percents and an example illustrating distribution of overhead costs **Assignments with Solutions Acetates** Includes the chapter's four assignments with solutions **Topic Review Video** • Base, Rate, Percentage • Percent of Increase or Decrease	**Test Bank**, also available in Word files on the IRCD **ExamView®** computerized version of the Test Bank **iLrn Business Math Homework** Includes the chapter's four assignments; grades are recorded in instructor's online grade book **WebTUTOR Advantage** Lecture notes, discussion threads	**Chapter Learning Objectives,** page 87 **The Bottom Line Summary of Learning Objectives,** page 97 **Self-Check Review Problems,** page 98 **Key Terms as Crossword Puzzles** http://deitz.swlearning.com **Internet Quizzes** http://deitz.swlearning.com **Topic Review Video** • Base, Rate, Percentage • Percent of Increase or Decrease **WebTUTOR Advantage** Lecture notes, discussion threads, flash cards and quizzes

CHAPTER 6: Commissions

Class Preparation / Lecture Tools	Testing Tools / Course Management	Student Mastery / Homework and Tutorials
Instructor's Annotated Edition with Teaching Notes and Solutions, pages 107–120 This material was included in Chapter 11 of the 13e. **PowerPoint® Teaching Slides** Slide 6-1 illustrates salesperson's commission **PowerPoint® Assignments with Solutions Slides** Includes the chapter's two assignments with solutions **Teaching Acetates** Acetate 6-1 illustrates salesperson's commission **Assignments with Solutions Acetates** Includes the chapter's two assignments with solutions	**Test Bank,** also available in Word files on the IRCD **ExamView®** computerized version of the Test Bank **iLrn Business Math Homework** Includes the chapter's assignments; grades are recorded in instructor's online grade book **WebTUTOR Advantage** Lecture notes, discussion threads	**Chapter Learning Objectives,** page 107 **The Bottom Line Summary of Learning Objectives,** page 114 **Self-Check Review Problems,** page 115 **Key Terms as Crossword Puzzles** http://deitz.swlearning.com **Internet Quizzes** http://deitz.swlearning.com **WebTUTOR Advantage** Lecture notes, discussion threads, flash cards and quizzes

CHAPTER 7: Discounts

Class Preparation / Lecture Tools	Testing Tools / Course Management	Student Mastery / Homework and Tutorials
Instructor's Annotated Edition with Teaching Notes and Solutions, pages 121–138 This material was included in Chapter 12 of the 13e. **PowerPoint® Teaching Slides** 7-1 on Trade Discounts 7-2 and 7-3 on Cash Discounts **PowerPoint® Assignments with Solutions Slides** Includes the chapter's two assignments with solutions **Teaching Acetates** 7-1 on Trade Discounts 7-2 and 7-3 on Cash Discounts **Assignments with Solutions Acetates** Includes the chapter's two assignments with solutions **Topic Review Video** • Cash Discounts	**Test Bank,** also available in Word files on the IRCD **ExamView®** computerized version of the Test Bank **iLrn Business Math Homework** Includes the chapter's assignments; grades are recorded in instructor's online grade book **WebTUTOR Advantage** Lecture notes, discussion threads	**Chapter Learning Objectives,** page 121 **The Bottom Line Summary of Learning Objectives,** page 132 **Self-Check Review Problems,** page 133 **Key Terms as Crossword Puzzles** http://deitz.swlearning.com **Internet Quizzes** http://deitz.swlearning.com **Topic Review Video** • Cash Discounts **WebTUTOR Advantage** Lecture notes, discussion threads, flash cards and quizzes

CHAPTER 8: Markup

Class Preparation / Lecture Tools	Testing Tools / Course Management	Student Mastery / Homework and Tutorials
Instructor's Annotated Edition with Teaching Notes and Solutions, pages 139–155 This material was included in Chapter 13 of the 13e. **PowerPoint® Teaching Slides** 8-1 Markup Based on Cost 8-2 Markup Based on Selling Price 8-3 Markup Conversion **PowerPoint® Assignments with Solutions Slides** Includes the chapter's two assignments with solutions **Teaching Acetates** 8-1 Markup Based on Cost 8-2 Markup Based on Selling Price 8-3 Markup Conversion **Assignments with Solutions Acetates** Includes the chapter's two assignments with solutions **Topic Review Video** • Markup Based on Cost, Markup Based on Selling Price	**Test Bank,** also available in Word files on the IRCD **ExamView®** computerized version of the Test Bank **iLrn Business Math Homework** Includes the chapter's two assignments; grades are recorded in instructor's online grade book **WebTUTOR Advantage** Lecture notes, discussion threads	**Chapter Learning Objectives,** page 139 **The Bottom Line Summary of Learning Objectives,** page 148 **Self-Check Review Problems,** page 149 **Key Terms as Crossword Puzzles** http://deitz.swlearning.com **Internet Quizzes** http://deitz.swlearning.com **Topic Review Video** • Markup Based on Cost, Markup Based on Selling Price **WebTUTOR Advantage** Lecture notes, discussion threads, flash cards and quizzes

PART 3: Accounting Applications
CHAPTER 9: Banking

Class Preparation / Lecture Tools	Testing Tools / Course Management	Student Mastery / Homework and Tutorials
Instructor's Annotated Edition with Teaching Notes and Solutions, pages 157–174 This material was included in Chapter 7 of the 13e. **PowerPoint® Teaching Slides** 9-1 Bank Statement Reconciliation **PowerPoint® Assignments with Solutions Slides** Includes the chapter's three assignments with solutions **Teaching Acetates** 9-1 Bank Statement Reconciliation **Assignments with Solutions Acetates** Includes the chapter's three assignments with solutions	**Test Bank,** also available in Word files on the IRCD **ExamView®** computerized version of the Test Bank **iLrn Business Math Homework** Includes the chapter's three assignments; grades are recorded in instructor's online grade book **WebTUTOR Advantage** Lecture notes, discussion threads	**Chapter Learning Objectives,** page 157 **The Bottom Line Summary of Learning Objectives,** page 167 **Self-Check Review Problems,** page 168 **Key Terms as Crossword Puzzles** http://deitz.swlearning.com **Internet Quizzes** http://deitz.swlearning.com **WebTUTOR Advantage** Lecture notes, discussion threads, flash cards and quizzes

CHAPTER 10: Payroll Records

Class Preparation / Lecture Tools	Testing Tools / Course Management	Student Mastery / Homework and Tutorials
Instructor's Annotated Edition with Teaching Notes and Solutions, pages 175–200 This material was included in Chapter 8 of the 13e. **PowerPoint® Teaching Slides** 10-1 Total earnings 10-2 Payroll register example **PowerPoint® Assignments with Solutions Slides** Includes the chapter's two assignments with solutions **Teaching Acetates** 10-1 Total earnings 10-2 Payroll register example **Assignments with Solutions Acetates** Includes the chapter's two assignments with solutions	**Test Bank,** also available in Word files on the IRCD **ExamView®** computerized version of the Test Bank **iLrn Business Math Homework** Includes the chapter's two assignments; grades are recorded in instructor's online grade book **WebTUTOR Advantage** Lecture notes, discussion threads	**Chapter Learning Objectives,** page 175 **The Bottom Line Summary of Learning Objectives,** page 191 **Self-Check Review Problems,** page 193 **Key Terms as Crossword Puzzles** http://deitz.swlearning.com **Internet Quizzes** http://deitz.swlearning.com **WebTUTOR Advantage** Lecture notes, discussion threads, flash cards and quizzes

CHAPTER 11: Taxes

Class Preparation / Lecture Tools	Testing Tools / Course Management	Student Mastery / Homework and Tutorials
Instructor's Annotated Edition with Teaching Notes and Solutions, pages 201–228 This material was included in Chapters 9 and 10 of the 13e. Coverage of corporate taxes has been deleted for the 14e. **PowerPoint® Teaching Slides** 11-1 Property Tax **PowerPoint® Assignments with Solutions Slides** Includes the chapter's three assignments with solutions **Teaching Acetates** 11-1 Property Tax **Assignments with Solutions Acetates** Includes the chapter's three assignments with solutions	**Test Bank,** also available in Word files on the IRCD **ExamView®** computerized version of the Test Bank **iLrn Business Math Homework** Includes the chapter's three assignments; grades are recorded in instructor's online grade book **WebTUTOR Advantage** Lecture notes, discussion threads	**Chapter Learning Objectives,** page 201 **The Bottom Line Summary of Learning Objectives,** page 218 **Self-Check Review Problems,** page 220 **Key Terms as Crossword Puzzles** http://deitz.swlearning.com **Internet Quizzes** http://deitz.swlearning.com **WebTUTOR Advantage** Lecture notes, discussion threads, flash cards and quizzes

CHAPTER 12: Insurance

Class Preparation / Lecture Tools	Testing Tools / Course Management	Student Mastery / Homework and Tutorials
Instructor's Annotated Edition with Teaching Notes and Solutions, pages 229–249 This material was included in Chapter 18 of the 13e. **PowerPoint® Teaching Slides** 12-1 Life Insurance definitions **PowerPoint® Assignments with Solutions Slides** Includes the chapter's three assignments with solutions **Teaching Acetates** 12-1 Life Insurance definitions **Assignments with Solutions Acetates** Includes the chapter's three assignments with solutions	**Test Bank,** also available in Word files on the IRCD **ExamView®** computerized version of the Test Bank **iLrn Business Math Homework** Includes the chapter's three assignments; grades are recorded in instructor's online grade book **WebTUTOR Advantage** Lecture notes, discussion threads	**Chapter Learning Objectives,** page 229 **The Bottom Line Summary of Learning Objectives,** page 240 **Self-Check Review Problems,** page 242 **Key Terms as Crossword Puzzles** http://deitz.swlearning.com **Internet Quizzes** http://deitz.swlearning.com **WebTUTOR Advantage** Lecture notes, discussion threads, flash cards and quizzes

PART 4: Interest Applications
CHAPTER 13: *Simple Interest*

Class Preparation / Lecture Tools	Testing Tools / Course Management	Student Mastery / Homework and Tutorials
Instructor's Annotated Edition with Teaching Notes and Solutions, pages 251–268 This material was included in Chapter 14 of the 13e. **PowerPoint® Teaching Slides** 13-1 Ordinary Interest vs Exact Interest 13-2 Interest and Time that equal 1% 13-3 Calculating Simple Interest **PowerPoint® Assignments with Solutions Slides** Includes the chapter's two assignments with solutions **Teaching Acetates** 13-1 Ordinary Interest vs Exact Interest 13-2 Interest and Time that equal 1% 13-3 Calculating Simple Interest **Assignments with Solutions Acetates** Includes the chapter's two assignments with solutions	**Test Bank**, also available in Word files on the IRCD **ExamView®** computerized version of the Test Bank **iLrn Business Math Homework** Includes the chapter's two assignments; grades are recorded in instructor's online grade book **WebTUTOR Advantage** Lecture notes, discussion threads	**Chapter Learning Objectives,** page 251 **The Bottom Line Summary of Learning Objectives,** page 260 **Self-Check Review Problems,** page 261 **Key Terms as Crossword Puzzles** http://deitz.swlearning.com **Internet Quizzes** http://deitz.swlearning.com **WebTUTOR Advantage** Lecture notes, discussion threads, flash cards and quizzes

CHAPTER 14: Installment Purchases

Class Preparation / Lecture Tools	Testing Tools / Course Management	Student Mastery / Homework and Tutorials
Instructor's Annotated Edition with Teaching Notes and Solutions, pages 269–294 This material was included in Chapter 17 of the 13e. The last section on home mortgages is new for the 14e. **PowerPoint® Assignments with Solutions Slides** Includes the chapter's three assignments with solutions **Assignments with Solutions Acetates** Includes the chapter's three assignments with solutions	**Test Bank,** also available in Word files on the IRCD **ExamView** ® computerized version of the Test Bank **iLrn Business Math Homework** Includes the chapter's three assignments; grades are recorded in instructor's online grade book **WebTUTOR Advantage** Lecture notes, discussion threads	**Chapter Learning Objectives,** page 269 **The Bottom Line Summary of Learning Objectives,** page 284 **Self-Check Review Problems,** page 285 **Key Terms as Crossword Puzzles** http://deitz.swlearning.com **Internet Quizzes** http://deitz.swlearning.com **WebTUTOR Advantage** Lecture notes, discussion threads, flash cards and quizzes

CHAPTER 15: Promissory Notes and Discounting

Class Preparation / Lecture Tools	Testing Tools / Course Management	Student Mastery / Homework and Tutorials
Instructor's Annotated Edition with Teaching Notes and Solutions, pages 295–314 This material was included in Chapters 15 and 16 of the 13e, and has been updated and condensed. **PowerPoint® Teaching Slides** 15-1 through 15-5 concern Notes and Discounting **PowerPoint® Assignments with Solutions Slides** Includes the chapter's three assignments with solutions **Teaching Acetates** Acetates 15-1 through 15-5 concern Notes and Discounting **Assignments with Solutions Acetates** Includes the chapter's three assignments with solutions **Topic Review Video** • Discounting Notes	**Test Bank,** also available in Word files on the IRCD **ExamView** ® computerized version of the Test Bank **iLrn Business Math Homework** Includes the chapter's three assignments; grades are recorded in instructor's online grade book **WebTUTOR Advantage** Lecture notes, discussion threads	**Chapter Learning Objectives,** page 295 **The Bottom Line Summary of Learning Objectives,** page 307 **Self-Check Review Problems,** page 308 **Key Terms as Crossword Puzzles** http://deitz.swlearning.com **Internet Quizzes** http://deitz.swlearning.com **Topic Review Video** • Discounting Notes **WebTUTOR Advantage** Lecture notes, discussion threads, flash cards and quizzes

CHAPTER 16: Compound Interest and Present Value

Class Preparation / Lecture Tools	Testing Tools / Course Management	Student Mastery / Homework and Tutorials
Instructor's Annotated Edition with Teaching Notes and Solutions, pages 315–341 This material was included in Chapter 25 of the 13e. The optional coverage of using business calculators is new for the 14e. **PowerPoint® Teaching Slides** 16-1 Compound Interest vs Simple Interest 16-2 Computing Compound Interest **PowerPoint® Assignments with Solutions Slides** Includes the chapter's two assignments with solutions **Teaching Acetates** 16-1 Compound Interest vs Simple Interest 16-2 Computing Compound Interest **Assignments with Solutions Acetates** Includes the chapter's two assignments with solutions	**Test Bank,** also available in Word files on the IRCD *ExamView*® computerized version of the Test Bank *iLrn* **Business Math Homework** Includes the chapter's two assignments; grades are recorded in instructor's online grade book *WebTUTOR Advantage* Lecture notes, discussion threads	**Chapter Learning Objectives,** page 315 **The Bottom Line Summary of Learning Objectives,** page 327 **Self-Check Review Problems,** page 328 **Key Terms as Crossword Puzzles** http://deitz.swlearning.com **Internet Quizzes** http://deitz.swlearning.com *WebTUTOR Advantage* Lecture notes, discussion threads, flash cards and quizzes

xxxiv **Resource Integration Guide**

PART 5: Business Applications
CHAPTER 17: (Comprehensive only) Inventory and Turnover

Class Preparation / Lecture Tools	Testing Tools / Course Management	Student Mastery / Homework and Tutorials
Instructor's Annotated Edition with Teaching Notes and Solutions, pages 343–362 This material was included in Chapter 19 of the 13e. **PowerPoint® Teaching Slides** 17-1 Inventory Costing Methods 17-2 Estimating Inventory Value **PowerPoint® Assignments with Solutions Slides** Includes the chapter's two assignments with solutions **Teaching Acetates** 17-1 Inventory Costing Methods 17-2 Estimating Inventory Value **Assignments with Solutions Acetates** Includes the chapter's two assignments with solutions	**Test Bank**, also available in Word files on the IRCD **ExamView®** computerized version of the Test Bank **iLrn Business Math Homework** Includes the chapter's two assignments; grades are recorded in instructor's online grade book **WebTUTOR Advantage** Lecture notes, discussion threads	**Chapter Learning Objectives,** page 343 **The Bottom Line Summary of Learning Objectives,** page 354 **Self-Check Review Problems,** page 356 **Key Terms as Crossword Puzzles** http://deitz.swlearning.com **Internet Quizzes** http://deitz.swlearning.com **WebTUTOR Advantage** Lecture notes, discussion threads, flash cards and quizzes

Resource Integration Guide

CHAPTER 18: (Comprehensive only) Depreciation

Class Preparation / Lecture Tools	Testing Tools / Course Management	Student Mastery / Homework and Tutorials
Instructor's Annotated Edition with Teaching Notes and Solutions, pages 363–382 This material was included in Chapter 20 of the 13e. **PowerPoint® Teaching Slides** 18-1 Straight-Line Depreciation 18-2 Double-Declining-Balance Depreciation 18-3 Sum-of-the-Years Digits Depreciation Method **PowerPoint® Assignments with Solutions Slides** Includes the chapter's two assignments with solutions **Teaching Acetates** 18-1 Straight-Line Depreciation 18-2 Double-Declining-Balance Depreciation 18-3 Sum-of-the-Years Digits Depreciation Method **Assignments with Solutions Acetates** Includes the chapter's two assignments with solutions	**Test Bank**, also available in Word files on the IRCD **ExamView®** computerized version of the Test Bank **iLrn Business Math Homework** Includes the chapter's two assignments; grades are recorded in instructor's online grade book **WebTUTOR Advantage** Lecture notes, discussion threads	**Chapter Learning Objectives,** page 363 **The Bottom Line Summary of Learning Objectives,** page 374 **Self-Check Review Problems,** page 375 **Key Terms as Crossword Puzzles** http://deitz.swlearning.com **Internet Quizzes** http://deitz.swlearning.com **WebTUTOR Advantage** Lecture notes, discussion threads, flash cards and quizzes

CHAPTER 19: (Comprehensive only) Financial Statements

Class Preparation / Lecture Tools	Testing Tools / Course Management	Student Mastery / Homework and Tutorials
Instructor's Annotated Edition with Teaching Notes and Solutions, pages 383–404 This material was included in Chapter 21 of the 13e. **PowerPoint® Teaching Slides** 19-1 Income statement analysis 19-2 and 19-3 Financial ratios 19-3 Rates of return **PowerPoint® Assignments with Solutions Slides** Includes the chapter's three assignments with solutions **Teaching Acetates** 19-1 Income statement analysis 19-2 and 19-3 Financial ratios 19-3 Rates of return **Assignments with Solutions Acetates** Includes the chapter's three assignments with solutions	**Test Bank,** also available in Word files on the IRCD **ExamView®** computerized version of the Test Bank **iLrn Business Math Homework** Includes the chapter's three assignments; grades are recorded in instructor's online grade book **WebTUTOR Advantage** Lecture notes, discussion threads	**Chapter Learning Objectives,** page 383 **The Bottom Line Summary of Learning Objectives,** page 393 **Self-Check Review Problems,** page 395 **Key Terms as Crossword Puzzles** http://deitz.swlearning.com **Internet Quizzes** http://deitz.swlearning.com **WebTUTOR Advantage** Lecture notes, discussion threads, flash cards and quizzes

CHAPTER 20: (Comprehensive only) International Business

Class Preparation / Lecture Tools	Testing Tools / Course Management	Student Mastery / Homework and Tutorials
Instructor's Annotated Edition with Teaching Notes and Solutions, pages 405-423 This material was included in Chapter 24 of the 13e. **PowerPoint® Assignments with Solutions Slides** Includes the chapter's two assignments with solutions **Assignments with Solutions Acetates** Includes the chapter's two assignments with solutions	**Test Bank,** also available in Word files on the IRCD **ExamView**® computerized version of the Test Bank **iLrn Business Math Homework** Includes the chapter's two assignments; grades are recorded in instructor's online grade book **WebTUTOR Advantage** Lecture notes, discussion threads	**Chapter Learning Objectives,** page 405 **The Bottom Line Summary of Learning Objectives,** page 414 **Self-Check Review Problems,** page 415 **Key Terms as Crossword Puzzles** http://deitz.swlearning.com **Internet Quizzes** http://deitz.swlearning.com **WebTUTOR Advantage** Lecture notes, discussion threads, flash cards and quizzes

PART 6: Corporate and Special Applications
CHAPTER 21: (Comprehensive only) Corporate Stocks

Class Preparation / Lecture Tools	Testing Tools / Course Management	Student Mastery / Homework and Tutorials
Instructor's Annotated Edition with Teaching Notes and Solutions, pages 425-444 This material was included in Chapter 22 of the 13e. **PowerPoint® Teaching Slides** 21-1 Dividends on preferred stock **PowerPoint® Assignments with Solutions Slides** Includes the chapter's two assignments with solutions **Teaching Acetates** 21-1 Dividends on preferred stock **Assignments with Solutions Acetates** Includes the chapter's two assignments with solutions	**Test Bank,** also available in Word files on the IRCD **ExamView**® computerized version of the Test Bank **iLrn Business Math Homework** Includes the chapter's two assignments; grades are recorded in instructor's online grade book **WebTUTOR Advantage** Lecture notes, discussion threads	**Chapter Learning Objectives,** page 425 **The Bottom Line Summary of Learning Objectives,** page 435 **Self-Check Review Problems,** page 436 **Key Terms as Crossword Puzzles** http://deitz.swlearning.com **Internet Quizzes** http://deitz.swlearning.com **WebTUTOR Advantage** Lecture notes, discussion threads, flash cards and quizzes

CHAPTER 22: (Comprehensive only) Corporate and Government Bonds

Class Preparation / Lecture Tools	Testing Tools / Course Management	Student Mastery / Homework and Tutorials
Instructor's Annotated Edition with Teaching Notes and Solutions, pages 445-460 This material was included in Chapter 23 of the 13e. **PowerPoint® Teaching Slides** 22-1 Determining the cost of bonds **PowerPoint® Assignments with Solutions Slides** Includes the chapter's two assignments with solutions **Teaching Acetates** 22-1 Determining the cost of bonds **Assignments with Solutions Acetates** Includes the chapter's two assignments with solutions	**Test Bank,** also available in Word files on the IRCD **ExamView®** computerized version of the Test Bank **iLrn Business Math Homework** Includes the chapter's two assignments; grades are recorded in instructor's online grade book **WebTUTOR Advantage** Lecture notes, discussion threads	**Chapter Learning Objectives,** page 445 **The Bottom Line Summary of Learning Objectives,** page 455 **Self-Check Review Problems,** page 456 **Key Terms as Crossword Puzzles** http://deitz.swlearning.com **Internet Quizzes** http://deitz.swlearning.com **WebTUTOR Advantage** Lecture notes, discussion threads, flash cards and quizzes

Resource Integration Guide xxxix

CHAPTER 23: (Comprehensive only) Annuities

Class Preparation / Lecture Tools	Testing Tools / Course Management	Student Mastery / Homework and Tutorials
Instructor's Annotated Edition with Teaching Notes and Solutions, pages 461–494 This material was included in Chapter 26 of the 13e. The optional coverage of using business calculators is new for the 14e. **PowerPoint® Teaching Slides** 23-1 Amortization Schedule **PowerPoint® Assignments with Solutions Slides** Includes the chapter's two assignments with solutions **Teaching Acetates** 23-1 Amortization Schedule **Assignments with Solutions Acetates** Includes the chapter's two assignments with solutions **Topic Review Video** • Annuities: Present Value, Annuities: Future Value	**Test Bank,** also available in Word files on the IRCD ExamView® computerized version of the Test Bank **iLrn Business Math Homework** Includes the chapter's two assignments; grades are recorded in instructor's online grade book WebTUTOR Advantage Lecture notes, discussion threads	**Chapter Learning Objectives,** page 461 **The Bottom Line Summary of Learning Objectives,** page 479 **Self-Check Review Problems,** page 480 **Key Terms as Crossword Puzzles** http://deitz.swlearning.com **Internet Quizzes** http://deitz.swlearning.com **Topic Review Video** • Annuities: Present Value, Annuities: Future Value WebTUTOR Advantage Lecture notes, discussion threads, flash cards and quizzes

CHAPTER 24: (Comprehensive only) Business Statistics

Class Preparation / Lecture Tools	Testing Tools / Course Management	Student Mastery / Homework and Tutorials
Instructor's Annotated Edition with Teaching Notes and Solutions, pages 495–518 This material was included in Chapter 27 of the 13e. **PowerPoint® Teaching Slides** 24-1 Mean **PowerPoint® Assignments with Solutions Slides** Includes the chapter's two assignments with solutions **Teaching Acetates** 24-1 Mean **Assignments with Solutions Acetates** Includes the chapter's two assignments with solutions **Topic Review Video** • How to Create Bar Charts, Line Graphs, and Pie Charts	**Test Bank**, also available in Word files on the IRCD **ExamView®** computerized version of the Test Bank **iLrn Business Math Homework** Includes the chapter's two assignments; grades are recorded in instructor's online grade book **WebTUTOR Advantage** Lecture notes, discussion threads	**Chapter Learning Objectives,** page 495 **The Bottom Line Summary of Learning Objectives,** page 509 **Self-Check Review Problems,** page 512 **Key Terms as Crossword Puzzles** http://deitz.swlearning.com **Internet Quizzes** http://deitz.swlearning.com **Topic Review Video** • How to Create Bar Charts, Line Graphs, and Pie Charts **WebTUTOR Advantage** Lecture notes, discussion threads, flash cards and quizzes

BRIEF CONTENTS

Part 1: Fundamental Review 2

1 Fundamental Processes 3
2 Fractions 29
3 Decimals 47
4 Word Problems and Equations 69

Part 2: Percentage Applications 86

5 Percents 87
6 Commissions 107
7 Discounts 121
8 Markup 139

Part 3: Accounting Applications 156

9 Banking 157
10 Payroll Records 175
11 Taxes 201
12 Insurance 229

Part 4: Interest Applications 250

13 Simple Interest 251
14 Installment Purchases 269
15 Promissory Notes and Discounting 295
16 Compound Interest and Present Value 315

Part 5: Business Applications 342

17 Inventory and Turnover 343
18 Depreciation 363
19 Financial Statements 383
20 International Business 405

Part 6: Corporate and Special Applications 424

21 Corporate Stocks 425
22 Corporate and Government Bonds 445
23 Annuities 461
24 Business Statistics 495

Appendix A Assignment Answers to Odd-Numbered Problems 519

Appendix B Answers to Self-Check Review Problems 528

Glossary 532

Index 538

Progress Report 543

CONTENTS

Part 1: Fundamental Review 2

1 Fundamental Processes 3
  Addition 4
    Number Combinations 4
    Repeated Digits 5
    Adding From Left To Right
    (Columns of Two-Digit
    Numbers) 5
    Checking Addition 5
    Horizontal Addition 6
  Subtraction 7
    Checking Subtraction 7
    Horizontal Subtraction 7
  Multiplication 8
    Checking Multiplication 9
    Multiplying Numbers Ending
    in Zero 9
    Multiplying When the
    Multiplier contains Zero
    Not On the End 9
    Multiplying the Product of
    two Factors 10
  Division 11
    Checking Division 12
    Dividing by 10 12
    Dividing by 100 13
    Dividing When the Divisor
    and Dividend End with
    Zeros 13
  Estimating 14
    Estimating when Multiplying .14
    Estimating when Dividing ...14

2 Fractions 29
  Vocabulary of Fractions 30
  Changing Improper Fractions
  and Mixed Numbers 30
  Changing Fractions to Lower
  and Higher Terms 32
  Adding Fractions and
  Mixed Numbers 33
  Subtracting Fractions and
  Mixed Numbers 34
    Borrowing 1 34

  Multiplying Fractions,
  Mixed Numbers, and Whole
  Numbers 36
  Canceling Common Factors in
  Numerators and Denominators ..37
  Dividing Fractions, Mixed
  Numbers, and Whole Numbers ..38

3 Decimals 47
  Fractions Versus Decimal
  Numbers 48
  Decimal Numbers and
  Electronic Displays 48
  Reading Decimal Numbers 49
    Reading Long Decimal
    Numbers 49
  Rounding Decimal Numbers ...50
    Rounding Up 50
  Whole Numbers, Decimal
  Numbers, and Arithmetic 51
  Adding Decimal Numbers 51
  Subtracting Decimal Numbers ..52
  Multiplying Decimal Numbers ..53
  Dividing Decimal Numbers 54
  Using Multipliers and Divisors
  that End with Zeros 57
  Approximating Products and
  Quotients 58

4 Word Problems and Equations .69
  Mental Computations 70
  Solving World Problems 70
  Solving Rate, Time, and
  Distance Problems 72
  Solving Simple Numeric
  Equations 74
  Numerical Relationships in
  a Series 76
  Making Quick Calculations by
  Rounding Numbers 77

Part 2: Percentage Applications 86

5 Percents 87
  Changing Percents to Decimals ..88

CONTENTS

Changing Decimals and
Fractions to Percents 89
Finding Base, Rate, and
Percentage 90
    Using Percents in Business . . .92
Using Percents to Measure
Increase and Decrease 92
    Computing Amounts of
    Increase and Decreases with
    a Calculator 94
Using Percents to Allocate
Overhead Expenses 94

6 Commissions **107**
Computing Sales Commissions
and Gross Pay 108
Computing Graduated Sales
Commissions 109
Computing Sales and
Purchases for Principles 111

7 Discounts **121**
Computing Trade Discounts 122
Computing a Series of Trade
Discounts 123
    Complement Method
    Shortcut 124
Computing the Equivalent
Single Discount Rate 125
Computing Cash Discounts
for Fully Paid Invoices 126
    Returned Merchandise and
    Freight Charges 127
Computing Cash Discounts for
Partially Paid Invoices 129

8 Markup **139**
Computing Markup Variables . .140
Computing Markup Based on
Cost 141
    Computing Selling Price
    Directly from Cost 141
    Computing Cost from
    Selling Price 142
Computing Markup Percent
Based on Cost 143

Computing Markup Based on
Selling Price 144
    Computing Cost Directly . . .144
    Computing Selling Price
    from Cost 145
Computing Markup Percent
Based on Selling Price 146

Part 3: Accounting Applications 156

9 Banking **157**
Using Deposit Slips and
Bank Checks 158
Using Checkbooks and Check
Registers 160
Reconciling Bank Statements . .161

10 Payroll Records **175**
Preparing a Payroll Register . . .176
Computing Federal Income
Tax Withholding Amounts 178
Computing Social Security,
Medicare, and Other
Withholdings 184
Completing an Employee's
Earnings Record 186
Computing an Employer's
Quarterly Federal Tax Return . .187
Computing an Employer's
Federal and State Unemployment
Tax Liability 189

11 Taxes **201**
Computing Sales Taxes 202
    Sales Tax as a Percent of
    Price 202
    Sales Tax as an Amount
    Per Unit 203
    Excise Tax as an Amount
    Per Unit 203
Computing Assessed Valuations
and Property Taxes 204
Computing Tax Rates in
Percents and Mills 205

CONTENTS

    Percents 205
    Mills .206
Computing Special Assessments,
Prorations, and Exemptions 207
Determining Taxable Income,
Using Standard Form 1040 209
    Computing Taxable Income .213
Determining Taxes Due, Using
Standard Form 1040 213
    Tax Credits and Net Tax 215

12 Insurance 229
Computing Auto Insurance
Costs 230
Computing Low-Risk and
High-Risk Rates 231
Computing Short Rates 232
Computing Coinsurance on
Property Losses 233
Computing Life Insurance
Premiums 235
Computing Cash Surrender
and Loan Values 236
Computing Medical Insurance
Contributions and
Reimbursements 237

Part 4: Interest Applications 250

13 Simple Interest 251
Computing Simple Interest 252
    Using Calculators 253
Computing Ordinary Interest . .254
Computing Exact Interest 254
Comparing Ordinary Interest
and Exact Interest 255
Estimating Exact Simple
Interest .256
    Combinations of Time and
    Interest that Yield 1% 256
    Other Rates and Times 256
    Estimating Exact Interest . . .256
Computing the Interest
Variables 257
    Finding the Interest Amount,
    Principal, Rate, or Time 258

14 Installment Purchases 269
Converting Interest Rates 270
Computing Simple Interest
on a Monthly Basis 271
Computing Finance Charges . . .271
Computing Costs of Installment
Purchases 273
Computing Effective Interest
Rates 275
    Increasing the Effective
    Rate 276
Amortizing a Loan 278
    Computing the Monthly
    Payment 278
    Loan Payment Schedule 280
Finding the Monthly Payment
of a Home Mortgage 281
    Amortization Schedule
    for a Mortgage 282

15 Promissory Notes and
Discounting 295
Promissory Notes 296
Computing the Number of
Interest Days of a Note 297
Determining the Due Date
of a Note 298
Computing the Maturity
Value of a Note 300
Discounting Promissory Notes . .300
    Non-Interest-Bearing
    Promissory Notes 302
Bank Discounting 303
    Comparing a Discount Rate
    to an Interest Rate 304
Borrowing Money to Take a
Cash Discount 305

16 Compound Interest and
Present Value 315
Computing Future Values and
Compound Interest 316
    Future Value Formula 317
    Various Compounding
    Periods 318
Calculators and Exponents 319

CONTENTS

Effective Rates 320
    Daily Compounding 321
Computing Present Values 322
Using Present Value Tables
and/or Formulas 323
    Present Value Formula 324
    Notes About the Future
    Value and Present Value
    Tables 325

Part 5: Business Applications 342

17 Inventory and Turnover 343
    Accounting for Inventory 344
        Inventory Sheets 344
        Perpetual Inventory Systems 344
    Computing Inventory, Using the
    Average Cost, FIFO, and LIFO
    Methods 346
        The Average Cost Method ..346
        The FIFO Method 346
        The LIFO Method 347
    Computing Inventory at the
    Lower of Cost or Market Value .347
    Estimating Inventory Value 349
    Computing Inventory
    Turnover 350

18 Depreciation 363
    Computing Depreciation with
    the Straight-Line Method 364
    Computing Depreciation with
    the Units-of-Production
    Method 365
    Computing Depreciation
    with the Declining-Balance
    Method 366
    Computing Depreciation with
    the Sum-of-the-Years-Digits
    Method 368
    Computing Depreciation with
    the Modified Accelerated Cost
    Recovery System 369
    Computing Partial-Year
    Depreciation 371

19 Financial Statements 383
    Analyzing Balance Sheets 384
    Analyzing Income Statements ..386
    Computing Business
    Operating Ratios 389
        Working Capital Ratio 389
        Acid Test Ratio 390
        Ratio of Accounts
        Receivable to Net Sales 390
        Inventory Turnover 391
        Relationship of Net Income
        to Net Sales 391
        Rate of Return Investment ..391

20 International Business 405
    Computing Currency Exchange
    Rates 406
    Computing the Effects of
    Exchange Rate Changes 408
    Computing Duties on Imports ..409
    Converting Between U.S.
    Weights and Measures and
    Metric Weights and Measures ..411

Part 6: Corporate and Special Applications 424

21 Corporate Stocks 425
    Computing the Costs and
    Proceeds of Stock Transactions .426
    Computing the Costs and
    Proceeds of Round and
    Odd Lots 429
    Computing the Rate of
    Yield and Gains or Losses 430
        The Rate of Yield 430
        Gain or Loss on Sale of
        Stock 430
    Computing Comparative
    Earning Potential 431

**22 Corporate and Government
Bonds** 445
    Computing Gains and Losses
    on Corporate Bonds 446

CONTENTS

Computing Annual Interest
on Corporate and Government
Bonds 447
    Newspaper Information on
    Bonds 448
    Commissions for Buying
    and Selling Bonds 449
Computing Accrued Interest
on Bond Transactions 449
Computing the Rate of Yield for
Bonds 450
Computing the Rate of
Yield to Maturity 451

23 Annuities 461
Computing the Future Value
of an Annuity 462
    Annuity Tables 463
    Future Value of an Annuity
    Formula 464
    Various Payment Periods ...464
    Using a Calculator to
    Compute Annuity Factors
    (Optional) 465
Computing Regular Payments
of an Annuity from the Future
Value 466
    Sinking Funds 467
Computing the Present
Value of an Annuity 468
    Present Value of an
    Annuity Formula 469
    Using a Calculator to
    Compute the Present
    Value of an Annuity 470
Computing Regular Payments
of an Annuity from the Present
Value 471
Computing the Payment to
Amortize a Loan 473

Creating a Loan Amortization
Schedule 474
    Using the Texas Instruments
    BA II Plus Business Calculator
    for Annuity Calculations
    (Optional) 475
    The Basic Annuity Keys 475
    Additional Annuity Keys ...476

24 Business Statistics 495
Statistical Averages:
Computing the Mean 496
Determining the Median 497
Determining the Mode 498
Constructing Frequency Tables .498
    Computing the Mean of
    Large Data Sets 499
    Charts and Graphs:
    Constructing Histograms ...500
    Constructing Bar Graphs ...501
    Comparative Bar Graph 502
    Component Bar Graph 503
Constructing Line Graphs 504
Constructing Pie Charts 507

**Appendix A   Assignment Answers
                to Odd-Numbered
                Problems** 519

**Appendix B   Answers to Self-Check
                Review Problems** ...528

Glossary 532

Index 538

Progress Report 543

14e

Contemporary Business Mathematics
for Colleges

Part 1

Fundamental Review

1. **Fundamental Processes**
2. **Fractions**
3. **Decimals**
4. **Word Problems and Equations**

Fundamental Processes

Learning Objectives
By studying this chapter and completing all assignments you will be able to:

Learning Objective 1 — Use shortcuts and simplifications to perform the fundamental process of addition rapidly and accurately.

Learning Objective 2 — Use shortcuts and simplifications to perform subtraction rapidly and accurately.

Learning Objective 3 — Use simplifications to perform the fundamental process of multiplication.

Learning Objective 4 — Use shortcuts and simplifications to perform division rapidly and accurately.

Learning Objective 5 — Estimate answers before performing operations.

Addition

Learning Objective 1

Use shortcuts and simplifications to perform the fundamental process of addition rapidly and accurately.

About half of all computations used in business involve addition. The more skilled you become in adding, the more rapidly you will get accurate answers. Addition is the process of finding the **sum** (total) of two or more **addends** (any of a set of numbers to be added).

NUMBER COMBINATIONS

Certain aids can help you add more accurately and rapidly. One of the most helpful is to combine any two numbers that total 10. The following combinations total 10. Practice the combinations until you can identify them instantly.

1	2	3	4	5	9	8	7	6	5
9	8	7	6	5	1	2	3	4	5

When these combinations are found sequentially in any column of numbers, you should add them as 10. In example A, by using the combinations of 10, you can simply add down the column by saying "9 plus 10 is 19, plus 10 is 29, plus 8 is 37" (or "9, 19, 29, 37").

The number 3 is carried over to the top of the next column and written in a small figure above the number 7. The combinations of 10 are used in adding the center column by simply saying "10, 20, 30."

In adding the left-hand column, you carry over the number 3 from the center column total. You can simply say "8, 18, 28, 32."

1.1 Emphasize to students that in business both speed and accuracy are important in making mathematical calculations. Although computers and calculators are used in business, they aren't always available, and a person working in business must be able to handle basic business calculations without them.

Combining vertical and horizontal addition (and subtraction) provides an excellent self-checking process, which is always encouraged in business applications.

EXAMPLE A

```
  3  3
  5  7  9
  4  2  4
  6  8  6
  9  0  3
  1  5  7
  4  5  8
3,2  0  7
```

Also learn to recognize the combinations of three numbers that total 10.

1	1	1	1	2	2	2	3
1	2	3	4	2	3	4	3
8	7	6	5	6	5	4	4

When three numbers totaling 10 appear in sequence in a column, you should combine them and add them as 10. In example B, you might add the numbers in the ones column as you add down the column, "10, 18, 28, 38, 41." Write the number 4, which is carried over as a small figure above the 1 in the tens column. Then use the combinations of 10 in adding the tens column by saying "5, 15, 25, 35, 43."

EXAMPLE B

```
              4
        (5)   1 │7│
              │6│2
       (15)  │4│1│ (10)
              │2│8 (18)
              2│5│
       (25)  │6│5│ (28)
              │5│4
              2│4
       (35)  │3│2│ (38)
       (43)   8 3 (41)
              ─────
              4 3 1
```

REPEATED DIGITS

When you're adding a column in which many of the digits are the same, it is often quicker to count the number of repeated digits and then multiply the digit by that number. In example C, the ones column totals 33: 10 + 10 + 13. The tens column shows five 4s, equaling 20: 5 × 4 = 20. The 3 that was carried over and the 5 are then added to the 20 for a total of 28 in the tens column. The total for the problem is 283.

EXAMPLE C

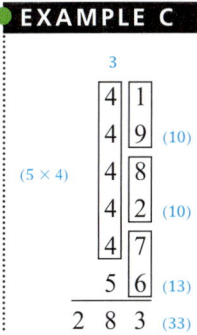

```
               3
              │4│ 1
              │4│ 9 (10)
      (5 × 4) │4│ 8
              │4│ 2 (10)
              │4│ 7
               5  6 (13)
              ─────
               2 8 3 (33)
```

ADDING FROM LEFT TO RIGHT (COLUMNS OF TWO-DIGIT NUMBERS)

When adding columns of two-digit numbers, you can easily count by tens and add the ones column to your total.

EXAMPLE D

Count:

```
 12    12
 24    22, 32 + 4 = 36
 51    46, 56, 66, 76, 86 +1 = 87
 43    97, 107, 117, 127 + 3 = 130
 32    140, 150, 160 + 2 = 162
───
162
```

CHECKING ADDITION

You should always check the accuracy of your addition. To do so, add the columns again in the opposite direction—that is, if you added down, add up for the check.

Chapter 1 Fundamental Processes

HORIZONTAL ADDITION

When using business records, you may need to add numbers horizontally. You may check several horizontal additions by adding the columns vertically and then adding these totals horizontally. This method is called **cross-checking.** The sums obtained by adding the totals horizontally and vertically should be the same.

EXAMPLE E

282	+	346	+	723	+	409	+	716	=	2,476
113	+	806	+	629	+	916	+	620	=	3,084
240	+	318	+	718	+	312	+	309	=	1,897
716	+	501	+	423	+	716	+	114	=	2,470
872	+	417	+	909	+	704	+	472	=	3,374
2,223	+	2,388	+	3,402	+	3,057	+	2,231	=	13,301

CONCEPT CHECK 1.1

Add horizontally and vertically; compare horizontal and vertical totals to verify accuracy. Use combinations to simplify addition.

1		1		1				
2 4	+	7 6	+	6 3	=	163	(4 + 6)	
3 6	+	2 4	+	2 5	=	85	(6 + 4)	(Note horizontal combinations)
2 7	+	4 3	+	1 2	=	82	(7 + 3)	
8 7	+	143	+	100	=	330		

COMPLETE ASSIGNMENT 1.1.

Subtraction

Subtraction is the process of finding the difference between the **minuend** (number from which subtraction is being made) and the **subtrahend** (number being subtracted); the result is the **difference.** When the subtrahend is greater than the minuend, the result is a negative difference. In business, a negative difference may be called a **credit balance.** A credit balance is frequently shown in parentheses.

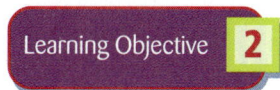

Use shortcuts and simplifications to perform subtraction rapidly and accurately.

EXAMPLE F

Positive Difference		Negative Difference (Credit Balance)
$18.88	Minuend	$12.00
−3.63	−Subtrahend	−13.50
$15.25	Difference	($ 1.50)

CHECKING SUBTRACTION

To check subtraction, use addition. If 209 is subtracted from 317, the difference is 108. You can check this result by adding the difference (108) to the subtrahend (209). The sum is 317. You can use the same procedure to check subtraction with a negative difference (credit balance).

EXAMPLE G

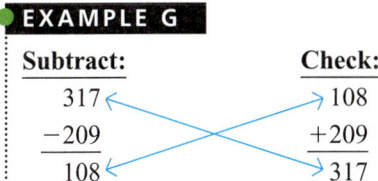

EXAMPLE H

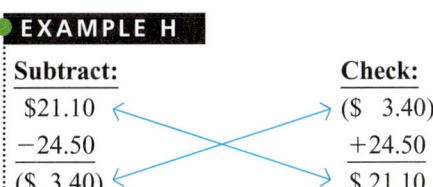

HORIZONTAL SUBTRACTION

When using certain business forms, you may have to subtract numbers horizontally. You can check a number of horizontal subtractions by adding the columns vertically and then subtracting these totals horizontally. This answer should equal the total of the differences in the column at the right.

EXAMPLE I

Minuend		Subtrahend		Difference
$ 120	−	$ 20	=	$100
283	−	10	=	273
440	−	110	=	330
$ 269	−	$149	=	$120
$1,112	−	$289	=	$823

✓ CONCEPT CHECK 1.2

Subtract:	Check:	Subtract horizontally: Check by comparing totals
276	142	27 − 13 = 14
−134	+134	24 − 11 = 13
142	276	36 − 10 = 26
		87 − 34 = 53

COMPLETE ASSIGNMENT 1.2.

Multiplication

Learning Objective 3

Use simplifications to perform the fundamental process of multiplication.

Multiplication, stated simply, is repeated addition. When two numbers (called **factors**) are multiplied, one number is repeated as many times as there are units in the other. The factor that is multiplied is called the **multiplicand**. The factor that indicates how many times to multiply is the **multiplier**. The result is the **product**.

STEPS to Multiply Two Numbers

1. Make the smaller factor the multiplier.
2. Multiply from right to left.
3. Add the products to get the final product.

EXAMPLE J

	456	(multiplicand)		In other words:
STEP 1	×237	(multiplier)	7 × 456 =	3,192
STEP 2	3 192	(product)	30 × 456 =	13,680
STEP 2	13 680	(product)	200 × 456 =	91,200
STEP 2	91 200	(product)	237 × 456 =	108,072
STEP 3	108,072	(final product)		

Part 1 Fundamental Review

CHECKING MULTIPLICATION

The best method of checking multiplication is to divide the product by the multiplier to obtain the multiplicand. Example K shows the relationship between multiplication and division.

EXAMPLE K

Multiplicand	22	→	22
Multiplier	×6	→	6)132
Product	132		↑

MULTIPLYING NUMBERS ENDING IN ZERO

To multiply a number by 10, simply add a zero to the end of the number. To multiply a number by 100, add two zeros to the end: $10 \times 46 = 460$; $7,689 \times 100 = 768,900$.

STEPS to Multiply Numbers with Zeros

1. Make the multiplier the factor with the smaller number of digits after ignoring zeros at the right-hand side of the number.
2. Ignore the right-hand zeros and multiply the remaining numbers.
3. Insert the zeros ignored in Step 2 to the right-hand side of the product.

EXAMPLE L

STEP 1 370×200: Make 2 the multiplier.
Ignored:
37   (1 zero)
×2   (2 zeros)

STEP 2   74   (3 zeros)
STEP 3   74 000 = 74,000

EXAMPLE M

STEP 1 $1,200 \times 160,800$: Make 12 the multiplier.
Ignored:
1,608   (2 zeros)
×12   (2 zeros)

3 216
16 08

STEP 2   19,296   (4 zeros)
STEP 3   19,296 0000 = 192,960,000

MULTIPLYING WHEN THE MULTIPLIER CONTAINS ZERO NOT ON THE END

Often a zero appears in the center of the multiplier rather than at the end. To multiply 42,674 by 401, first multiply the multiplicand by 1 and write down the product. Then multiply by 4 (which is really 400) and write the result two places, instead of one, to the left. In other words, one extra place is left for each zero in the multiplier.

EXAMPLE N

```
    42,674
  × 401
    42 674
 17 069 6      (2 places)
 17,112,274
```

1.2 237×456 can be illustrated by the addition process.

7×456:
456
456
456
456
456
456
456

30×456:   4,560
4,560
4,560

200×456:   45,600
45,600
108,072

We use three-factor multiplication later in such business applications as chain discounts and computing cubic volume for storage and room size (i.e., length × width × height).

Here is an example of a business-type calculation requiring multiplying the product of two factors: Total sales for a car wash that washes 200 cars per day at $6 per car for 7 days is $200 \times \$6 \times 7 = \$8,400$.

Whenever more than one zero appears in the multiplier, the multiplication process is similar. To multiply 33,222 by 2,004, as in Example O, first multiply 33,222 by 4. Then multiply 33,222 by 2, writing the answer three places to the left. Remember, extra places must be left for the two zeros (1 place + 2 extra places = 3 places).

EXAMPLE O

```
   33,222
 × 2,004
  132 888
 66 444        (3 places)
 66,576,888
```

MULTIPLYING THE PRODUCT OF TWO FACTORS

Sometimes in business you will need to multiply two factors and then multiply the product of those factors by a third factor. As shown in example P, you begin by multiplying the first two factors and then multiply that product by the third factor.

EXAMPLE P

$21 \times 30 \times 15 = 9{,}450$

```
   21            630
  ×30           ×15
  630          3150
                630
               9,450
```

MULTIPLYING BY 25

A shortcut for multiplying by 25 is to multiply by 100 (increase by two zeros) and divide by 4.

EXAMPLE Q
321 × 25
32,100 ÷ 4 = 8,025

EXAMPLE R
828 × 25
82,800 ÷ 4 = 20,700

MULTIPLYING BY 50

A shortcut for multiplying by 50 is to multiply by 100 (increase by two zeros) and divide by 2.

EXAMPLE S
732 × 50
73,200 ÷ 2 = 36,600

✓ CONCEPT CHECK 1.3

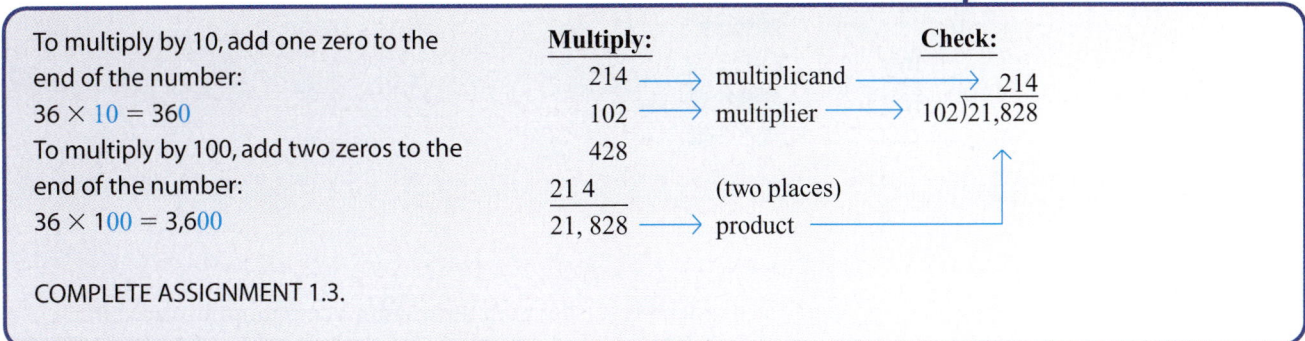

To multiply by 10, add one zero to the end of the number:
36 × 10 = 360

To multiply by 100, add two zeros to the end of the number:
36 × 100 = 3,600

COMPLETE ASSIGNMENT 1.3.

Division

Division is the process of finding how many times one number (the **divisor**) is contained in another (the **dividend**). The result is called the **quotient**. If anything remains after the division is completed, it is called the **remainder**. In example T, 47 ÷ 2 = 23 (with 1 left over), 47 is the dividend, 2 is the divisor, 1 is the remainder, and 23 with a remainder of (1) is the quotient.

Learning Objective 4

Use shortcuts and simplifications to perform division rapidly and accurately.

● EXAMPLE T

$$\begin{array}{r} 23\ (1) \\ 2\overline{)47} \\ \underline{4} \\ 7 \\ \underline{6} \\ 1 \end{array}$$

1.3 Note that in the example the remainder is enclosed in parentheses rather than shown as remainder over divisor. The reason is that we have not introduced fractions at this time. If you feel that it is better to introduce the concept of remainder over divisor at this time, feel free to do so.

Emphasize the placement of the zeros in the quotient, as shown in example V.

STEPS in Long Division

1. Write the divisor in front of and the dividend inside of a division bracket ($\overline{)\ \ \ }$).
2. As the first partial dividend, use only as many digits at the left of the dividend as you need in order to have a number that is equal to or larger than the divisor.
3. Write the number of times the divisor will go into the partial dividend selected in Step 2.
4. Multiply the divisor by this answer, write the product under the partial dividend, and subtract.
5. Next to the remainder thus obtained, bring down the next digit of the dividend to form the second partial dividend.
6. Divide as before, and repeat the process until all the digits of the dividend have been used.

Chapter 1 Fundamental Processes

EXAMPLE U

```
           174
    164)28,536
        16 4
        12 13
        11 48
           656
           656
             0
```

STEP 3
STEPS 1 & 2
STEP 4
STEP 5
STEP 6

When the partial dividend is smaller than the divisor, a zero must be placed in the quotient above that digit. This process is continued until the partial dividend is at least as large as the divisor. Then continue the long division steps, as shown in example V.

EXAMPLE V

```
        20,108
    34)683,672
       68
        3 6
        3 4
          272
          272
            0
```

CHECKING DIVISION

To check division, simply multiply the quotient by the divisor and add any remainder to the product. The result will equal the original dividend. (Examples W and X provide checks for examples U and V.)

EXAMPLE W

```
       174
     ×164
      696
    10 44
    17 4
    28,536
```

EXAMPLE X

```
      20,108
        ×34
      80 432
     603 24
     683,672
```

Note: Division is the reverse process of multiplication.

DIVIDING BY 10

To divide by 10, drop the digit at the extreme right of the dividend; the dropped digit will be the remainder.

EXAMPLE Y

79**0** ÷ 10 = 79 (0 remainder)

EXAMPLE Z

3,65**2** ÷ 10 = 365 (2 remainder)

DIVIDING BY 100

To divide by 100, drop the two right-hand digits of the dividend—they will be the remainder.

EXAMPLE AA

81,400 ÷ 100 = 814 (0 remainder)

EXAMPLE BB

257,948 ÷ 100 = 2,579 (48 remainder)

DIVIDING WHEN THE DIVISOR AND DIVIDEND END WITH ZEROS

When a divisor and dividend both end with zeros, a division shortcut is to delete the ending zeros common to both and then divide.

EXAMPLE CC

Both Divisor and Dividend End with Zeros	Zeros Common to Divisor and Dividend Have Been Dropped	Answer
8,400 ÷ 200	84 ÷ 2	42
46,000 ÷ 2,300	460 ÷ 23	20
42,000 ÷ 100	420 ÷ 1	420
20,000,000 ÷ 4,000	20,000 ÷ 4	5,000
2,760 ÷ 270	276 ÷ 27	10 (6 remainder)
3,200 ÷ 1,000	32 ÷ 10	3 (2 remainder)

✓ CONCEPT CHECK 1.4

Divide:

```
                    21     quotient
divisor ⟶ 32)683    ⟵ dividend
                    64
                    ---
                    43
                    32
                    ---
                    11     ⟵ remainder
```

Check:

divisor × quotient + remainder = dividend
↓ ↓ ↓ ↓
32 × 21 + 11 = 683

Dividing by 10: 860 ÷ 10 = 86
863 ÷ 10 = 86 (3 remainder)

Dividing by 100: 19,300 ÷ 100 = 193
19,346 ÷ 100 = 193 (46 remainder)

COMPLETE ASSIGNMENT 1.4.

Estimating

Learning Objective 5

Estimate answers before performing operations.

1.4 Emphasize that students should round numbers to zeros and mentally arrive at an approximate answer in *all* business calculations.

ESTIMATING WHEN MULTIPLYING

Is estimating important? *Yes*, it is! In using a calculator to make computations, you may possibly omit keystrokes, accidentally repeat keystrokes, or incorrectly shift/omit decimal points. There is a great deal of difference between 3 times $14.87 and 3 times $1,487. When working with calculations in any manner—such as entering items into a spreadsheet, a cash register, or a calculator—you should always have a mental estimate of the final product.

Mentally estimating an answer provides a good method for checking whether your product is a reasonable answer.

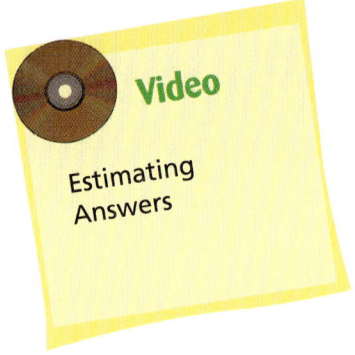

Video — Estimating Answers

STEPS to Estimate a Multiplication Answer

1. Round both the multiplicand and multiplier to the nearest 10 for two-digit numbers, the nearest 100 for three-digit numbers, the nearest 1,000 for four-digit numbers, etc.
2. Drop the zeros to the right of the nonzero numbers.
3. Mentally multiply the nonzero numbers to determine the base product.
4. Reinsert *all* zeros dropped in Step 2.

EXAMPLE DD

Problem	Round to	Drop Zeros	Base Product	Reinsert Zeros Estimated Answer	Real Answer
68 × 21	70 × 20	7 × 2	14	1,400	1,428
693 × 1,957	700 × 2,000	7 × 2	14	1,400,000	1,356,201
7,869 × 43,242	8,000 × 40,000	8 × 4	32	320,000,000	340,271,298
9 × 511,739	9 × 500,000	9 × 5	45	4,500,000	4,605,651
891 × 39 × 104	900 × 40 × 100	9 × 4 × 1	36	3,600,000	3,613,896

ESTIMATING WHEN DIVIDING

Before doing long division problems, estimate a whole-number answer. The process of mentally estimating whole-number answers helps to avoid major and embarrassing errors.

STEPS to Estimate a Long Division Answer

1. Round both the divisor and dividend to the nearest 10 for two-digit numbers, the nearest 100 for three-digit numbers, the nearest 1,000 for four-digit numbers, etc.
2. Drop the number of zeros common to both.
3. Mentally divide the remaining divisor into the remaining dividend.

EXAMPLE EE

Problem	Round to	Drop Zeros	Estimated Answer	Real Answer
77 ÷ 39	80 ÷ 40	8 ÷ 4	2	1.97
196 ÷ 63	200 ÷ 60	20 ÷ 6	3*	3.11*
2,891 ÷ 114	3,000 ÷ 100	30 ÷ 1	30	25.36
592 ÷ 29	600 ÷ 30	60 ÷ 3	20	20.41
18,476 ÷ 384	20,000 ÷ 400	200 ÷ 4	50	48.11
917 ÷ 186	900 ÷ 200	9 ÷ 2	4*	4.93*
21,716,412 ÷ 40,796	20,000,000 ÷ 40,000	2,000 ÷ 4	500	532.32
99,624 ÷ 476	100,000 ÷ 500	1,000 ÷ 5	200	209.29
29,200 ÷ 316	30,000 ÷ 300	300 ÷ 3	100	92.41

*Because 20 ÷ 6 and 9 ÷ 2 would result in remainders we can reasonably assume that the real number will be *larger*.

CONCEPT CHECK 1.5

ESTIMATING MULTIPLICATION ANSWERS

			Reinsert Zeros		
Problem	Round to	Drop Zeros	Base Product	Estimated Answer	Real Answer
47 × 31	50 × 30	5 × 3	15	1,500	1,457
498 × 221	500 × 200	5 × 2	10	100,000	110,058

ESTIMATING DIVISION ANSWERS

Problem	Round to	Drop Zeros	Estimated Answer	Real Answer
88 ÷ 29	90 ÷ 30	9 ÷ 3	3	3.03
9,811 ÷ 394	10,000 ÷ 400	100 ÷ 4	25	24.90

COMPLETE ASSIGNMENT 1.5.

Chapter Terms for Review

- addend
- credit balance
- cross-checking
- difference
- dividend
- divisor
- factors
- minuend
- multiplicand
- multiplier
- product
- quotient
- remainder
- subtrahend
- sum

Try Microsoft® Excel

Using the Student CD found in your textbook, read the Introduction file in the folder Excel Templates and try the Problems for Chapter 1.

THE BOTTOM LINE

Summary of chapter learning objectives:

Learning Objective	Example

1.1
 	
 Use shortcuts and simplifications to perform the fundamental process of addition rapidly and accurately
	Add the following, using the technique indicated.
	
	 Number combinations **Repeated digits** **Counting by tens**
	 1. 8 2. 18 3. 52 4. 23
	      2 62 58 41
	      3 43 57 37
	      2 27 52 56
	   +5 +80 +51 +42
	
	 Add and then check by adding both vertically and horizontally.
	
	 5. 22 + 54 + 63 + 37 = _____
	      27 + 82 + 44 + 19 = _____
	      83 + 39 + 72 + 12 = _____
	      91 + 71 + 21 + 84 = _____
	      __ + __ + __ + __ = _____

1.2
 	
 Use shortcuts and simplifications to perform subtraction rapidly and accurately
	Subtract the following and then check by addition.
	
	 6. 228 _____ 7. 335 _____
	    −134 +134 −217 +217
	
	 Subtract horizontally Subtract by
	 and check. changing numbers.
	
	 8. 245 − 130 = _____ 9. 53
	      432 − 212 = _____ −18
	      381 − 270 = _____
	      183 − 111 = _____
	      __ − __ = _____

1.3
 	
 Use simplifications to perform the fundamental process of multiplication
	Multiply. **Multiplying by numbers ending in zero**
	 10. 227 11. 437 12. 879
	      ×143 ×100 ×10
	
	 Multiplying by 25 **Multiplying by 50**
	 13. 354 14. 846
	      ×25 ×50

Answers: 1. 20   2. 230   3. 270   4. 199   5. 821   6. 94   7. 118   8. 518   9. 35   10. 32,461   11. 43,700   12. 8,790   13. 8,850   14. 42,300

THE BOTTOM LINE

Summary of chapter learning objectives:

Learning Objective	Example
1.4 Use shortcuts and simplifications to perform division rapidly and accurately	Divide and check the answer by multiplication. 15. 27)1,512 27 × _____ **Dividing by numbers ending in 0** 16. 8,430 ÷ 10 = _____ 17. 127,400 ÷ 100 = _____ **Dividing when both divisor and dividend end with zeros** 18. 7,400 ÷ 200 = _____ 19. 53,200 ÷ 400 = _____ 20. 140,000 ÷ 2,000 = _____
1.5 Estimate answers before performing operations	Estimate these multiplication answers. Show your rounding, dropping of zeros with base product, estimated answer, and real answer. See table below.

Problem	Round to	Dropped Zeros and Base Product	Estimated Answer	Real Answer
21. 47 × 31	_____	_____	_____	_____
22. 498 × 221	_____	_____	_____	_____

Estimate these division answers. Show your rounding, dropping of zeros, estimated answer, and real answer.

Problem	Round to	Drop Zeros	Estimated Answer	Real Answer
23. 88 ÷ 29	_____	_____	_____	_____
24. 9,811 ÷ 394	_____	_____	_____	_____

Answers: 15. 56 16. 843 17. 1,274 18. 37 19. 133 20. 70 21. 50 × 30; 5 × 3 = 15; 1,500; 1,457 22. 500 × 200; 5 × 2 = 10; 100,000; 110,058 23. 90 ÷ 30; 9 ÷ 3; 3; 3.03 24. 10,000 ÷ 400; 100 ÷ 4; 25; 24.90

Chapter 1   Fundamental Processes

SELF-CHECK

Review Problems for Chapter 1

1. $8 + 9 + 3 + 12 + 6 =$ _____
2. $32 + 47 + 36 + 12 =$ _____
   $17 + 22 + 17 + 11 =$ _____
   $14 + 98 + 47 + 81 =$ _____
   $77 + 62 + 21 + 44 =$ _____
   ___ ___ ___ ___ = _____
3. $9,078$
   $-6,382$
4. $717 \div 14 =$ _____
5. $98 \times 13 =$ _____
6. $789 \div 36 =$ _____
7. $842 \times 200 =$ _____
8. $974 \div 12 =$ _____
9. $27\overline{)876}$
10. $2,006 \times 304$
11. $395 \div 79$

12. $800 \div 25 =$ _____
13. $4,000,000 \div 400 =$ _____
14. $\$370 - \$148 =$ _____
    $\$422 - \$109 =$ _____
    $\$982 - \$777 =$ _____
    $\$\_\_\_ - \$\_\_\_ =$ _____
15. $1,472$
    $\times 28$
16. $704 \times 1,002 =$ _____
17. $704 \div 25 =$ _____
18. $16,000 \div 25 =$ _____
19. $6,000,006 \div 300$
20. $77,777 \div 707$

Estimate answers for each of the following.

21. $78 \times 29 =$ _____
22. $103 \times 19 =$ _____
23. $397 \times 200 =$ _____
24. $3,982 \times 99 =$ _____
25. $1,503 \times 600 =$ _____

26. $396 \div 79 =$ _____
27. $892 \div 29 =$ _____
28. $9,891 \div 480 =$ _____
29. $3,111 \div 59 =$ _____
30. $6,219 \div 3,114 =$ _____

Assignment 1.1: Addition

Name _____

Date _____ Score _____

Learning Objective 1

A (10 points) Add the following. Where possible, use combinations of 10. (1 point for each correct answer)

1. 18	2. 41	3. 19	4. 34	5. 97	6. 50	7. 72	8. 82	9. 38	10. 92
52	29	54	33	44	54	99	43	39	37
35	17	14	43	33	54	99	47	22	51
42	36	81	37	76	47	89	93	45	36
43	44	28	36	32	59	47	58	47	24
16	15	11	34	72	54	63	34	25	21
22	56	43	32	34	55	40	22	13	19
58	62	51	38	76	55	62	46	29	25
14	66	76	32	27	35	68	73	79	63
300	**366**	**377**	**319**	**491**	**463**	**639**	**498**	**337**	**368**

Score for A (10)

B (10 points) Add the following. (1 point for each correct answer)

11. 209	12. 782	13. 127	14. 920	15. 347	16. 852	17. 251	18. 885	19. 275	20. 438
301	280	145	751	399	428	271	115	342	412
116	438	665	359	354	112	244	316	342	200
214	473	818	822	334	238	234	584	898	415
375	655	682	807	192	959	589	736	505	315
1,215	**2,628**	**2,437**	**3,659**	**1,626**	**2,589**	**1,589**	**2,636**	**2,362**	**1,780**

Score for B (10)

C (10 points) Add the following. (1 point for each correct answer)

21. 248.28	22. 201.22	23. 234.81	24. 238.69	25. 326.52
820.14	513.14	371.60	982.30	117.38
306.80	250.54	271.37	376.48	267.34
521.98	2,647.55	408.55	728.90	118.66
1,897.20	**3,612.45**	**1,286.33**	**2,326.37**	**829.90**

26. 703.91	27. 126.92	28. 442.71	29. 535.13	30. 233.48
422.38	32.15	71.93	44.78	607.22
721.05	873.19	416.90	208.17	211.25
446.21	872.52	236.19	6,481.29	211.25
2,293.55	**1,904.78**	**1,167.73**	**7,269.37**	**1,263.20**

Score for C (10)

Chapter 1 Fundamental Processes 19

Assignment 1.1 Continued

D (10 points) Add the following. Use the count-by-10s-and-add-the-1s method. (1 point for each correct answer)

31.	32.	33.	34.	35.	36.	37.	38.	39.	40.
10.76	20.43	33.79	45.86	33.27	11.43	88.71	94.32	55.93	22.79
31.43	82.76	42.56	22.18	98.21	27.43	56.32	74.23	10.70	43.28
88.33	30.42	12.70	33.81	90.01	11.51	83.70	21.44	30.46	12.48
33.08	64.22	21.20	10.04	11.33	21.48	44.12	63.01	47.05	53.20
12.33	56.03	22.19	80.31	33.04	11.80	23.51	34.20	80.11	30.22
175.93	253.86	132.44	192.20	265.86	83.65	296.36	287.20	224.25	161.97

Score for D (10)

E (30 points) Business Application. The following is the first part of a weekly sales summary—the Weekly Sales Report for the computer department. Complete the totals, both horizontal and vertical, and verify your addition by comparing the vertical and horizontal grand totals. (2 points for each column/row; 4 points for grand total)

DEPARTMENT SALES REPORT
Week of December 11–17, 20XX

Department: COMPUTERS

SALESPERSON	SUN	MON	TUE	WED	THU	FRI	SAT	TOTAL
Whalen	3,443	—	—	8,643	3,176	7,885	9,378	32,525
Tsao	—	8,772	—	9,483	7,339	8,113	9,771	43,478
Culver	8,722	2,443	3,114	5,729	6,193	—	—	26,201
Hernandez	6,117	8,783	—	—	5,685	9,473	11,492	41,550
Ingake	—	3,114	8,492	7,652	3,994	14,119	12,378	49,749
Greenberg	—	—	5,141	2,739	8,941	2,836	10,242	29,899
Total	18,282	23,112	16,747	34,246	35,328	42,426	53,261	223,402

Score for E (30)

F (30 points) Business Application. The following is the second part of the weekly sales summary—the Consolidated Sales Report for the entire store. Fill in the figures from the Department Sales Report and complete the totals, both horizontal and vertical. Verify your addition by comparing the horizontal and vertical grand totals. (2 points for each column/row; 2 points for grand total)

STORE SALES REPORT
Week of December 11–17, 20XX

DEPARTMENT	SUN	MON	TUE	WED	THU	FRI	SAT	TOTAL
Home Audio	3,465	1,147	1,523	2,403	1,773	2,873	3,432	16,616
Auto Audio	1,278	1,785	1,713	2,117	2,563	3,499	9,971	22,926
Video/TV	15,230	12,377	10,429	9,384	8,773	11,245	13,486	80,924
Computers	18,282	23,112	16,747	34,246	35,328	42,426	53,261	223,402
Telecomm	849	722	531	733	1,012	1,239	1,375	6,461
Games	882	248	379	287	415	978	1,015	4,204
Repairs	732	892	384	658	981	1,043	1,774	6,464
Total	40,718	40,283	31,706	49,828	50,845	63,303	84,314	360,997

Score for F (30)

Assignment 1.2: Subtraction

Name

Date   Score

Learning Objective 2

A (18 points) Subtract the following. (One point for each correct answer)

1. 77 −16 61	2. 90 −17 73	3. 72 −25 47	4. 63 −29 34	5. 84 −48 36	6. 38 −49 (11)	7. 92 −16 76	8. 83 −65 18	9. 80 −20 60
10. 39 −36 3	11. 20 −13 7	12. 13 −26 (13)	13. 73 −14 59	14. 63 −19 44	15. 68 −39 29	16. 99 −27 72	17. 57 −43 14	18. 96 −39 57

Score for A (18)

B (12 points) Subtract the following. Then check your subtraction by adding the subtrahend and the difference and comparing your total to the minuend. (2 points for each correct answer)

19. 584 −173 411 584	20. 963 −874 89 963	21. 103 −310 (207) 103	22. 714 −30 684 714	23. 616 −333 283 616	24. 9003 −3116 5887 9003

Score for B (12)

C (6 points) Subtract the following. (1 point for each correct answer)

25. $97.17 −23.19 $73.98	26. $15.67 −0.88 $14.79	27. $71.69 −10.87 $60.82	28. $43.21 −47.18 ($3.97)	29. $80.41 −41.80 $38.61	30. $99.32 −18.66 $80.66

Score for C (6)

D (9 points) Subtract the following. ($1\frac{1}{2}$ points for each correct answer)

31. $8,042.88 −3,400.07 $4,642.81	32. $964.38 −201.83 $762.55	33. $9,011.09 −795.08 $8,216.01	34. $7,430.29 −2,597.73 $4,832.56	35. $3,385.03 −233.42 $3,151.61	36. $1,029.27 −89.27 $940.00

Score for D (9)

E (15 points) Sometimes a double subtraction is necessary. The following problems are of this type. (3 points for each correct final answer)

37. $7,672.18 −564.27 $7,107.91 −124.13 $6,983.78	38. $11,739.93 −3,142.18 8,597.75 −1,694.25 $6,903.50	39. $734.12 −672.18 61.94 −13.14 $48.80	40. $745.89 −250.15 495.74 −224.13 $271.61	41. $1,837,042.03 −6,218.18 1,830,823.85 −39,917.16 $1,790,906.69

Score for E (15)

Assignment 1.2 Continued

F (20 points) Business Application. In many cases, multiple subtractions are required to complete a business transaction. (1 point for each intermediate answer; 2 points for each final answer)

WINTER CATALOG CLEARANCE SALE ON SOFTWARE AND GAMES
10% REDUCTIONS ON CATALOG ORDERS
10% PREFERRED CUSTOMER DISCOUNTS
MAIL-IN REBATE OFFERS

Item	Sierra Half-Life	The Sims 2	Grand Theft Auto	Street Legal	Zoo Tycoon
List price	$43.95	$45.70	$42.25	$49.95	$53.75
Less 10% catalog rate	−4.40	−4.57	−4.23	−5.00	−5.38
	39.55	41.13	38.02	44.95	48.37
Less 10% preferred customer rate	−3.96	−4.11	−3.80	−4.50	−4.84
	35.59	37.02	34.22	40.45	43.53
Mail−in rebate	−7.50	−6.25	−7.50	−6.75	−5.75
Your price	$28.09	$30.77	$26.72	$33.70	$37.78

Score for F (20)

G (20 points) Business Application. Maintaining a budget involves both addition and subtraction. Keeping a budget sometimes involves a continuous record of cash income and expenses. Study the example and then complete the balances. (2 points for each balance)

Date 2/1/98	To	Subtract Expenses	Add Income	Balance
				$1,475.38
2/2/98	Salary income		$700.00	2,175.38
2/3/98	Hinson Real Estate	$550.00		1,625.38
2/5/98	PG&E	23.22		1,602.16
2/6/98	Pacific Bell	18.76		1,583.40
2/6/98	Macy's	43.22		1,540.18
2/10/98	Chevron	15.75		1,524.43
2/16/98	Salary income		$700.00	2,224.43
2/17/98	Fitness USA	25.00		2,199.43
2/18/98	John Simms, D.D.S.	30.00		2,169.43
2/23/98	Prudential Insurance	17.73		2,151.70
2/25/98	Visa	85.42		2,066.28
2/27/98	General Motors Finance	257.87		1,808.41

Score for G (20)

Assignment 1.3: Multiplication

Name _____

Date _____ Score _____

Learning Objective **3**

A (12 points) Multiply the following. ($\frac{1}{2}$ point for each correct answer)

1. $2 \times 12 =$ 24
2. $8 \times 16 =$ 128
3. $13 \times 40 =$ 520
4. $14 \times 48 =$ 672
5. $9 \times 10 =$ 90
6. $5 \times 15 =$ 75
7. $15 \times 16 =$ 240
8. $60 \times 7 =$ 420
9. $8 \times 9 =$ 72
10. $6 \times 12 =$ 72
11. $12 \times 12 =$ 144
12. $55 \times 9 =$ 495
13. $6 \times 8 =$ 48
14. $8 \times 12 =$ 96
15. $4 \times 20 =$ 80
16. $62 \times 70 =$ 4,340
17. $6 \times 6 =$ 36
18. $7 \times 22 =$ 154
19. $8 \times 11 =$ 88
20. $14 \times 700 =$ 9,800
21. $2 \times 14 =$ 28
22. $9 \times 22 =$ 198
23. $8 \times 17 =$ 136
24. $70 \times 70 =$ 4,900

Score for A (12) _____

B (24 points) Find the products. (2 points for each correct answer)

	25.	26.	27.	28.	29.	30.
	1,728	3,026	38,246	5,017	3,600	8,179
	× 42	× 372	× 8,297	× 201	× 300	× 81
	72,576	1,125,672	317,327,062	1,008,417	1,080,000	662,499

	31.	32.	33.	34.	35.	36.
	8,222	67,406	1,236	27,000	8,125	3,716
	× 509	× 3,006	× 444	× 420	× 279	× 418
	4,184,998	202,622,436	548,784	11,340,000	2,266,875	1,553,288

Score for B (24) _____

C (10 points) Multiply by using shortcuts. (2 points for each correct answer)

	37.	38.	39.	40.	41.
	3,684	4,999	6,642	3,212	1,500
	× 50	× 50	× 25	× 50	× 25
	184,200	249,950	166,050	160,600	37,500

Score for C (10) _____

D (18 points) Multiply the three factors. (2 points for each final product)

42. $23 \times 22 \times 21 =$ 10,626
43. $47 \times 16 \times 70 =$ 52,640
44. $44 \times 44 \times 44 =$ 85,184
45. $14 \times 100 \times 7 =$ 9,800
46. $915 \times 40 \times 20 =$ 732,000
47. $10 \times 10 \times 10 =$ 1,000
48. $30 \times 30 \times 30 =$ 27,000
49. $17 \times 34 \times 1,013 =$ 585,514
50. $1,500 \times 9 \times 3 =$ 40,500

Score for D (18) _____

Chapter 1 Fundamental Processes

Assignment 1.3 Continued

E (12 points) Complete the five multiplication problems and then add the five products. (1 point for each correct answer)

51. 12 × 12.00 = 144.00
52. 27 × 8.16 = 220.32
53. 104 × 3.52 = 366.08
54. 6 × 92.92 = 557.52
55. 55 × 32.50 = 1,787.50
56. Total = 3,075.42

57. 21 × 7 × 16 = 2,352
58. 13 × 101 × 22 = 28,886
59. 33 × 14 × 7 = 3,234
60. 99 × 11 × 100 = 108,900
61. 3 × 88 × 100 = 26,400
62. Total = 169,772

Score for E (12)

F (24 points) Business Application. Complete the merchandise inventory TOTAL column. (1 point for each correct total; 8 points for correct grand total)

MERCHANDISE INVENTORY
JUNE 30, 20xx

Stock Number	Description	Price	# in Stock	Total
G473-2	Linspire 4.5	$39.99	58	$ 2,319.42
G763-4	Spysweeper	$39.99	172	$ 6,878.28
G865-A	Encarta	$49.95	98	$ 4,895.10
G2238-1	Turbo Tax	$34.99	225	$ 7,872.75
G873-2	Ever Quest 2	$42.75	88	$ 3,762.00
S876-3	Microsoft Word	$98.77	178	$17,581.06
S4433	Uninstaller 4	$32.59	85	$ 2,770.15
S887-32	Doom 3	$45.79	110	$ 5,036.90
S4536	Netscape Navigator	$38.79	100	$ 3,879.00
S1322	Norton Utilities 7.0	$67.85	68	$ 4,613.80
S458-2	Quicken	$27.75	205	$ 5,688.75
S5382	City of Heros	$95.69	80	$ 7,655.20
E5673-E	Typing Tutor	$26.59	108	$ 2,871.72
E82-18	Atari Atar	$52.49	25	$ 1,312.25
E2442	Adobe 6	$45.29	307	$13,904.03
E3578-1	Perfect Spanish	$44.79	80	$ 3,583.20
			TOTAL	$94,623.61

Score for F (24)

Assignment 1.4: Division

Name

Date   Score

Learning Objective 4

A (10 points) Divide the following problems mentally. ($\frac{1}{2}$ point for each correct quotient)

1. 72 ÷ 6 = 12
2. 90 ÷ 6 = 17
3. 66 ÷ 22 = 3
4. 110 ÷ 5 = 22
5. 126 ÷ 3 = 42
6. 188 ÷ 21 = 9
7. 88 ÷ 22 = 4
8. 144 ÷ 12 = 12
9. 360 ÷ 20 = 18
10. 135 ÷ 9 = 15
11. 990 ÷ 33 = 30
12. 361 ÷ 19 = 19
13. 156 ÷ 12 = 13
14. 900 ÷ 15 = 60
15. 1,782 ÷ 18 = 99
16. 84 ÷ 12 = 7
17. 104 ÷ 2 = 52
18. 561 ÷ 17 = 33
19. 119 ÷ 7 = 17
20. 225 ÷ 15 = 15

Score for A (10)

B (10 points) Divide by shortcut methods. Express remainders in parentheses. (1 point for each correct answer)

21. 1,818 ÷ 333 = 5 (153)
22. 107,300 ÷ 100 = 1,073
23. 97,600 ÷ 100 = 976
24. 2,200 ÷ 100 = 22
25. 7,800 ÷ 20 = 390
26. 6,450 ÷ 320 = 20 (50)
27. 9,005 ÷ 100 = 90 (5)
28. 387 ÷ 10 = 38 (7)
29. 7,600 ÷ 1,000 = 7 (600)
30. 3,250,000 ÷ 10,000 = 325

Score for B (10)

C (50 points) Divide. Show the remainder in parentheses after the whole number in the quotient. (2 points for each correct answer)

31. 21)478 = 22 (16)
32. 13)2,795 = 215
33. 23)14,076 = 612
34. 7)4,919 = 702 (5)
35. 36)6,436 = 178 (28)
36. 23)478 = 20 (18)
37. 271)50,001 = 184 (137)
38. 33)97,382 = 2,950 (32)
39. 926)926,007 = 1,000 (7)
40. 77)12,770 = 165 (65)
41. 506)10,238 = 20 (118)
42. 9)818,173 = 90,908 (1)
43. 700)362,497 = 517 (597)
44. 111)34,173 = 307 (96)
45. 88)97,817 = 1,111 (49)
46. 13)$67,209 = $5,169 (12)
47. 6)$13.20 = $2.20
48. 54)78,540 = 1,454 (24)
49. 51)100 = 1 (49)
50. 26)111,013 = 4,269 (19)
51. 66)73,428 = 1,112 (36)
52. 1,014)20,016 = 19 (750)
53. 66)17,209 = 260 (49)
54. 65)372,000 = 5,723 (5)
55. 29)58,004,316 = 2,000,148 (24)

Score for C (50)

Chapter 1 Fundamental Processes 25

Assignment 1.4 Continued

D (10 points) Divide and check the following problems. (2 points for each correct answer)

56. 22)1,364 = 62
57. 31)1,395 = 45
58. 92)7,284 = 79 (16)
59. 21)2,214 = 105 (9)
60. 31)642 = 20 (22)

Check:

	56	57	58	59	60
	62	45	92	105	31
×	22	31	79	21	20
			7,268	2,205	620
+			16	9	22
=	1,364	1,395	7,284	2,214	642

Score for D (10)

E (20 points) Business Applications. As an estimator for a printing company, you must estimate the paper costs for printing jobs. Paper is priced by the ream, which is 500 pages. Compute the paper costs of the jobs. (1 point for each correct computation)

No. of Booklets	No. of Pages	Total Pages	Reams of Paper	Cost per Ream	Total Paper Cost
250	66	16,500	33	$2.00	$66.00
120	150	18,000	36	$4.25	153.00
75	220	16,500	33	$4.83	159.39
110	250	27,500	55	$3.75	206.25
25	280	7,000	14	$3.15	44.10
30	250	7,500	15	$4.10	61.50
		Total reams	186	Total paper cost	$690.24

Score for E (20)

Part 1 Fundamental Review

Assignment 1.5: Estimating

Name

Date   Score

Learning Objective 5

A (60 points) Estimate an answer for each of the following problems. Show your rounding, dropping of zeros with base product, and final estimate. (1 point for each correct answer)

Problem	Round to	Dropped Zeros and Base Product	Estimated Answer
1. 1,095 × 427	1,000 × 400	1 × 4 = 4	400,000
2. 78,221 × 6,099	80,000 × 6,000	8 × 6 = 48	480,000,000
3. 34,007 × 80	30,000 × 80	3 × 8 = 24	2,400,000
4. 56 × 1,528	60 × 2,000	6 × 2 = 12	120,000
5. 18 × 2,855 × 93	20 × 3,000 × 90	2 × 3 × 9 = 54	5,400,000
6. 20 × 17 × 19	20 × 20 × 20	2 × 2 × 2 = 8	8,000
7. 2,997 × 13	3,000 × 10	3 × 1 = 3	30,000
8. 41 × 19 × 3	40 × 20 × 3	4 × 2 × 3 = 24	2,400
9. 212 × 101 × 99	200 × 100 × 100	2 × 1 × 1 = 2	2,000,000
10. 23 × 10,322	20 × 10,000	2 × 1 = 2	200,000
11. 777 × 777	800 × 800	8 × 8 = 64	640,000
12. 29,301 × 21	30,000 × 20	3 × 2 = 6	600,000
13. 72,111 × 108	70,000 × 100	7 × 1 = 7	7,000,000
14. 13 × 100 × 6	10 × 100 × 6	1 × 1 × 6 = 6	6,000
15. 99 × 99 × 99	100 × 100 × 100	1 × 1 × 1 = 1	1,000,000
16. 28 × 42	30 × 40	3 × 4 = 12	1,200
17. 111 × 39	100 × 40	1 × 4 = 4	4,000
18. 7 × 99	7 × 100	7 × 1 = 7	700
19. 204 × 17	200 × 20	2 × 2 = 4	4000
20. 11 × 12 × 13	10 × 10 × 10	1 × 1 × 1 = 1	1,000

Score for A (60)

Chapter 1 Fundamental Processes

Assignment 1.5 Continued

B (20 points) Estimate an answer for each of the following problems. Show your rounding, dropping of zeros with base product, estimated answer, and real answer. (1 point for each correct answer)

Problem	Round to	Dropped Zeros and Base Product	Estimated Answer	Real Answer
21. 883 × 294	900 × 300	9 × 3 = 27	270,000	259,602
22. 42,100 × 412	42,000 × 400	42 × 4 = 168	16,800,000	17,345,200
23. 19,965 × 492	20,000 × 500	2 × 5 = 10	10,000,000	9,822,780
24. 89 × 33	90 × 30	9 × 3 = 27	2,700	2,937
25. 793 × 199	800 × 200	8 × 2 = 16	160,000	157,807

Score for B (20)

C (20 points) Estimate an answer for each of the following division problems. Show your rounding, dropping of zeros, estimated answer, and real answer. Round to two decimal places. (1 point for each correct answer)

Problem	Round to	Drop Zeros	Estimated Answer	Real Answer
26. 123 ÷ 41	120 ÷ 40	12 ÷ 4	3	3
27. 612 ÷ 12	600 ÷ 10	60 ÷ 1	60	51
28. 4,836 ÷ 78	4,800 ÷ 80	480 ÷ 8	60	62
29. 19,760 ÷ 95	20,000 ÷ 100	200 ÷ 1	200	208
30. 21,033 ÷ 690	21,000 ÷ 700	210 ÷ 7	30	30 (333)

Score for C (20)

Fractions

2

Learning Objectives
By studying this chapter and completing all assignments you will be able to:

Learning Objective 1 — Change improper fractions and mixed numbers.

Learning Objective 2 — Change fractions to lower and higher terms.

Learning Objective 3 — Add fractions and mixed numbers.

Learning Objective 4 — Subtract fractions and mixed numbers.

Learning Objective 5 — Multiply fractions, mixed numbers, and whole numbers.

Learning Objective 6 — Divide fractions, mixed numbers, and whole numbers.

Fractions are a natural part of cultures around the world. Very young children who cannot yet read learn simple fractions such as one half and one third when their parents teach them about sharing a candy bar or a pizza. Before the development of inexpensive handheld calculators, fractions were more important than today because they permitted shortcuts in arithmetic. However, fractions are still important in some industries. Moreover, the rules of fractions will always remain very important in algebra and higher mathematics.

Vocabulary of Fractions

2.1 The line separating the numerator and denominator is often called the "fraction bar." Encourage students to write the fraction bar as a horizontal line instead of a slanted line. Because of printing fonts, even in this book, fractions are sometimes written with a slanted line, such as 1/4. However, when someone is writing by hand, using a dull pencil, it is not uncommon to confuse 2 1/4 with 21/4.

A restaurant cuts its medium-sized pizzas into six pieces. Each piece is "*one sixth*" of the pizza. If you take two pieces of pizza, you have "*two sixths*" of the pizza. With numbers, two sixths is written as $\frac{2}{6}$. The 2 is called the **numerator,** and the six is called the **denominator**. $\frac{2}{6}$ is called a **proper fraction** because its numerator (2) is smaller than its denominator (6). If you buy two medium-sized pizzas and cut each into six pieces, you will have twelve pieces, or twelve sixths, written as $\frac{12}{6}$. $\frac{12}{6}$ is called an **improper fraction** because its numerator (12) is larger than its denominator (6). If you eat one of the twelve slices of pizza, eleven pieces remain, or $\frac{11}{6}$, or one whole pizza and $\frac{5}{6}$ of the other pizza. We can write this result as $1\frac{5}{6}$, which is called a **mixed number**. $1\frac{5}{6}$ is simply another way to write $\frac{11}{6}$. Figure 2-1 illustrates these concepts.

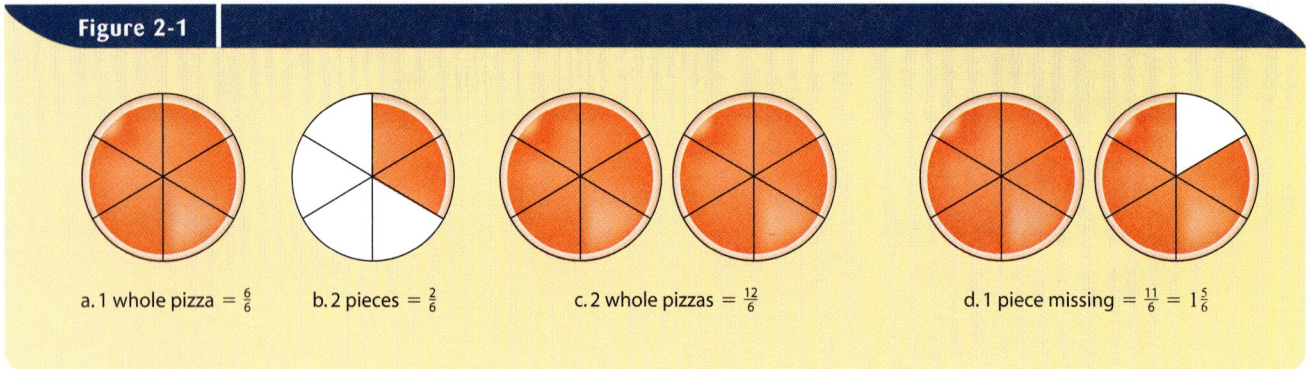

Figure 2-1
a. 1 whole pizza = $\frac{6}{6}$ b. 2 pieces = $\frac{2}{6}$ c. 2 whole pizzas = $\frac{12}{6}$ d. 1 piece missing = $\frac{11}{6}$ = $1\frac{5}{6}$

Changing Improper Fractions and Mixed Numbers

With simple arithmetic, we can change improper fractions to mixed numbers and mixed numbers to improper fractions.

Change improper fractions and mixed numbers.

STEPS to Change an Improper Fraction to a Mixed Number

1. Divide the numerator by the denominator.
2. The quotient is the whole-number part of the mixed number.
3. The remainder is the numerator of the fraction part.
4. The original denominator is the denominator of the fraction part.

EXAMPLE A

Change $\frac{11}{8}$ to a mixed number.

STEP 1

STEPS 2, 3, & 4

$$\frac{11}{8} = 8\overline{)11}^{1\ R3} \qquad \text{Thus,} \quad \frac{11}{8} = 1\frac{3}{8}$$
$$\phantom{\frac{11}{8} = 8\overline{)11}}\frac{8}{3}$$

Note: Refer to Point A in Figure 2-2 to see where this mixed number appears on a ruler.

Figure 2-2

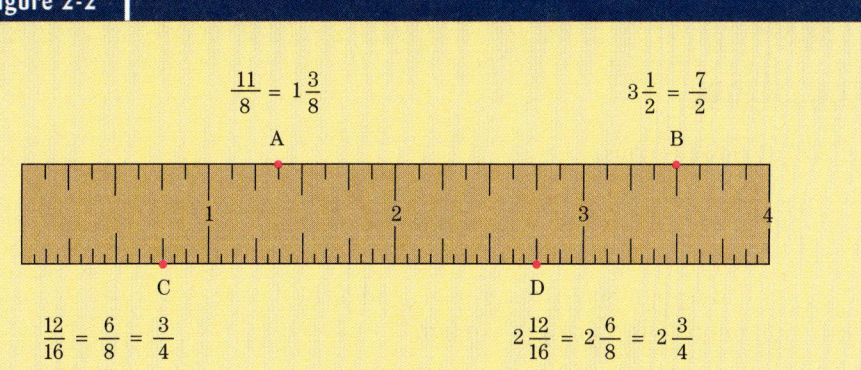

$\frac{11}{8} = 1\frac{3}{8}$ at A

$3\frac{1}{2} = \frac{7}{2}$ at B

$\frac{12}{16} = \frac{6}{8} = \frac{3}{4}$ at C

$2\frac{12}{16} = 2\frac{6}{8} = 2\frac{3}{4}$ at D

STEPS to Change a Mixed Number to an Improper Fraction

1. Multiply the denominator of the fraction by the whole number.
2. Add the numerator of the fraction to the product of Step 1. The sum is the numerator of the improper fraction.
3. The denominator of the fraction of the mixed number is the denominator of the improper fraction.

2.2 Emphasize that $3\frac{1}{2}$ means $3 + \frac{1}{2}$.

EXAMPLE B

Change $3\frac{1}{2}$ to an improper fraction.

STEP 1

STEPS 2, 3

$$2 \times 3 = 6 \qquad \text{Thus,} \ 3\frac{1}{2} = \frac{6 + 1}{2} = \frac{7}{2}$$

See Point B in Figure 2-2.

Changing Fractions to Lower and Higher Terms

Learning Objective 2

Change fractions to lower and higher terms.

Read Point C on the measuring tape shown in Figure 2-2. Point C marks the distance $\frac{12}{16}$ of an inch, but it could also be read as $\frac{6}{8}$ or $\frac{3}{4}$ of an inch. Thus $\frac{12}{16}$, $\frac{6}{8}$, and $\frac{3}{4}$ are three ways to write the same value. We say that $\frac{6}{8}$ is in **lower terms** and $\frac{12}{16}$ is in **higher terms** because 8 is a smaller denominator than 16. We also say that $\frac{3}{4}$ is in **lowest terms** because it cannot be changed to any lower terms. When we change a fraction to lower terms, we say that we are *reducing* the fraction to lower terms. If we change a mixed number such as $2\frac{12}{16}$ to $2\frac{3}{4}$, we say that we have reduced the mixed number to its lowest terms. When we change a fraction to higher terms, we say that we are *raising* the fraction to higher terms.

2.3 The word *cancel* is commonly used, as in "cancel a common factor." Emphasize that the word *cancel* means to *divide* the common factor. The word *cancel* is not wrong, but some students simply cross out common numbers without thinking. A useful illustration might be $\frac{16}{64}$.

If the two 6s are "crossed out" (as opposed to canceled) the result is $\frac{16}{64} = \frac{1}{4}$, which is the correct answer. Naturally, however, that method of "canceling" does not work with most other fractions—such as $\frac{16}{96}$, for instance. You can even challenge students to find other examples that do work.

> **STEPS** to Reduce a Fraction to Lowest Terms
>
> 1. Divide both the numerator and the denominator by a common divisor greater than 1 to arrive at a reduced fraction.
> 2. If necessary, repeat Step 1 until the fraction cannot be reduced any further.
>
> Note: If a fraction's numerator and denominator have no common divisor greater than 1, the fraction is already in lowest terms.

● **EXAMPLE C**

Reduce $\frac{12}{16}$ to lowest terms.

$$\frac{12}{16} = \frac{12 \div 2}{16 \div 2} = \frac{6}{8} = \frac{6 \div 2}{8 \div 2} = \frac{3}{4} \quad \text{or} \quad \frac{12}{16} = \frac{12 \div 4}{16 \div 4} = \frac{3}{4}$$

Note that dividing by 4 once is faster than dividing by 2 twice. Always try to use the greatest common divisor that you can find.

> **STEPS** to Raise a Fraction to Higher Terms
>
> 1. Divide the new denominator by the old denominator. The quotient is the *common multiplier*.
> 2. Multiply the old numerator by the common multiplier.
> 3. Multiply the old denominator by the common multiplier.

● **EXAMPLE D**

Video

Reducing and Raising Fractions

Raise $\frac{3}{4}$ to twenty-fourths.

STEP 1

$$\frac{3}{4} = \frac{?}{24} \quad 24 \div 4 = 6$$

STEPS 2 & 3

So, $\frac{3}{4} = \frac{3 \times 6}{4 \times 6} = \frac{18}{24}$

Adding Fractions and Mixed Numbers

Fractions and mixed numbers are all numbers—they can be added and subtracted just like whole numbers. However, **fractions and mixed numbers cannot be added or subtracted until they have the same denominators called a common denominator.**

When you add fractions and/or mixed numbers, you must first find a **common denominator,** which is a denominator shared by all of the fractions and it will be the denominator of the fraction part of the answer. The smallest common denominator possible is called the **least common denominator.** However, if the least common denominator is not easily apparent, it may be quicker to use the first common denominator that you can discover and then reduce the answer to lowest terms. The product of all of the denominators will always be a common denominator, but very often there will be a smaller common denominator.

Learning Objectives 3

Add fractions and mixed numbers.

Video
Adding and Subtracting Fractions and Mixed Numbers

STEPS to Add Two or More Fractions and/or Mixed Numbers

1. If necessary, change the fraction parts to fractions with common denominators. The common denominator is the denominator in the fraction part of the answer.
2. Add the numerators to make the numerator of the fraction part of the answer. If there are any whole-number parts, add them to make the whole-number part of the answer.
3. If necessary, reduce the fraction part to a mixed number in lowest terms and mentally combine any whole-number parts to make a final mixed-number answer.

EXAMPLE E

Add $2\frac{7}{8}$ and $2\frac{5}{8}$.
The fractions already have a common denominator of 8.

$$\begin{array}{r} 2\frac{7}{8} \\ + 2\frac{5}{8} \\ \hline 4\frac{12}{8} = 4 + 1\frac{4}{8} = 5\frac{1}{2} \end{array}$$

EXAMPLE F

Add $\frac{5}{6}$ and $\frac{3}{4}$.
A common denominator is $6 \times 4 = 24$.

$$\begin{array}{r} \frac{5}{6} = \frac{5 \times 4}{6 \times 4} = \frac{20}{24} \\ + \frac{3}{4} = \frac{3 \times 6}{4 \times 6} = \frac{18}{24} \\ \hline \frac{38}{24} = 1\frac{14}{24} = 1\frac{7}{12} \end{array}$$

2.4 If time permits you can experiment. Give each student two problems: Solve the first one by using the common denominator that is the product of the three denominators and then reducing the answer to lowest terms; solve the second one by first finding the least common denominator (no credit for using any other denominator). Have the students time themselves. Students will probably not be uniformly quicker with one method or the other. As an alternative, you could let half the class solve problem 1 and the other half solve problem 2. (You can simplify the two problems if you want to by having the students add only the first two fractions in each problem.)

1. $\frac{7}{15} + \frac{5}{6} + \frac{9}{10}$

A common denominator is $15 \times 6 \times 10$.

2. $\frac{5}{14} + \frac{7}{10} + \frac{4}{35}$

Find the least common denominator; then add.

Chapter 2 Fractions **33**

EXAMPLE G

Add $3\frac{3}{8}$, $7\frac{5}{6}$, and $\frac{1}{4}$.
The least common denominator is 24.

STEP 1

$$3\frac{5}{8} = 3\frac{15}{24}$$

$$7\frac{5}{6} = 7\frac{20}{24}$$

$$+\frac{1}{4} = +\frac{6}{24}$$

STEP 2

$$\text{Sum} = 10\frac{41}{24}$$

STEP 3

$$10 + 1\frac{17}{24} = 11\frac{17}{24}$$

✓ CONCEPT CHECK 2.1

a. Add $\frac{3}{5}$, $\frac{2}{3}$, and $\frac{7}{9}$.
The least common denominator is 45.

$$\frac{3}{5} = \frac{3 \times 9}{5 \times 9} = \frac{27}{45}$$

$$\frac{2}{3} = \frac{2 \times 15}{3 \times 15} = \frac{30}{45}$$

$$+\frac{7}{9} = \frac{7 \times 5}{9 \times 5} = +\frac{35}{45}$$

$$\frac{92}{45} = 2\frac{2}{45}$$

b. Add $1\frac{5}{6}$ and $2\frac{5}{9}$.
A common denominator is $6 \times 9 = 54$.

$$1\frac{5}{6} = 1\frac{45}{54}$$

$$+2\frac{5}{9} = +2\frac{30}{54}$$

$$3\frac{75}{54} = 4\frac{21}{54} = 4\frac{7}{18}$$

Subtracting Fractions and Mixed Numbers

Learning Objective 4

Subtract fractions and mixed numbers.

The procedure for subtracting one fraction from another is essentially the same as the procedure for adding one fraction to another. When you calculate the difference $3\frac{1}{4} - \frac{3}{4}$, $3\frac{1}{4}$ is called the *minuend* and $\frac{3}{4}$ is called the *subtrahend*, as in the subtraction of whole numbers.

BORROWING 1

2.5 Point out that one reason for learning the words *subtrahend* and *minuend* is to be able to write and read explanations such as the one for "borrowing."

Sometimes, as with $3\frac{1}{4} - \frac{3}{4}$, the fraction part of the minuend is smaller than the fraction part of the subtrahend. To make the fraction part of the minuend larger than the fraction part of the subtrahend, you have to "borrow 1" from the whole-number part of the minuend. Actually, you're just rewriting the minuend. Remember that $3\frac{1}{4}$ means $3 + \frac{1}{4}$, or the same as $2 + 1 + \frac{1}{4}$, $2 + \frac{4}{4} + \frac{1}{4}$, or $2\frac{5}{4}$. These are simply four different ways to express the same quantity. Figure 2-3 is useful in understanding borrowing.

Figure 2-3

a. 3 whole units plus $\frac{1}{4}$ of a unit

a. 2 "whole units" plus $\frac{5}{4}$ of a unit

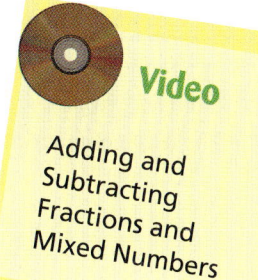

Video

Adding and Subtracting Fractions and Mixed Numbers

STEPS to Subtract One Fraction or Mixed Number from Another

1. If necessary, change the fractions so that all fractions have a common denominator. The common denominator is the denominator in the fraction part of the answer.
2. If necessary, "borrow 1" from the whole-number part of the minuend so that the fraction part of the minuend is at least as large as the fraction part of the subtrahend.
3. Subtract the numerators in the fractions to make the numerator in the fraction part of the answer.
4. If there are any whole-number parts, subtract them to make the whole-number part of the answer.
5. If necessary, reduce the fraction in the answer to lowest terms.

2.6 In Step 2, "borrow 1" means to "subtract 1 from the whole-number part, convert it to a fraction with the same denominator as the fraction part, and add it to the fraction part." That will be clear to most students, but it is a cumbersome and potentially confusing instruction to include in the "steps."

EXAMPLE H

STEP 3   STEP 5

$$\frac{7}{8} - \frac{3}{8} = \frac{7-3}{8} = \frac{4}{8} = \frac{1}{2}$$

EXAMPLE I

STEP 1   STEP 3

$$\frac{3}{4} - \frac{1}{5} = \frac{3 \times 5}{4 \times 5} - \frac{1 \times 4}{5 \times 4} = \frac{15}{20} - \frac{4}{20} = \frac{11}{20}$$

EXAMPLE J

STEP 1

$$\begin{array}{r} 5\frac{3}{4} = 5\frac{9}{12} \\ -\,2\frac{1}{3} = -\,2\frac{4}{12} \\ \hline \end{array}$$

STEPS 3 & 4

$$3\frac{5}{12}$$

EXAMPLE K

STEP 1   STEP 2

$$\begin{array}{r} 4\frac{4}{9} = 4\frac{8}{18} = 3\frac{18}{18} + \frac{8}{18} = 3\frac{26}{18} \\ -\,1\frac{5}{6} = -\,1\frac{15}{18} = -\,1\frac{15}{18} \phantom{+ \frac{8}{18}} = -\,1\frac{15}{18} \\ \hline \end{array}$$

STEPS 3 & 4

$$2\frac{11}{18}$$

© ROSE ALCORN/THOMSON

✓ CONCEPT CHECK 2.2

a. Subtract $\frac{5}{6}$ from $\frac{7}{8}$.

The least common denominator is 24.

$$\frac{7}{8} = \frac{21}{24}$$
$$-\frac{5}{6} = -\frac{20}{24}$$
$$\frac{1}{24}$$

b. Subtract $2\frac{7}{10}$ from $5\frac{4}{15}$.

The least common denominator is 30.

$$5\frac{4}{15} = 5\frac{14}{30} = 4\frac{44}{30}$$
$$-2\frac{7}{10} = -2\frac{21}{30} = -2\frac{21}{30}$$
$$2\frac{23}{30}$$

COMPLETE ASSIGNMENT 2.1.

Multiplying Fractions, Mixed Numbers, and Whole Numbers

Learning Objective 5

Multiply fractions, mixed numbers, and whole numbers.

In fractions, multiplication is the simplest operation and division is the next simplest. The reason is that multiplication and division do not require common denominators like addition and subtraction do. Recall that any mixed number can be changed to an improper fraction. Also, a whole number can be written as an improper fraction by writing the number in the numerator with a denominator of 1. For example, the whole number 5 can be written as the fraction $\frac{5}{1}$.

Video

Multiplication and Division of Mixed Numbers

STEPS to Multiply Fractions, Mixed Numbers, and Whole Numbers

1. If necessary, change any mixed or whole numbers to improper fractions.
2. Multiply all the numerators to get the numerator of the product.
3. Multiply all the denominators to get the denominator of the product.
4. Write the product as a fraction or mixed number in lowest terms.

EXAMPLE L

STEP 1   STEPS 2 & 3   STEP 4

$$1\frac{2}{3} \times \frac{4}{5} = \frac{5}{3} \times \frac{4}{5} = \frac{5 \times 4}{3 \times 5} = \frac{20}{15} = 1\frac{5}{15} = 1\frac{1}{3}$$

EXAMPLE M

STEPS 2 & 3   STEP 4

$$\frac{2}{3} \times \frac{4}{5} \times \frac{5}{6} = \frac{2 \times 4 \times 5}{3 \times 5 \times 6} = \frac{40}{90} = \frac{4}{9}$$

Note: The word *of* often means *multiply* when it is used with fractions. For example, you know that "$\frac{1}{2}$ of 6 bottles" is 3 bottles. And $\frac{1}{2}$ of $6 = \frac{1}{2} \times \frac{6}{1} = \frac{6}{2} = 3$. For this reason, in this age of calculators, multiplication may be the most important arithmetic operation with fractions. In verbal communication, we will always be using expressions like "$\frac{1}{2}$ of 6."

Canceling Common Factors in Numerators and Denominators

As the last step in example M, we reduced the fraction $\frac{40}{90}$ to its lowest terms, $\frac{4}{9}$. Recall that reducing this fraction means that we divide both the numerator and the denominator by 10. As an option, we can do the division in advance, before doing any multiplication. Examining the three numerators and three denominators we discover that they have common factors of 2 and 5 ($2 \times 5 = 10$). Divide out, or **cancel,** both common factors in the numerators and denominators as shown in example N. This division of the common factors is often called **cancellation.** Canceling common factors is an option; it is not required to calculate the correct product.

EXAMPLE N

Multiply the three fractions, using cancellation.

$$\frac{2}{3} \times \frac{4}{5} \times \frac{5}{6} = \frac{2}{3} \times \frac{4}{\overset{1}{\underset{1}{5}}} \times \frac{\overset{1}{5}}{6} = \frac{2}{3} \times \frac{\overset{2}{4}}{\underset{1}{5}} \times \frac{\overset{1}{5}}{\underset{3}{6}} = \frac{2 \times 2 \times 1}{3 \times 1 \times 3} = \frac{4}{9}$$

EXAMPLE O

Multiply the fraction and the whole number, using cancellation.

$$12 \times \frac{3}{4} = \frac{12}{1} \times \frac{3}{4} = \frac{\overset{3}{12}}{\underset{1}{1}} \times \frac{3}{4} = \frac{3 \times 3}{1 \times 1} = \frac{9}{1} = 9$$

EXAMPLE P

Multiply the fraction and the mixed number, using cancellation.

    STEP 1   STEPS 2 & 3

$$\frac{2}{5} \times 2\frac{3}{4} = \frac{2}{5} \times \frac{11}{4} = \frac{\overset{1}{2}}{5} \times \frac{11}{\underset{2}{4}} = \frac{1 \times 4}{5 \times 2} = \frac{11}{10} = 1\frac{1}{10}$$

✓ CONCEPT CHECK 2.3

Multiply the fraction, whole number, and mixed number, using cancellation.

    STEP 1   STEPS 2 & 3   STEP 4

$$\frac{1}{8} \times 4 \times 2\frac{1}{3} = \frac{1}{8} \times \frac{4}{1} \times \frac{7}{3} = \frac{1}{8} \times \frac{\overset{1}{4}}{\underset{2}{1}} \times \frac{7}{3} = \frac{1 \times 1 \times 7}{2 \times 1 \times 3} = \frac{7}{6} = 1\frac{1}{6}$$

Dividing Fractions, Mixed Numbers, and Whole Numbers

Learning Objective 6

Divide fractions, mixed numbers, and whole numbers.

Recall that with whole numbers, division is the *inverse* of multiplication. You can check a multiplication problem by division. With fractions, you actually perform a division problem by doing multiplication. That is, you *invert the divisor and multiply*.

STEPS to Divide Fractions, Mixed Numbers, and Whole Numbers

1. If necessary, change the dividend and/or the divisor from mixed or whole numbers to improper fractions.
2. Invert the divisor (that is, exchange the numerator and denominator).
3. Change the division symbol to a multiplication symbol.
4. Multiply the two factors (canceling where possible, if desired).
5. Write the quotient as a proper fraction or mixed number in lowest terms.

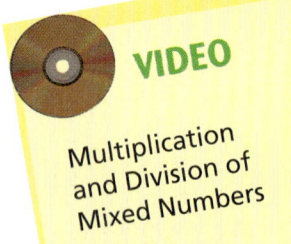

VIDEO
Multiplication and Division of Mixed Numbers

EXAMPLE Q

STEPS 2 & 3 — STEP 4

$$\frac{3}{10} \div \frac{2}{5} = \frac{3}{10} \times \frac{5}{2} = \frac{3}{\underset{2}{10}} \times \frac{\overset{1}{5}}{2} = \frac{3 \times 1}{2 \times 2} = \frac{3}{4}$$

EXAMPLE R

STEP 1 — STEPS 2 & 3 — STEP 4 — STEP 5

$$6 \div 1\frac{3}{5} = \frac{6}{1} \div \frac{8}{5} = \frac{6}{1} \times \frac{5}{8} = \frac{\overset{3}{6}}{1} \times \frac{5}{\underset{4}{8}} = \frac{3 \times 5}{1 \times 4} = \frac{15}{4} = 3\frac{3}{4}$$

✓ CONCEPT CHECK 2.4

Divide $3\frac{3}{4}$ by $1\frac{1}{2}$.

Change both mixed numbers to improper fractions: $\frac{15}{4} \div \frac{3}{2}$.
Invert the divisor $\frac{3}{2}$ to $\frac{2}{3}$ and multiply:

$$\frac{15}{4} \times \frac{2}{3} = \frac{\overset{5}{15}}{\underset{2}{4}} \times \frac{\overset{1}{2}}{\underset{1}{3}} = \frac{5 \times 1}{2 \times 1} = \frac{5}{2} = 2\frac{1}{2}$$

COMPLETE ASSIGNMENT 2.2.

Chapter Terms for Review

- cancel
- cancellation
- common denominator
- denominator
- fractions
- higher terms
- improper fraction
- least common denominator
- lower terms
- lowest terms
- mixed number
- numerator
- proper fraction

THE BOTTOM LINE

Summary of chapter learning objectives:

Learning Objective	Example
2.1 Change improper fractions and mixed numbers	1(a). Change $\frac{18}{5}$ to a mixed number. 1(b). Change $3\frac{2}{5}$ to an improper fraction.
2.2 Change fractions to lower and higher terms	2(a). Reduce $\frac{24}{60}$ to lowest terms. 2(b). Raise $\frac{7}{12}$ to sixtieths; that is, $\frac{7}{12} = \frac{?}{60}$.
2.3 Add fractions and mixed numbers	3. Add $\frac{7}{8}, \frac{5}{6},$ and $2\frac{3}{4}$.
2.4 Subtract fractions and mixed numbers	4. Subtract $1\frac{3}{4}$ from $4\frac{2}{5}$.
2.5 Multiply fractions, mixed numbers, and whole numbers	5. Multiply: $\frac{2}{9} \times \frac{6}{7}$.
2.6 Divide fractions, mixed numbers, and whole numbers	6. Divide: $1\frac{4}{5} \div \frac{3}{4}$.

Answers: 1(a). $3\frac{3}{5}$ 1(b). $\frac{17}{5}$ 2(a). $\frac{2}{5}$ 2(b). $\frac{35}{60}$ 3. $4\frac{11}{24}$ 4. $2\frac{13}{20}$ 5. $\frac{4}{21}$ 6. $2\frac{2}{5}$

SELF-CHECK

Review Problems for Chapter 2

Write all answers as proper fractions or mixed numbers in lowest terms.

1. Change $2\frac{5}{6}$ to an improper fraction _____

2. Change $\frac{90}{12}$ to a mixed number _____

3. Reduce $\frac{54}{63}$ to lowest terms _____

4. Raise $\frac{10}{14}$ to 56ths _____

5. Add $\frac{2}{3}, \frac{3}{5},$ and $\frac{3}{10}$ _____

6. Add $\frac{5}{8}$ and $1\frac{1}{6}$ _____

7. Add $\frac{3}{4}, 2\frac{4}{5},$ and 4 _____

8. Subtract $\frac{1}{3}$ from $\frac{4}{5}$ _____

9. Subtract $\frac{8}{9}$ from $2\frac{5}{6}$ _____

10. Subtract $2\frac{4}{9}$ from $4\frac{1}{5}$ _____

11. Multiply $\frac{5}{6}$ by $\frac{9}{25}$ _____

12. Multiply $\frac{9}{16}$ by $1\frac{13}{15}$ _____

13. Multiply $2\frac{1}{10}, \frac{8}{15},$ and $2\frac{1}{12}$ _____

14. Divide $\frac{15}{16}$ by $\frac{5}{12}$ _____

15. Divide $1\frac{11}{25}$ by $\frac{24}{35}$ _____

16. Divide $1\frac{5}{7}$ by $1\frac{13}{14}$ _____

17. JoAnn Brandt decided to use an expensive, but effective, herbicide to kill weeds and brush on a client's land. For one part of the land, she needed $3\frac{2}{3}$ quarts of herbicide; for a second part, she needed $2\frac{3}{4}$ quarts; and for the third part, she needed $1\frac{5}{6}$ quarts. In total, how many quarts of herbicide did JoAnn need for this client? _____

18. Cabinetmaker Dave Smith needs to make a cabinet door. The cabinet drawing shows an opening $24\frac{1}{16}$ inches wide. Dave wants a space of $\frac{1}{8}$ inch on each side of the cabinet door. How wide should he make the door? _____

19. The Central Hotel just hired a new chef. This chef makes a hot sauce that uses $1\frac{3}{4}$ tablespoons of chili powder, but he needs to increase the recipe by $3\frac{1}{2}$ times. How many tablespoons of chili powder should he use? _____

20. How many whole pieces of copper $2\frac{5}{8}$ inches long can be cut out of one piece that is $24\frac{1}{2}$ inches long? _____ How long is the shorter piece that is left over? _____

Notes

Assignment 2.1: Addition and Subtraction of Fractions

Name _____

Date _____ Score _____

Learning Objectives 1 2 3 4

A (12 points) Change the improper fractions to whole numbers or to mixed numbers. Change the mixed numbers to improper fractions. (1 point for each correct answer)

1. $\dfrac{13}{6}$ $2\dfrac{1}{6}$
2. $\dfrac{32}{10}$ $3\dfrac{1}{5}$
3. $\dfrac{18}{6}$ 3
4. $\dfrac{25}{15}$ $1\dfrac{2}{3}$
5. $\dfrac{11}{7}$ $1\dfrac{4}{7}$
6. $\dfrac{25}{8}$ $3\dfrac{1}{8}$
7. $3\dfrac{7}{10}$ $\dfrac{37}{10}$
8. $2\dfrac{11}{12}$ $\dfrac{35}{12}$
9. $2\dfrac{5}{8}$ $\dfrac{21}{8}$
10. $3\dfrac{3}{4}$ $\dfrac{15}{4}$
11. $6\dfrac{3}{5}$ $\dfrac{33}{5}$
12. $1\dfrac{3}{5}$ $\dfrac{24}{15}$

Score for A (12) _____

B (15 points) In problems 13–20, reduce each fraction to lowest terms. In problems 21–27, raise each fraction to higher terms, as indicated. (1 point for each correct answer)

13. $\dfrac{10}{25}$ $\dfrac{2}{5}$
14. $\dfrac{9}{24}$ $\dfrac{3}{8}$
15. $\dfrac{10}{12}$ $\dfrac{5}{6}$
16. $\dfrac{12}{20}$ $\dfrac{3}{5}$
17. $\dfrac{32}{48}$ $\dfrac{2}{3}$
18. $\dfrac{24}{42}$ $\dfrac{4}{7}$
19. $\dfrac{42}{60}$ $\dfrac{7}{10}$
20. $\dfrac{16}{32}$ $\dfrac{1}{2}$
21. $\dfrac{1}{6} = \dfrac{13}{18}$
22. $\dfrac{3}{4} = \dfrac{15}{20}$
23. $\dfrac{5}{8} = \dfrac{15}{24}$
24. $\dfrac{7}{12} = \dfrac{21}{36}$
25. $\dfrac{11}{6} = \dfrac{88}{48}$
26. $\dfrac{2}{3} = \dfrac{10}{15}$
27. $\dfrac{4}{5} = \dfrac{36}{45}$

Score for B (15) _____

C (24 points) Add the following fractions and mixed numbers. Write the answers as fractions or mixed numbers, with fractions in lowest terms. (3 points for each correct answer)

28. $\dfrac{5}{8}$
    $+\dfrac{3}{8}$
    $\overline{\dfrac{8}{8} = 1}$

29. $\dfrac{3}{10}$
    $+\dfrac{3}{10}$
    $\overline{\dfrac{6}{10} = \dfrac{3}{5}}$

30. $\dfrac{9}{16}$
    $+2\dfrac{11}{16}$
    $\overline{2\dfrac{20}{16} = 3\dfrac{1}{4}}$

31. $1\dfrac{2}{3} = 1\dfrac{8}{12}$
    $+2\dfrac{3}{4} = +2\dfrac{9}{12}$
    $\overline{3\dfrac{17}{12} = 4\dfrac{5}{12}}$

32. $1\dfrac{1}{4} = 1\dfrac{6}{24}$
    $\dfrac{5}{8} = \dfrac{15}{24}$
    $+4\dfrac{11}{12} = +4\dfrac{22}{24}$
    $\overline{5\dfrac{43}{24} = 6\dfrac{19}{24}}$

33. $4\dfrac{1}{2} = 4\dfrac{3}{6}$
    $3\dfrac{2}{3} = 3\dfrac{4}{6}$
    $+\dfrac{5}{6} = +\dfrac{5}{6}$
    $\overline{7\dfrac{12}{6} = 9}$

34. $\dfrac{4}{5} = \dfrac{24}{30}$
    $3\dfrac{5}{6} = 3\dfrac{25}{30}$
    $+5\dfrac{1}{3} = +5\dfrac{10}{30}$
    $\overline{8\dfrac{59}{30} = 9\dfrac{29}{30}}$

35. $2\dfrac{5}{9} = 2\dfrac{25}{45}$
    $3\dfrac{8}{15} = 3\dfrac{24}{45}$
    $+1\dfrac{1}{5} = +1\dfrac{9}{45}$
    $\overline{6\dfrac{58}{45} = 7\dfrac{13}{45}}$

Score for C (24) _____

Chapter 2 Fractions 43

Assignment 2.1 Continued

D (24 points) Subtract the following fractions and mixed numbers. Write the answers as proper fractions or mixed numbers, with fractions in lowest terms. (3 points for each correct answer)

36. $\dfrac{5}{8}$
    $-\dfrac{3}{8}$
    $\dfrac{2}{8} = \dfrac{1}{4}$

37. $2\dfrac{7}{12}$
    $-1\dfrac{1}{12}$
    $1\dfrac{6}{12} = 1\dfrac{1}{2}$

38. $\dfrac{3}{4} = \dfrac{12}{16}$
    $-\dfrac{5}{16} = -\dfrac{5}{16}$
    $\dfrac{7}{16}$

39. $2\dfrac{3}{4} = 2\dfrac{9}{12}$
    $-1\dfrac{1}{12} = -1\dfrac{1}{12}$
    $1\dfrac{8}{12} = 1\dfrac{2}{3}$

40. $3\dfrac{2}{3} = 3\dfrac{4}{6} = 2\dfrac{10}{6}$
    $-2\dfrac{5}{6} = -2\dfrac{5}{6} = -2\dfrac{5}{6}$
    $\dfrac{5}{6}$

41. $3\dfrac{3}{5} = 3\dfrac{12}{20} = 2\dfrac{32}{20}$
    $-1\dfrac{3}{4} = -1\dfrac{15}{20} = -1\dfrac{15}{20}$
    $1\dfrac{17}{20}$

42. $6\dfrac{7}{8} = 6\dfrac{21}{24}$
    $-2\dfrac{2}{3} = -2\dfrac{16}{24}$
    $4\dfrac{5}{24}$

43. $4\dfrac{2}{5} = 4\dfrac{12}{30} = 3\dfrac{42}{30}$
    $-1\dfrac{5}{6} = -1\dfrac{25}{30} = -1\dfrac{25}{30}$
    $2\dfrac{17}{30}$

Score for D (24)

E (25 points) Business Applications and Critical Thinking. Solve the following. Write your answers as fractions or mixed numbers in lowest terms. (5 points for each correct answer)

44. A restaurant sells three different hamburgers, based on the amount of meat used: "The $\tfrac{1}{4}$ Pounder," "The $\tfrac{1}{3}$ Pounder," and a giant—"The $\tfrac{1}{2}$ Pounder." Students bought one of each to compare them. What was the total amount of meat used in the three hamburgers? $1\tfrac{1}{12}$ pounds

    $\tfrac{1}{4} + \tfrac{1}{3} + \tfrac{1}{2} = \tfrac{3}{12} + \tfrac{4}{12} + \tfrac{6}{12} = \tfrac{13}{12} = 1\tfrac{1}{12}$

45. Judy Mihalyi specialized in custom painting, but for the first coat she could combine leftover paints when the colors were relatively the same. She had three containers of different shades of white: $2\tfrac{2}{3}$ gallons, $2\tfrac{2}{5}$ gallons, and $2\tfrac{1}{2}$ gallons. If Judy combined the contents of all the containers, how much paint did she have? $7\tfrac{17}{30}$ gallons

    $2\tfrac{2}{3} + 2\tfrac{2}{5} + 2\tfrac{1}{2} = 2\tfrac{20}{30} + 2\tfrac{12}{30} + 2\tfrac{15}{30} = 6\tfrac{47}{30} = 7\tfrac{17}{30}$

46. Contractor Don Fleming has a top board that is $\tfrac{13}{16}$ inch thick. Don wants to use wood screws to attach it to a bottom board. If a wood screw is $1\tfrac{1}{2}$ inches long, how much of the screw will be left over to go into the bottom board? $\tfrac{11}{16}$ in.

    $1\tfrac{1}{2} - \tfrac{13}{16} = 1\tfrac{8}{16} - \tfrac{13}{16} = \tfrac{24}{16} - \tfrac{13}{16} = \tfrac{11}{16}$ in.

47. Robert Landles is planning to attach a plywood panel to the wall with nails that are $1\tfrac{3}{4}$ inches long. The panel is $\tfrac{3}{8}$ inch thick. Beneath the panel will be a layer of sheetrock that is $\tfrac{1}{2}$ inch thick. How many inches of the nail will go into the wood frame that is underneath the sheetrock? $\tfrac{7}{8}$ in.

    $\tfrac{3}{8} + \tfrac{1}{2} = \tfrac{3}{8} + \tfrac{4}{8} = \tfrac{7}{8}$ in. (plywood plus sheetrock) $1\tfrac{3}{4} - \tfrac{7}{8} = 1\tfrac{6}{8} - \tfrac{7}{8} = \tfrac{14}{8} - \tfrac{7}{8} = \tfrac{7}{8}$ in. to go into wood frame

48. Paris Fabric Center sold four pieces of wool fabric to a tailor. The pieces measure $3\tfrac{1}{4}$ yards, $2\tfrac{1}{3}$ yards, $1\tfrac{3}{4}$ yards, and $4\tfrac{1}{2}$ yards. How many yards of wool did the tailor purchase? $11\tfrac{5}{6}$

    $3\tfrac{1}{4} + 2\tfrac{1}{3} + 1\tfrac{3}{4} + 4\tfrac{1}{2} = 3\tfrac{3}{12} + 2\tfrac{4}{12} + 1\tfrac{9}{12} + 4\tfrac{6}{12} = 10\tfrac{22}{12} = 11\tfrac{10}{12} = 11\tfrac{5}{6}$ yd purchased

Score for E (25)

Assignment 2.2: Multiplication and Division of Fractions

Name

Date   Score

Learning Objectives **1 2 5 6**

A (32 points) Change whole or mixed numbers to improper fractions and multiply. Cancel if possible. Where the word *of* appears, replace it by the multiplication symbol. Write the answers as mixed numbers or proper fractions in lowest terms. (4 points for each correct answer)

1. $\dfrac{5}{6} \times \dfrac{8}{15} = \dfrac{4}{9}$

   $\dfrac{5}{6} \times \dfrac{8}{15} = \dfrac{1 \times 4}{3 \times 3} = \dfrac{4}{9}$

2. $\dfrac{3}{10} \times \dfrac{6}{7} \times \dfrac{5}{6} = \dfrac{3}{14}$

   $\dfrac{3}{10} \times \dfrac{6}{7} \times \dfrac{5}{6} = \dfrac{3 \times 1 \times 1}{2 \times 7 \times 1} = \dfrac{3}{14}$

3. $\dfrac{3}{4}$ of $\dfrac{5}{6} = \dfrac{5}{8}$

   $\dfrac{3}{4} \times \dfrac{5}{6} = \dfrac{1 \times 5}{4 \times 2} = \dfrac{5}{8}$

4. $\dfrac{5}{18} \times \dfrac{4}{9} \times \dfrac{3}{10} = \dfrac{1}{27}$

   $\dfrac{5}{18} \times \dfrac{4}{9} \times \dfrac{3}{10} = \dfrac{1 \times 1 \times 1}{9 \times 3 \times 1} = \dfrac{1}{27}$

5. $4\dfrac{1}{2} \times 1\dfrac{5}{9} = 7$

   $\dfrac{9}{2} \times \dfrac{14}{9} = \dfrac{1 \times 7}{1 \times 1} = \dfrac{7}{1} = 7$

6. $\dfrac{5}{8}$ of $10 = 6\dfrac{1}{4}$

   $\dfrac{5}{8} \times \dfrac{10}{1} = \dfrac{5 \times 5}{4 \times 1} = \dfrac{25}{4} = 6\dfrac{1}{4}$

7. $1\dfrac{7}{8} \times 12 \times \dfrac{3}{10} = 6\dfrac{3}{4}$

   $\dfrac{15}{8} \times \dfrac{12}{1} \times \dfrac{3}{10} = \dfrac{3 \times 3 \times 3}{2 \times 1 \times 2} = \dfrac{27}{4} = 6\dfrac{3}{4}$

8. $1\dfrac{1}{3} \times 1\dfrac{7}{8} \times 1\dfrac{4}{5} = 4\dfrac{1}{2}$

   $\dfrac{4}{3} \times \dfrac{15}{8} \times \dfrac{9}{5} = \dfrac{1 \times 1 \times 9}{1 \times 2 \times 1} = \dfrac{9}{2} = 4\dfrac{1}{2}$

Score for A (32)

B (32 points) Change the mixed numbers to improper fractions and divide. Cancel where possible. Write the quotients as mixed numbers or proper fractions in lowest terms. (4 points for each correct answer)

9. $\dfrac{7}{8} \div \dfrac{3}{4} = 1\dfrac{1}{6}$

   $\dfrac{7}{8} \times \dfrac{4}{3} = \dfrac{7 \times 1}{2 \times 3} = \dfrac{7}{6} = 1\dfrac{1}{6}$

10. $\dfrac{7}{10} \div \dfrac{4}{15} = 2\dfrac{5}{8}$

    $\dfrac{7}{10} \times \dfrac{15}{4} = \dfrac{7 \times 3}{2 \times 4} = \dfrac{21}{8} = 2\dfrac{5}{8}$

11. $\dfrac{3}{4} \div \dfrac{7}{8} = \dfrac{6}{7}$

    $\dfrac{3}{4} \times \dfrac{8}{7} = \dfrac{3 \times 2}{1 \times 7} = \dfrac{6}{7}$

12. $\dfrac{7}{10} \div 2\dfrac{4}{5} = \dfrac{1}{4}$

    $\dfrac{7}{10} \div \dfrac{14}{5} = \dfrac{7}{10} \times \dfrac{5}{14} = \dfrac{1 \times 1}{2 \times 2} = \dfrac{1}{4}$

13. $6\dfrac{1}{4} \div 4\dfrac{3}{8} = 1\dfrac{3}{7}$

    $\dfrac{25}{4} \div \dfrac{35}{8} = \dfrac{25}{4} \times \dfrac{8}{35} = \dfrac{5 \times 2}{1 \times 7} = \dfrac{10}{7} = 1\dfrac{3}{7}$

14. $3\dfrac{5}{6} \div 1\dfrac{7}{12} = 2\dfrac{8}{19}$

    $\dfrac{23}{6} \div \dfrac{19}{12} = \dfrac{23}{6} \times \dfrac{12}{19} = \dfrac{23 \times 2}{1 \times 19} = \dfrac{46}{19} = 2\dfrac{8}{19}$

Chapter 2 Fractions 45

Assignment 2.2 Continued

15. $3\frac{1}{3} \div \frac{4}{5} = 4\frac{1}{6}$

    $\frac{10}{3} \div \frac{4}{5} = \frac{10}{3} \times \frac{5}{4} = \frac{5 \times 5}{3 \times 2} = \frac{25}{6} = 4\frac{1}{6}$

16. $2\frac{1}{3} \div 1\frac{3}{4} = 1\frac{1}{3}$

    $\frac{7}{3} \div \frac{7}{4} = \frac{7}{3} \times \frac{4}{7} = \frac{1 \times 4}{3 \times 1} = \frac{4}{3} = 1\frac{1}{3}$

    Score for B (32)

C (36 points) Business Applications and Critical Thinking. Use fractions and mixed numbers to solve each of the following. State the answers as whole numbers, proper fractions, or mixed numbers in lowest terms. (6 points for each correct answer)

17. Last week, East Shore Concrete Co. built a small driveway that required $5\frac{1}{3}$ cubic yards of concrete. This week, the company must build another one that is $2\frac{1}{2}$ times larger. How much concrete will be required? $13\frac{1}{3}$ cu yd

    $2\frac{1}{2} \times 5\frac{1}{3} = \frac{5}{2} \times \frac{16}{3} = \frac{5 \times 8}{1 \times 3} = \frac{40}{3} = 13\frac{1}{3}$ cu yd

18. Athena Nguyen bought eight pieces of copper tubing that were each $6\frac{3}{4}$ inches long. What was the total length of tubing that Athena bought? (Give the answer in inches.) 54 in.

    $8 \times 6\frac{3}{4} = \frac{8}{1} \times \frac{27}{4} = \frac{2 \times 27}{1 \times 1} = \frac{54}{1} = 54$ in.

19. Linda Johanssen had $2\frac{1}{4}$ quarts of liquid fertilizer in a container. Her supervisor asked her to mix $\frac{2}{3}$ of the fertilizer with water and save the remainder. How many quarts of fertilizer did Linda mix with water? $1\frac{1}{2}$ qt

    $\frac{2}{3} \times 2\frac{1}{4} = \frac{2}{3} \times \frac{9}{4} = \frac{1 \times 3}{1 \times 2} = \frac{3}{2} = 1\frac{1}{2}$ qt

20. Landscaper Ron Benoit needs several pieces of PVC irrigation pipe, each 6 feet 8 inches long. PVC pipe comes in 20-foot lengths. How many pieces can Ron cut out of one length of pipe? (*Hint:* 8 inches equals $\frac{2}{3}$ foot.) 3 pieces

    $20 \div 6\frac{2}{3} = \frac{20}{1} \div \frac{20}{3} = \frac{20}{1} \times \frac{3}{20} = \frac{1 \times 3}{1 \times 1} = \frac{3}{1} = 13$

21. Robert Burke has a diesel-powered generator on his ranch. The generator has a tank that holds $3\frac{3}{4}$ gallons of diesel fuel. He stores the diesel fuel in 55-gallon drums (barrels). How many times can Robert refill his generator from one drum of fuel? $14\frac{2}{3}$ times

    $55 \div 3\frac{3}{4} = \frac{55}{1} \div \frac{15}{4} = \frac{55}{1} \times \frac{4}{15} = \frac{11 \times 4}{1 \times 3} = \frac{44}{3} = 14\frac{2}{3}$

22. Home builders Bill and John Walter are planning a narrow stairway to an attic. The stairs will each be 2 feet 4 inches long. They will cut the stairs from boards that are 8 feet long. How many whole stairs can they cut from one 8-foot board? (*Hint:* 4 inches is $\frac{1}{3}$ foot.) 3 stairs

    $8 \div 2\frac{1}{3} = \frac{8}{1} \div \frac{7}{3} = \frac{8}{1} \times \frac{3}{7} = \frac{24}{7} = 3\frac{3}{7}$, or 3 whole stairs

    Score for C (36)

Part 1 Fundamental Review

Decimals

3

Learning Objectives
By studying this chapter and completing all assignments you will be able to:

Learning Objective 1 — Read decimal numbers.

Learning Objective 2 — Round decimal numbers.

Learning Objective 3 — Add two or more decimal numbers.

Learning Objective 4 — Subtract one decimal number from another.

Learning Objective 5 — Multiply two decimal numbers.

Learning Objective 6 — Divide one decimal number by another decimal number.

Learning Objective 7 — Multiply and divide by decimal numbers that end with zeros.

Learning Objective 8 — Approximate products and quotients.

Fractions Versus Decimal Numbers

McDonald's restaurant sells a hamburger sandwich called the Quarter Pounder. The sandwich is named for the amount of meat: one-quarter pound of ground beef. McDonald's—or anyone—can describe the same amount of meat in four different ways: 4 ounces, $\frac{1}{4}$ pound, 0.25 pound, or 25% of a pound. To express less than 1 pound, McDonald's could use smaller units, fractions, decimals, or percents.

All four expressions are useful, but which one is best? It depends on what you're doing: whether you're buying or selling, whether you're speaking or writing, whether you're just estimating or making accurate financial records, or whether you're working with large volumes of something cheap or small quantities of something very expensive. For McDonald's, a Four Ouncer wouldn't sell as well as a Quarter Pounder, but Bloomingdale's sells perfume by the (fluid) ounce rather than by the gallon, quart, pint, or even cup.

Verbal expressions such as "half of a candy bar" or "a third of the pizza" are so common that children learn them before they can even read. We reviewed fractions in Chapter 2. Because of calculators, most calculations are now performed using decimal numbers. We review decimals here in Chapter 3. Percents are a combination of decimal numbers and a few common fractions. Percents are as easy to use as decimals and also allow simple verbal expressions. We review percents in Chapter 5.

Chapter 3 has three main concepts: vocabulary, calculating, and estimating. Calculating with decimals is the same as with whole numbers except that there is a decimal point. Thus, calculating with decimals is actually "managing the decimal point," which your calculator does automatically. Estimating, which is important to check your calculator, still requires that you must "manage the decimal point."

Decimal Numbers and Electronic Displays

A customer in a delicatessen might ask for "a quarter of a pound of salami, please" or perhaps "four ounces of salami." However, the food scale in the delicatessen probably has an electronic display and is calibrated only in pounds. It will likely display "0.25" or 0.250." As a fraction, a quarter of a pound is written as $\frac{1}{4}$ pound; three quarters of a pound is $\frac{3}{4}$ pound. In the U.S. monetary system, a quarter is the name of the coin whose value is twenty-five cents. And three quarters are worth seventy-five cents. When we write these monetary amounts we write either whole numbers or decimals: Either 25¢ and 75¢, or $0.25 and $0.75. It is highly unlikely that anyone would ever write $$\frac{1}{4}$ or $$\frac{3}{4}$.

Almost all business transactions and record keeping are best done in decimals rather than fractions. The calculations are usually more straightforward and more accurate. Today, specialized calculators, computers, and measurement instruments have electronic displays that are calibrated in decimals, not fractions.

Modern gasoline pumps used in the United States are calibrated in gallons and typically measure the volume of gasoline sold accurate to three decimal places. Suppose that an automobile owner buys gasoline and the display shows 12.761 gallons. 12.761 is a number; it is called a **mixed decimal.** The 12 is the whole number part of the number; the 761 is the **pure decimal** part. The period (or dot) that separates the 12 from the 761 is the **decimal point.** We say that the number 12.761 has three **decimal places** because there are three digits to the right of the decimal point.

Most calculators and computer spreadsheets permit you to change the number of decimal places that are displayed. A new calculator will often be preset to display exactly two decimal places because that is how the money system is designed. Divide 1 by 3 with your calculator. The correct answer is 0.3333333333 . . . ., a repeating number that never stops. Count the number of 3s that appear in the calculator. That is the number of decimal places your calculator is set to display. Read the instruction manual. Perhaps you can change the display to show more or fewer decimal places. *Note:* Your calculator also displays a zero (0) to the left of the decimal point. We will follow that same convention in this book. Every pure decimal number will be preceded by a zero (0).

3.1 The U.S system for weight is not a decimal system. Post office scales are typically in pounds and ounces. However, grocery store scales are typically in pounds and tenths of pounds. Therefore, some persons may be temporarily confused as they move from scale to scale.

Reading Decimal Numbers

Reading decimal numbers, both mixed and pure, is like reading whole numbers: Each "place," or column, represents a different value. Starting at the decimal point and reading to the *left*, the places represents ones, tens, hundreds, thousands, and so on. Starting at the decimal point and reading to the *right*, the vocabulary is new: The places represent *tenths, hundredths, thousandths,* and so on.

Recall the vocabulary words *tenths, hundredths,* and *thousandths* from your review of decimals in Chapter 2. As money, the decimal $0.10 represents 10¢, but also $\frac{10}{100} \cdot \frac{10}{100}$ is pronounced as "ten *hundredths.*" But $\frac{10}{100}$ can be reduced to $\frac{1}{10}$ which is "one tenth." Like the fraction, the decimal 0.10 is read as "ten *hundredths;*" the decimal 0.1 is "one *tenth.*" At the gasoline pump, the display showed 12.761. As a fraction, it is written $12\frac{761}{1000}$. Both numbers are pronounced "twelve *and* seven hundred sixty-one *thousandths.*" The decimal point is read as the word "*and.*"

Figure 3-1 illustrates the place values of the number system on both sides of the decimal point for the number 607,194.35824. The pure decimal part of the number in Figure 3-1 is 0.35824, which is pronounced "thirty-five thousand eight hundred twenty-four *hundred-thousandths.*" The decimal 0.0582 is pronounced "five hundred eighty-two *ten-thousandths.*"

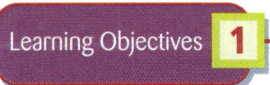

Read decimal numbers.

3.2 When numbers are written, the word *and* is used only to indicate a decimal point. But in spoken English, people commonly use the word *and* in other ways, as in "one hundred and fifty dollars." Often the word *and* is slurred so that the phrase sounds like "one hundred'n fifty." This inconsistency of usage is why, when accuracy is important, it makes sense to read numbers orally by saying each digit and using the word *point* to indicate the decimal point.

3.3 Remind students that commas are not used to the right of the decimal point.

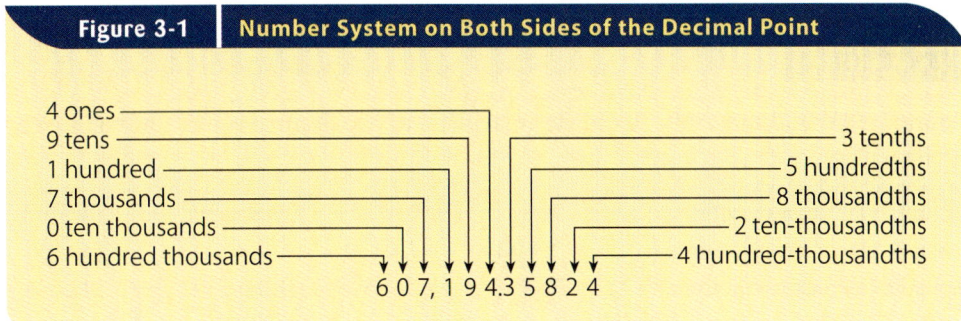

Figure 3-1 Number System on Both Sides of the Decimal Point

READING LONG DECIMAL NUMBERS

The entire number in Figure 3-1—607,194.35824—is read as "six hundred seven thousand one hundred ninety-four and thirty-five thousand eight hundred twenty-four hundred-thousandths." For a long number, reciting it orally is inefficient and might be confusing to the listener. For such a number, it may be better simply to read the digits and commas, from left to right. The word *point* is used for the decimal point.

EXAMPLE A

Recite orally the number 607,194.35824.

Number	Oral Recitation
607,194.35824	"six zero seven comma one nine four point three five eight two four"

CONCEPT CHECK 3.1

a. Write 37.062 using words: Thirty-seven and sixty-two thousandths
b. Write fifteen and seven hundredths using digits: 15.07

Rounding Decimal Numbers

Learning Objective 2

Round decimal numbers.

In the preceding section, you reviewed how to read and write decimal numbers such as 148.65392. However, in many business situations, if the whole number part is as large as 148, the digits on the extreme right may not be very important. Maybe only the digit in the tenths or hundredths column is significant. **Rounding off** such a number to make it simpler is common. You rounded off whole numbers in Chapter 1. The procedure is the same with decimal numbers.

STEPS to Round Decimal Numbers

1. Find the last place, or digit, to be retained.
2. Examine the digit to the right of the last digit to be retained.
3. a. If it is equal to or greater than 5, increase the digit to be retained by 1. Drop all digits to the right of the ones retained.
   b. If it is less than 5, leave the digit to be retained unchanged. Drop all digits to the right of the ones retained.

3.4 At the post office, weights are essentially rounded up. Currently, a 1-ounce letter costs 37¢, and a 2-ounce letter costs 63¢. Any letter between 1 and 2 ounces also costs 63¢. A weight of only 1.05 ounces costs the same as a weight of 2 ounces.

When lumber is sold by the linear foot, a 7′11″ board cannot be sold as an 8′ board. It could be "rounded down" and sold as a 7′ board. If you need 8′ boards to build a house, 7′11″ is not long enough. Rounding down is also called *truncating*.

Mention that many calculators will round off to a specified number of decimal places. The calculator will not round off whole numbers, however. For example, it will not round to the nearest hundred or thousand. Some calculators may truncate rather than round off.

EXAMPLE B

Round 7.3951 and 148.65392 to one decimal place, to two decimal places, and to three decimal places.

Round to the nearest tenth	7.3951 → 7.4	148.65392 → 148.7
Round to the nearest hundredth	7.3951 → 7.40	148.65392 → 148.65
Round to the nearest thousandth	7.3951 → 7.395	148.65392 → 148.654

ROUNDING UP

Retail businesses, such as grocery stores, often use a different method of rounding to a whole number of cents. Suppose that a grocery store has lemons priced at 3 for $1.00. Usually the store will charge $0.34 for one lemon, even though $1.00 divided by 3 is $0.3333 (to four places). The store has rounded up to the next larger whole cent. To round up monetary amounts, always increase any partial cent to the next whole cent. For example, $27.842 would round up to $27.85.

CONCEPT CHECK 3.2

Round 3.4681 to the nearest hundredth (that is, to two decimal places).

STEP 1	Find the hundredths digit.	3.4681	(The 6)
STEP 2	Examine the digit to the right of the 6.	3.4681	(It is greater than 5.)
STEP 3a	Increase the 6 to a 7 and drop the digits 81 at the right.	3.47	(The answer)

Round up 8.5014 to the nearest tenth (that is, to one decimal place).

| STEP 1 | Find the tenths digit. | 8.5014 | (The 5) |
| STEP 2 | Increase the 5 to a 6 and drop the digits 014 at the right. | 8.6 | (The answer) |

Whole Numbers, Decimal Numbers, and Arithmetic

In Chapter 1, we reviewed arithmetic with whole numbers. There were also some problems involving money in which the numbers contained decimal points. A whole number is simply a mixed decimal where the pure decimal part is zero. For simplicity, the zeros and the decimal point are omitted. In the examples that follow, when you see a whole number, you may need to place a decimal point at the right end and maybe even write one or more zeros after it. As you calculate, "manage the decimal point" as described in the following sections.

Adding Decimal Numbers

To add two or more decimal numbers, follow these steps.

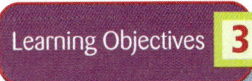

Add two or more decimal numbers.

STEPS to Add Decimal Numbers

1. Arrange the numbers in columns, with the decimal points in a vertical line.
2. Add each column, from right to left, as with whole numbers. Insert the decimal point.

   Option: You may want to write zeros in some of the right-hand columns of decimal numbers so that each number has the same number of decimal places.

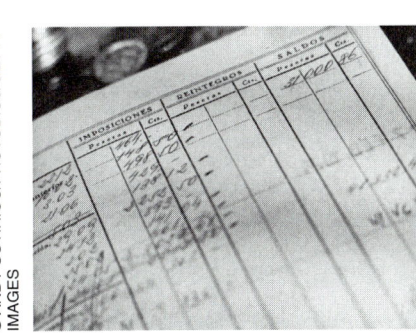

Chapter 3   Decimals   51

EXAMPLE C

Add 4.326, 218.6004, 7.09, 15, and 0.87782.

STEP 1	STEP 2		STEP 2 WITH OPTION
4.326	4.326		4.32600
218.6004	218.6004		218.60040
7.09	7.09	or	7.09000
15.	15.		15.00000
0.87782	+ 0.87782		+ 0.87782
	245.89422		245.89422

✓ CONCEPT CHECK 3.3

Add these decimal numbers: 8.95, 13.791, and 0.6.

First align:	Then add:	Or, write zeros and add:
8.95	8.95	8.950
13.791	13.791	13.791
0.6	+ 0.6	+ 0.600
	23.341	23.341

Subtracting Decimal Numbers

Learning Objective 4

Subtract one decimal number from another.

Subtracting one decimal number from another is similar to subtracting whole numbers. When you aren't using a calculator, you should write enough zeros so that both numbers have the same number of places. To subtract one decimal number from another, follow these steps.

STEPS to Subtract Decimal Numbers

1. Arrange the numbers in columns, with the decimal points in a vertical line.
2. If necessary, write enough extra zeros so that both numbers have the same number of decimal places.
3. Subtract each column, from right to left, as with whole numbers. Insert the decimal point.

EXAMPLE D

Subtract 4.935 from 12.8.

STEP 1	STEPS 2 & 3
12.8	12.800
− 4.935	− 4.935
	7.865

EXAMPLE E

Subtract 9.4 from 82.113.

STEP 1	STEPS 2 & 3
82.113	82.113
− 9.4	− 9.400
	72.713

CONCEPT CHECK 3.4

Subtract 53.784 from 207.6.

Align: Write zeros and subtract:
207.6   207.600
 53.784   − 53.784
            ─────────
             153.816

COMPLETE ASSIGNMENT 3.1.

Multiplying Decimal Numbers

To multiply one decimal number by another, follow these steps.

Learning Objectives 5

Multiply two decimal numbers.

STEPS to Multiply Decimal Numbers

1. Multiply the two numbers as if they were whole numbers.
2. Count the *total* number of decimal places in the two original numbers.
3. a. In the product, place the decimal point so that the number of decimal places is the same as the number in Step 2. (Count from right to left.)
   b. If necessary, insert zeros in front of the left-hand digit to provide enough decimal places. (See example G.)

EXAMPLE F
3.764 × 21

STEP 1:
```
    3.764    (3 places)
  ×   2.1    (1 place)
  ───────
    3 764
    7 5 28         STEP 2
  ───────
    7.9 044   (3 + 1 = 4 places)
```
STEP 3

EXAMPLE G
3.764 × 0.0021

STEP 1:
```
      3.764    (3 places)
  ×  0.0021    (4 places)
  ─────────
       3764
       7528         STEP 2
  ─────────
  0.0079044   (3 + 4 = 7
              places; insert
              2 zeros)
```
STEP 3

In business applications, zeros that come at the right end of the decimal part of the product are often omitted (example H). Do not omit zeros that come at the end of the whole-number part (example I). When the product is written in dollars and cents, two decimal places are written, including zeros at the end (example J). Please be aware that some calculators may not display any zeros at the right end.

3.5 Review with students how to set the number of decimal places on their calculators. This setting will dictate how many zeros the calculator displays at the end of a decimal.

Chapter 3 Decimals 53

EXAMPLE H
0.76 × 0.5 = 0.380   (3 places)
May be written as 0.38

EXAMPLE I
12.5 × 1.6 = 20.00   (2 places)
May be written as 20

EXAMPLE J
$8.40 × 6.5 = $54.600   (3 places)
Should be written as $54.60

✓ CONCEPT CHECK 3.5

a. Multiply 2.36 × 3.4

    2.36   (2 places)
 × 3.4   (1 place)
    944
 7 08
  8.024   (3 places)

b. Multiply 0.236 × 0.34

  0.236   (3 places)
× 0.34   (2 places)
  9 44
 7 0 8
0.08 0 24   (5 places; insert 1 zero)

Dividing Decimal Numbers

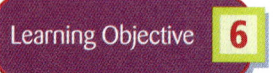

Learning Objective 6
Divide one decimal number by another decimal number.

When dividing decimal numbers, remember that a whole number will have its decimal point immediately to the right of the units digit.

To divide one decimal number by another, follow these steps.

STEPS to Divide one Decimal Number by Another

1. Arrange the divisor, dividend, and division bracket ($\overline{)}$) as in whole-number long division.
2. Move the decimal point in the divisor to the right until the divisor is a whole number. (You won't have to move it if the divisor is already a whole number.)
3. Move the decimal point in the dividend to the right exactly the same number of decimal places as you did in Step 2. If necessary, add more zeros to the right end of the dividend. (See example K.)
4. Write the decimal point in the quotient directly above the new decimal point in the dividend.
5. If necessary, write zeros in the quotient between the decimal point and the first nonzero digit. (See example L.)
6. Divide as you would for whole numbers.

EXAMPLE K

	STEP 1	STEP 2 STEP 3	STEP 4	STEP 6

$2.7 \div 0.15$ is $0.15\overline{)2.7}$ = $0.15\overline{)2.70.}$ = $15.\overline{)270.}$ = $15.\overline{)270.}$
$\phantom{2.7 \div 0.15 \text{ is } 0.15\overline{)2.7} = 0.15\overline{)2.70.} = 15.\overline{)270.} = }\underline{-15}$
$\phantom{2.7 \div 0.15 \text{ is } 0.15\overline{)2.7} = 0.15\overline{)2.70.} = 15.\overline{)270.} = }120$
$\phantom{2.7 \div 0.15 \text{ is } 0.15\overline{)2.7} = 0.15\overline{)2.70.} = 15.\overline{)270.} = }\underline{-120}$
$\phantom{2.7 \div 0.15 \text{ is } 0.15\overline{)2.7} = 0.15\overline{)2.70.} = 15.\overline{)270.} = }0$

Answer: 18.

EXAMPLE L

	STEP 1	STEPS 2, 3, & 4	STEPS 5 & 6

$0.096 \div 4$ is $4\overline{)0.096}$ = $4.\overline{)0.096}$ = $4.\overline{)0.096}$

Answer: 0.024 with subtractions −8, 16, −16, 0.

Recall from Chapter 1 that, in long division with two whole numbers, you write a *remainder* when the division doesn't come out evenly, for example, $17 \div 8 = 2$ with a remainder of 1. In division with decimals, you don't write remainders. You simply keep dividing until you have some required number of decimal places. To get the required number of decimal places, you may have to keep adding zeros to the right end of the dividend. (See example M.)

3.6 Remind students that "calculators do not do remainders;" only humans do.

EXAMPLE M

Calculate $17 \div 8$ to three decimal places.

	STEP 1	STEPS 2, 3, & 4	STEP 6

$17 \div 8$ is $8\overline{)17}$ = $8.\overline{)17.}$ = $8.\overline{)17.}$ = $8.\overline{)17.000}$

Answer: 2.125 with subtractions −16, 10, −8, 20, −16, 40, −40, 0.

Chapter 3 Decimals 55

✓ CONCEPT CHECK 3.6

Divide 1.026 by 15.

STEPS 1 & 4   STEPS 5 & 6

$$15\overline{)1.026} \;=\; 15\overline{)1.0260}$$
$$\underline{-\;90}$$
$$126$$
$$\underline{-\;120}$$
$$60$$
$$\underline{-\;60}$$
$$0$$

Divide 0.009 by 0.4.

STEPS 1, 2, 3, & 4   STEPS 5 & 6

$$0.4\overline{)0.0.09} \;=\; 4\overline{)0.0900}$$
$$\underline{-\;8}$$
$$10$$
$$\underline{-\;8}$$
$$20$$
$$\underline{-\;20}$$
$$0$$

In example M, $17 \div 8 = 2.125$. But recall that $17 \div 8$ can also be written as the fraction $\frac{17}{8}$. 2.125 is called the **decimal equivalent** of $\frac{17}{8}$. Decimal equivalents can be useful when you are working with fractions and have a calculator available. Even with simple fractions, and no calculator, it is often simpler to use decimal equivalents because you don't need a common denominator.

● EXAMPLE N

Compute $\frac{1}{2} + \frac{3}{4} - \frac{2}{5}$. This requires that all fractions have a common denominator of 20. But $\frac{1}{2} = 0.5, \frac{3}{4} = 0.75$, and $\frac{2}{5} = 0.4$. Therefore, we have $\frac{1}{2} + \frac{3}{4} - \frac{2}{5} = 0.05 + 0.75 - 0.4 = 0.85$.

For difficult fractions, use a calculator to convert the fractions to their decimal equivalents. Then use the calculator to perform the required operation. (If possible, you should use the memory of your calculator to store the intermediate answers.)

● EXAMPLE O

Compute $\frac{8}{15} + \frac{7}{12} - \frac{3}{7}$.

[8] [÷] [15] [=] gives   0.53333333
[7] [÷] [12] [=] gives   0.58333333
[3] [÷] [7] [=] gives   0.42857143
                                      1.54523809

The preceeding example assumes that your calculator is displaying eight decimal places. Also, if you use the memory to store the intermediate answers, your calculator may round off the intermediate answers and give you a final answer of 1.54523810 or 1.5452381. Some calculators make it even easier to compute fractions using decimal equivalents. A few have an "algebraic operating system" that automatically does multiplication and division before addition and subtraction. For those calculators, you might use keystrokes like these:

[8] [÷] [15] [+] [7] [÷] [12] [−] [3] [÷] [7] [=] **1.5452380,** or possibly **1.5452381**

Many calculators that do not have an "algebraic operating system" will have parentheses, permitting this type of calculation:

[(] [8] [÷] [15] [)] [+] [(] [7] [÷] [12] [)] [−] [(] [3] [÷] [7] [)] [=] **1.5452380,** or possibly **1.5452381**

Using Multipliers and Divisors that End with Zeros

In Chapter 1, we showed simple multiplication and division shortcuts when the multiplier or the divisor is a whole number ending in zeros (e.g., 30, 200, or 1,000). The same shortcuts may be used with decimal numbers. We just "manage the decimal point."

If the multiplier is 10, 100, 1,000, and so on, there is just one step.

Step 1   Move the decimal point in the multiplicand to the *right* the same number of places as the number of zeros in the multiplier. (See example P.) If necessary, add zeros to the *right* end of the multiplicand before multiplying. (See example Q.)

Learning Objectives 7

Multiply and divide by decimal numbers that end with zeros.

EXAMPLE P
0.56 × 10 = 0.5.6 = 5.6

(1 place)

EXAMPLE Q
4.73 × 1,000 − 4.730 = 4,730

(3 places)

If the multiplier ends in zeros but has a first digit that is not 1 (for example, 300 or 2,000), there are two steps.

Step 1   Multiply the multiplicand by the nonzero part of the multiplier.

Step 2   Move the decimal point in the product from Step 1 to the *right* the same number of places as the number of zeros in the multiplier.

EXAMPLE R
Multiply 3.431 by 2,000

Multiply by 2: 3.431 × 2 = 6.862

Move the decimal point three places to the right: 6.862. ⟶ 6,862.

If the divisor is 10, 100, 1000, and so on, there is just one step.

Step 1   Move the decimal point in the dividend to the *left* the same number of places as the number of zeros in the divisor. (See example S.) If necessary, add zeros to the *left* end of the dividend. (See example T.)

EXAMPLE S
735.1 ÷ 100
735.1 ÷ 100 = 7.35.1 = 7.351

(2 places)

EXAMPLE T
9.64 ÷ 1,000
9.64 ÷ 1,000 = .009.64 = 0.00964

(3 places)

Chapter 3   Decimals

If the divisor ends in zeros (for example, 300 or 2,000) but has a first digit that is not 1, there are two steps.

Step 1   Divide the dividend by the nonzero part of the divisor.

Step 2   Move the decimal point in the quotient from Step 1 to the *left* the same number of places as the number of zeros in the divisor.

EXAMPLE U

Divide 615.24 by 300

Divide by 3: $615.24 \div 3 = 205.08$

Move the decimal point two places to the left: $2.05.08 \longrightarrow 2.0508$

CONCEPT CHECK 3.7

a. Multiply 0.413 by 300

$0.413 \times 3 = 1.239$

Move the decimal point two places to the right:

$1.23.9 \longrightarrow 123.9$

b. Divide 4.375 by 10

Move the decimal point one place to the left:

$4.375 \div 10 = .4.375 \longrightarrow 0.4375$

COMPLETE ASSIGNMENT 3.2

Approximating Products and Quotients

Learning Objective 8

Approximate products and quotients.

Business people today almost always use calculators or computers to do important computations. But calculators are perfect only if every single key is pressed correctly. Often, you can discover a calculator error by doing some simple mental approximations. The objective is to determine whether the answer is approximately the right size—that is, whether the decimal point is in the correct position. To do so, we round each decimal number to only one nonzero digit and all the rest to zeros. Follow these steps:

STEPS   **to Approximate a Multiplication Problem**

1. In each factor, round the first nonzero digit from the left end. (How does the digit to its right compare to 5?)
2. Change all the digits to the right of the first nonzero digit to zero.
3. Multiply the two new factors.
4. Place the decimal point in the product.

EXAMPLE V

Approximate 3.764 × 7.1

	STEP 1	STEPS 2 & 3
3.764	⟶ 4.000	4
× 7.1	⟶ × 7.0	× 7
		28

EXAMPLE W

Approximate 0.089 × 61.18

	STEP 1	STEPS 2 & 3
0.089	⟶ 0.090	0.09
× 61.18	⟶ × 60.00	× 60
		5.40

The actual answers are 26.7244 and 5.44502.

In division, the mental approximation will be easier if you change the decimal numbers so that the division will end evenly after one step. To do this, first round the divisor to one nonzero digit and then round the dividend to two nonzero digits, evenly divisible by the new divisor.

> **STEPS to Approximate a Division Problem**
>
> 1. Round the divisor to a *single nonzero digit* at the left, followed by all zeros.
> 2. Round the dividend to a *two-digit number* at the left, followed by all zeros. Select the two-digit number so that it is evenly divisible by the new divisor.
> 3. Divide the new dividend by the new divisor.
> 4. Place the decimal point correctly in the quotient.

EXAMPLE X

Approximate 4.764 ÷ 8.1

	STEP 1	STEP 2	STEPS 3 & 4
8.1)4.764	⟶ 8.0)4.764	⟶ 8.)4.800	⟶ 0.6 8.)4.8 −4.8 0

EXAMPLE Y

Approximate 61.18 ÷ 0.089

	STEP 1	STEP 2	STEPS 3 & 4
0.089)61.18	⟶ 0.090)61.18	⟶ 0.09)63.00	⟶ 700. 9.)6300. 63 0

3.7 In example Y, we say that 0.089 has two *significant digits*, 0.090 has one *significant digit*, and 63.00 has two *significant digits*. Significant digits are nonzero digits that are either preceded or followed by all zeros.

The actual answers are 0.5882 and 687.4157 (to four decimal places).

CONCEPT CHECK 3.8

a. Approximate 6.891×0.614

$6.891 \longrightarrow 7.000$
$0.614 \longrightarrow 0.600$

$$\begin{array}{r} 0.6 \quad \text{(1 place)} \\ \times\ 7 \quad \text{(0 places)} \\ \hline 4.2 \quad \text{(1 place)} \end{array}$$

Compare with $6.891 \times 0.614 = 4.231074$

b. Approximate $0.0738 \div 92.65$
Remember to round off the divisor first.

$92.65 \longrightarrow 90.00$
$0.0738 \longrightarrow 0.0720$

$$90\overline{)0.072} \longrightarrow 90\overline{)0.0720}\ \ \ \begin{array}{r}.0008\\ \hline 720 \\ \hline 0\end{array}$$

Compare with $0.0738 \div 92.65 = 0.000796546$

COMPLETE ASSIGNMENT 3.3

Chapter Terms for Review

decimal equivalent
decimal places
decimal point

mixed decimal
pure decimal
rounding off

Try Microsoft® Excel

1. Set up and complete the following tables using the appropriate Excel formulas. Refer to your Student CD template for solutions.

Date	Auto Sales	Part Sales	Total Sales
6/4/04	$ 36,628.14	$ 1,782.28	
6/5/04	$ 42,789.40	$ 2,047.33	
6/6/04	$ 58,334.98	$ 1,132.48	
6/7/04	$ 96,782.04	$ 3,006.04	
6/8/04	$ 29,765.55	$ 2,333.33	
Total			

Date	Total Receipts	Total Cash	Cash Short
7/15/04	$ 974.58	$ 969.30	
7/16/04	$ 888.07	$ 888.02	
7/17/04	$ 1,384.17	$ 1,350.23	
Total			

Date	Units Sold	Price Per Unit	Total Sales
5/24/04	47	$ 107.16	
5/25/04	63	$ 107.16	
5/26/04	72	$ 107.16	
5/27/04	39	$ 107.16	
Total			

Date	Total Sale	Price Per Unit	Units Sold
5/24/04	$ 5,036.52	107.16	
5/25/04	$ 6,751.08	107.16	
5/26/04	$ 7,715.52	107.16	
5/27/04	$ 4,179.24	107.16	
Total			

THE BOTTOM LINE

Summary of chapter learning objectives:

Learning Objective	Example
3.1 Read decimal numbers	1. Write 8.427, using words. 2. Write forty-one and eleven ten-thousandths, using digits.
3.2 Round decimal numbers	3. Round 0.506489 to the nearest thousandth (that is, to three decimal places). 4. Round up 13.26012 to the next hundredth (that is, to two decimal places).
3.3 Add two or more decimal numbers	5. Add 82.9, 14.872, and 2.09.
3.4 Subtract one decimal number from another	6. Subtract 14.5977 from 19.34.
3.5 Multiply two decimal numbers	7. Multiply: 4.68 × 3.5 _____
3.6 Divide one decimal number by another decimal number	8. Divide: 0.084 ÷ 4 _____ 9. Divide: 0.064 ÷ 2.5 _____
3.7 Multiply and divide by decimals that end with zeros	10. Multiply: 0.069782 × 1000 _____ 11. Divide: 9.462 by 100 _____ 12. Multiply: 0.0623 × 20 _____ 13. Divide: 84.6 by 300 _____
3.8 Approximate products and quotients	14. Approximate 48.79 × 0.47 _____ 15. Approximate 0.2688 ÷ 0.713 _____

Answers: 1. eight and four hundred twenty-seven ten-thousandths 2. 41.0011 3. 0.506 4. 13.27 5. 99.862 6. 4.7423 7. 16.38 8. 0.021 9. 0.0256 10. 69.782 11. 0.09462 12. 1.246 13. 0.282 14. 25 15. 0.4

SELF-CHECK

Review Problems for Chapter 3

1. Write "one hundred sixteen and fourteen ten-thousandths" as a number _____

2. Write 6,431.719, using words _____

3. Round 3.475 feet to the nearest tenth _____

4. Round $12.667 to the nearest cent _____

5. Add 3.79475 and 739.85 _____

6. Add 12.42, 0.087, and 8.3 _____

7. Subtract 8.693 from 11.41 _____

8. Subtract 162.78 from 341.2494 _____

9. Multiply 3.722 by 0.483 (do not round off) _____

10. Multiply $17.75 by 14.62 (round off to the nearest cent) _____

In problems 11 and 12, divide to three places and round to the nearest hundredth.

11. Divide 45.88 by 14.2 _____

12. Divide $6.25 by 8.41 _____

In problems 13 and 14, use shortcuts to solve each problem and round to the nearest hundredth.

13. Multiply 86.493 by 100 _____

14. Divide $2,762.35 by 1,000 _____

In problems 15 and 16, pick the best approximate answers from the possible answers.

15. Multiply 48.98 by 11.2 _____   (a) 0.5   (b) 5   (c) 50   (d) 500   (e) 5,000

16. Divide $6.65 by 8.21 _____   (a) $0.008   (b) $0.08   (c) $0.80   (d) $8.00   (e) $80.0

17. DeLois McBryde owns a chain of very large, upscale bookstores. She decides to start selling coffee drinks such as espresso and cappuccino at one of her stores. During the first day, the store has total sales of $4,188.25. Of the total, $362.50 was from coffee drinks. How much of the total was from books and other items? _____

18. Gary Gehlert operates tennis and golf shops at a desert resort. Last year, he started selling on the Internet as well. He had the following profits last year: Tennis (shop), $52,418.12; Golf (shop), $168,078.51; Tennis (Internet), $8,993.84; and Golf (Internet), $18,745.49. What were the total profits from these sources? _____

19. Dean Treggas, a landscape contractor, needed to plant 226 1-gallon plants and 164 5-gallon plants. Dean uses about 0.8 cubic foot of planting soil for each 1-gallon plant and 2.5 cubic feet of soil for each 5-gallon plant. How many cubic feet of planting soil will Dean need for all these plants? _____

20. Planting soil is sold by the cubic yard. How many cubic yards of planting soil will Dean Treggas need to do his planting in question 19? (Round the answer to two decimal places.) _____

Assignment 3.1: Addition and Subtraction of Decimal Numbers

Name

Date   Score

Learning Objectives 1 2 3 4

A (13 points) Use digits to write each number that is expressed in words. Use words to write each number that is expressed in digits. (1 point for each correct answer)

1. Six hundred thirteen ten-thousandths 0.0613
2. Nineteen thousandths 0.019
3. Sixty-four hundredths 0.64
4. Seventy-six and seventy-one ten-thousandths 76.0071
5. Eight hundred sixty and ninety-eight hundred-thousandths 860.00098
6. Eighteen and six thousandths 18.006
7. 26.085 twenty-six and eighty-five thousandths
8. 0.0004 four ten-thousandths
9. 492.3 four hundred ninety-two and three tenths
10. 0.081 eighty-one thousandths
11. 42.0481 forty-two and four hundred eighty-one ten-thousandths
12. 6.018 six and eighteen thousandths
13. 1,007.4 one thousand seven and four tenths

Score for A (13)

B (24 points) Round as indicated. (1 point for each correct answer)

Nearest Tenth

14. 6.3517 qt   6.4 qt
15. 48.77 mi   48.8 mi
16. 3.824 gal   3.8 gal
17. 374.29 lb   374.3 lb
18. 7.35 ft   7.4 ft
19. 6.375 oz   6.4 oz

Nearest Cent

20. $6.425 $6.43
21. $0.098 $0.10
22. $942.3449 $942.34
23. $8.1047 $8.10
24. $0.0449 $0.04
25. $51.375 $51.38

Nearest Thousandth

26. 5.37575 pt   5.376 pt
27. 0.00549 gal 0.005 gal
28. 14.6445 oz   14.645 oz
29. 5.040603 ft 5.041 ft
30. 8.9989 mi   8.999 mi
31. 0.200499 lb 0.200 lb

UP to the *Next* Cent

32. $9.681 $9.69
33. $0.159 $0.16
34. $72.535 $72.54
35. $2.0917 $2.10
36. $11.4485 $11.45
37. $0.6545 $0.66

Score for B (24)

Chapter 3 Decimals

Assignment 3.1 Continued

C **(27 points) Write the following numbers in columns, and then add. (3 points for each correct answer)**

38. 3.84, 42.81, 747.114
    3.84
    42.81
    747.114
    ———
    793.764

39. 0.7323, 4.084, 17.42
    0.7323
    4.084
    17.42
    ———
    22.2363

40. 15.4, 32.574, 9.51, 74.0822
    15.4
    32.574
    9.51
    74.0822
    ———
    131.5662

41. 24.78, 71.402, 8.3176
    24.78
    71.402
    8.3176
    ———
    104.4996

42. 6.084, 107.4, 48.2007
    6.084
    107.4
    48.2007
    ———
    161.6847

43. 6.4, 3.211, 12.6, 7.07
    6.4
    3.211
    12.6
    7.07
    ———
    29.281

44. 337.51, 6.1761, 16.078
    337.51
    6.1761
    16.078
    ———
    359.7641

45. 36.7, 208.51, 3.992
    36.7
    208.51
    3.992
    ———
    249.202

46. 0.592, 1.82, 0.774, 6.5
    0.592
    1.82
    0.774
    6.5
    ———
    9.686

Score for C (27) _____

D **(36 points) Subtract the following. (3 points for each correct answer)**

47.   0.734
     − 0.37
     ———
       0.364

48.   0.04264
     − 0.00497
     ————
       0.03767

49.   26.04
     − 8.625
     ———
      17.415

50.   0.7212
     − 0.034
     ———
       0.6872

51.   12.
     − 4.37
     ———
        7.63

52.   804.07
     − 167.1
     ———
       636.97

53.   3.2525
     − 2.843
     ———
       0.4095

54.   708.932
     − 419.058
     ————
       289.874

55.   0.365
     − 0.189
     ———
       0.176

56.   4.37
     − 1.9055
     ———
       2.4645

57.   7.624
     − 5.947
     ———
       1.677

58.   1.0045
     − 1.003
     ———
       0.0015

Score for D (36) _____

Part 1 Fundamental Review

Assignment 3.2: Multiplication and Division of Decimal Numbers

Name

Date   Score

Learning Objectives **5 6 7 8**

A (32 points) Multiply the following. Round monetary products to the nearest cent. Do not round nonmonetary products. (4 points for each correct answer)

1. $16.75
    × 64
    67 00
    1 005 0
    $1,072.00

    $1,072.00

2. $24.60
    × 4.5
    12 300
    98 40
    $110.700

    $110.70

3. $420.00
    × 0.806
    2 52000
    00 0000
    336 000
    $338.52000

    $338.52

4. $57.80
    × 0.35
    2 8900
    17 340
    $20.2300

    $20.23

5. 107.21
    × 0.74
    4 2884
    75 047
    79.3354

    79.3354

6. 52.93
    × 0.45
    2 6465
    21 172
    23.8185

    23.8185

7. 285.70326
    × 0.28
    22 8562608
    57 140652
    79.9969128

    79.9969128

8. 816.04
    × 0.403
    2 44812
    00 0000
    326 416
    328.86412

    328.86412

Score for A (32)

B (24 points) Divide the following. Round monetary quotients to the nearest cent. Round nonmonetary quotients to two decimal places. (4 points for each correct answer)

9. $1.85
    7)$12.95
    7
    59
    56
    35
    35

    $1.85

10. $18.75
    0.36)$6.7500
    36
    315
    288
    270
    252
    180
    180

    $18.75

11. $45.25
    1.2)$54.300
    48
    63
    60
    30
    24
    60
    60

    $45.25

Chapter 3 Decimals 65

Assignment 3.2 Continued

```
         1.713                    6.122                    8.764
12. 1.5)2.5 700          13. 0.11)0.67350        14. 0.09)0.78880
        15                        66                       72
        107                       13                       68
        105                       11                       63
         20                       25                       58
         15                       22                       54
         50                       30                       40
         45                       22                       36

        1.71                     6.12                     8.76
```

Score for B (24)

C (12 points) Multiply and/or divide by just moving the decimal point or by doing some simple multiplication/division and moving the decimal point. Round monetary answers to the nearest cent. Do not round nonmonetary answers. (1 point for each correct answer)

15. $0.0625 \times 1,000$ = 62.5
16. 50.708×100 = 5,070.8
17. $0.047 \times 10,000$ = 470
18. $763 \div 100$ = 7.63
19. $6.32 \div 10$ = 0.632
20. $27.469 \div 1,000$ = 0.027469

21. $\$72.41 \times 300$ = $21,723.00
22. $\$32.25 \times 20$ = $645.00
23. $\$0.07 \times 4,000$ = $280.00
24. $\$2.50 \times 40$ = $100.00
25. $\$86.50 \div 200$ = $0.43
26. $\$9,612 \div 40$ = $240.30

Score for C (12)

D (32 points) For each of the following problems, underline the estimate that is most nearly correct. (2 points for each correct answer)

#	Problem	(a)	(b)	(c)	(d)
27.	0.077×0.52	4.0	0.4	<u>0.04</u>	0.004
28.	76.7×0.8477	0.064	0.64	6.4	<u>64</u>
29.	0.38×71.918	0.28	2.8	<u>28</u>	280
30.	0.00907×6.12	<u>0.054</u>	0.54	5.4	54
31.	0.0782×0.5503	0.0048	<u>0.048</u>	0.48	4.8
32.	0.0417×0.0957	0.04	<u>0.004</u>	0.0004	0.00004
33.	268.25×0.9175	27,000	2,700	<u>270</u>	27
34.	0.00487×0.0059	0.000003	<u>0.00003</u>	0.0003	0.003
35.	19.1×6104	120	1,200	12,000	<u>120,000</u>
36.	$7.958 \div 0.514$	<u>16</u>	160	1,600	16,000
37.	$3.575 \div 893.12$	<u>0.004</u>	0.04	0.4	4
38.	$0.0614 \div 0.00398$	0.15	1.5	<u>15</u>	150
39.	$0.8397 \div 6.12$	0.14	1.4	<u>14</u>	140
40.	$0.5379 \div 0.591$	900	90	9	<u>0.9</u>
41.	$5.112 \div 0.0692$	<u>70</u>	7	0.7	0.07
42.	$2.671 \div 0.0926$	300	<u>30</u>	3	0.3

Score for D (32)

Assignment 3.3: Decimal Numbers in Business

Name

Date   Score

Learning Objectives **3** **4** **5** **6**

A (36 points) Business Applications and Critical Thinking. Solve the following. Do not round your final answers. (6 points for each correct answer)

1. Gary Floyd had 21.5 feet of rope. He cut off a piece 14.75 feet long. How much did he have left?

    <u>6.75 ft</u>   21.5
                         −14.75
                          6.75

2. Cho Jewelers had only 12.7 ounces of gold on hand, so Mr. Cho bought 22.5 ounces more to make Christmas items. He used 18.7 ounces for gold rings. How much gold did he have left?

    <u>16.5 oz</u>   12.7   35.2
                         +22.5   −18.7
                          35.2   16.5

3. Judy Taylor reads meters for the gas and electric company. She walked 3.6 miles on Monday; 3.7 miles on Tuesday, 2.9 miles on Wednesday, 3.25 miles on Thursday, and 3.4 miles on Friday. What was her total distance for the week?

    <u>16.85 mi</u>   3.6
                          3.7
                          2.9
                          3.25
                        + 3.4
                         16.85

4. Four messenger service drivers need gasoline for their cars. Individually, they buy 12.4, 8.9, 13.8, and 13.9 gallons. How much did they purchase all together?

    <u>49.0 gal</u>   12.4
                          8.9
                         13.8
                         13.9
                         49.0

5. A retail customer owes a total of $226.54 on her department store account. She visits the store to return an item that cost $47.79. While there, she buys two items that cost $55.88 and $67.50, respectively. What is her new account balance at the store?

    <u>$302.13</u> $226.54 $178.75
                       − 47.79   55.88
                       $178.75   + 67.50
                                    $302.13

6. Parker Paving Co. delivered 6.2 tons of asphalt. It used 4.7 tons for a driveway and 1.2 tons for a walkway. How much asphalt was left?

    <u>0.3 t</u>   6.2   1.5
                         −4.7   −1.2
                          1.5   0.3

Score for A (36)

Chapter 3 Decimals 67

Assignment 3.3 Continued

B (64 points) Business Applications and Critical Thinking. Solve the following business problems. Use shortcuts where possible. If necessary, round answers to two decimal places. (8 points for each correct answer)

7. Bill Wells Hardware sells a large-diameter plastic pipe for $0.07 per foot and copper pipe for $1.02 per foot. How much will Katy Cruz save by using plastic pipe if she needs 300 feet of pipe? $285

   $1.02
   −0.07
   $0.95

   $0.95
   × 3
   $2.85

   Move decimal point 2 places right
   $2.85 ⟶ $285

8. Benoit Landscaping sent three truckloads of topsoil to a job. The soil cost $21.50 per cubic yard. Two trucks carried 7.25 cubic yards each; the third carried 6.75 cubic yards. What was the total cost of all the topsoil? $456.88

   7.25
   7.25
   6.75
   21.25

   $21.50
   ×21.25
   10750
   4300
   2150
   0 4300
   $456.8750

9. Wholesale, 1,000 2-ounce plastic bottles cost 3.5 cents each, and 2,000 4-ounce bottles cost 4.5 cents each. What is the total cost of all 3,000 bottles? $125

   3.5¢ = $0.035   4.5¢ = $0.045
   $0.035 × 1,000 = $35 $0.045 × 2,000 = $90
   $35 + $90 = $125

10. Evelyn Haynes uses her car as a delivery vehicle. On Monday, she bought 14.62 gallons of regular gasoline at $2.179 per gallon. On Thursday, she bought 15.52 gallons at $2.239. How much did she pay for gasoline that week? $66.61

    14.62 × $2.179 = $31.85698 or $31.86
    15.52 × $2.239 = $34.74928 or $34.75
    $66.61

11. Electrician Tom Stewart paid $95.50 for 500 feet of multistrand electrical wire. What was the cost per foot for this particular wire?

    $0.08
    $95.50 ÷ 5 = $19.10
    Move decimal point 2 places left
    $19.10 ⟶ $0.191 or $0.19

12. A pizza chef has 24 pounds of flour on hand. He needs 3.75 pounds of flour for one large recipe of pizza dough. How many recipes can he make with the flour on hand? (Round to the nearest tenth.) 6.4 recipes

    ```
           6.4
    3.75)24.000
         2250
         1500
         1500
            0
    ```

13. Paint thinner costs $1.29 per gallon. How many gallons can a painting contractor buy for $10? (Round to the nearest tenth.) 7.8 gal

    ```
           7.75
    1.29)10.00.00
         9 03
          970
          903
           670
           645
           250
    ```

14. Jackie Barner earns $22.60 per hour. How many hours did she work during a partial day for which her pay was $152.55? 6.75 hr

    ```
            6.75
    22.60)152.55.00
          135 60
           16950
           15820
            11300
            11300
                0
    ```

Score for B (64)

Word Problems and Equations 4

Learning Objectives
By studying this chapter and completing all assignments you will be able to:

Learning Objective 1 — Use mental computations in simple addition, subtraction, multiplication, and division.

Learning Objective 2 — Use a systematic approach to solve word problems.

Learning Objective 3 — Apply formulas to solve rate, time, and distance problems.

Learning Objective 4 — Solve simple numerical equations.

Learning Objective 5 — Recognize numerical relationships in a series.

Learning Objective 6 — Do quick mental calculations through a process of rounding numbers.

Mental Computations

Learning Objective 1

Use mental computations in simple addition, subtraction, multiplication, and division.

Simple computations need to be made quickly in business. Practicing mental computation drills will improve your speed and accuracy in using the four fundamental math processes.

In example A you should be able to obtain the ten answers without using pencil, paper, or an electronic calculator. Mentally compute each problem. Each computation is done from left to right. In these problems, addition, subtraction, multiplication, and division are done in the sequence in which they appear.

4.1 Dictating to students simple problems incorporating the four fundamental processes will help them develop this skill. For example, you might dictate, slowly and evenly: "Seven plus nine minus four times two plus one divided by five equals …?"

EXAMPLE A

$7 + 3 + 8 + 4$ = 22
$27 - 2 - 5 + 8 + 2$ = 30
$60 \div 2 \div 3 \div 5$ = 2
$3 + 4 + 2 + 10 - 4$ = 15
$3 \times 4 \times 2 \times 10$ = 240
$28 \div 4 \times 5 \times 2$ = 70
$26 \div 2 + 2 \times 2 \times 2 \div 6 + 10$ = 20
$180 \times 2 \div 6 - 20 \div 8 \times 5$ = 25
$100 \times 5 - 20 - 80 - 40 \div 6$ = 60
$4{,}000 \div 2 + 100 \div 7 - 299$ = 1

✓ CONCEPT CHECK 4.1

Practice computations until you can do them mentally without extra copying or writing. Use the simplification techniques in Chapter 1 whenever possible: number combinations, repeated digits, counting by 10s and adding 1s, subtraction by changing numbers, multiplying numbers ending in zeros, and dividing numbers ending in zeros. Do the following computations mentally.

$7 + 3 + 6 + 6 + 6 \times 20 \div 10 + 31 = 87$

Think: $10 + (3 \times 6 = 18) 28 (2 \times 28) + 0 560 \, (560) - 0 56 \, (66, 76, 86, 87) = 87$

$78 - 29 + 7 + 7 + 7 \times 40 = 2{,}800$

Think: $(79 - 30 = 49) + (3 \times 7 = 21) 70 (7 \times 4) + 00 = 2{,}800$

Solving Word Problems

Learning Objective 2

Use a systematic approach to solve word problems.

You might have little difficulty with computations expressed in numbers only. In example B you would quickly answer 350.

EXAMPLE B

$15 + 15 + 10 \times 10 - 50 = 350$

70 Part 1 Fundamental Review

However, you might not answer $350 as quickly when the business problem in example C appears, even though it uses the same numerical elements as example B.

EXAMPLE C

A company orders carpeting for three offices measuring 15 square yards, 15 square yards, and 10 square yards, respectively. A carpet dealer sells the carpet for $10 a square yard and gives a $50 discount when the sale is for three or more offices. How much would the company pay to have the three offices carpeted?

15 sq yd + 15 sq yd + 10 sq yd = 40 sq yd
40 sq yd × $10 = $400 gross price
$400 − $50 discount = $350 net price

Business problems involving computations simply require addition, subtraction, multiplication, and division.

> **STEPS** to Solve Word Problems
>
> Read the entire problem carefully and then:
> 1. Determine exactly what is being requested.
> 2. Determine the processes you will use to solve the problem.

We use these steps to solve the word problem in example C.

STEP 1 What is requested: How much money would the company pay?

STEP 2 What process will be used:
Add square yards in the 3 offices: 15 + 15 + 10 = 40.
Multiply the $10 per square yard cost by total square yards: 40 × $10 = $400.
Subtract the $50 discount: $400 − $50 = $350.

Some word problems will involve all four fundamental processes: addition, subtraction, multiplication, and division.

EXAMPLE D

Phoebe Elias owns half of a small bakery. Last week she baked 6 cakes on Monday, 9 on Tuesday, 11 on Wednesday, 8 on Thursday, and 6 on Friday. She sold all cakes for $9 each. It cost Phoebe $5 to make each cake; the rest was her profit on each cake. Phoebe split her profit evenly with her partner. How much did her partner receive from last week's cakes?

STEP 1 What is requested: How much money did Phoebe's partner receive?

STEP 2 What process will be used:
Add the cakes baked: 6 + 9 + 11 + 8 + 6 = 40.
Subtract the cost from the sales price: $9 − $5 = $4 profit per cake.
Multiply the $4 profit per cake by the number of cakes sold: 40 × $4 = $160.
Divide the total profit by 2: $160 ÷ 2 = $80 received by the partner.

CONCEPT CHECK 4.2

Summary of steps for solving word problems:
1. Determine what is being requested.
2. Determine the processes you will use to solve the problem.

Problem: Maria wants to upholster three chairs. Two chairs will require 4 yards of material each; the third will require 3 yards. One material costs $32 per yard; the other is $24 per yard. What is the difference between the costs of the two materials for upholstering the chairs?

STEP 1 What is requested: Difference in cost between the two materials.

STEP 2 The process to be used:
Add amount of material needed: 4 yd + 4 yd + 3 yd = 11 yd.
Cost of material for three chairs, first material: 11 yd × $32 per yd = $352.
Cost of material for three chairs, second material: 11 yd × $24 per yd = $264.
Difference in cost between the two materials: $352 − $264 = $88 difference in cost.

Solving Rate, Time, and Distance Problems

Learning Objective 3

Apply formulas to solve rate, time, and distance problems.

In some business word problems, you must compute how much is done in a given amount of time at a specific speed. These rate, time, and distance problems are solved with a simple formula: Rate (speed) × Time = Distance (amount done). If you are given any two factors, it is easy, by formula, to find the third.

Rate × Time = Distance
Distance ÷ Time = Rate
Distance ÷ Rate = Time

4.3 Some students may be familiar with the function hierarchy (multiplication and division before addition and subtraction when no parentheses are shown). For uniformity and simplicity, this hierarchy is ignored in these early chapters.

EXAMPLE E

Jan traveled at 35 miles per hour for 5 hours. How far did Jan travel?
35 mph × 5 hr = 175 mi
(Rate × Time = Distance)

EXAMPLE F

Jan traveled 175 miles in 5 hours. How fast was Jan traveling?
175 mi ÷ 5 hr = 35 mph
(Distance ÷ Time = Rate)

EXAMPLE G

At 35 miles per hour, how long would it take Jan to travel a total of 175 miles?
175 mi ÷ 35 mph = 5 hr
(Distance ÷ Rate = Time)

72 Part 1 Fundamental Review

EXAMPLE H

Jan and Ahmed start traveling toward each other from 300 miles apart. Jan is traveling at 35 miles per hour; Ahmed is traveling at 40 miles per hour. How much time will elapse before they meet?

Distance = 300 mi

Total rate = 35 mph (Jan) + 40 mph (Ahmed) = 75 mph

300 mi ÷ 75 mph = 4 hr

(Distance ÷ Rate = Time)

EXAMPLE I

Jan and Ahmed start traveling toward each other from 300 miles apart. Jan is traveling at 35 miles per hour; Ahmed is traveling at 40 miles per hour. How much distance will Jan travel before they meet?

Total rate = 35 mph (Jan) + 40 mph (Ahmed) = 75 mph

Time = 300 mi ÷ 75 mph = 4 hr

Jan's distance = 35 mph (Jan's Rate) × 4 hr (Time) = 140 mi

EXAMPLE J

Mary needs to type a term paper that will be 30 pages long. Each page contains about 200 words. If Mary can type 40 words per minute, how many minutes will it take her to complete the paper?

Choose a formula: We know distance (amount done) and speed (rate). Therefore, we choose the formula for time.

Distance (amount done) ÷ Rate (speed) = Time

30 pages × 200 words = 6,000 words ÷ 40 wpm = 150 min

EXAMPLE K

Flora also had a paper to type, but hers was 9,000 words in length. She was able to type it in 150 minutes. How fast did she type?

Choose a formula: We know distance (amount done) and time. Therefore, we choose the formula for rate.

Distance (amount done) ÷ Time = Rate (speed)

9,000 words ÷ 150 min = 60 wpm

EXAMPLE L

It is approximately 400 miles from San Francisco to Los Angeles. Roy's friends tell him that he can make the trip in 6 hours if he averages 60 miles per hour. Is this true?

Choose a formula: We know the rate and the time, so we choose the formula for distance.

Rate (speed) × Time = Distance (amount done)

60 mph × 6 hr = 360 mi

Can he get there in 6 hours? *No.*

CONCEPT CHECK 4.3

The basic formulas:

a. Rate (speed) × Time = Distance (amount done)
   If you know any *two* factors, you can find the *third*.
b. Distance (amount done) ÷ Time = Rate (speed)
c. Distance (amount done) ÷ Rate (speed) = Time

Apply the appropriate formula to answer the following question: A machine that produces tortillas at the Baja Restaurant can produce 200 tortillas per hour, or 1,600 tortillas in an 8-hour day. A new machine can produce 3,000 tortillas in 6 hours. How many more tortillas per hour can the new machine produce than the old one?

Distance (amount done) ÷ Time = Rate
1,600 tortillas ÷ 8 hr   = 200 per hr
3,000 tortillas ÷ 6 hr   = 500 per hr
Difference: 500 − 200 = 300 more tortillas per hr

Solving Simple Numerical Equations

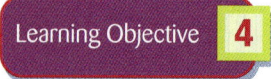

Solve simple numerical equations.

A **numerical sentence** in which both sides of an equal sign contain calculations is called an **equation.** For example, five plus five equals twelve minus two ($5 + 5 = 12 - 2$) is an equation, as is seven minus one equals thirty divided by five ($7 - 1 = 30 \div 5$).

For an equation to be true, the numbers on the left of the equal sign must always compute to the same answer as the numbers on the right of the equal sign. Moving a number from one side of the equation to the other changes its sign. A plus sign will change to minus; a minus sign will change to plus. A multiplication sign will change to division; a division sign will change to multiplication.

● EXAMPLE M   Addition—Subtraction

$6 + 4 + 5 = 17 - 2$

Change only the -2:
$6 + 4 + 5 + 2 = 17$

Change only the $+5$:
$6 + 4 = 17 - 2 - 5$
$6 + 4 = 10$ and $17 - 2 - 5 = 10$

Change the $+5$ and the -2:
$6 + 4 + 2 = 17 - 5$
Check: $6 + 4 + 2 = 12$
$17 - 5 = 12$

● EXAMPLE N   Multiplication—Division

$3 \times 8 = 48 \div 2$

Change only the $\div 2$:
$3 \times 8 \times 2 = 48$

Change only the $\times 8$:
$3 = 48 \div 2 \div 8$

Change the $\times 8$ and $\div 2$:
$3 \times 2 = 48 \div 8$
Check: $3 \times 2 = 6$
$48 \div 8 = 6$

A numerical equation may be incomplete, with one factor missing, but provide enough information to be completed.

EXAMPLE O
$6 + 2 = 5 + ?$
$6 + 2 = 8$ so $5 + ? = 8$
Therefore, $? = 3$
Or change a number
$6 + 2 - ? = 5$
Therefore, $? = 3$

EXAMPLE P
$15 - 3 = 2 + ?$
$15 - 3 = 12$ so $2 + ? = 12$
Therefore, $? = 10$
Or change a number
$15 - 3 - ? = 7$
Therefore, $? = 5$

EXAMPLE Q
$7 + 3 + 6 = 4 + 4 + ?$
$7 + 3 + 6 = 16$ so $4 + 4 + ? = 16$
Therefore, $? = 8$
Or change a number
$7 + 3 + 6 - ? = 4 + 4$
Therefore, $? = 8$

EXAMPLE R
$20 \div 5 = 2 \times ?$
$20 \div 5 = 4$ so $2 \times ? = 4$
Therefore, $? = 2$
Or change a number
$20 \div 5 \div ? = 2$
Therefore, $? = 2$

In business, numerical sentences with equations frequently compare items. Note the following examples:

EXAMPLE S
4 items at $0.50 each = 10 items at ? each
4 items at $0.50 each = $2.00
10 items at ? each = $2.00
$2.00 \div 10$ items = $0.20
Therefore, $? = \$0.20$
Or change a number
$4 \times 0.50 \div ? = 10$
Therefore, $? = 0.20$

EXAMPLE T
6 tickets at $5 each = 15 tickets at ? each
6 tickets at $5 each = $30
15 tickets at ? each = $30
$30 \div 15$ tickets = $2
Therefore, $? = \$2$
Or change a number
$6 \times 5 \div ? = 15$
Therefore, $? = \$2$

EXAMPLE U
A company had sales of $25,000 and $20,000 for January and February of last year, respectively. If January sales this year were $30,000, what is the amount needed for February in order to equal last year's sales for the two months?

January LY $25,000 + February LY $20,000 = $45,000
January $30,000 + February (?)   = $45,000
$45,000 - $30,000 = $15,000
Therefore, $? = \$15,000$

Chapter 4 Word Problems and Equations

CONCEPT CHECK 4.4

Both sides of a true equation are equal. Each side may contain calculations.
$7 + 5 = 14 - 2$
$2 \times 9 = 36 \div 2$
A number may be moved from one side of an equation to the other by reversing its sign.

$8 = 6 + 2$	$8 - 2 = 6$	$7 + 3 = 10$	$7 = 10 - 3$
$12 = 4 \times 3$	$12 \div 3 = 4$	$24 \div 12 = 2$	$24 = 2 \times 12$

Numerical Relationships in a Series

Learning Objective 5

Recognize numerical relationships in a series.

Relationships in a series of numbers may be found by comparing the first three or four terms in a series and then extrapolating the numbers that would most logically come next. For example, examining the series 320, 160, 80, 40 indicates that each term is found by dividing the preceding number by 2. The next two numbers in the series would logically be 20 and 10—that is, $40 \div 2 = 20$ and $20 \div 2 = 10$.

Examining the series 7, 14, 21, 28 suggests the addition of 7 to each preceding number. The next two numbers in this series would logically be 35 and 42 ($28 + 7 = 35$ and $35 + 7 = 42$).

In the series 5, 15, 35, 75, 155, seeing a relationship is difficult; however, a relationship does exist. Each number results from multiplying the preceding number by 2 and then adding 5. In this series, the next number would logically be 315 ($155 \times 2 + 5 = 315$).

Recognizing numerical and series relationships can be important in analyzing, communicating, and computing numbers. These relationship series are also used frequently in initial employment tests.

CONCEPT CHECK 4.5

In studying relationships in a numerical series, look for patterns. Patterns most commonly fall into categories:

Addition	2, 7, 12, 17, 22, 27	(+ 5, or 32)
Alternating addition/subtraction	12, 24, 18, 30, 24, 36, 30	(+ 12, − 6, or 42, 36)
Subtraction	39, 32, 25, 18, 11, 4	(− 7, or − 3)
Alternating subtraction/addition	64, 59, 61, 56, 58, 53, 55	(− 5, + 2, or 50, 52)
Multiplication	4, 12, 36, 108, 324, 972	(× 3, or 2,916)
Division	384, 192, 96, 48, 24	(÷ 2, or 12)

You can also devise patterns such as multiplication with addition or subtraction, division with addition or subtraction, and many other combinations.

Making Quick Calculations by Rounding Numbers

Quick calculations are beneficial when working in business situations. *Rounding* odd and difficult-to-compute amounts to even whole numbers that are easier to compute is a technique often used in business. By rounding, you will be able to get quick and accurate answers without having to write out the computations.

Learning Objective 6

Do quick mental calculations through a process of rounding numbers.

EXAMPLE V

How much would 5 items at $2.99 each cost?

To make this computation easily, think "$2.99 is $0.01 less than $3.00." Then think "5 times $3 equals $15." Finally, think "$15.00 less $0.05 (5 × $0.01) is $14.95," which is the correct answer.

4.4 Making quick calculations by rounding numbers is a natural sequel to estimating and approximating, covered in Chapter 1. Student review of and practice in estimating will develop and enrich this thinking process.

EXAMPLE W

The total cost of 3 equally priced dresses is $119.85. How much does each dress cost?

To figure out this problem easily, think "$119.85 is $0.15 less than $120.00." Then think "$120 divided by 3 = $40, and $40.00 less $0.05 ($0.15 ÷ 3) is $39.95," the correct answer.

EXAMPLE X

At 19 miles per gallon, how many miles would a car go on 9 gallons of gas?

To figure out this problem easily, think "19 is just 1 mile less than 20." Then think "9 times 20 = 180, and 180 minus 9 (9 × 1) is 171," the correct answer.

✓ CONCEPT CHECK 4.6

You may have noticed that making quick calculations is quite similar to making estimations, which you did in Chapter 1. In fact, quick calculation is only an additional step. After estimating an answer, you determine the degree to which the estimated, or rounded, answer differs from the actual answer by mentally correcting for the amount of the estimation or rounding.

COMPLETE ASSIGNMENTS 4.1 AND 4.2

Chapter Terms for Review

equation

numerical sentence

Chapter 4 Word Problems and Equations 77

THE BOTTOM LINE

Summary of chapter learning objectives:

Learning Objective	Example
4.1 Use mental computations in simple addition, subtraction, multiplication, and division	Use mental computations. 1. Add: $\quad 4 + 3 + 8 + 11 + 9 + 2 + 3 =$ _____ 2. Add by combining numbers: $\quad 4 + 6 + 8 + 8 + 8 + 30 + 10 =$ _____ 3. Subtract: $\quad 84 - 7 - 12 - 23 =$ _____ 4. Subtract and add: $\quad 9 + 4 - 2 - 8 + 4 =$ _____ 5. Multiply and divide: $\quad 4 + 4 \times 2 \div 4 + 14 =$ _____ 6. Multiply and divide: $\quad 18 \div 3 + 10 - 5 \times 3 =$ _____
4.2 Use a systematic approach to solve word problems involving basic math processes	Use the two-step process to solve the word problem. 7. Martha is preparing to make two dresses. One will require 3 yards of material; the other will require 4 yards of material. The material for the first dress costs $12.00 per yard; the material for the second costs $15.00 per yard. Buttons and trimming will cost $8.00 for each dress. What will be the total cost? Determine what is being requested. Determine the processes to be used to solve the problem. Answer: _____
4.3 Apply formulas to solve rate, time, and distance problems	8. At an average rate of 50 miles per hour, how long would it take to drive 650 miles? _____ 9. At an average rate of 60 miles per hour, how far could you drive in 6 hours? _____ 10. If you drove 70 miles per hour and covered 280 miles, how much time did it take? _____
4.4 Solve simple numerical equations	11. $7 + 8 - 2 = 5 + 9 - 1$ 12. $5 \times 12 = 120 \div 2$ Change the 12 to the opposite side and test the equation.

Answers: 1. 40 2. 74 3. 42 4. 7 5. 18 6. 33 7. $112 8. 13 hr 9. 360 mi 10. 4 hr 11. $7 - 2 = 5 + 9 - 1 - 8$ 12. $5 = 120 \div 2 \div 12$

THE BOTTOM LINE

Summary of chapter learning objectives:

Learning Objective	Example
4.5 Recognize numeric relationships in a series	Insert the next two numbers. 13. 4, 7, 6, 9, 8, 11, _____, _____ Pattern: _____ 14. 12, 48, 24, 96, 48, _____, _____ Pattern: _____
4.6 Do quick mental calculations through a process of rounding numbers	15. What is the cost of 8 items at $3.99 each? 16. At 59 miles per hour, how far would a car go in 20 hours?

Answers: 13. (+3, −1) 10, 13   14. (× 4, ÷ 2) 192, 96   15. (8 × $4.00) = $32.00 − 0.08 = $31.92   16. (60 × 20) = 1200 − 20 = 1180 mi

Chapter 4   Word Problems and Equations

SELF-CHECK

Review Problems for Chapter 4

1. Add: $7 + 9 + 4 + 8 + 2 = $ _____

2. Subtract: $70 - 7 - 4 - 8 - 3 - 6 = $ _____

3. Multiply: $4 \times 2 \times 3 \times 2 \times 2 = $ _____

4. Divide: $120 \div 2 \div 3 \div 5 \div 2 = $ _____

5. In the first four months of the year, a corporation had monthly earnings of $12,493, $6,007, $3,028, and $9,728. What was its total earnings in the four months? _____

6. If the corporation in question 5 had earnings of $74,500 at the end of the year, how much did it earn in the last eight months of the year? _____

7. If a tour bus gets 7 miles per gallon of gas and used 61 gallons in a week, how many miles did it travel in the week? _____

8. An employer earned $4,000. Half the earnings went into an employee bonus pool. The pool was split among 5 employees. How much did each employee receive? _____

9. A delivery firm bought 21 gallons of gas on Monday, 15 on Tuesday, 24 on Wednesday, 34 on Thursday, and 11 on Friday. If gas cost $2.15 per gallon, how much did the delivery firm pay for the week's gas? _____

10. A store owner planned to give away $1,200 at Christmas. The owner gave $150 to each of 5 full-time employees and $50 to each of 4 part-time employees. The remainder was given to a local charity. How much did the charity receive? _____

11. How long would it take to travel 1,265 miles at 55 miles per hour? _____

12. Bob and Mary start traveling toward each other from 1,330 miles apart. Bob is traveling at 30 miles per hour, Mary at 40 miles per hour. How many hours elapse before they meet? _____

13. Bob and Mary start traveling toward each other from 960 miles apart. Bob is traveling at 25 miles per hour, Mary at 55 miles per hour. How many hours elapse before they meet? _____

14. $41 - 6 = 27 + $ _____

15. $72 + 72 = 300 - $ _____

16. $10 \times 3 = 90 \div $ _____

17. Four items at $9 each = _____ items at $12 each

18. What is the next number in the series 3, 7, 8, 12, ? _____

19. What is the next number in the series 5, 20, 10, 40, ? _____

20. To find the price of 7 items at $1.99 you would think: 7 times $_____ less 7 times $_____ = $13.93

Assignment 4.1: Word Problems, Equations, and Series

Name

Date   Score

Learning Objectives **1** **2** **5**

A (20 points) Do the steps in the order in which they occur. Do not use scratch paper or an electronic calculator. (1 point for each correct answer)

1. $14 + 5 + 3 + 6 =$ 28
2. $6 \times 6 - 4 \div 8 \times 2 =$ 8
3. $12 - 3 - 2 - 5 =$ 2
4. $14 \div 2 \times 5 \times 2 + 5 =$ 75
5. $40 \div 4 \div 2 \div 5 =$ 1
6. $9 \times 2 + 2 \times 6 - 20 \div 4 =$ 25
7. $3 \times 2 \times 5 \times 2 =$ 60
8. $(4 - 3) \times 5 \times 5 \times 5 - 3 =$ 122
9. $25 \div 5 \times 3 + 1 + 11 + 2 - 6 =$ 23
10. $(12 + 12 + 12 + 14) \div 5 \times 3 + 8 =$ 38
11. $100 \times 5 - 50 \div 9 + 5 \div 11 \times 3 =$ 15
12. $(36 \div 3 \div 4 + 10 + 5 - 3) \times 5 =$ 75
13. $(15 \div 3 \times 2 + 8 - 3 + 12) \div 3 =$ 9
14. $(5 \times 8) + (20 \times 3 \div 6 \div 5) + 4 =$ 46
15. $9 \div 3 \times 7 + 4 + 5 \times 4 - 6 =$ 114
16. $680 \div 2 \div 2 + 10 \div 6 \times 2 + 8 =$ 68
17. $32 \times 2 \div 8 \times 100 + 200 \div 4 + 3 =$ 253
18. $12 + 10 + 3 + 26 + 29 \div 4 \times 3 =$ 60
19. $1{,}000 \times 4 \times 2 - 5{,}000 \div 3 =$ 1,000
20. $3 + 4 + 5 + 6 + 7 \div 5 \times 800 =$ 4,000

Score for A (20)

B (10 points) Do these problems without using scratch paper or an electronic calculator. (2 points for each correct answer)

21. How much would you pay for 8 gallons of gasoline selling at $2.05 per gallon? $16.40
22. How many items would you have if you had 3 books, 7 cards, and 21 pencils? 31
23. If six people divided three pizzas so that each person got one piece, how many slices would each pizza have? 2
24. How much would you have if you received $7.00 from one person, $23.00 from a second, $12.00 from a third, and $4.00 from a fourth? $46.
25. If 27 people were divided into three equal groups and each group added 2 additional members, how many members would be in each group? 11

Score for B (10)

C (10 points) Do the steps in the order in which they occur. Do these problems without using scratch paper or an electronic calculator. (1 point for each correct answer)

26. 12 items at $3 each plus $2 tax = $38
27. 15 watches at $30 each less a $50 discount = $400
28. 3 lamps at $22 each plus 7 bulbs at $2 each = $80
29. 100 belts at $4 each less discounts of $60 and $30 = $310
30. 3 dozen scissors at $11.20 per dozen plus a $4 shipping charge = $37.60
31. 8 pounds of pears at $3 per pound plus 50¢ per pound for packaging = $28

Assignment 4.1 Continued

32. $38 sale price plus $3 tax less a $11 discount plus a $5 delivery charge = __$35__
33. 6 bath towels at $8 each and 4 hand towels at $3 each plus $2.50 tax = __$62.50__
34. 4 dozen brushes at $25 per dozen plus $5 tax plus $7 shipping charge = __$112.00__
35. 2 shirts at $30 each, 4 ties at $10 each, and 7 pairs of socks at $2 each = __$114__

Score for C (10)

D (40 points) Complete the following equations by supplying the missing items. (2 points for each correct answer)

36. $27 + 3 =$ __22__ $+ 8$
37. $13 +$ __22__ $= 7 + 28$
38. __16__ $+ 4 = 4 + 16$
39. $400 = 17 - 2 +$ __385__
40. $22 - 9 =$ __19__ $- 6$
41. $36 -$ __11__ $= 17 + 8$
42. $9 + 17 - 3 = 4 \times 7$ ____ $- 5$
43. $160 \div 4 + 2 = 7 \times 7 -$ __7__
44. $13 - 11 \times 40$ ____ $= 8 \times 8 + 16$
45. __3__ $\times 3 \times 3 = 9 \div 3 \times 9$
46. $4 \times 20 =$ __76__ $+ 4$
47. __16__ $\div 2 = 9 - 1$
48. $64 \div 32 = 900 \div$ __450__
49. $15 - 9 - 2 = 25 -$ __21__
50. __25__ $+ 6 = 43 - 12$
51. $(7 \times 8) - 6 =$ __50__
52. $15 \times 2 \times 2 =$ __60__
53. $13 \times$ __5__ $= 77 - 12$
54. __10__ $\times 9 = 99 - 9$
55. $6 \times$ __15__ $= 10 \times 9$

Score for D (40)

E (20 points) In each of the following problems, a definite relationship exists among the numbers in each series. Extend each series two items by following the correct process. (8 points for each problem; 1 point for each correct line)

56. Extend each series below through addition.
    a. 4, 8, 12, 16, __20, 24__
    c. 2, 4, 7, 11, 13, __16, 20__
    b. 1, 4, 5, 8, __9, 12__

57. Extend each series below through subtraction.
    a. 50, 45, 40, 35, __30, 25__
    c. 100, 90, 81, 73, __66, 60__
    b. 50, 45, 43, 38, __36, 31__

58. Extend each series below through multiplication.
    a. 4, 8, 16, 32, __64, 128__
    c. 2, 4, 20, 40, __200, 400__
    b. 5, 25, 125, __625, 3, 125__

59. Extend each series below through division.
    a. 15,625, 3,125, 625, 125, __25; 5__
    c. 10,000, 2,000, 1,000, 200, __100, 20__
    b. 729, 243, 81, 27, __9, 3__

60. Extend each series below through combinations of the four processes above.
    a. 72, 75, 69, 72, __66, 69__
    e. 7, 4, 8, 5, __9, 6__
    b. 200, 100, 300, 150, __450, 225__
    f. 30, 10, 60, 20, __120, 40__
    c. 6, 9, 18, 21, 42, __45, 90__
    g. 10, 40, 20, 80, __40, 160__
    d. 240, 120, 600, 300, 1,500, __750, 3,750__
    h. 100, 50, 40, 20, __10, 5__

Score for E (20)

Assignment 4.2: Word Problems, Formulas, and Equations

Name

Date   Score

Learning Objectives **1 2 3 4 6**

A **(40 points) Solve the following word problems. (5 points for each correct answer)**

1. A store regularly sold 2 cans of soup for $1.28. It advertised a special sale of 6 cans for $3.12. A customer bought 12 cans at the sale. How much did the customer save over the regular price? $0.72
   $1.28 ÷ 2 = $0.64 each $0.64 − $0.52 = $0.12
   $3.12 ÷ 6 = $0.52 each $0.12 × 12 = $1.44 saved

2. A sales representative's car gets 18 miles to a gallon of gas. It was driven 120 miles each day for 30 days. Gas cost an average of $2.27 per gallon. What was the sales representative's total 30-day cost for gas? $254.00
   120 × 30 = 3,600 mi
   3,600 ÷ 18 = 200 gal
   200 × $2.27 = $454.00

3. A store clerk sold a customer a ruler for $1.67, three pencils for $0.29 each, notebook paper for $0.99, and an eraser for $0.35 and was given $10.00 in payment. How much change did the clerk give the customer from the $10.00? (All prices include tax.) $6.12
   $1.67 + (3 × $0.29) + $0.99 + $0.35 = $3.88
   $10.00 − $3.88 = $6.12

4. A college student worked at a local store for $9.00 per hour, as his class schedule permitted. The student worked 3 hours each Monday, Tuesday, Wednesday, and Thursday. He also worked 2 hours each Friday and 8 hours each Saturday. How many weeks did the student have to work to earn $792 for a new bicycle?
   4 weeks
   (3 × 4) + 2 + 8 = 22 hr
   $9.00 × 22 = $198 per week
   $792 ÷ $198 = 4 weeks

5. A box, a crate, and a trunk weigh a total of 370 pounds. The crate weighs 160 pounds. The trunk weighs 4 pounds more than the box. What does the box weigh? 103 lb
   370 − 160 − 4 ÷ 2 = 103 lb

6. A hotel has 12 floors. Each floor has 30 *single-person* rooms and 40 *two-person* rooms. What is the total *guest* capacity of the hotel? 1,320
   30 + (40 × 2) × 12 = 1,320 guest capacity

7. A department store offers its customers socks for $1.50 per pair or $15.00 per dozen. If two customers buy 1 dozen together and each pays half the cost, how much will each customer save by paying the quantity price? $1.50
   $1.50 × 12 = $18.00
   $18.00 − $15.00 = $3.00
   $3.00 ÷ 2 = $1.50 each saved

8. Supply Clerk A ordered 5 staplers for $27.50 total and 2 large boxes of staples for $1.75 each. Supply Clerk B ordered a box of computer disks for $8.50 and a box of computer paper for $39.95. How much more did Clerk B spend than Clerk A? (All prices include tax.) $17.45
   A: $27.50 + ($1.75 × 2) = $31.00
   B: $8.50 + $39.95 = $48.45
   $48.45 − $31.00 = $17.45

Score for A (40)

Assignment 4.2 Continued

B **(10 points) Solve the following time, rate, distance problems. (5 points for each correct answer)**

9. Wendy leaves St. Paul to travel the 2,000 miles to Los Angeles, driving at a speed of 55 miles per hour. Mark leaves Los Angeles to travel the same 2,000-mile route to St. Paul, driving at a speed of 45 miles per hour. How many miles will Mark have traveled when they meet? __900 mi__
   $2,000 \div 100 = 20$ hr
   $20 \times 45 = 900$ mi

10. Car A traveled to a destination 840 miles away at 60 miles per hour. Car B traveled to a destination 660 miles away at 55 miles per hour. How much longer did Car A travel than Car B? __2 hr__
    A: $840 \div 60 = 14$ hr
    B: $660 \div 35 = 12$ hr
    14 hr $- 12$ hr $= 2$ hr

Score for B (10)

C **(40 points) Solve each of the problems without writing any computations on paper and without using a calculator or a computer. (2 points for each correct answer)**

11. 5 items at $1.99 = __$9.95__
12. 2 items at $7.98 = __$15.96__
13. 4 items at $19.98 = __$79.92__
14. 2 items at $49.96 = __$99.92__
15. 15 items at $0.99 = __$14.85__
16. 10 items at $9.99 = __$99.90__
17. 6 items at $3.95 = __$23.70__
18. 5 items at $1.02 = __$5.10__
19. 19 items at $40 = __$760__
20. 3 items at $19.99 = __$59.97__
21. 20 items at $40.05 = __$801__
22. 30 items at $1.99 = __$59.70__
23. 20 items at $39.98 = __$799.60__
24. 2 items at $5.99 = __$11.98__
25. 48 items at $5 = __$240__
26. 5 items at $1.97 = __$9.85__
27. 7 items at $7.97 = __$55.79__
28. 2 items at $99.98 = __$199.96__
29. 30 items at $2.98 = __$89.40__
30. 99 items at $1.90 = __$188.10__

Score for C (40)

D **(10 points) In each of the following equations, rewrite the equation by moving the last number on each side of the equal sign to the other side and making appropriate sign changes so that the equation is still true. (Example: Given $13 + 7 + 2 = 10 + 12$; Answer $13 + 7 - 12 = 10 - 2$) (1 point for each correct equation)**

31. $6 + 4 + 5 = 17 - 2$
    $6 + 4 + 2 = 17 - 5$
32. $6 \times 2 \div 3 = 8 \div 4 \times 2$
    $6 \times 2 \div 2 = 8 \div 4 \times 3$
33. $9 - 3 - 3 = 2 + 1$
    $9 - 3 - 1 = 2 + 3$
34. $8 \div 2 \times 4 = 24 \div 3 \times 2$
    $8 \div 2 \div 2 = 24 \div 3 \div 4$
35. $20 + 1 - 7 = 16 - 2$
    $20 + 1 + 2 = 16 + 7$
36. $3 \times 3 \times 3 = 18 \div 2 \times 3$
    $3 \times 3 \div 3 = 18 \div 2 \div 3$
37. $12 + 3 - 5 = 7 + 3$
    $12 + 3 - 3 = 7 + 5$
38. $7 \times 4 \div 2 = 28 \times 2 \div 4$
    $7 \times 4 \times 4 = 28 \times 2 \times 2$
39. $64 - 32 - 16 = 8 + 8$
    $64 - 32 - 8 = 8 + 16$
40. $63 \div 7 \times 2 = 3 \times 2 \times 3$
    $63 \div 7 \div 3 = 3 \times 2 \div 2$

Score for D (10)

Notes

Part 2

Percentage Applications

5 Percents
6 Commissions
7 Discounts
8 Markup

Percents

5

Learning Objectives
By studying this chapter and completing all assignments you will be able to:

Learning Objective 1   Change percents to decimals.

Learning Objective 2   Change fractions and decimals to percents.

Learning Objective 3   Find Base, Rate, and Percentage.

Learning Objective 4   Use percents to measure increase and decrease.

Learning Objective 5   Use percents to allocate overhead expenses.

Percents and percentages are used extensively in various business and nonbusiness applications. Airlines are required to publish the "on time percentage" for each of their flights. Every bank publishes its loan rates as percents. The Food and Drug Administration (FDA) says that packaged foods must contain labels with nutritional information, much of which is written in percents. Colleges and universities often describe the ethnic diversity of their student bodies and faculty using percents.

Changing Percents to Decimals

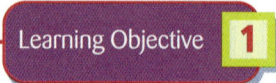

Change percents to decimals.

We use percents because the word *percent* makes verbal and written communication easier. Suppose that we have a 5% sales tax. Which of these phrases sounds better: (a) "five percent," (b) "five-hundredths," (c) "one-twentieth," or even (d) "point zero five"? Imagine how complicated the latter three phrases would be if the sales tax rate were 5.25%. But by using the word *percent*, we can just say "five point two five percent."

Percents themselves are actually not used in arithmetic. Before you can do any calculation with a percent, you must change the percent to a decimal. If you use a calculator with a percent key [%], the calculator will first convert the percent to a decimal. Take a calculator with a percent key and observe the display closely. Enter **75%**; that is, press these three keys: [7] [5] [%]. After pressing the [%] key, the display shows **0.75**. There is no percent symbol and the decimal point has moved two places to the *left*. The calculator will use the 0.75 in all of its calculations that involve 75%.

Sometimes a percent has a fractional part. For example, we might have a tax rate that is stated as $5\frac{1}{2}$%. Even using a calculator, first we must write the fraction as a decimal to get 5.5%. Using the calculator, press these keys: [5] [.] [5] [%]. After pressing [%], the display shows **0.055**. Notice that to move two places to the left, the calculator had to insert an extra zero.

5.1 Ask students how to say percents in other languages. For example, in Italian, the word for *100* is "cento," the word for *20* is "venti," and the words for *percent* are "per cento." So 100% is "cento per cento," and 20% is "venti per cento."

In Spanish, the word for *100* is "ciento" (or sometimes "cien"), the word for *20* is "veinte," and the words for *percent* are "por ciento." So 100% is "ciento por ciento," and 20% is "veinte por ciento."

5.2 Remind students again that in percents, when a decimal number has no preceding whole number, we will write a zero to the left of the decimal point.

STEPS to Change a Percent to a Decimal
1. If the percent has a fractional part, convert the fraction to its decimal equivalent.
2. Remove the percent symbol.
3. Move the decimal point two places to the *left* (insert zeros if needed).

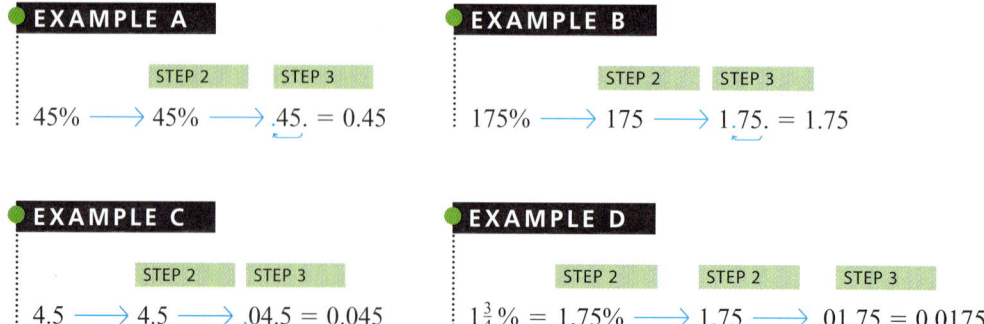

(Note: Check the answers to these examples with the percent key on your calculator.)

Part 2 Percentage Applications

✓ CONCEPT CHECK 5.1

a. Change 250% to a decimal.

$250\% \longrightarrow 250 \longrightarrow 2.50. = 2.50$ or 2.5

b. Change $\frac{1}{4}\%$ to a decimal.

$\frac{1}{4}\% = 0.25\% \longrightarrow 0.25 \longrightarrow .00.25 = 0.0025$

Changing Decimals and Fractions to Percents

Changing a decimal to a percent is exactly the opposite from changing a percent to a decimal: Move the decimal point two places to the *right*, and then write a percent symbol. If you have a fraction or a mixed number, first change it to a decimal as you did in Chapter 3. Then change the decimal to a percent. (A decimal point at the extreme right end of the percent is omitted. Examine example E below.)

Learning Objective 2

Change fractions and decimals to percents.

STEPS to change a Fraction or a Decimal to a Percent

1. If the number is a fraction, or a mixed number, convert it to its decimal equivalent.
2. Move the decimal point two places to the *right* (insert zeros if needed).
3. Write a percent symbol at the *right* end of the new number.

EXAMPLE E

STEP 1   STEP 2   STEP 3

$\frac{4}{5} = 0.8 \longrightarrow 0.80. \longrightarrow 80\%$ or 80%

EXAMPLE F

STEP 1   STEP 2   STEP 3

$1\frac{3}{8} = 1.375 \longrightarrow 1.37.5 \longrightarrow 137.5\%$

EXAMPLE G

STEP 2   STEP 3

$0.4 \longrightarrow 0.40. \longrightarrow 40\%$

EXAMPLE H

STEP 2   STEP 3

$1.1875 \longrightarrow 1.18.75 \longrightarrow 118.75\%$

EXAMPLE I

STEP 2   STEP 3

$2.5 \longrightarrow 2.50. \longrightarrow 2.50\%$

EXAMPLE J

STEP 2   STEP 3

$1 = 1. \longrightarrow 1.00. \longrightarrow 100\%$

(Note: To check these examples with your calculator, you can multiply the decimal number by 100 and write the percent symbol at the right end of the answer.)

✓ CONCEPT CHECK 5.2

a. Change $2\frac{7}{10}$ to a percent.

$$2\frac{7}{10} = 2.7 \longrightarrow 2.70. \longrightarrow 270\%$$

b. Change 0.075 to a percent.

$$0.075 \longrightarrow 0.07.5 \longrightarrow 7.5\%$$

Finding Base, Rate, and Percentage

Learning Objective 3

Find Base, Rate, and Percentage.

© STEVE MASON/PHOTODISC/GETTY IMAGES

Suppose that you have $5 and spend $4 for breakfast. Example E showed that the fraction $\frac{4}{5}$ equals 80%. You can say that "you spent 80% of your money ($5) for your breakfast ($4)." Without the context of your breakfast, you have simply "80% of $5 = $4." In this book we call 80% the **Rate** (**R**), $5 the **Base** (**B**) amount, and $4 the **Percentage** (**P**) amount. The Base and the Percentage amounts will always have the same units (e.g., dollars, feet, or pounds). The Rate is the percent. (The word *rate* comes from the word *ratio*—in this case, $\frac{4}{5}$.) It may make sense for you to think of the Base amount as the denominator in the rate (that is, ratio = $\frac{4}{5}$) because the denominator is the "base" (i.e., bottom) of the fraction.

Note: In practice, the terms *percent* and *percentage* are often used interchangeably. Sometimes, you will see the word *percentage* used to mean a rate and the word *percent* used to mean an amount. You will even see the two words *percentage rate* to mean the rate. In this book, however, we use only the one meaning for each word.

● **EXAMPLE K**

80% of $5 = $4
80% is the Rate
$5 is the Base
$4 is the Percentage

● **EXAMPLE L**

25% of 20 ft = 5 ft
25% is the Rate
20 ft is the Base
5 ft is the Percentage

● **EXAMPLE M**

50% of 60 gal = 30 gal
50% is the Rate
60 gal is the Base
30 gal is the Percentage

5.3 Students may understandably be confused by the way in which we distinguish between *percent* and *percentage* because the terms often are not used consistently. Ask students to look up the definitions and then find examples on television or in the newspaper. They will quickly discover the inconsistent usage.

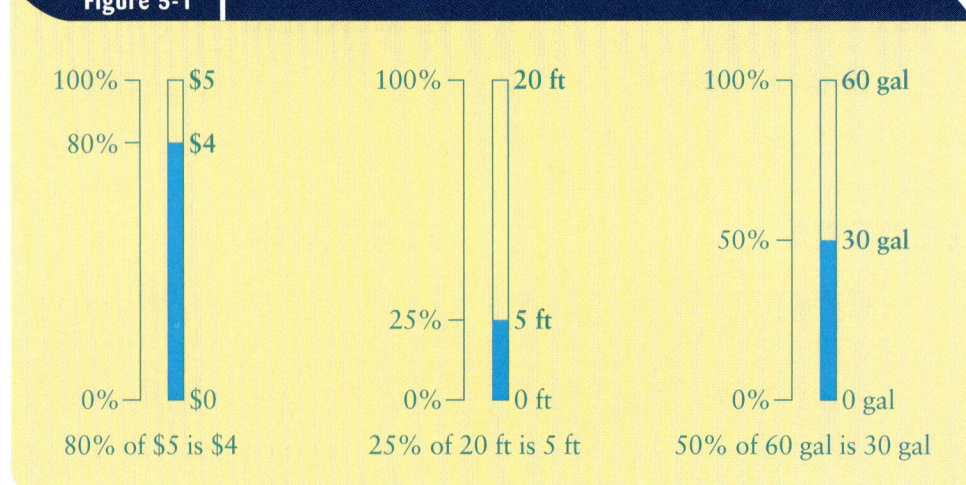

Figure 5-1

80% of $5 is $4   25% of 20 ft is 5 ft   50% of 60 gal is 30 gal

Figure 5-1 shows three diagrams, one each for examples K, L, and M. In each diagram, the Rate (or percent) is shown in the left-hand column. Each Percentage is represented by the shaded portion of the right-hand column. Each Base is represented by the entire height of the right-hand column.

The word *of* often appears in problems that involve percents. Recall from Chapter 2 that with fractions **of** means **multiply**. We just showed that $80\% = \frac{\$4}{\$5}$. Also recall that you can "check" a division problem by multiplication. We would get $80\% \times \$5 = \$4$. In words, we say that "80% *of* $5 is $4."

Rule: The number that follows the word *of* is the Base (and is the denominator in the fraction); the number that follows the word *is* is the Percentage amount.

The preceding examples illustrate the basic relationship among the Rate, Base, and Percentage: Rate $\times$ Base = Percentage. As a formula, it is written as $R \times B = P$ or as $P = R \times B$.

When you know any two of these three numbers, you can calculate the third by changing the formula:

If you want to find B, the formula becomes $B = P \div R$ or $P \div R = B$.
If you want to find R, the formula becomes $R = P \div B$ or $P \div B = R$.

5.4 Looking at examples L and M, we could say that "$\frac{1}{4}$ of 20 feet is 5 feet" and "$\frac{1}{2}$ of 60 gallons is 30 gallons." The mathematics are $\frac{1}{4} \times 20 = 5$ and $\frac{1}{2} \times 60 = 30$.

EXAMPLE N
Find P when
R = 50% and B = 300 yd

EXAMPLE O
Find R when
B = 30 lb and P = 6 lb

EXAMPLE P
Find B when
P = $45 and R = 75%

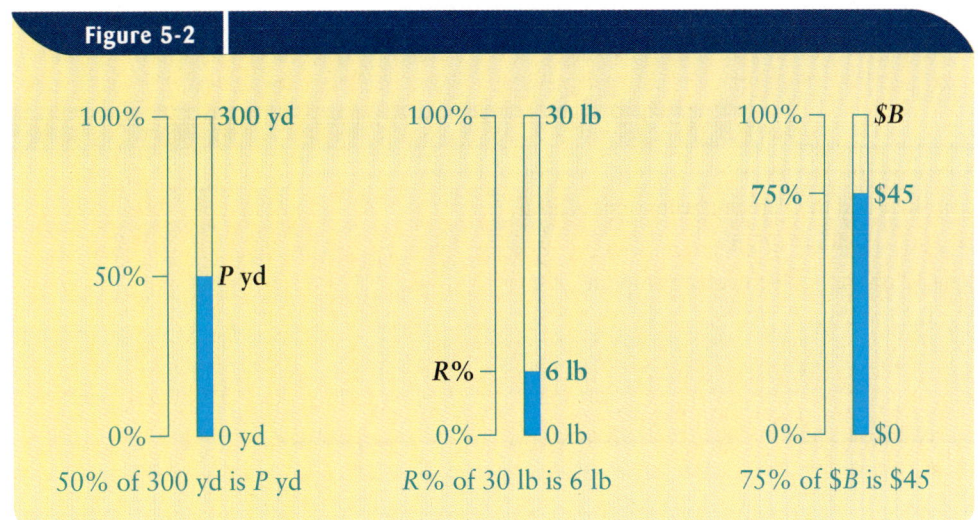

Figure 5-2

50% of 300 yd is P yd   R% of 30 lb is 6 lb   75% of $B is $45

Figure 5.2 illustrates these relationships, which are calculated as follows.

$P = R \times B$	$R = P \div B$	$B = P \div R$
$P = 50\% \times 300$ yd	$R = 6$ lb $\div 30$ lb	$B = \$45 \div 75\%$
$P = 0.50 \times 300$ yd	$R = 0.20$	$B = \$45 \div 0.75$
$P = 150$ yd	$R = 20\%$	$B = \$60$

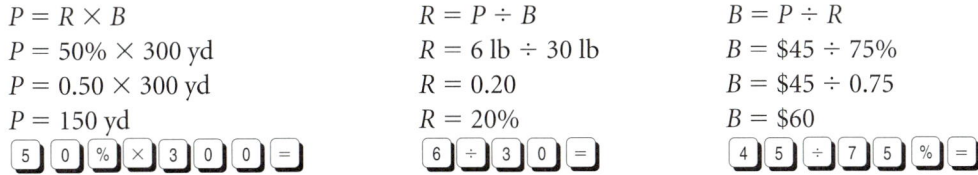

Note that in example O, the calculator cannot automatically "move" the decimal point two places to the right. If you want the calculator to do it, you "multiply by 100." It is faster to just move the decimal point places without a calculator.

USING PERCENTS IN BUSINESS

Percent problems occur frequently in business. Examples Q and R are typical fundamental applications, in which we solve for the Base (*B*) amount and the Rate (*R*), respectively.

EXAMPLE Q

Lena Hoover is a financial analyst. In December, she received a $600 bonus, which equaled 15% of her monthly salary. What was her monthly salary?

P = amount of bonus = $600
R = rate of bonus = 15%
B = monthly salary = ?
As $P \div R = B$,
$P \div R = \$600 \div 15\% = \$600 \div 0.15 = \$4,000$ monthly salary

EXAMPLE R

Last year Bayside Coffee Shop had total expenses of $300,000. Of that total, $210,000 was the expense for employee salaries. At Bayside, employee salary expense is what percent of total expenses?

P = employee salaries = $210,000
R = ?
B = total expenses = $300,000
Since $P \div B = R$,
$P \div B = \$210,000 \div \$300,000 = 0.70 = 70\%$

✓ CONCEPT CHECK 5.3

a. Find the Base when the Rate is 40% and the Percentage amount is 50 ft.
   $B = P \div R = 50\text{ft} \div 40\% = 50\text{ft} \div 0.40 = 125\text{ ft}$
b. Find the Rate when the Base is 12 oz and the Percentage amount 3 oz.
   $R = P \div B = 3\text{ oz} \div 12\text{ oz} = 0.25 = 25\%$

COMPLETE ASSIGNMENT 5.1.

Using Percents to Measure Increase and Decrease

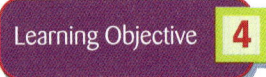

Use percents to measure increase and decrease.

In business, percents are used to measure change from one year to the next or from one month to the next. Real estate firms compare the number of homes sold this year with the number of homes sold last year. Read and carefully compare the following four statements about home sales last year and this year:

Joslin Realty sold 40% more homes this year than it did last year, when it sold 135 homes.

Part 2 Percentage Applications

Rossi & Shanley Real Estate sold 25 more homes this year than last year, which represents 20% more homes this year than last year.

Real estate agent Nancy Lo sold 5 fewer homes this year than she did last year, when she sold 40 homes.

Charles Peterson, a real estate broker, sold 30 homes last year; this year he sold 36 homes.

The number of homes sold last year is the Base (B) amount (last year is called the *base year*). The change in homes sold can be reported as a number, which would be the Percentage amount (P), or as a percent, which would be the Rate (R). If any two of the three values are given, the third can be determined using one of the three formulas in this chapter.

5.5 Point out that, if you increase by an amount and then decrease by the same amount, the percents are different. Suppose that you start with 100 and increase by 25 to 125. That is a 25% increase. If you then decrease by 25, back to 100, that is only a 20% decrease. Similarly, suppose that you start with 100 and increase by 10% to 110. If you then decrease by 10%, you go to 99 instead of 100. A similar difference will occur in Chapter 8 when we encounter markup based on cost and markup based on selling price.

EXAMPLE S

Find the number of additional homes (P) that Joslin Realty sold this year.

$B = 135$ and $R = 40\%$. Since $P = R \times B$,
$P = 40\% \times 135 = 0.40 \times 135 = 54$ more homes this year

EXAMPLE T

Find the number of homes that Rossi & Shanley Real Estate sold last year (B).

$P = 25$ and $R = 20\%$. Since $B = P \div R$,
$B = 25 \div 20\% = 25 \div 0.20 = 125$ homes sold last year

EXAMPLE U

Find Nancy Lo's rate of decrease (R) from last year's sales.

$P = 5$ and $B = 40$. Since $R = P \div B$,
$R = 5 \div 40 = 0.125 = 12.5\%$ decrease

To find the percent change when the only numbers reported are the amounts (B) for last year and this year, the first step is to find the **amount of increase** or the **amount of decrease**. P is the difference between the amounts for the two years. Then use $R = P \div B$ to find the **rate of increase** or the **rate of decrease**.

EXAMPLE V

Find Charles Peterson's rate of change (R).

Charles sold 30 homes last year (B) and 36 this year. The amount of change is

$P = 36 - 30 = 6$ more homes this year

The rate of change is

$R = P \div B = 6 \div 30 = 0.20 = 20\%$ increase

COMPUTING AMOUNTS OF INCREASE AND DECREASE WITH A CALCULATOR

Review example S. Now consider a variation of example S that says, "Find the total number of homes that Joslin Realty sold this year." Last year it sold 135 homes. There was a 40% increase, which means 54 more homes were sold this year. The total number of homes sold this year was 135 + 54 = 189 homes. Many calculators allow you to calculate 189 with the following keystrokes: [1][3][5][+][4][0][%][=]. The display will show the answer, 189.

If you need to know the actual amount of the increase, it will usually show in the calculator display immediately after you press the [%] key, but before you press the [=] key.

Similarly, suppose the original example had said, "The real estate agency sold 40% *fewer* homes this year than it did last year, when it sold 135 homes. Find the total number of homes that it sold this year." The amount of the *decrease* is 54 homes. Therefore, the total number sold this year is 135 − 54 = 81 homes. On the calculator, you would use the following keystrokes: [1][3][5][−][4][0][%][=]. The display will show the answer, 81.

✓ CONCEPT CHECK 5.4

A company had sales of $200,000 this month and $160,000 last month (*B*). Find both the amount of increase (*P*) and the rate of increase (*R*).

The amount of increase is
$P = \$200{,}000 - \$160{,}000 = \$40{,}000$

The rate of increase is
$R = P \div B = \$40{,}000 \div \$160{,}000 = 0.25 = 25\%$

COMPLETE ASSIGNMENTS 5.2 AND 5.3.

Using Percents to Allocate Overhead Expenses

Learning Objective 5

Use percents to allocate overhead expenses.

5.6 Ask students to develop other bases for allocating overhead in different businesses. Several of them will likely be familiar with various types of businesses with different attributes that could be used to allocate overhead. Even very obvious measurements, such as number of employees in different departments, might be suggested.

Many businesses are organized into divisions or departments. Suppose Cotton's Clothing is a retailer of sportswear. It has three departments: women's clothes, men's clothes, and children's clothes. Management and owners of Cotton's need to measure the profitability of each department. Cotton's also knows the amounts it paid for the merchandise sold and the salaries of employees in each department. Cotton's can subtract these departmental costs from the departmental revenues.

But what about rent and other general costs such as electricity? These costs that are not directly related to the types of merchandise sold are called **overhead costs**. For example, Cotton's monthly rental expense might be $12,000 for the entire building. How should that single amount be divided among the three departments? Should each department be assigned $\frac{1}{3}$, or $4,000, of the total rent?

Businesses can *allocate*, or distribute, the rent based on a measurement related to the total cost. Rent is a cost of using the building; it could be allocated on the basis of floor space, since each department occupies some of that space.

THE BOTTOM LINE

Summary of chapter learning objectives:

Learning Objective	Example
5.1 Change percents to decimals	1. Change 4.25% to a decimal.
5.2 Change fractions and decimals to percents	2. Change 0.45 to a percent. 3. Change $\frac{7}{8}$ to a percent.
5.3 Find Base, Rate, and Percentage	4. Find the Percentage: 35% of 40 = P 5. Find the Rate: R% of 140 = 28 6. Find the Base: 80% of B = 220
5.4 Use percents to measure increase and decrease	7. Increase a $4,000 salary by 15%. 8. From 300 to 240 is a decrease of what percent?
5.5 Use percents to allocate overhead expenses	9. A company has three stores, A, B and C, with 4, 6, and 10 employees, respectively. Based on the number of employees, allocate a $3,000 expense among the stores.

Answers: 1. 0.0425 2. 45% 3. 87.5% 4. 14 5. 20% 6. 275 7. the increase is $600 8. 20% 9. Store A: $600; Store B: $900; Store C: $1,500

Chapter 5 Percents

SELF-CHECK

Review Problems for Chapter 5

1. Change 17.1% to a decimal _____
2. Change 0.625 to a percent _____
3. Change 150% to a decimal _____
4. Change 0.0075 to a percent _____
5. Change 0.06% to a decimal _____
6. Change $\frac{2}{5}$ to a percent _____
7. 14% of 50 = _____
8. 250% of 60 = _____
9. 25% of _____ = 45
10. 100% of _____ = 70
11. _____% of 40 = 35
12. _____% of 90 = 144

13. Sales were $100,000 two months ago and increased by 20% last month. How much were sales last month? _____

14. Sales were $120,000 last month and decreased by 20% this month. How much were sales this month? _____

15. Expenses were $200,000 two years ago and $400,000 last year. What was the percent increase last year? _____

16. Expenses were $400,000 last year and $200,000 this year. What was the percent decrease this year? _____

17. Peggy Covey owns a nursery. This year she sold 195 more rose bushes than she did last year. This represents a 12% increase over the previous year. How many rose bushes did Peggy's nursery sell last year? _____

18. Jim Dukes manages Internet sales for a company that started selling its product over the Internet two years ago. Last year, company sales over the Internet were only about $500,000. This year, sales were $1,625,000. Calculate the company's percent increase in Internet sales this year. _____

19. Ken Chard is a bank teller. When he started this morning, his cash drawer had coins worth $86. The coins represented only 2.5% of all the money that Ken had in his cash drawer. What was the total value of all this money? _____

20. Nancy McGraw is an orthopedic surgeon. Last winter, Dr. McGraw performed 50 emergency surgeries. Thirty-two of those surgeries were the result of ski injuries. What percent of Dr. McGraw's emergency surgeries were the result of ski injuries? _____

Assignment 5.1: Base, Rate, and Percentage

Name

Date   Score

Learning Objectives 1 2 3

A (20 points) Change the percents to decimals. Change the nonpercents to percents. (1 point for each correct answer)

1. 31% = __0.31__
   31%

2. 100% = __1__
   100%

3. $3\frac{1}{3}$% = __0.0333__
   $03\frac{1}{3}$%

4. 0.875 = __87.5%__
   0.875

5. 3 = __300%__
   3.00

6. $33\frac{2}{3}$% = __0.3367__
   $33\frac{2}{3}$%

7. 0.15 = __15%__
   0.15

8. 0.3 = __30%__
   0.30

9. $1\frac{3}{4}$ = __175%__
   $1\frac{3}{4}$ = 1.75

10. 5.2% = __0.052__
    05.2

11. 224.5% = __2.245__
    224.5%

12. 0.0003% = __0.000003__
    00.0003%

13. 0.52 = __52%__
    0.52

14. 350% = __3.5__
    350%

15. $0.08\frac{1}{4}$ = __8.25%__
    0.0825

16. $\frac{1}{2}$ = __50%__
    $\frac{1}{2}$ = 0.50

17. 4.0 = __400%__
    4.00

18. 0.000025 = __0.0025%__
    0.000025

19. 0.1% = __0.001__
    00.1%

20. 1,000% = __10__
    1000%

Score for A (20)

B (30 points) In the following problems, find each Percentage amount. (2 points for each correct answer)

21. 0.375% of 56 = __0.21__
    0.00375 × 56

22. 0.25% of 1,600 = __4__
    0.0025 × 1,600

23. 100% of 11.17 = __11.17__
    1 × 11.17

24. 62.5% of 24 = __15__
    0.625 × 24

25. 40% of 0.85 = __0.34__
    0.40 × 0.85

26. 250% of $66 = __$165__
    2.5 × $66

27. 25% of $1.16 = __$0.29__
    0.25 × $1.16

28. 120% of $45 = __$54__
    1.2 × $45

29. 2.5% of $66 = __$1.65__
    0.025 × $66

30. 50% of $162 = __$81__
    0.5 × $162

31. 8% of 200 = __16__
    0.08 × 200

32. 15% of 0.08 = __0.012__
    0.15 × 0.08

33. 187.5% of 40 = __75__
    1.875 × 40

34. 1.5% of $86 = __$1.29__
    0.015 × $86

35. 0.2% of 480 = __0.96__
    0.002 × 480

Score for B (30)

Chapter 5 Percents 99

Assignment 5.1 Continued

C (50 points) In each of the following problems, find the Percentage amount, the Rate, or the Base amount. Write rates as percents. Round dollars to the nearest cent. (2 points for each correct answer)

	(R)	(B)	(P)		(R)	(B)	(P)

36. 35% of $40.00 = $14
    14 ÷ 0.35

37. 20% of $35 = $7
    $7 ÷ $35 = 0.20

38. 200% of 0.12 = 0.24
    0.24 ÷ 0.12 = 2.00

39. 200% of 14.2 = 28.4
    28.4 ÷ 14.2 = 2.00

40. 3.5% of 400 = 14
    14 ÷ 400 = 0.035

41. 80% of $1.20 = $0.96
    $0.96 ÷ 0.8

42. 1.25% of 128 = 1.6
    1.6 ÷ 0.0125

43. 150% of 80 = 120
    120 ÷ 80 = 1.50

44. 25% of 0.056 = 0.014
    0.014 ÷ 0.056 = 0.25

45. 175% of $48 = $84
    $84 ÷ 1.75

46. 2.5% of $2,820 = $70.50
    0.025 × $2,820

47. 0.25% of $8,000 = $20
    $20 ÷ 0.0025

48. 250% of 9.76 = 24.4
    24.4 ÷ 2.5

49. 62.5% of 56 = 35
    35 ÷ 0.625

50. 0.025% of $16,400 = $4.10
    0.00025 × $16,400

51. 140% of 480 = 672
    672 ÷ 1.4

52. 120% of $42.50 = $51
    $51 ÷ 1.20

53. 40% of 5.4 = 2.16
    2.16 ÷ 5.4 = 0.40

54. 2,000% of $2,340 = $46,800
    $46,800 ÷ $2,340 = 20.00

55. 15% of $140 = $21.00
    0.15 × $140

56. 180% of $90 = $162
    1.80 × $90

57. 160% of 85 = 136
    136 ÷ 85 = 1.60

58. 125% of $416 = $520
    $520 ÷ 1.25

59. 12% of 25 = 3
    3 ÷ 0.12

60. 16% of 2.1 = 0.336
    0.336 ÷ 2.1 = 0.16

Score for C (50)

Assignment 5.2: Rate of Increase and Rate of Decrease

Name

Date   Score

Learning Objective **4**

A (40 points) Calculate the missing values. ($2\frac{1}{2}$ points for each correct answer)

1. Decreasing the base value of 280 by 25% gives the new value <u>210</u>.
   0.25 × 280 = 70
   280 − 70 = 210

2. Increasing the base value of 240 by 40% gives the new value <u>336</u>.
   0.40 × 240 = 96
   240 + 96 = 336

3. Start with 75, decrease it by 60%, and end up with <u>30</u>.
   0.60 × 75 = 45
   75 − 45 = 30

4. Start with 80, increase it by 14%, and end up with <u>91.2</u>.
   0.14 × 80 = 11.20
   80 + 11.2 = 91.2

5. Sales were $8,000 last month and increased by 4% this month. Sales were <u>$8,320</u> this month.
   0.04 × $8,000 = $320
   $8,000 + $320 = $8,320

6. Profits were $44,000 last month, but decreased by 2% this month. Profits were <u>$43,120</u> this month.
   0.02 × $44,000 = $880
   $44,000 − $880 = $43,120

7. Base value = 272; increase = 100%; new (final) value = <u>544</u>
   1.00 × 272 = 272
   272 + 272 = 544

8. Base value = 250; decrease = 100%; new (final) value = <u>0</u>
   1.00 × 250 = 250
   250 − 250 = 0

9. A $17 increase is 10% of the base value of <u>$170</u>.
   $17 ÷ 0.10 = $170

10. A decrease of 45 units is 15% of the base value of <u>300</u> units.
    45 ÷ 0.15 = 300

11. The price decreased from $450 to $378; the percent decrease was <u>16%</u>.
    $450 − $378 = $72
    $72 ÷ $450 = 0.16, or 16%

12. Production increased from 8,000 units to 10,000 units; the percent increase was <u>25%</u>.
    10,000 − 8,000 = 2,000 increase
    2,000 ÷ 8,000 = 0.25, or 25% increase

13. $300 is what percent less than $400?
    <u>25%</u>
    $400 − $300 = $100
    $100 ÷ $400 = 0.25, or 25%

14. 320 is what percent greater than 160? <u>100%</u>
    320 − 160 = 160
    160 ÷ 160 = 1.00, or 100%

15. Sales were $500,000 in June but only $400,000 in July. The rate of decrease was <u>20%</u>.
    $500,000 − $400,000 = $100,000
    $100,000 ÷ $500,000 = 0.20, or 20%

16. Profits were $11,000 last month and $10,000 the previous month. The rate of increase was <u>10%</u>.
    $11,000 − $10,000 = $1,000
    $1,000 ÷ $10,000 = 0.10, or 10%

Score for A (40)

Assignment 5.2 Continued

B (30 points) The following table shows the volumes of various items sold by Thrift's Speed Shop during the past two years. Compute the amount of change and the rate of change between this year and last year. Compute the rates to the nearest tenth of a percent. If the amount and rate are increases, write a + in front of them; if they are decreases, enclose them in parentheses (). (1 point for each correct amount; 2 points for each correct rate)

Thrift's Speed Shop
Volume Sold (number of units)

Description of Item	This Year	Last Year	Amount of Change	Rate of Change
17. Batteries	516	541	(25)	(4.6%)
18. Brake fluid (pints)	1,781	1,602	+179	+11.2%
19. Coolant (gallons)	2,045	1,815	+230	+12.7%
20. Headlight lamps	4,907	4,084	+823	+20.2%
21. Oil (quarts)	13,428	14,746	(1,318)	(8.9%)
22. Mufflers	639	585	+54	+9.2%
23. Shock absorbers	895	1,084	(189)	(17.4%)
24. Tires, auto	6,742	5,866	+876	+14.9%
25. Tires, truck	2,115	1,805	+310	+17.2%
26. Wiper blades	1,927	2,342	(415)	(17.7%)

Score for B (30)

C (30 points) During May and June, Hillman's Paint Store had sales in the amounts shown in the following table. Compute the amount of change and the rate of change between May and June. Compute the rates of change to the nearest tenth of a percent. If the amount and rate are increases, write a + in front of them; if they are decreases, then enclose them in parentheses (). (1 point for each correct amount; 2 points for each correct rate)

Hillman's Paint Store
Volume Sold (in dollars)

Description of Item	June	May	Amount of Change	Rate of Change
27. Brush, 2" wide	$611.14	$674.67	($63.53)	(9.4%)
28. Brush, 3" wide	564.20	512.51	+51.69	+10.1%
29. Brush, 4" wide	429.87	374.27	+55.60	+14.9%
30. Drop cloth, 9 × 12	143.50	175.66	(32.16)	(18.3%)
31. Drop cloth, 12 × 15	174.29	151.55	+22.74	+15%
32. Paint, latex (gal)	38,506.24	36,382.13	+2,124.11	+5.8%
33. Paint, latex (qt)	5,072.35	4,878.96	+193.39	+4.0%
34. Paint, oil (gal)	7,308.44	7,564.27	(255.83)	(3.4%)
35. Paint, oil (qt)	4,358.35	4,574.96	(216.61)	(4.7%)
36. Paint scraper	274.10	238.82	+35.28	+14.8%

Score for C (30)

Assignment 5.3: Business Applications

Name

Date   Score

Learning Objectives **3** **4**

A (50 points) Solve the following problems. Round dollar amounts to the nearest cent. Round other amounts to the nearest tenth. Write rates as percents to the nearest tenth of a percent. (5 points for each correct answer)

1. Walter Electric shipped 5,500 capacitors in May. Clients eventually returned 4% of the capacitors. How many of the May capacitors were eventually returned? 220
   $P = R \times B = 0.04 \times 5,500 = 220$

2. Jim Walter, CEO of Walter Electric, wants the company to reduce the percent of capacitors that customers return. In June, the company shipped 5,000 capacitors, and 150 were eventually returned. What percent of the June shipment was eventually returned? 3%
   $R = P \div B = 150 \div 5,000 = 0.03$, or 3%

3. By July of the following year, Walter Electric had reduced the percent of capacitors returned to 2% of the number shipped. If 130 capacitors were returned from that month's shipment, how many had been shipped? 6,500
   $B = P \div R = 130 \div 0.02 = 6,500$

4. A food importer, Fontaine's Food Expo, imports 60% of its vinegars from France, 30% from Italy, and 10% from Spain. The total value of all the vinegars that it imports is $920,000. What is the value of the vinegars that are *not* imported from France? $368,000
   30% + 10% = 40%, or 0.40
   $P = R \times B = 0.40 \times \$920,000 = \$368,000$

5. Next year, Fontaine's is planning to import $640,000 worth of vinegars from France, $300,000 worth of vinegars from Italy, and $260,000 worth of vinegars from Spain. If next year's imports occur as currently being planned, what percent of the total imports will be from Italy? 25%
   $640,000 + $300,000 + $260,000 = $1,200,000
   $R = P \div B = \$300,000 \div \$1,200,000 = 0.25$, or 25%

6. Rigik Parka Products, Inc., manufactures only parkas for adults and children. Last year, Rigik manufactured all its children's parkas in Asia. Those children's parkas represented 35% of all the Rigik production. If the company made a total of 240,000 parkas, how many children's parkas did it produce? 84,000
   $P = R \times B = 0.35 \times 240,000 = 84,000$

7. This year, Rigik again plans to manufacture all its children's parkas in Asia, and Rigik will expand the children's product line to 40% of the total number of parkas produced. If Rigik plans to produce 112,000 children's parkas, how many parkas does the company plan to produce in total? 280,000
   $B = P \div R = 112,000 \div 0.40 = 280,000$

8. Next year, Rigik plans to keep the percent of children's parkas at 40% but increase the number of children's parkas produced to 125,000. How many parkas does the company plan to produce for adults? (*Hint:* First you need to calculate the total number of all parkas to be produced next year.) 187,500
   $B = P \div R = 125,000 \div 0.40 = 312,500$ total
   312,500 − 125,000 = 187,500 for adults

9. Manuel Sosa is a single father. He tries to save 15% of his monthly salary for his son's education. In August, Manuel's salary was $4,800. How much should he save to meet his objective? $720
   $P = R \times B = 0.15 \times \$4,800 = \$720$

10. In September, Manuel Sosa got a promotion and a raise. Because his monthly expenses did not increase very much, Manuel was able to save more dollars. He saved $1,350, which was 25% of his new salary. How much was Manuel's new salary? $5,400
    $B = P \div R = \$1,350 \div 0.25 = \$5,400$

Score for A (50)

Assignment 5.3 Continued

B (50 points) Solve the following problems. Round dollar amounts to the nearest cent. Round other amounts to the nearest tenth. Write rates as percents to the nearest tenth of a percent. (5 points for each correct answer)

11. Norman Brewer, a paralegal, will receive a 4% salary increase this month. Hence he will receive $130 more salary this month than he received last month. What was Norman's salary last month? $3,250
    $B = P \div R = \$130 \div 0.04 = \$3,250$

12. Roberta Coke works in the marketing research department of a soft-drink company. Yesterday Roberta received a raise of $375 per month. Roberta now earns 6% more than she did before the raise. How much does she earn now? $6,625
    $B = P \div R = \$375 \div 0.06 = \$6,250$
    $\$6,250 + \$375 = \$6,625$

13. A farmers' market is held downtown every Saturday. The volume has been increasing by about 3% every week. If the volume was $51,400 this week, what should the volume be next week? $52,942
    $P = R \times B = 0.03 \times \$51,400 = \$1,542$
    $\$51,400 + \$1,542 = \$52,942$

14. Marcia Almeida works as a sales analyst for a toy manufacturer. She predicts that toy sales will decrease by 5% between May and June. If the amount of the sales decrease is $175,000, what level of sales is she predicting for June? $3,325,000
    $B = P \div R = \$175,000 \div 0.05 = \$3,500,000$ sales in May
    $\$3,500,000 - \$175,000 = \$3,325,000$ sales in June

15. Last month, Fred Gerhardt started working as an apprentice machinist. One of his first projects was to reduce the diameter of a metal shaft from 0.180 inch to 0.162 inch. By what percent did he reduce the diameter of the shaft? 10%
    0.180 in. − 0.162 in. = 0.018 in. decrease
    $R = P \div B = 0.018$ in. $\div 0.180$ in. $= 0.10$, or 10%

16. Judy Gregory, a mechanical engineer, was able to increase the efficiency of a manufacturing facility. By doing so, she decreased the cost to manufacture a commercial quality lawn mower by $18, which was 15% of the former cost. What will be the new reduced cost to manufacture the lawn mower? $102
    $B = P \div R = \$18 \div 0.15 = \$120$ former cost
    $\$120 - \$18$ decrease $= \$102$ new cost

17. Richard Phipps is the purchasing manager for a janitorial service. He orders all the supplies used by his company. Because of new contracts to clean three new office buildings, Richard ordered an additional $5,000 worth of supplies this month. This was an 8% increase from last month. What was the value of the supplies that Richard ordered last month? $62,500
    $B = P \div R = \$5,000 \div 0.08 = \$62,500$

18. Nancy Yamamoto owns a gift shop that had sales of $175,000 in November. Because of the Christmas holiday season, Nancy predicts that the shop will have a 200% increase in sales in December. What total sales is Nancy predicting for December? $525,000
    $P = R \times B = 200 \times \$175,000 = \$350,000$
    $\$175,000 + \$350,000 = \$525,000$

19. Suppose that Yamamoto's Gift Shop had sales of $175,000 in November and then doubled its sales in December. What would be the percent increase for December over November? 100%
    $B = \$175,000$ in November; $P =$ increase $=$ additional $\$175,000$ in December
    $R = P \div B = \$175,000 \div \$175,000 = 1.0$, or 100% increase
    Note: Remind students that doubling the base is always a 100% increase.

20. Because of Father's Day, Martin's Men's Store had sales of $350,000 in June. Sales decreased by 50% in July. What were Martin's sales in July? $175,000
    $P = R \times B = 0.50 \times \$350,000 = \$175,000$ decrease
    $\$350,000 - \$175,000$ decrease $= \$175,000$ in July
    Note: Remind students that problems like 19 and 20, which have the same numbers, have different rates because the bases change.

Score for B (50)

Assignment 5.4: Allocation of Overhead

Name

Date Score

Learning Objective 5

A (20 points) Complete the square feet, percent, and distribution columns below. Round percents to the nearest whole number. (1 point for each correct answer in column 1; 2 points for each correct answer in columns 2 and 3)

1. Maye Chau owns small restaurants in four different towns: (a) Alleghany, (b) Delwood, (c) Bangor, and (d) Lakeside. She manages all four restaurants from central office that she maintains at the Alleghany restaurant. Monthly office expenses are distributed among the four restaurants based on the floor space of each. In the following table, complete the distribution table for monthly expenses of $16,000.

Store	Space Occupied	Square Feet	Percent of Total	Distribution of Expense
(a) Alleghany	60 ft × 40 ft	2,400	32%	$ 5,120
(b) Delwood	40 ft × 45 ft	1,800	24%	3,840
(c) Bangor	70 ft × 30 ft	2,100	28%	4,480
(d) Lakeside	30 ft × 40 ft	1,200	16%	2,560
Total		7,500	100%	$16,000

(a) $\frac{2,400}{7,500} = 0.32; 0.32 \times \$16,000 = \$5,120$ (c) $\frac{2,100}{7,500} = 0.28; 0.28 \times \$16,000 = \$4,480$

(b) $\frac{1,800}{7,500} = 0.24; 0.24 \times \$16,000 = \$3,840$ (d) $\frac{1,200}{7,500} = 0.16; 0.16 \times \$16,000 = \$2,560$

Score for A (20)

B (16 points) Complete the percent and distribution columns in the following table. Before computing the distribution, round each percent to the nearest whole number. (2 points for each correct answer)

2. Diane Kingsley owns a temporary services company. She employs four types of employees whom she places into temporary positions: (a) bookkeepers, (b) secretaries, (c) food service people, and (d) hotel service people. Diane rents office space for $5,200 per month. She distributes the rent among the four labor groups, according to the number of people employed in each group. Calculate the percents and the resulting distributions.

	Number of Employees	Percent of Total	Distribution of Rent
(a) Bookkeepers	18	15%	$ 780
(b) Secretaries	36	30%	1,560
(c) Food Service	42	35%	1,820
(d) Hotel Service	24	20%	1,040
Total	120	100%	$5,200

(a) $\frac{18}{120} = 0.15; 0.15 \times \$5,200 = \$780$ (c) $\frac{42}{120} = 0.35; 0.35 \times \$5,200 = \$1,820$

(b) $\frac{36}{120} = 0.30; 0.30 \times \$5,200 = \$1,560$ (d) $\frac{24}{120} = 0.20; 0.20 \times \$5,200 = \$1,040$

Score for B (16)

Chapter 5 Percents

Assignment 5.4 Continued

C (64 points) The following situations provide practice in distributing monthly overhead expenses at a central office. From the information given in the following table, complete the distributions indicated in problems 3 through 6. Remember: Answers for each problem should sum to the total monthly overhead expense. (4 points for each correct answer)

Monthly Overhead Expense		Basis of Distribution	Location				
			East	West	North	South	TOTAL
Insurance	$20,000	Square feet	19,200	9,600	14,400	16,800	60,000
Utilities	15,000	Machine hours worked	18,000	14,400	10,800	28,800	72,000
Rent	26,000	Units produced	10,200	7,800	5,700	6,300	30,000
Maintenance	12,000	Number of employees	30	75	105	90	300

3. Distribute insurance expense based on the number of square feet at each location.

East $6,400 ; West $3,200 ; North $4,800 ; South $5,600 Check.

$\dfrac{19,200}{60,000} = 0.32$ $\dfrac{9,600}{60,000} = 0.16$ $\dfrac{14,400}{60,000} = 0.24$ $\dfrac{16,800}{60,000} = 0.28$ $ 6,400
3,200
4,800

0.32 × $20,000   0.16 × $20,000   0.24 × $20,000   0.28 × $20,000   5,600
$20,000

4. Distribute utilities expense based on the number of machine hours worked in each location.

East $3,750 ; West $3,000 ; North $2,250 ; South $6,000 Check.

$\dfrac{18,000}{72,000} = 0.25$ $\dfrac{14,400}{72,000} = 0.20$ $\dfrac{10,800}{72,000} = 0.15$ $\dfrac{28,800}{72,000} = 0.40$ $ 3,750
3,000
2,250

0.25 × $15,000   0.20 × $15,000   0.15 × $15,000   0.40 × $15,000   6,000
$15,000

5. Distribute rent expense based on the units produced at each location.

East $8,840 ; West $6,760 ; North $4,940 ; South $5,460 Check.

$\dfrac{10,200}{30,000} = 0.34$ $\dfrac{7,800}{30,000} = 0.26$ $\dfrac{5,700}{30,000} = 0.19$ $\dfrac{6,300}{30,000} = 0.21$ $ 8,840
6,760
4,940

0.34× $26,000   0.26 × $26,000   0.19 × $26,000   0.21 × $26,000   5,460
$26,000

6. Distribute maintenance expense based on the number of employees at each location.

East $1,200 ; West $3,000 ; North $4,200 ; South $3,600 Check.

$\dfrac{30}{300} = 0.10$ $\dfrac{75}{300} = 0.25$ $\dfrac{105}{300} = 0.35$ $\dfrac{90}{300} = 0.30$ $ 1,200
3,000
4,200

0.10 × $12,000   0.25 × $12,000   0.35 × $12,000   0.30 × $12,000   3,600
$12,000

Score for C (64)

Commissions 6

Learning Objectives
By studying this chapter and completing all assignments you will be able to:

Learning Objective 1 — Compute sales commissions and gross pay.

Learning Objective 2 — Compute graduated sales commissions.

Learning Objective 3 — Compute sales and purchases for principals.

A **commission** is a payment to an employee or to an agent for performing a business transaction or service. The most familiar type of commission is that received by a salesperson. Many companies have employees who are paid either totally or partially on a commission basis. People who sell insurance, real estate, and automobiles typically are in this category.

For a business owner, one advantage of using the commission method to pay employees is that the commission is an incentive. Employees are paid on the basis of the volume of business they produce for the company. They can earn more by being more productive.

Besides typical salespeople, other businesspeople provide selling and buying services. These include commission merchants, agents, and brokers, all of whom are paid a commission for their services. The person for whom the services are provided is called the **principal**. A commission merchant will normally take actual possession of the merchandise and make the sales transaction in his or her name. A **broker,** however, will usually make the transaction in the principal's name and does not take possession of the merchandise.

6.1 Students should be able to discuss the advantages and disadvantages of commission-based pay to both employees and employers and in different types of businesses.

Computing Sales Commissions and Gross Pay

Learning Objective 1

Compute sales commissions and gross pay.

A sales commission paid to a salesperson is usually a stated percent of the dollar value of the goods or services sold. Whether the commission is based on the wholesale or retail value of the goods will depend on the type of business and merchandise sold. The rate used to calculate the commission also will vary among different businesses. In some companies, the salesperson receives both a salary and a commission.

STEPS to Compute Commission and Total Pay

1. Multiply the commission rate by the amount sold to get the commission amount.
2. If there is a salary, add it to the commission amount to get the total gross pay.

EXAMPLE A

Kay Schiff sells yachts and marine equipment for Delta Marine Sales. She receives a base salary of $3,000 per month and earns a commission that is 2% of the value of all boating equipment that she sells during the month. Find her commission and total pay during September, a month in which she sold $132,000 worth of equipment.

STEP 1   2% × $132,000 = 0.02 × $132,000 = $2,640 commission
STEP 2 $2,640 commission + $3,000 base salary = $5,640 total pay

Commissions normally are paid only on actual sales. Thus goods that are returned or orders that are canceled are not subject to commission. The reason for this policy is to protect the business owner. Suppose that Delta Marine Sales in example A pays the 2% commission whether or not the goods are returned. When Kay Schiff got an order for $20,000, her commission would be 2% × $2,000 = $400. If the goods were all returned but the commission were still paid, the owner would have to pay her $400. Because no goods were sold, the owner actually would lose $400 on this transaction.

STEPS to Compute Commission When a Sale Involves Returned Goods

1. Subtract the value of the returned goods from the total ordered to determine the amount sold.
2. Multiply the commission rate by the amount sold to get the commission amount.

EXAMPLE B

Hobart Hamilton is a salesperson for Aggie Office Supply. He works on commission-only basis—he receives a commission of 2.5% on his monthly sales, but no base salary. What are his commission and total pay during a month when he sells $166,000 worth of office products, but one of his customers cancels an order for $25,000 and returns the merchandise that had already been delivered?

STEP 1 $168,000 − $25,000 = $141,000
STEP 2   2.5% × $141,500 = 0.025 × $141,000 = $3,525 commission
              Total Pay = $3,525, as he is paid on a commission-only basis

✓ CONCEPT CHECK 6.1

Compute the commission and gross pay for a salesperson who is paid a $1,800 salary and earns a 4% commission. Total sales were $88,000, but there were returns of $6,000.

$88,000 − $6,000 = $82,000 net sales
4% × $82,000 = 0.04 × $82,000 = $3,280   commission
                             + 1,800   salary
                             $5,080   gross pay

Computing Graduated Sales Commissions

Commission plans provide incentives for employees because they can earn more money by selling more products. A company can provide additional incentives for even greater productivity by using **graduated commission rates**. As the level of sales increases, so does the commission rate.

Learning Objectives 2

Compute graduated sales commissions.

STEPS to Compute Commission Under a Graduated Rates Plan

1. Compute the dollar amount at each rate level by using subtraction.
2. Multiply each level's commission rate by the level's sales dollars.
3. Add the products computed in Step 2 to determine the total commission.

Chapter 6   Commissions   109

6.2 The text doesn't mention that salespeople are often paid a higher rate for new customers or new accounts than for "add-ons" or repeat orders, sometimes even when the margins are lower on the initial purchase. In a discussion, students should be able to think of reasons why—maybe to get new business or to "lock in" customers to more profitable future business. Getting new business is usually harder than soliciting subsequent sales.

EXAMPLE C

Donna Chin has a monthly commission plan under which she receives 2% on the first $40,000 of sales during the month and 3% on sales above $40,000 for the month. If Donna has sales of $75,000 during a month, compute her commission for that month.

STEP 1

$75,000 total sales
− 40,000 at 2%
$35,000 at 3%

STEP 2

$40,000 × 0.02 = $ 800
  35,000 × 0.03 = 1,050

STEP 3

Total commission = $1,850

EXAMPLE D

Assume that Donna has a monthly commission plan under which she receives 2% on the first $40,000 of sales during the month, 3% on sales from $40,000 to $80,000, and 4% on all sales over $80,000. If Donna has sales of $126,000 during a month, compute her commission for that month.

STEP 1

$126,000 total sales
− 40,000 at 2%
$86,000
− 40,000 at 3%
$46,000 at 4%

STEP 2

$40,000 × 0.02 = $ 800
  40,000 × 0.03 = 1,200
  46,000 × 0.04 = 1,840

STEP 3

Total commission = $3,840

The same graduated incentive plan can be defined in terms of bonus rates. The calculations are similar.

EXAMPLE E

Dale Crist has a monthly commission plan under which he receives 2% on all sales during the month. If Dale has sales over $40,000, he receives a bonus of 1% of everything over $40,000. If he sells more than $80,000, he receives a "super bonus" of an additional 1% of everything over $80,000. What is Dale's commission for a month during which he sold $112,000?

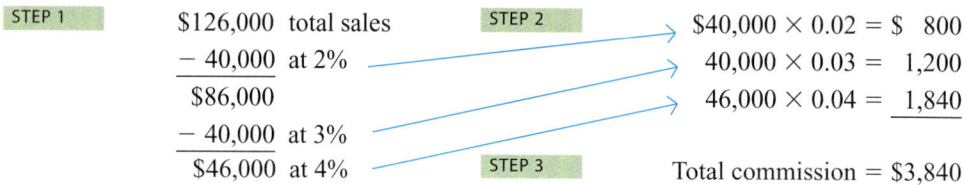

	0	$40,000	$80,000	$96,500
Base		$40,000		0.02 × $112,000 = $2,240
Bonus			$112,000 − $40,000 = $72,000	0.01 × $ 72,000 = 720
Super Bonus				$112,000 − $80,000 = $32,000	0.01 × $ 32,000 = 320

Total commission (add the three commission amounts) = $3,280

✓ CONCEPT CHECK 6.2

Compute the total commission on sales of $184,000. The commission is graduated: 1% on sales to $50,000, 2% on sales from $50,000 to $100,000, and 3% on sales above $100,000.

1% × $50,000 = 0.01 × $50,000	=	$ 500
2% × $50,000 = 0.02 × $50,000	=	1,000
3% × $84,400 = 0.03 × $84,000	=	2,520
Total commission	=	$4,020

Computing Sales and Purchases for Principals

A producer may send goods to an agent, often called a **commission merchant,** for sale at the best possible price. Such a shipment is a **consignment.** The party who sends the shipment is the **consignor;** the party to whom it is sent—that is, the commission merchant—is the **consignee.**

Whatever amount the commission merchant gets for the consignment is the **gross proceeds.** The commission is generally a certain percent of the gross proceeds. Sometimes it is a certain amount per unit of weight or measure of the goods sold. The commission and any other sales expenses (e.g., transportation, advertising, storage, and insurance) are the **charges.** The charges are deducted from the gross proceeds. The resulting amount, which is sent to the consignor, is the **net proceeds.**

Learning Objectives 3

Compute sales and purchases for principals.

6.3 Only the basics of sales and purchases for principals are discussed here. Enormous sales volumes are handled by manufacturers' representatives and various brokers who are only alluded to here. The person earning the commission is in the middle, and sometimes it is difficult to distinguish the real "principal." For example, a home decorator might be working directly for and in close cooperation with a homeowner. However, the decorator might earn all of his or her pay from the manufacturer or dealer, who gives the decorator a lower price. In many cases, the manufacturer may not even sell directly to the public. Students may want to discuss ethical ramifications of this built-in conflict of interest.

EXAMPLE F

Jack Phelps, owner of Willowbrook Farms, has been trying to sell a used livestock truck and a used tractor. Unsuccessful after 3 months, Phelps consigns the items to Acme Equipment Brokers. They agree on commission rates of 6% on the gross proceeds from the truck and 9% on the gross proceeds from the tractor. Acme sells the truck for $42,500 and the tractor for $78,600. Acme also pays $610 to deliver the truck and $835 to deliver the tractor. What are the net proceeds due Willowbrook Farms from the sale of the equipment?

Truck:	Commission: 0.06 × $42,500 =	$2,550		Gross proceeds:	$42,500
	Freight:	+ 610		less charges	− 3,160
	Total charges	$3,160		Net Proceeds:	$39,340
Tractor:	Commission: 0.09 × $78,600 =	$7,074		Gross proceeds:	$78,600
	Freight:	+ 835		less charges	− 7,909
	Total charges	$7,909		Net proceeds:	$70,691

$39,340 + $70,691 = $110,031 Total Net Proceeds

Along with the net proceeds, the commission merchant sends the consignor a form known as an **account sales.** It is a detailed statement of the amount of the sales and the various deductions. Figure 6-1 shows a typical account sales.

Chapter 6 Commissions 111

Figure 6-1 Account Sales

ACME EQUIPMENT BROKERS

August 16, 20-- NO. 67324
309 Sule Road, Wilbraham, MA 01095-2073

NAME Willowbrook Farms
ADDRESS 127 N. Kaye
Albany, GA 31704-5606

BELOW ARE ACCOUNT SALES OF Consignment No. 76
RECEIVED August 1, 20--
and sold for account of Same

DATE	CHARGES	AMOUNT	DATE	SALES	AMOUNT
Aug. 1	Freight (truck)	$ 610	Aug. 10	Truck	$42,500
16	6% Commission (truck)	2,550			
	Net proceeds (truck)	39,340	13	Tractor	78,600
				Gross proceeds	$121,100
	Freight (tractor)	835			
	9% Commission (tractor)	7,074			
	Net proceeds (tractor)	70,691			
	Total	$121,100			

When commission merchants purchase goods for their principals, the price they pay for the merchandise is the **prime cost.** The prime cost and all charges are the **gross cost,** or the cost the principal pays.

EXAMPLE G

Asia-Pacific Tours commissioned Specialty Marketing Group to purchase 10,000 vinyl travel bags that will be labeled with Asia-Pacific's logo and used as promotional items. For this size order, Specialty Marketing purchased the bags for $4.29 each. Charges include the commission, which is 6% of the prime cost; storage, $125; and freight, $168. What is the gross cost that Asia-Pacific should pay to Specialty Marketing?

$ 4.29
× 10,000 units
$ 42,900 prime cost

$42,900 prime cost
× 0.06
$ 2,574 commission

$2,574 commission + $125 storage + $168 freight = $2,867 charges
$42,900 prime cost + $2,867 charges = $45,767 gross cost

An **account purchase** is a detailed statement from the commission merchant to the principal. It shows the cost of goods purchased, including charges. Figure 6-2 shows a typical account purchase, for the transaction in example G.

Figure 6-2 | Account Purchase

SPECIALTY MARKETING GROUP

4445 Mission Street
San Francisco, CA 94112

ACCOUNT PURCHASE NO. 1311

October 26 20 __

Bought on Consignment for

Asia-Pacific Tours
7300 Harbor Place
San Francisco, CA 94104

DATE	DESCRIPTION	CHARGES	AMOUNT
Oct. 23	10,000 units stock #T805 @ $4.29		$42,900.00
23	6% commission	$2,574.00	
	Storage	125.00	
	Freight	168.00	2,867.00
	Gross Cost		$45,767.00

✓ CONCEPT CHECK 6.3

a. Compute the commission and the net proceeds on a consignment sale of $6,500. The commission rate is 5%, local delivery charges are $328.16, and storage charges are $125.
5% × $6,500 = 0.05 × $6,500 = $325 commission
$6,500 − $325 − $328.16 − $125 = $5,721.84 net proceeds

b. Compute the commission and gross cost on a $12,500 purchase for a principal. The commission rate is 7%, air freight is $138.70, and local delivery charges are $64.60.
7% × $12,500 = 0.07 × $12,500 = $875 commission
$12,500 + $875 + $138.70 + $64.60 = $13,578.30 gross cost

COMPLETE ASSIGNMENTS 6.1 AND 6.2.

Chapter Terms for Review

account purchase
account sales
broker
charges
commission
commission merchant
consignee
consignment

consignor
graduated commission rates
gross cost
gross proceeds
net proceeds
prime cost
principal

THE BOTTOM LINE

Summary of chapter learning objectives:

Learning Objective	Example
6.1 Compute sales commissions and gross pay	1. A salesperson gets a $2,240 salary and a 2% commission. Find the commission and the gross pay when sales are $58,200 and returns are $6,500.
6.2 Compute graduated sales commissions	2. A salesperson has a graduated commission rate: 1% on sales up to $100,000; 2% on sales from $100,000 to $200,000; and 2.5% on sales above $200,000. Find the commission when sales are $255,000.
6.3 Compute sales and purchases for principals	3. A broker sells a principal's merchandise at a gross sales price of $15,600 and a commission rate of 3.5%. There are sales costs of $300 for storage and $119 for delivery. Find the commission and net proceeds. 4. A commission merchant purchases merchandise for a principal at a prime cost of $8,400. The commission rate is 8%, air freight is $139, and local delivery is $75. Find the commission and gross cost.

Answers: 1. Commission: $1,034; Gross pay: $3,274   2. $4,375   3. Commission: $546; Net proceeds: $14,635   4. Commission: $672; Gross costs: $9,286

SELF-CHECK

Review Problems for Chapter 6

In problems 1–4, compute both the commission and the total pay based on the information given.

1. Salary, $3,000; commission rate, 6%; total sales, $58,000; returns, $0

    a. Commission _____   b. Total pay _____

2. Salary, $2,500; commission rate, 5%; total sales, $91,000; returns, $5,000

    a. Commission _____   b. Total pay _____

3. Salary, $4,500; commission rate, 4%; total sales, $74,000; returns, $8,975

    a. Commission _____   b. Total pay _____

4. Salary, $0; commission rate, 8%; total sales, $98,000; returns, $11,425

    a. Commission _____   b. Total pay _____

5. Compute the total commission on sales of $160,000 if the commission rates are 3% on the first $100,000 and 5% on everything above $100,000. _____

6. Compute the total commission on sales of $85,000 if the commission rates are 3% on the first $100,000 and 5% on everything above $100,000. _____

7. Compute the total commission on sales of $250,000 if the commission rates are 2% on the first $75,000; then 3% on the next $75,000; and 4% on everything above $150,000. _____

8. Compute the total commission on sales of $135,000 if the commission rates are 2% on the first $75,000; then 3% on the next $75,000; and 4% on everything above $150,000. _____

9. Compute the total commission on sales of $70,000 if the commission rates are 2% on the first $75,000; then 3% on the next $75,000; and 4% on everything above $150,000. _____

10. Compute the total commission on sales of $115,000 if the commission rates are 4% on the first $35,000; then 6% on the next $45,000; and 8% on everything above $80,000. _____

11. Larry Leong is paid 3% on all sales. He is also paid a bonus of an additional 1% on any sales above $75,000. Calculate Larry's total commission on sales of $125,000. _____

12. Gloria Alvares is paid 4% on all sales. She is also paid a bonus of an additional 2% on any sales above $40,000. Calculate Gloria's total commission on sales of $105,000. _____

13. Charles White sells used logging equipment on consignment. He charges 20% plus expenses. Calculate Charles's commission on a log truck he sold for $42,750. _____

14. For the sale in problem 13, Charles also paid an additional $290 to deliver the truck to the new owner. Calculate the net proceeds that Charles's principal should receive. _____

15. Sue Lyon is a designer who purchases furniture for clients. She charges 15% of the price, plus expenses. Calculate Sue's commission on furniture priced at $21,400. _____

16. For the sale in problem 15, calculate the gross cost to the client if Sue also had expenses of $646. _____

Notes

Assignment 6.1: Commission

Name

Date   Score

Learning Objectives 1 2 3

A (24 points) Find the commission and the total gross pay. (2 points for each correct answer)

Employee	Monthly Salary	Commission Rate	Monthly Sales	Commission	Gross Pay
1. Li, Walter	$ 0	8%	$45,000	$3,600	$3,600
$0.08 \times \$45,000 = \$3,600 + 0 = \$3,600$					
2. Starr, Karen	2,000	3%	36,000	1,080	3,080
$0.03 \times \$36,000 = \$1,080 + \$2,000 = \$3,080$					
3. Aguire, Luis	1,500	5%	42,000	2,100	3,600
$0.05 \times \$42,000 = \$2,100 + \$1,500 = \$3,600$					
4. Gupta, Rajeev	3,000	2%	40,000	800	3,800
$0.02 \times \$40,000 = \$800 + \$3,000 = \$3,800$					
5. Rogerro, George	1,800	6.5%	64,000	3,840	5,640
$0.06 \times \$64,000 = \$3,840 + \$1,800 = \$5,640$					
6. Tang, Suzanne	2,500	4%	57,000	2,280	4,780
$0.04 \times \$57,000 = \$2,280 + \$2,500 = \$4,780$					

Score for A (24)

B (36 points) Compute the total commission for the following commission payment plans. (6 points for each correct answer)

Graduated Commission Rates	Sales	Commission	
7. 2% on sales to $60,000 4% on sales above $60,000	$106,000	$3,040	$0.02 \times \$60,000 = \$1,200$ $0.04 \times 46,000 = \underline{1,840}$ $\$3,040$
8. 1% on sales to $150,000 2% on sales above $150,000	$188,000	2,260	$0.01 \times \$150,000 = \$1,500$ $0.02 \times 38,000 = \underline{760}$ $\$2,260$
9. 3% on sales to $50,000 5% on sales above $50,000	$ 94,400	3,720	$0.03 \times \$50,000 = \$1,500$ $0.05 \times 44,400 = \underline{2,220}$ $\$3,720$
10. 1% on sales to $75,000 2% on sales from $75,000 to $150,000 3% on sales above $150,000	$240,000	4,950	$0.01 \times \$75,000 = \$\ 750$ $0.02 \times 75,000 = 1,500$ $0.03 \times 90,000 = \underline{2,700}$ $\$4,950$
11. 3% on sales to $50,000 4% on sales from $50,000 to $100,000 5% on sales above $100,000	$128,000	4,900	$0.03 \times \$50,000 = \$1,500$ $0.04 \times 50,000 = 2,000$ $0.05 \times 28,000 = \underline{1,400}$ $\$4,900$

Chapter 6 Commissions

Assignment 6.1 Continued

12. 2% on sales to $65,000 $124,800   3,094 $0.02 \times \$65,000 = \$1,300$
    3% on sales from $65,000 to $130,000 $0.03 \times 59,800 = \underline{1,794}$
    4% on sales above $150,000 $\$3,094$

Score for B (36)

C (20 points) Janet Cronin is a commission merchant. She charges different commission rates to sell different types of merchandise. During May, she completed the following consignment sales for consignors. Find Janet's commission on each sale and the net proceeds sent to each consignor. (2 points for each correct answer)

	Gross Sales	Comm. Rate	Commission	Local Delivery	Storage	Air Freight	Net Proceeds
13.	$38,400	3%	$1,152	$68.75	$0	$183.50	$36,995.75
	$38,400 × 0.03			$38,400 − $1,152 − $68.75 − $183.50			
14.	1,600	4.5%	$72	88.50	65.00	0	$1,374.50
	$1,600 × 0.045			$1,600 − $72 − $88.50 − $65			
15.	8,400	6%	$504	284.00	0	0	$7,612.00
	$8,400 × 0.06			$8,400 − $504 − $284			
16.	12,880	5%	$644	0	0	148.00	$12,088.00
	$12,880 × 0.05			$12,880 − $644 − $148			
17.	5,600	3.5%	$196	0	85.00	112.00	$5,207.00
	$5,600 × 0.035			$5,600 − $196 − $85 − $112			

Score for C (20)

D (20 points) Alvin Guiterez, a commission merchant in Dallas, buys merchandise exclusively for principals. Listed below are five recent transactions. Compute Alvin's commission on each purchase and the gross cost. (2 points for each correct answer)

	Prime Cost	Comm. Rate	Commission	Local Delivery	Storage	Air Freight	Gross Cost
18.	$16,600	5%	$830	$89.50	$88.00	$0	$17,607.50
	$16,600 × 0.05			$16,600 + $830 + $89.50 + $88			
19.	4,900	11%	$539	0	0	195.00	$5,634.00
	$4,900 × 0.11			$4,900 + $539 + $195			
20.	8,400	6%	$504	30.00	58.00	196.00	$9,188.00
	$8,400 × 0.06			$8,400 + $504 + $30 + $58 + $196			
21.	4,850	8%	$388	0	110.00	108.00	$5,456.00
	$4,850 × 0.08			$4,850 + $388 + $110 + $108			
22.	19,000	7%	$1,330	50.00	0	0	$20,380.00
	$19,000 × 0.07			$19,000 + $1,330 + $50			

Score for D (20)

Assignment 6.2: Applications with Commission

Name

Date   Score

Learning Objectives **1** **2** **3**

A (56 points) Solve each of the following business application problems involving salespeople who are paid partly or entirely on a commission basis. Solve the problems in order, because some of the questions are sequential. (8 points for each correct answer)

1. Pat Endicot sells memberships to an athletic club. He receives a monthly salary of $1,200 plus a commission of 12% on new membership fees. What was Pat's monthly pay for May, when he sold new memberships valued at $34,500? $5,340

    0.12 × $34,500 = $4,140   commission
                            +1,200   salary
                            $5,340   monthly pay

2. Roberta Reavis sells commercial restaurant supplies and equipment. She is paid on a commission-only basis. She receives 2% for her sales up to $60,000. For the next $90,000 of sales, she is paid 3%, and for any sales above $150,000 she is paid 4%. How much commission would Roberta earn in a month when her sales were $175,000? $4,900

    0.02 × $60,000 = $1,200
    0.03 ×   90,000 = 2,700
    0.04 ×   25,000 = 1,000
                          $4,900   commission

3. Roberta Reavis (problem 2) is not paid commission on any restaurant supplies or equipment that are later returned. If an item is returned, its price is deducted from Roberta's total sales to get her net sales. The commission-only rate is applied to her net sales. Suppose that Roberta sold merchandise worth $175,000 but that $40,000 of that was later returned. What would be Roberta's commission on net sales? $3,450

    $175,000   total sales   0.02 × $60,000 = $1,200
    − 40,000   returns   0.03 × 75,000 = +2,250
    $135,000   net sales $3,450   commission

4. Dana Kline works for Southwest Appliance Depot. She receives a monthly salary of $2,500 for which she must sell $20,000 worth of appliances. She also receives a commission of 4% on net sales above $20,000. What will be Dana's pay for October when her net appliance sales were $42,000? $3,380

    $42,000   net sales   0.04 × $22,000 = $ 880   commission
    − 20,000   +2,500   salary
    $22,000   commission sales $3,380   monthly pay

5. Southwest Appliance Depot (problem 4) offers service contracts with all appliance sales. To encourage salespeople such as Dana to sell more service contracts, the company pays a commission of 20% on all service contracts. What will be her total pay for a month if she sells $42,000 worth of appliances and $1,500 worth of service contracts? $3,680

    0.20 × $1,500 = $ 300   commission on service contracts
                          880   commission on appliances
                      2,500   salary
                    $3,680   total pay

6. Stockbrokers for companies such as PaineWebber are normally paid a commission on the stocks that they buy and sell for their clients. Suppose that the commission rate is 0.5% of the value of the stock. What will the commission be on 5,000 shares of General Motors stock that is selling for $67.31 per share? $1,682.75

    5,000 × $67.31 = $336,550   0.005 × $336,550 = $1,682.75   commission

Chapter 6 Commissions 119

Assignment 6.2 Continued

7. Joni Lopez works in telemarketing. Her job is to make telephone calls from a computerized list of names and try to convince people to make an appointment with a life insurance salesperson. Joni receives 30¢ for each completed telephone call, $6.00 for each appointment made and kept, and 0.75% of any initial revenue that results from the appointment. How much would Joni earn if she completed 868 calls, 137 persons made and kept appointments, and $28,500 in revenue resulted from the appointments? $1,298.15

868 × $0.30 =	$ 260.40	telephone calls
137 × $6.00 =	822.00	appointments
0.0075 × $28,500 =	213.75	revenue percentage
	$1,296.15	total earnings

   Score for A (56)

B (24 points) Solve each of the following business applications about consignment sales and commission merchants. (8 points for each correct answer)

8. Teresa Fowler is a commission merchant who charges a 15% commission to sell antique furniture from her showroom. Henry Marshal owns antique furniture, which he transports to the showroom where Theresa sells it for $9,600. Henry agrees to pay Theresa $488 to have the furniture delivered to the buyer from the showroom. What will be Henry's net proceeds from the sale? $7,672

   Commission: 0.15 × $9,600 = $1,440
   Net proceeds: $9,600 − $1,440 − $488 = $7,672

9. Suppose, in problem 8, that payment of the $488 delivery expense was Theresa's responsibility instead of Henry's. Then what would be Theresa's net earnings from the sale? $952

   Commission: 0.15 × $9,600 = $1,440
   Less delivery expense: −488
   $ 952

10. Sandy McCulloch makes artistic weavings that are used as wall hangings. She sells her weavings primarily at open-air art shows and street fairs through her agent, Ruth Danielson. Ruth charges 20% on all sales, plus the fees to operate a sales booth and transportation expenses. What will be Sandy's net proceeds if Ruth sold weavings worth $32,400 at four different art shows? Each art show charged a booth fee of $500, and Ruth's total transportation expenses were $425. $23,495

    Booth fees: 4 × $500 = $2,000 $32,400
    Transportation: 425   − 8,905
    Commission: 0.20 × $32,400 = +6,480 $23,495
    Total costs: $8,905

   Score for B (24)

C (20 points) The following problems involve the purchase of a home. (10 points for each correct answer)

11. JoAnn Ednie has a house that she would like to sell and she asks real estate broker Gene Jenkins to sell it. Gene owns Jenkins/Weekly Real Estate, which advises JoAnn that she should be able to sell her house for $180,000. The commission rate for selling a house is 6%. If the house sells for the expected price, what will be the total commission amount that JoAnn pays? $10,800

    Commission: 0.06 × $180,000 = $10,800

12. See problem 11. To sell her home, JoAnn Ednie must pay some additional fees for three home inspections and title insurance, as well as fees to the county to record the transaction. These fees total $3,500 and are added to the 6% commission. What will JoAnn's net proceeds be from the sale of her $180,000 home? $165,700

    Commission: $10,800 $180,000
    Fees   + 3,500   − 14,300
    Total costs: $14,300 $165,700   net proceeds

    Score for C (20)

Discounts

7

Learning Objectives
By studying this chapter and completing all assignments you will be able to:

Learning Objective **1** — Compute trade discounts.

Learning Objective **2** — Compute a series of trade discounts.

Learning Objective **3** — Compute the equivalent single discount rate for a series of trade discounts.

Learning Objective **4** — Compute cash discounts and remittance amounts for fully paid invoices.

Learning Objective **5** — Compute cash discounts and remittance amounts for partially paid invoices.

When one business sells merchandise to another business, the seller often offers two types of discounts: trade discounts and cash discounts. Trade discounts affect the agreed-upon selling price *before* the sale happens. Cash discounts affect the amount actually paid *after* the transaction.

Computing Trade Discounts

Learning Objective 1

Compute trade discounts.

Businesses that sell products want to attract and keep customers who make repeated, large-volume purchases. Manufacturers, distributors, and wholesalers frequently offer **trade discounts** to buyers "in the trade," generally based on the quantity purchased. For example, Eastern Restaurant Supply gives a 40% discount to Regal Meals, a local chain of 34 sidewalk sandwich carts that sell hot dogs and sausage sandwiches. Another Eastern customer is Suzi Wilson, founder and owner of Suzi's Muffins. Suzi's business is still small. She bakes her muffins between 11 P.M. and 2 A.M. in oven space that she leases from a bakery. Eastern gives Suzi only a 25% discount because she doesn't do as much business with Eastern as Regal Meals does. Eastern also sells to people who are not "in the trade." These retail customers pay the regular **list price**, or full price without a discount.

Large restaurant chains such as McDonald's or Burger King can go directly to the manufacturer for most items or even do their own manufacturing. They can have items manufactured to their exact specifications for a contracted price. They reduce their costs by eliminating the distributors (the "middle men").

The two traditional methods for computing trade discounts are the discount method and the complement method. You can use both to find the **net price** that a distributor will charge a customer after the discount. The discount method is useful when you want to know both the net price and the actual amount of the trade discount. The **complement method** is used to find only the net price. It gets its name because you use the **complement rate**, which is 100% minus the discount rate. Each method has only two steps.

7.1 Students may be more familiar with cash discounts than with trade discounts. Trade discounts are discussed first only because the trade discount is awarded by the seller before the sale, to encourage the sale, whereas the cash discount is awarded after the sale, to encourage the payment.

7.2 Ask students which method might be more useful to businesses. If the buyer is comparing different sellers, what is of primary interest to the buyer is the ultimate net price. It doesn't make financial sense to base the decision on a larger discount if the net price isn't lower. Compare the situation to that of retailers that have "big sales" at various times throughout the year. Some that have reputations for large discounts also have high original prices.

> **STEPS** to Compute Net Price with the Discount Method
>
> 1. Multiply the discount rate by the list price to get the discount amount:
>   Discount = Trade discount rate × List price
> 2. Subtract the discount from the list price to get the net price:
>   Net price = List price − Discount

EXAMPLE A

Eastern Restaurant Supply sells a set of stainless steel trays to Suzi's Muffins. The list price is $120, and Suzi qualifies for a 25% trade discount. Compute the net price using the discount method.

STEP 1   Discount = 0.25 × $120 = $30

STEP 2   Net price = $120 − $30 = $90

STEPS to Compute Net Price with the Complement Method

1. Subtract the discount rate from 100% to get the complement rate:
   Complement rate = 100% − Trade discount rate
2. Multiply the complement rate by the list price to get the net price:
   Net price = Complement rate × List price

EXAMPLE B

Using the data in example A, compute the net price, using the complement method.

STEP 1 Complement rate = 100% − 25% = 75%

STEP 2 Net price = 0.75 × $240 = $180

CONCEPT CHECK 7.1

a. Compute the trade discount amount and the net price, using the discount method.

   List price = $240   Trade discount = 30%
   Discount amount = 0.30 × $240 = $72
   Net price = $240 − $72 = $168

b. Compute the complement rate and the net price, using the complement method.

   List price = $240   Trade discount = 30%
   Complement rate = 100% − 30% = 70%
   Net price = 0.70 × $240 = $168

Computing a Series of Trade Discounts

A distributor or manufacturer may give additional discounts to customers who actually buy the largest volumes. Suppose that Eastern Restaurant Supply gives all food preparation businesses a 25% discount for being in the trade. However, if one business buys twice as much from Eastern, it may be rewarded with additional discounts. For example, Suzi's Muffins may receive its first discount of 25% automatically. Then, Suzi's gets an additional 20% discount if its accumulated purchases were between $10,000 and $25,000 during the previous year and another 10% if accumulated purchases were more than $25,000 during the previous year. Therefore, Suzi's Muffins could have discounts of 25%, 20%, and 10%, called a **series of discounts.**

Both the discount method and the complement method can be used to compute the net price for a series of discounts. *The two methods are the same as shown previously, except that the steps are repeated for each discount in the series.* For example, if there are three discounts, repeat the steps three times. Apply the first **discount rate** to the list price. For the second and third discounts, compute intermediate prices and then apply the discount rates to them.

Learning Objectives 2

Compute a series of trade discounts.

EXAMPLE C

Eastern Restaurant Supply sells a set of mixing bowls with a list price of $200. Suzi's Muffins qualifies for the series of discounts: 25%, 20%, 10%. Compute the net price using the discount method.

	1st discount	2nd discount	3rd discount
STEP 1	0.25 × $200 = $50	0.20 × $150 = $30	0.10 × $120 = $12
STEP 2	$200 − $50 = $150	$150 − $30 = $120	$120 − $12 = $108

EXAMPLE D

Using the data in example C, calculate the net price using the complement method.

	1st discount	2nd discount	3rd discount
STEP 1	100% − 25% = 75%	100% − 20% = 80%	100% − 10% = 90%
STEP 2	0.75 × $200 = $150	0.80 × $150 = $120	0.90 × $120 = $108

COMPLEMENT METHOD SHORTCUT

When you use complement rates, you may not need to write all of the intermediate prices. If not, an efficient shortcut is

Multiply the list price by all of the complement rates successively.

EXAMPLE E

Repeat example D, using the shortcut. The list price is $200, and the discounts are 25%, 20%, and 10%. The complement rates are 75%, 80%, and 90%.
Net price = $200 × 0.75 × 0.80 × 0.90 = $108

Note: Remember that there should be *no rounding* until you reach the final net price. Then round it to the nearest cent.

✓ CONCEPT CHECK 7.2

a. A wholesaler offers a series of trade discounts: 30%, 25%, and 10%. Find each of the discount amounts and the final net price on a $1,500 purchase.

First discount amount:	$1,500 × 0.30 = $450
Second discount amount:	$1,500 − $450 = $1,050; $1,050 × 0.25 = $262.50
Third discount amount:	$1,050 − $262.50 = $787.50; $787.50 × 0.10 = $78.75
Net price:	$787.50 − $78.75 = $708.75

b. A series of trade discounts is 30%, 25%, and 10%. Find each of the complement rates, and use the shortcut to calculate the final net price on a purchase of $1,500.

First complement rate:	100% − 30% = 70%
Second complement rate:	100% − 25% = 75%
Third complement rate:	100% − 10% = 90%
Net price:	$1,500 × 0.70 × 0.75 × 0.90 = $708.75

Computing the Equivalent Single Discount Rate

Suppose that an Eastern competitor, United Food Services, offers a single discount of 45% to Suzi's Muffins. How does that rate compare to the series of discounts from Eastern, 25%, 20%, and 10%? Suzi or her accountant could check by calculating the **equivalent single discount rate**, which is the single discount rate that can be used in place of two or more trade discount rates to determine the same discount amount.

The most efficient way to find the single discount rate that is equivalent to a series of discounts is similar to the shortcut used in example E.

Learning Objectives 3

Compute the equivalent single discount rate for a series of trade discounts.

> **STEPS to Compute the Equivalent Single Discount Rate**
>
> 1. Compute the complement of each rate.
> 2. Multiply all the complement rates (as decimals), and then write the product as a percent.
> 3. Subtract the product (Step 2) from 100% to get the equivalent single discount rate.

EXAMPLE F

Find the equivalent single discount rate for Eastern's series of discounts: 25%, 20%, and 10%.

STEP 1
- 1st complement rate = 100% − 25% = 75%
- 2nd complement rate = 100% − 20% = 80%
- 3rd complement rate = 100% − 10% = 90%

STEP 2
- Product of complements = 0.75 × 0.80 × 0.90 = 54%

STEP 3
- Equivalent single discount = 100% − 54% = 46%

7.3 Point out that the series of discounts in example F is just slightly better than the single discount of 45%. If the series were 25%, 20%, and 5%, the single discount would be better.

✓ CONCEPT CHECK 7.3

A series of trade discounts is 50%, 30%, and 10%. Find the three complement rates and then find the equivalent single trade discount rate.

Complement rates:	100% − 50% = 50%, 100% − 30% = 70%, 100% − 10% = 90%
Product of the complement rates:	0.50 × 0.70 × 0.90 = 0.315, or 31.5%
Equivalent single discount rate:	100% − 31.5% = 68.5%

COMPLETE ASSIGNMENT 7.1.

Computing Cash Discounts for Fully Paid Invoices

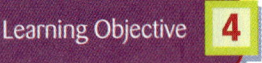

Compute cash discounts and remittance amounts for fully paid invoices.

When a seller sends merchandise to a buyer, the seller usually wants to get its payment quickly and some buyers often try to delay payment as long as possible. Sellers can encourage early payment by offering a **cash discount;** they can discourage late payment by assessing an extra interest payment; or they can do both. These stipulations are called the **terms of payment,** or simply the *terms*. The terms describe details about cash discounts and/or penalty periods.

After shipping merchandise to a buyer, the seller usually sends a document called an invoice, requesting payment. An **invoice** lists each item, its cost (including packaging and freight), and the total cost. The invoice also states the terms of payment. The amount the buyer pays is called the **remittance.** The **net purchase amount** is the price of the merchandise actually purchased, including allowances for returns and excluding handling and other costs.

> **STEPS to Compute the Remittance**
>
> 1. Multiply the discount rate (expressed as a decimal) by the net purchase amount to get the cash discount:
>   Cash discount = Discount rate × Net purchase amount
> 2. Subtract the cash discount from the net purchase amount to get the remittance:
>   Remittance = Net purchase amount − Cash discount

7.4 For large purchases, the amount of the cash discount may be large enough that the buyer will consider borrowing money for a few days just to get the discount. (See Chapter 15.)

7.5 The term *net purchase amount* isn't really needed here, but it will be useful when returned merchandise and/or freight charges are considered later in the chapter.

Figure 7-1 shows an invoice from National Automotive Supply, which sold car wax to Broadway Motors for $528. The wax will be shipped via UPS, and National will pay for the shipping. The invoice lists terms of 2/10, n/30. The **invoice date,** or the beginning of the discount period, is May 23.

Video — Cash Discounts

Figure 7-1 Sales Invoice

NATIONAL AUTOMOTIVE SUPPLY

INVOICE NO. 782535

SOLD TO: Broadway Motors
730 W. Columbia Dr.
Peoria, IL 62170-1184

DATE: May 23, 200−
TERMS: 2/10, n/30
SHIP VIA: UPS

QUANTITY	DESCRIPTION	UNIT PRICE	GROSS AMOUNT	NET AMOUNT
24 gals.	Car wax	$22.00	$528.00	$528.00

126 Part 2 Percentage Applications

The expression 2/10, n/30 means that Broadway Motors can get a 2% discount if it pays the full invoice within 10 days of the invoice date. Ten days after May 23 is June 2, which is called the **discount date.** The 10-day period between May 23 and June 2 is called the **discount period.** The n/30 is short for net 30, which means that if Broadway Motors does not pay within 30 days, National will charge an interest penalty. Thirty days after May 23 is June 22, which is called the **due date.** (See Figure 7-2)

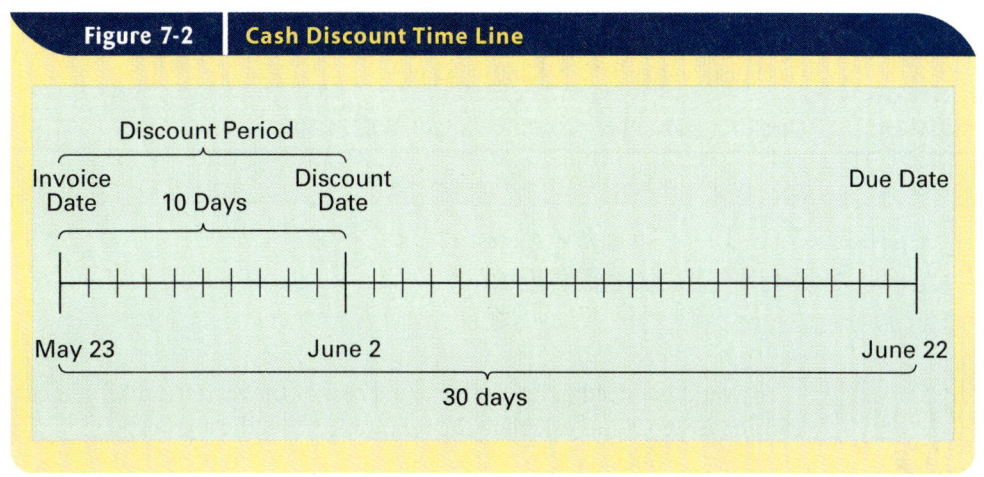

Figure 7-2 Cash Discount Time Line

EXAMPLE G

Compute the remittance due if Broadway Motors pays National within the 10-day discount period.

STEP 1   Cash discount = 2% of $528 = 0.02 × $528 = $10.56

STEP 2   Remittance   = $528 − $10.56 = $517.44

All companies do not use exactly the same notation for writing their terms; 2/10, n/30 is also written as 2/10, net 30 or as 2-10, n-30. Likewise, there can be more than one discount rate and discount period. For example 2/5, 1/15, n/30 means that the seller gets a 2% discount by paying within 5 days, gets a 1% discount by paying between 6 and 15 days, and must pay a penalty after 30 days.

7.6 As with trade discounts, the method to be used depends on the information wanted. With computers and calculators, time and accuracy are no longer considerations.

RETURNED MERCHANDISE AND FREIGHT CHARGES

The seller gives a discount only on merchandise that is actually purchased—the net purchases. For example, there is no discount on returned items. Likewise, there is no discount on charges from a third party, such as freight.

STEPS to Compute the Remittance When There Are Merchandise Returns and/or Freight Charges

1. Net purchase = Invoice amount − Merchandise returns − Freight
2. Cash discount = Discount rate × Net purchase
3. Cost of merchandise = Net purchase − Cash discount
4. Remittance = Cost of merchandise + Freight, if any

EXAMPLE H

National Automotive Supply sells merchandise to Broadway Motors. The invoice amount is $510, which includes $30 in freight charges. The invoice date is August 13, and the terms are 2/10, n/30. Broadway Motors returns $200 worth of merchandise and pays the rest of the invoice before the discount date. Compute the cash discount and the remittance. Also, determine the discount date and due date.

STEP 1	Net purchase = $510 − $200 − $30 = $280
STEP 2	Cash discount = 0.02 × $280 = $5.60
STEP 3	Cost of merchandise = $280 − $5.60 = $274.40
STEP 4	Remittance = $274.40 + $30 = $304.40

Discount date = August 13 + 10 days = August 23
Due date = August 13 + 30 days = September 12

If you don't need to know the actual cost of the merchandise, you can eliminate Step 3 and calculate the remittance directly:

Remittance = $280.00 − $5.60 + $30.00 = $304.40

There is also a complement method for cash discounts. However, it isn't used as often as the discount method because most businesses want to know the amount of the cash discount before deciding whether to pay the invoice early. In the complement method for cash discounts, only Steps 2 and 3 change.

STEPS to Compute the Remittance with the Complement Method

1. Net purchase = Invoice amount − Merchandise returns − Freight
2. Complement rate = 100% − Cash discount rate
3. Cost of merchandise = Net purchase × Complement rate
4. Remittance = Cost of merchandise + Freight, if any

EXAMPLE I

Solve example H by using the complement method for cash discounts. The invoice amount is $510, merchandise returns are $200, and freight is $30.

STEP 1	Net purchase = $510 − $200 − $30 = $280
STEP 2	Complement rate = 100% − 2% = 98%
STEP 3	Cost of merchandise = 0.98 × $280 = $274.40
STEP 4	Remittance = $274.40 + $30 = $304.40

CONCEPT CHECK 7.4

a. Use the given information to calculate the discount date, due date, cash discount, and remittance.

Terms:	1/10, n/60	Discount date = August 24 + 10 days = September 3
Invoice date:	August 24	Due date = August 24 + 60 days = October 23
Invoice amount:	$852.43	
Returned goods:	$187.23	Net purchases = $852.43 − $187.23 − $47.20 = $618.00
Freight:	$47.20	Cash discount = 0.01 × $618 = $6.18
		Remittance = $618 − $6.18 + $47.20 = $659.02

b. Calculate the remittance for the problem in part (a), using the complement method.

Net purchases = $852.43 − $187.23 − $47.20 = $618.00
Complement rate = 100% − 1% = 99%
Cost of merchandise = 0.99 × $618 = $611.82
Remittance = $611.82 + $47.20 = $659.02

Computing Cash Discounts for Partially Paid Invoices

Sometimes a buyer wants to take advantage of a cash discount but can afford to pay only part of the invoice within the discount period. The invoice will be reduced by the amount paid (remittance) plus the amount of the discount. The total of the amount paid plus the amount of cash discount is called the **amount credited** to the buyer's account. To compute the amount credited, you need to know the complement rate: 100% − Discount rate.

Learning Objectives 5

Compute cash discounts and remittance amounts for partially paid invoices.

STEPS to Compute the Unpaid Balance

1. Compute the complement of the discount rate (100% − Discount rate).
2. Compute the amount credited by dividing the amount paid (remittance) by the complement rate.
3. Compute the unpaid balance by subtracting the amount credited (Step 2) from the invoice amount.

EXAMPLE J

Larry Eickworth operates a shop called Space Savers, a do-it-yourself center for closets and storage. Larry buys shelving materials with an invoice price of $484 and terms of 2/10, net 60. Within the 10-day discount period, he sends in a check for $300. How much credit should Larry receive, and what is his unpaid balance?

STEP 1 Complement rate = 100% − 2% = 98%

STEP 2 Amount credited = $300 ÷ 0.98 = $306.1224, or $306.12

STEP 3 Unpaid balance = $484.00 − $306.12 = $177.88

Note that, in example J, Larry receives $1.00 credit for every $0.98 paid. In other words, the $300 actually remitted is 98% of the total amount credited. We check this result with multiplication:

Cash discount = 0.02 × $306.12 = $6.1224, or $6.12
Remittance = $306.12 − $6.12 = $300.00

A slightly different situation, which arises less frequently, is when a buyer decides in advance the total amount that he or she wants to have credited to the account. This problem is exactly like the original cash discount problems.

EXAMPLE K

Larry Eickworth buys $484 worth of shelving materials for use in his closet and storage shop. The terms are 2/10, net 60. Larry wants to pay enough within the 10-day discount period to reduce his unpaid balance by exactly $300. What amount should he remit to the seller? What will be his unpaid balance?

STEP 1 Cash discount = 2% × $300 = $6

STEP 2 Remittance = $300 − $6 = $294

STEP 3 Unpaid balance = $484 − $300 = $184

✓ CONCEPT CHECK 7.5

a. An invoice for $476 has terms of 1/15, net 25. How much is the unpaid balance after a $350 remittance is made within the discount period?
Complement rate = 100% − 1% = 99%
Amount credited = $350 ÷ 0.99 = $353.54
Unpaid balance = $476.00 − $353.54 = $122.46

b. An invoice for $476 has terms of 1/15, net 25. What size remittance should be made in order to have a total of $350 credited to the account?
Cash discount = $350 × 0.01 = $3.50
Remittance = $350.00 − $3.50 = $346.50

COMPLETE ASSIGNMENT 7.2.

Chapter Terms for Review

- amount credited
- cash discount
- complement method
- complement rate
- discount date
- discount period
- discount rate
- due date
- equivalent single discount rate
- invoice
- invoice date
- list price
- net price
- net purchase amount
- remittance
- series of discounts
- terms of payment
- trade discounts

Try Microsoft® Excel

1. Find the required remittance for goods with a list price of $240, a trade discount of 25% and a cash discount of 5%.

    The formula is List Price × (1 − Trade Discount %) × (1 − Cash Discount %) = Remittance. Enter the values in the columns as labeled, and enter the formula in the Remittance cell. Format the remittance cell for Currency with 2 digits after the decimal point.

List Price	Trade Discount	Cash Discount	Remittance

2. What is the remittance amount for goods with a list price of $2200, a trade discount of 40%, and another discount of 25%?

List Price	Trade Discount	Cash Discount	Remittance

3. What is the remittance amount for goods with a list price of $1650, a trade discount of 30%, and another discount of 20%?

List Price	Trade Discount	Cash Discount	Remittance

Refer to your Student CD template for solutions.

THE BOTTOM LINE

Summary of chapter learning objectives:

Learning Objective	Example
7.1 Compute trade discounts	1. Find the net price on a list price of $280 with a 25% trade discount, using the discount and the complement methods.
7.2 Compute a series of trade discounts	2. Find the net price on a list price of $800 with a series of trade discounts of 25% and 10%. Use both the discount method and the complement method.
7.3 Compute the equivalent single discount rate for a series of trade discounts	3. A series of trade discounts is 25%, 20%, 15%. Use complement rates to find the equivalent single discount rate.
7.4 Compute cash discounts and remittance amounts for fully paid invoices	An invoice is dated December 26 and has terms of 2/10, net 25. The total amount is $964.24, with $141.34 of returned goods and $82.90 freight. 4. Compute the discount date, due date, cash discount, and remittance. 5. Compute the remittance using the complement rate.
7.5 Compute cash discounts and remittance amounts for partially paid invoices	An invoice for $500 has terms of 3/5, net 45. 6. Compute the unpaid balance after a $400 payment within the discount period. 7. Compute the remittance required within the discount period in order to have $400 credited to the account.

Answers: 1. Discount method: $280 − $70 = $210; complement method: 0.75 × $280 = $210
2. Discount method: $800 − $200 = $600, $600 − $60 = $540; complement method: 0.75 × 0.90 × $800 = $540
3. 49% 4. Discount date: Jan. 5; due date: Jan. 20; cash discount: $14.80; remittance: $808.10 5. $808.10
6. $87.63 7. $388.00

SELF-CHECK

Review Problems for Chapter 7

In problems 1 and 2, use the discount method to compute the missing terms.

1 List price, $650; trade discount, 20%

  a. Discount amount _____

  b. Net price _____

2 List price, $1,200; trade discounts, 30% and 20%

  a. First discount amount _____

  b. Second discount amount _____

  c. Net price _____

In problems 3 and 4, use the complement method to compute the missing terms.

3 List price, $875; trade discount, 40%

  a. Complement rate _____

  b. Net price _____

4 List price, $1,600; trade discounts, 25% and 10%

  a. First complement rate _____

  b. Second complement rate _____

  c. Net price _____

5 Patty Duncan is a broker of hotel rooms in Europe. To tour directors, she offers a standard trade discount of 40% off the list price. She has additional discounts of 20% and 10%, which are based on the number of tours in a season and the total number of tourists. Compute the equivalent single discount rate for tour organizer Kristi Atchison who qualifies for all three discounts.

  a. First complement rate _____

  b. Second complement rate _____

  c. Third complement rate _____

  d. Equivalent single discount rate _____

Use the invoice information given in problems 6 and 7 to compute the missing terms.

6 Terms: 2/10, n/30
Invoice Date: July 25
Invoice Amount: $874.55
Freight: 0
Returned Goods: 0

  a. Discount date _____

  b. Due date _____

  c. Discount amount _____

  d. Remittance _____

7 Terms: 3/5, net 45
Invoice Date: December 28
Invoice Amount: $2,480
Freight: $143
Returned Goods: $642

  a. Discount date _____

  b. Due date _____

  c. Complement rate _____

  d. Remittance _____

8 Joyce Thompson purchased some new pieces of office furniture for her Internet consulting firm. The invoice amount was $16,540 with terms of 2/10, net 60 and the discount would apply to any partial payment made within the discount period. Joyce sent in a check for $10,000 by the discount date. Find: (a) the amount credited to Joyce's account _____; and (b) the unpaid balance _____.

Notes

Assignment 7.1: Trade Discounts

Name

Date   Score

Learning Objectives 1 2 3

A (24 points) Problems 1–3: Find the dollar amount of the trade discount and the net price, using the discount method. Problems 4–6: Find the complement rate and the net price, using the complement method. (2 points for each correct answer)

	Trade Discount	List Price	Discount Amount	Net Price	
1.	35%	$1,260	$441	$819	$0.35 \times \$1,260 = \$441$ $\$1,260 - \$441 = \$819$
2.	30%	$6,470	$1,941	$4,529	$0.30 \times \$6,470 = \$1,941$ $\$6,470 - \$1,941 = \$4,529$
3.	25%	$8,480	$2,120	$6,360	$0.25 \times \$8,480 = \$2,120$ $\$8,480 - \$2,120 = \$6,360$

	Trade Discount	List Price	Complement Rate	Net Price	
4.	30%	$1,670	70%	$1,169	$100\% - 30\% = 70\%$ $0.70 \times \$1,670 = \$1,169$
5.	40%	$3,750	60%	$2,250	$100\% - 40\% = 60\%$ $0.60 \times \$3,750 = \$2,250$
6.	35%	$4,720	65%	$3,068	$100\% - 35\% = 65\%$ $0.65 \times \$4,720 = \$3,068$

Score for A (24)

B (16 points) Find the amount of each discount in the given series of trade discounts. Then find the net price. Where a discount doesn't exist, enter a dash. (2 points for each correct answer)

	List Price	Trade Discounts	Trade Discount Amounts			Net Price
			First	Second	Third	
7.	$2,400	30%, 25%	$720	$420	—	$1,260
	$0.30 \times \$2,400 = \$720$		$0.25 \times \$1,680 = \$420$			
	$\$2,400 - \$720 = \$1,680$		$\$1,680 - \$420 = \$1,260$			
8.	$1,600	40%, 25%, 20%	$640	$240	$144	$576
	$0.40 \times \$1,600 = \$640$		$0.25 \times \$960 = \$240$		$0.20 \times \$720 = \$144$	
	$\$1,600 - \$640 = \$960$		$\$960 - \$240 = \$720$		$\$720 - \$144 = \$576$	

Score for B (16)

Assignment 7.1 Continued

C (20 points) Find the complement rate for each discount in the given series of trade discounts. Then find the net price, using the complement method. Where a complement rate doesn't exist, place a dash. (2.5 points for each correct answer)

List Price	Trade Discounts	Complement Rates			Net Price
		First	Second	Third	
9. $1,800	30%, 15%	70%	85%	—	$1,071

$100\% - 30\% = 70\%$
$100\% - 15\% = 85\%$
$0.70 \times 0.85 \times \$1,800 = \$1,071$

10. $2,000	40%, 20%, 10%	60%	80%	90%	$864

$100\% - 40\% = 60\%$
$100\% - 20\% = 80\%$
$100\% - 10\% = 90\%$
$0.60 \times 0.80 \times 0.90 \times \$2,000 = \$864$

Score for C (20)

D (20 points) Find the complement rate for each discount in the given series of trade discounts. Then find the equivalent single discount rate, to the nearest $\frac{1}{10}$ of a percent. (2.5 points for each correct answer)

Trade Discounts	Complement Rates			Equivalent Single Discount Rates
	First	Second	Third	
11. 30%, 20%, 5%	70%	80%	95%	46.8%

$100\% - 30\% = 70\%$ $100\% - 20\% = 80\%$ $100\% - 5\% = 95\%$
$0.70 \times 0.80 \times 0.95 = 0.532$ $1.000 - 0.532 = 0.468,$ or 46.8%

12. 20%, 10%, 5%	80%	90%	95%	31.6%

$100\% - 20\% = 80\%$ $100\% - 10\% = 90\%$ $100\% - 5\% = 95\%$
$0.80 \times 0.90 \times 0.95 = 0.684$ $1.000 - 0.684 = 0.316,$ or 31.6%

Score for D (20)

E (20 points) Solve each of the following business applications about trade discounts. Use either the discount method or the complement method. (10 points for each correct answer)

13. Gifford Landscaping, Inc., purchased $425 worth of plants and $180 worth of soil and fertilizer from a garden supply wholesaler. The wholesaler gives Gifford a 20% trade discount on the plants and a 30% trade discount on the other items. Compute the net price that Gifford Landscaping will be required to pay. $466

$0.20 \times \$425 = \$85$ $\$425 - \$85 = \$340$ plants
$0.30 \times \$180 = \$54$ $\$180 - \$54 = +\ 126$ soil, etc.
                                                   $\$466$ total

14. Hackett Roofing is purchasing redwood shakes to reroof a house. The shakes have a list price of $15,600. The Pacific Roofing Supply Company gives Hackett the normal trade discount of 25%. In addition, Pacific gives Hackett two further trade discounts of 20% and 10% because of the large volume of business that the company has done with Pacific so far this year. What is Hackett's net price on the order of redwood shakes? $8,424

$100\% - 25\% = 75\%$
$100\% - 20\% = 80\%$
$100\% - 10\% = 90\%$
$0.75 \times 0.80 \times 0.90 \times \$15,600 = \$8,424$

Score for E (20)

Part 2 Percentage Applications

Assignment 7.2: Cash Discounts

Name

Date   Score

Learning Objectives 4 5

A (64 points) For the following problems, find the discount date, the due date, the amount of the cash discount, and the amount of the remittance. (2 points for each correct date and 6 points for each correct amount)

1. Terms: 3/5, n/25   Discount date: June 1
   Invoice date: May 27   Due date: June 21
   Invoice amount: $622.56   Discount amount: $18.68
                                       Remittance: $603.88

   Disc. date: May 27 + 5 = June 1
   Due date: May 27 + 25 = June 21
   0.03 × $622.56 = $18.6768, or $18.68
   $622.56 − $18.68 = $603.88

2. Terms: 2/10, n/30   Discount date: Aug. 2
   Invoice date: July 23   Due date: Aug. 22
   Invoice amount: $484.86   Discount amount: $8.80
   Freight: $45.00   Remittance: $476.06

   Disc. date: July 23 + 10 = Aug. 2
   Due date: July 23 + 30 = Aug. 22
   $484.86 − $45 = $439.86
   0.02 × $439.86 = $8.7972, or $8.80
   $439.86 − $8.80 + $45 = $476.06

3. Terms: 1.5/15, net 45   Discount date: Sept. 4
   Invoice date: Aug. 20   Due date: Oct. 4
   Invoice amount: $692.00   Discount amount: $6.75
   Returned goods: $242.00   Remittance: $443.25

   Disc. date: Aug. 20 + 15 = Sept. 4
   Due date: Aug. 20 + 45 = Oct. 4
   $692 − $242 = $450
   0.015 × $450 = $6.75
   $450 − $6.75 = $443.25

4. Terms: 2.5/20, N/60   Discount date: Jan. 17
   Invoice date: Dec. 28   Due date: Feb. 26
   Invoice amount: $1,245.55   Discount amount: $19.42
   Returned goods: $398.75   Remittance: $827.38
   Freight: $70.00

   Disc. date: Dec. 28 + 20 = Jan. 17
   Due date: Dec. 28 + 60 = Feb. 26
   $1,245.55 − $398.75 − $70 = $776.80
   0.025 × $776.80 = $19.42
   $776.80 − $19.42 + $70 = $827.38

Score for A (64)

Chapter 7 Discounts 137

Assignment 7.2 Continued

B (16 points) For the following problems, find the discount date, the complement rate, and the amount of the remittance. (2 points for each date and rate; 4 points for each correct remittance)

5. Terms: 2/10, n/25
   Invoice date: March 29
   Invoice amount: $582.50

   Discount date: Apr. 8
   Complement rate: 98%
   Remittance: $570.85

   Disc. date: Mar. 29 + 10 = Apr. 8
   Comp rate: 100% − 2% = 98%
   0.98 × $582.50 = $570.85

6. Terms: 1/25, net 55
   Invoice date: July 9
   Invoice amount: $684.92
   Returned goods: $171.12
   Freight: $45.00

   Discount date: Aug. 3
   Complement rate: 99%
   Remittance: $509.11

   Disc. date: July 9 + 25 = Aug. 3 $464.11 + $45 = $509.11
   Comp. rate: 100% − 1% = 99%
   $684.92 − $171.12 − $45.00 = $468.80
   0.99 × $468.80 = $464.112, or $464.11

Score for B (16)

C (20 points) The following problems involve partial payments made within the discount period. Solve for the items indicated. (5 points for each correct answer)

7. Terms: 3/7, n/45
   Invoice date: Feb. 27
   Invoice amount: $664.27

   Amount credited: $412.37
   Remittance: $400
   Unpaid balance: $251.90

   Comp. rate: 3% = 97%
   $400 ÷ 0.97 = $412.3711, or $412.37
   $664.27 − $412.37 = $251.90

8. Terms: 2/15, net 35
   Invoice date: Feb. 15
   Invoice amount: $832.90
   Returned goods: $186.00

   Amount credited: $510.20
   Remittance: $500
   Unpaid balance: $136.70

   Comp. rate: 100% − 2% = 98%
   $500 ÷ 0.98 = $510.2041, or $510.20
   $832.90 − $186 − $510.20 = $136.70

Score for C (20)

Markup

8

Learning Objectives
By studying this chapter and completing all assignments you will be able to:

Learning Objective 1 — Compute the variables in the basic markup formula.

Learning Objective 2 — Compute the markup variables when the markup percent is based on cost.

Learning Objective 3 — Compute markup percent based on cost.

Learning Objective 4 — Compute the markup variables when the markup percent is based on selling price.

Learning Objective 5 — Compute markup percent based on selling price.

Computing Markup Variables

Learning Objective 1

Compute the variables in the basic markup formula.

8.1 You may want to mention that these numbers are used only for purposes of illustration. The store buys shoes in quantities, and the prices and the freight are also for quantities. Also, taxes are not included in the discussion. Finally, not all shoes will sell at a price that gives the desired profit. Some sizes and styles left at the end of the season may sell if the price is marked low enough.

Some businesses manufacture products and sell them. Other businesses buy products from someone else and then resell them. Both types of businesses must sell their products for more than it costs to produce or purchase them. This price increase is called the **markup**.

Athletes' World is a chain of retail stores that sells athletic equipment and athletic clothing. The store buys shoes directly from a manufacturer. Suppose that the manufacturer charges $43.00 per pair for one particular type of athletic shoe. The prorated amount to deliver one pair to the store is $0.50. The total cost of the shoes, with delivery, is $43.50. $43.50 is called the **cost of goods sold**, or just the cost.

If Athletes' World sells the shoes for exactly the cost, $43.50, it will actually lose money on the sale. The store has many other expenses—such as rent, utilities, and salaries—that are not part of the cost of acquiring the shoes. Athletes' World must mark up the selling price far enough above the cost of the shoes to cover all these additional costs—and also leave some profit for the owners.

The total amount that Athletes' World marks up the selling price is called the **dollar markup**. (*Note*: Markup is expressed both in dollars and in percents. To eliminate confusion, in this book we use two separate terms: *dollar markup* and *markup percent*.)

Suppose that Athletes' World accountants estimate that $18.80 of additional expenses should be allocated to each pair of athletic shoes. Also, suppose that the store would like a profit of $16.00 on each pair of shoes. Then the total dollar markup that it should give the shoes is $18.80 + $16.00 = $34.80.

To determine the selling price of the shoes, Athletes' World adds the dollar markup to the cost of goods sold (cost), using the basic markup formula:

Selling price = Cost + Dollar markup = $43.50 + $34.80 = $78.30

Because the dollar markup is the difference between the selling price and the cost of goods sold, it is often useful to rewrite the formula as

Dollar markup = Selling price − Cost = $78.30 − $43.50 = $34.80

Likewise, cost is the difference between selling price and dollar markup. Thus,

Cost = Selling price − Dollar markup = $78.30 − $34.80 = $43.50

✔ CONCEPT CHECK 8.1

Compute the missing terms in the basic markup formula:
Selling price = Cost + Dollar markup

a. Cost = $417.82; Dollar markup = $204.20

    Selling price = Cost + Dollar markup
              = $417.82 + $204.20 = $622.02

b. Cost = $154.40; Selling price = $392.12

    Dollar markup = Selling price − Cost
               = $392.12 − $154.40 = $237.72

c. Dollar markup = $41.26; Selling price = $93.20

    Cost = Selling price − Dollar markup
           = $93.20 − $41.26 = $51.94

Computing Markup Based on Cost

In the example, Athletes' World computed its markup directly by determining its expenses and the desired profit. However, this method isn't practical when a business has hundreds or thousands of items. Allocating expenses and profit to each item would be too tedious. A more practical method is for the owner, an employee, or an accountant to analyze prior sales of the company or a similar company. The analyst can look at the costs of goods, additional expenses, and desired profit to determine a percent to use to mark up various items, called the **markup percent**.

One company may use different markup percents for different types of items. For example, an appliance store typically also performs repair services and sells replacement parts for the appliances it sells. The store may have one markup percent for the actual appliance, a second markup percent for repair services, and a third markup percent for replacement parts.

In Chapter 5 on percents, we introduced three terms: rate, base, and percentage. In this chapter, rate is the *markup percent*, or **markup rate.** Percentage is the *dollar markup*. Determining the base is more challenging because sometimes *cost* is the base and sometimes *selling price* is the base. For some businesses, cost may be the more logical base for calculating dollar markup. However, calculating dollar markup based on selling price is an advantageous method for many retail stores.

The accountant for Athletes' World says that, in order to pay all expenses and have a reasonable profit, and based upon a cost of $43.50, the company should have an 80% markup. When the cost and the markup percent are known, the dollar markup and the selling price can be computed.

Learning Objective 2

Compute the markup variables when the markup percent is based on cost.

8.2 Most students will already be aware that the percent of markup varies with the type of business and also the type of product. A high-volume supermarket may have low markups on many staples but higher markups on snack foods and even higher markups on nongrocery items such as hardware and paper products. A jewelry store, especially one that makes much of its own jewelry, has a relatively low number of sales each week and will therefore have much higher markups.

Video — Markup Based on Cost/Selling Price

STEPS to Compute the Selling Price Based on Cost

1. Multiply the cost by the markup percent to get the dollar markup.
2. Add the dollar markup to the cost to get the selling price.

For the Athletes' World's athletic shoes:

STEP 1   Dollar markup = Markup percent × Cost = 0.80 × $43.50 = $34.80

STEP 2   Selling price = Cost + Dollar markup = $43.50 + $34.80 = $78.30

8.3 These equations do not lend themselves easily to the use of letters instead of words. In speech, which tends to be less exact, "markup" may refer to the amount of dollar markup in one sentence and then to the markup percent in a subsequent sentence.

EXAMPLE A

Using markup based on cost, what are the dollar markup and the selling price on merchandise that costs $60 and has a 35% markup?

STEP 1   Dollar markup = Markup percent × Cost = 0.35 × $60 = $21

STEP 2   Selling price = Cost + Dollar markup = $60 + $21 = $81

COMPUTING SELLING PRICE DIRECTLY FROM COST

You can compute the selling price directly from the cost, without computing the dollar markup.

> **STEPS** to Compute the Selling Price Directly from the Cost
>
> 1. Add 100% to the markup percent.
> 2. Multiply this sum by the cost to get the selling price.

EXAMPLE B

What is the selling price of an item that has a cost of $250 and a markup percent of 40% based on cost?

STEP 1   Markup percent + 100% = 40% + 100% = 140%

STEP 2   Selling price = (Markup percent + 100%) × Cost = 1.40 × $250 = $350

COMPUTING COST FROM SELLING PRICE

When you know the selling price and the markup percent, the procedure for computing cost is just the reverse of that for computing selling price.

> **STEPS** to Compute the Cost from the Markup Percent
>
> 1. Add the markup percent to 100%.
> 2. Divide the selling price by this sum to get the cost.

EXAMPLE C

The selling price of a pair of shoes is $75. The markup percent based on cost is 25%. Find the cost.

STEP 1   100% + Markup percent = 100% + 25% = 125%

STEP 2   Cost = Selling price ÷ (100% + Markup percent) = $75 ÷ 1.25 = $60

You can always check your work in markup problems.
Cost is $60, and markup percent is 25%.
Dollar markup = Cost × Markup percent = $60 × 0.25 = $15
Selling Price = Cost + Dollar markup = $60 + $15 = $75
It checks!

✔ CONCEPT CHECK 8.2

Compute the required values when the markup percent is based on cost.

a. Cost = $1,240; Markup percent = 40%
   Find dollar markup, and then find selling price.
   Dollar markup = 0.40 × $1,240 = $496
   Selling price = $1,240 + $496 = $1,736

b. Cost = $330; Markup percent = 50%
   Find 100% + Markup percent, and then find selling price directly.
   100% + Markup percent = 100% + 50% = 150%
   Selling price = 1.50 × $330 = $495

c. Selling price = $780; Markup percent = 25%
   Find 100% + Markup percent, and then find cost directly.
   100% + Markup percent = 100% + 25% = 125%
   Cost = $780 ÷ 1.25 = $624

Computing Markup Percent Based on Cost

In the illustration for Athletes' World, the accountant determined that the markup percent needed to be 80% of cost, which meant that the selling price needed to be $78.30. However, management may want to price the shoes at $79.95. Now, the markup is no longer 80% of cost. The **markup percent based on cost** can be computed in two steps.

Learning Objective 3

Compute markup percent based on cost.

STEPS to Compute the Markup Percent Based on Cost

1. Subtract the cost from the selling price to get the dollar markup.
2. Divide the dollar markup by the cost to get the markup percent.

For the athletic shoes from Athletes' World, priced at $79.95:

STEP 1   Dollar markup = Selling price − Cost = $79.95 − $43.50 = $36.45

STEP 2   Markup percent = Dollar markup ÷ Cost = $36.45 ÷ $43.50 = 0.838, or 83.8% (rounded to one decimal place)

EXAMPLE D

What is the markup percent based on cost when the selling price is $120 and the cost is $80?

STEP 1   Dollar markup = Selling price − Cost = $120 − $80 = $40

STEP 2   Markup percent = Dollar markup ÷ Cost = $40 ÷ $80 = 0.50, or 50%

EXAMPLE E

What is the markup percent based on cost when the dollar markup is already known to be $30 and the cost is $75? (Step 1 is not necessary.)

STEP 2   Markup percent = Dollar markup ÷ Cost = $30 ÷ $75 = 0.40, or 40%

✓ CONCEPT CHECK 8.3

Cost = $1,600; Selling price = $2,560
Find the markup percent based on cost.

Dollar markup = $2,560 − $1,600 = $960
Markup percent = $960 ÷ $1,600 = 0.60, or 60%

COMPLETE ASSIGNMENT 8.1.

Computing Markup Based on Selling Price

Learning Objective 4

Compute the markup variables when the markup percent is based on selling price.

Although many businesses base their markup on cost, many others, often retailers, commonly use a percent of selling price—that is, they use **markup based on selling price**. That doesn't mean that selling price is determined without considering cost or even before considering cost. It merely means that the dollar markup is computed by multiplying the markup percent by the selling price.

Many individuals start their own business when they observe another successful business selling a product. New owners believe that they can acquire the product, pay all expenses, and still sell it for less than the existing business is selling its product. Instead of basing the selling price on costs, expenses, and satisfactory profit, the new owners may price their product just under the competition's price. They base their selling price on the competition's selling price rather than marking up from their own costs.

Basing markup calculations on selling price can be an advantage in a retail store where the salesperson or sales manager has the authority to lower the sales price immediately in order to make a sale.

8.4 Emphasize that the same markup percent will result in different amounts of dollar markup, depending on whether it is based on cost or selling price.

Likewise, the same dollar markup will result in different markup percents. Here are the formulas to convert the markup percents from one base to the other:

Markup % (on SP) = Markup % (on C) ÷ [100% + Markup % (on C)]

Markup % (on C) = Markup % (on SP) ÷ [100% − Markup % (on SP)]

Possible discussion questions:

Can markup percent based on cost equal 100%? Yes

Can markup percent based on cost exceed 100%? Yes

Can markup percent based on selling price equal 100%? Not really, because then the cost would be zero and the markup percent based on cost would be infinite.

Can markup percent based on selling price exceed 100%? Not really, because then the cost would be negative (maybe the "seller" was paid by someone to take the item away).

STEPS to Compute the Dollar Markup and Cost from the Markup Percent

1. Multiply the selling price by the markup percent to get the dollar markup.
2. Subtract the dollar markup from the selling price to get the cost.

EXAMPLE F

Roy Brainard enters Floyd's Appliance Store to buy a washing machine. He finds one with a selling price of $400. He knows that he can buy it for $375 at another store, but he prefers this store because of its reputation for good service. He tells the sales manager, "I would buy it for $375." The manager, Jesse Cullen, knows that the markup percent is 40% based on selling price. What is the cost of the washing machine?

STEP 1 Dollar markup = Markup percent × Selling price = 0.40 × $400 = $160

STEP 2 Cost = Selling price − Dollar markup = $400 − $160 = $240

Jesse can then decide whether she prefers no sale or one for which she gets a $135 markup. Although it would be helpful if Jesse knew how much markup she would need to pay for expenses, at least she would know the cost.

EXAMPLE G

Find the dollar markup and the cost of an item that sells for $120 and has a markup percent that is 30% based on selling price.

STEP 1 Dollar markup = Markup percent × Selling price = 0.30 × $120 = $36

STEP 2 Cost = Selling price − Dollar markup = $120 − $36 = $84

COMPUTING COST DIRECTLY

You can compute the cost directly from the selling price, without computing the dollar markup.

STEPS to Compute the Cost from the Markup Percent and Selling Price

1. Subtract the markup percent from 100%.
2. Multiply this difference by the selling price to get the cost.

Video
Markup Based on Cost/Selling Price

EXAMPLE H

What is the cost of an item that has a selling price of $240 and a markup percent of 60% based on selling price?

STEP 1   100% − Markup percent = 100% − 60% = 40%

STEP 2   Cost = (100% − Markup percent) × Selling price = 0.40 × $240 = $96

COMPUTING SELLING PRICE FROM COST

When you know the cost and the markup percent, the procedure for computing cost is just the reverse of that for computing selling price.

STEPS to Compute the Selling Price from the Cost

1. Subtract the markup percent from 100%.
2. Divide the cost by this difference to get the selling price.

EXAMPLE I

The cost of a mountain bike is $120. The markup percent based on selling price is 40%. Find the selling price.

STEP 1   100% − Markup percent = 100% − 40% = 60%

STEP 2   Selling price = Cost ÷ (100% − Markup percent) = $120 ÷ 0.60 = $200

You can always check your work in markup problems:
Selling price is $200, and markup percent is 40% based on selling price.
Dollar markup = Markup percent × Selling price = 0.40 × $200 = $80
Cost = Selling price − Dollar markup = $200 − $80 = $120

It checks!

✔ CONCEPT CHECK 8.4

Compute the required values when the markup percent is based on selling price.

a. Selling price = $750; Markup percent = 50%
   Find dollar markup, and then find cost.
   Dollar markup = 0.50 × $750 = $375
   Selling price = $750 − $375 = $375

b. Selling price = $40; Markup percent = 30%
   Find 100% − Markup percent, and then find cost directly.
   100% − Markup percent = 100% − 30% = 70%
   Cost = 0.70 × $40 = $28

c. Cost = $150; Markup percent = 40%
   Find 100% − Markup percent, and then find selling price directly.
   100% − Markup percent = 100% − 40% = 60%
   Cost = $150 ÷ 0.60 = $250

Chapter 8 Markup

Computing Markup Percent Based on Selling Price

Learning Objective 5

Compute markup percent based on selling price

In the illustration for Athletes' World, the pair of athletic shoes had a cost of $43.50. The store owner decided that the selling price of the athletic shoes would be $79.95. The markup percent based on selling price can be calculated in two steps.

> **STEPS to Compute the Markup Percent from the Selling Price**
> 1. Subtract the cost from the selling price to get the dollar markup.
> 2. Divide the dollar markup by the selling price to get the markup percent.

For Athletes' World's athletic shoes,

STEP 1 Dollar markup = Selling price − Cost = $79.95 − $43.50 = $36.45

STEP 2 Markup percent = Dollar markup ÷ Selling price = $36.45 ÷ $79.95 = 0.456, or 45.6% (rounded to one decimal place)

EXAMPLE J

What is the markup percent based on selling price when the selling price is $80 and the cost is $50?

STEP 1 Dollar markup = Selling price − Cost = $80 − $50 = $30

STEP 2 Markup percent = Dollar markup ÷ Selling price = $30 ÷ $80 = 0.375, or 37.5%

EXAMPLE K

What is the markup percent based on selling price when the dollar markup is already known to be $150 and the selling price is $375? (Step 1 is not necessary.)

STEP 2 Markup percent = Dollar markup ÷ Selling price = $150 ÷ $375 = 0.40, or 40%

CONCEPT CHECK 8.5

Cost = $1,600; Selling price = $2,560
Find the markup percent based on selling price.

Dollar markup = $2,560 − $1,600 = $960
Markup percent = $960 ÷ $2,560 = 0.375, or 37.5%

COMPLETE ASSIGNMENT 8.2.

Chapter Terms for Review

cost of goods sold

dollar markup

markup

markup based on selling price

markup percent

markup percent based on cost

markup rate

THE BOTTOM LINE

Summary of chapter learning objectives:

Learning Objective	Example
8.1 Compute the variables in the basic markup formula	Find the missing variables in the basic formula: Selling price = Cost + Dollar markup 1. Cost = $231.50; Dollar markup = 109.12 2. Cost = $34.20; Selling price = $59.95 3. Dollar markup = $475; Selling price = $900
8.2 Compute the markup variables when the markup percent is based on cost	4. Cost = $800; Markup percent = 35% Find the dollar markup and then find the selling price. Find 100% + Markup percent, and then find selling price. 5. Selling price = $2,100; Markup percent = 40% Find 100% + Markup percent and then find cost.
8.3 Compute the markup percent based on cost	6. Cost = $80; Selling price = $108 Find the markup percent based on cost.
8.4 Compute the markup variables when the markup percent is based on selling price	7. Selling price = $820; Markup percent = 25% Find the dollar markup and then find the cost. Find 100% − Markup percent and then find the cost. 8. Cost = $1,350; Markup percent = 40% Find 100% − Markup percent, and then find the selling price.
8.5 Compute the markup percent based on selling price	9. Cost = $288; Selling price = $640 Find the markup percent based on the selling price.

Answers: 1. Selling price = $340.62 2. Dollar markup = $25.75 3. Cost = $425 4. $280, $1,080, 135%; $1,080 5. 140%, $1,500 6. 35% 7. $205, $615, 75%, $615 8. 60%, $2,250 9. 55%

SELF-CHECK

Review Problems for Chapter 8

1 Find the missing terms.

	Cost of Goods Sold	Dollar Markup	Selling Price		Cost of Goods Sold	Dollar Markup	Selling Price
a.	$28.90	$14.45	_____	c.	_____	$1,405	$2,975
b.	$188.12	_____	$399.95	d.	$426.25	_____	$998.88

In problems 2–9, the markup percent is based on cost. Find the missing terms. Round all percents to the nearest one tenth of a percent.

	Cost	Markup Percent	Dollar Markup	Selling Price		Cost	Markup Percent	100% + Markup Percent	Selling Price
2	$500	50%	a. _____	b. _____	**4**	$225	60%	a. _____	b. _____
3	$36	65%	a. _____	b. _____	**5**	$165	40%	a. _____	b. _____

	Selling Price	Markup Percent	100% + Markup Percent	Cost		Selling Price	Cost	Dollar Markup	Markup Percent
6	$840	100%	a. _____	b. _____	**8**	$480	$240	a. _____	b. _____
7	$98	40%	a. _____	b. _____	**9**	$2,000	$1,600	a. _____	b. _____

In problems 10–13 the markup percent is based on selling price. Find the missing terms. Round all percents to the nearest one tenth of a percent.

	Selling Price	Markup Percent	Dollar Markup	Cost		Selling Price	Markup Percent	100% − Markup Percent	Cost
10	$240	30%	a. _____	b. _____	**12**	$1,240	40%	a. _____	b. _____
11	$144	25%	a. _____	b. _____	**13**	$528	75%	a. _____	b. _____

	Cost	Markup Percent	100% − Markup Percent	Selling Price		Selling Price	Cost	Dollar Markup	Markup Percent
14	$960	60%	a. _____	b. _____	**16**	$800	$480	a. _____	b. _____
15	$36	25%	a. _____	b. _____	**17**	$3,750	$1,500	a. _____	b. _____

18 Carol Wilson sells high-end toys, specializing in all wooden toys for preschool children. She pays $40 for a toy truck. Carol sells the toy truck for $50.   a. Find the dollar markup. _____   b. Find the markup percent based on cost. _____   c. Find the markup percent based on selling price. _____

Notes

Assignment 8.1: Markup Based on Cost

Name

Date Score

Learning Objectives 1 2 3

A (12 points) Calculate the missing terms. (2 points for each correct answer)

	Cost	Dollar Markup	Selling Price		Cost	Dollar Markup	Selling Price
1.	$480.70	$175.25	$655.95	2.	$48.51	$21.44	$69.95
	$480.70 + $175.25 = $655.95				$69.95 − $48.51 = $21.44		
3.	$455.48	$374.50	$829.98	4.	$175.50	$57.50	$233.00
	$829.98 − $374.50 = $455.48				$175.50 + $57.50 = $233.00		
5.	$629.00	$280.99	$909.99	6.	$397.00	$352.49	$749.49
	$909.99 − $629.00 = $280.99				$749.49 − $352.49 = $397.00		

Score for A (12)

B (32 points) In the following problems, the markup percent is based on *cost*. Find the missing terms. (2 points for each correct answer)

	Cost	Markup Percent	Dollar Markup	Selling Price		Cost	100% + Markup Percent	Markup Percent	Selling Price
7.	$850	40%	$340	$1,190	8.	$160	125%	225%	$360
	0.40 × $850 = $340					100% + 125% = 225%			
	$850 + $340 = $1,190					2.25 × $160 = $360			
9.	$1,500	70%	$1,050	$2,550	10.	$240	100%	200%	$480.00
	0.70 × $1,500 = $1,050					100% + 100% = 200%			
	$1,500 + $1,050 = $2,550					2.0 × $240 = $480.			
11.	$640	75%	$480	$1,120	12.	$800	30%	130%	$1,040
	0.75 × $640 = $480					$100 + 30% = 130%			
	$640 + $480 = $1,120					1.3 × $800 = $1,040			
13.	$1,500	150%	$2,250	$3,750	14.	$120	200%	300%	$360
	1.5 × $1,500 = $2,250					100% + 200% = 300%			
	$1,500 + $2,250 = $3,750					3.0 × $120 = $360			

Score for B (32)

Chapter 8 Markup 151

Assignment 8.1 Continued

C (32 points) In the following problems, the markup percent is based on cost. Find the missing terms. Round all percents to the nearest tenth of a percent. (2 points for each correct answer)

	Selling Price	Markup Percent	100% + Markup Percent	Cost		Selling Price	Cost	Dollar Markup	Markup Percent
15.	$1,240	60%	160%	$775	16.	$48	$30	$18	60%
	100% + 60% = 160%; $1,240 ÷ 1.60 = $775					$48 − $30 = $18; $18 ÷ $30 = 0.6			
17.	$110	100%	200%	$55	18.	$1,922	$1,240	$682	55%
	100% + 100% = 200%; $110 ÷ 2 = $55					$1,922 − $1,240 = $682; $682 ÷ $1,240 = 0.55			
19.	$594	35%	135%	$440	20.	$679	$388	$291	75.0%
	100% + 35% = 135%; $594 ÷ 1.35 = $440					$679 − $388 = $291; $291 ÷ $388 = 0.75			
21.	$1,050	150%	250%	$420	22.	$216	$96	$120	125%
	100% + 150% = 250%; $1,050 ÷ 2.50 = $420					$216 − $96 = $120; $120 ÷ $96 = 1.25			

Score for C (32)

D (24 points) Business Applications. In the following problems, the markup percent is based on cost. Round all percents to the nearest tenth of a percent. (3 points for each correct answer)

23. Susan Chin owns a firm that sells office furniture to local businesses. One set of six matched pieces costs Susan $2,100. To cover her own business expenses and allow a reasonable profit, Susan marks up this set by 75% of the cost. Find the dollar markup and the selling price.

    Dollar markup $1,575   0.75 × $2,100 = $1,575
    Selling price $3,675 $2,100 + $1,575 = $3,675

24. Stan Wegner manufactures a handheld heart monitoring device. He sells it for $840, which represents a markup of 275% on his production cost. Stan marks it up this much to cover additional business expenses and profit as well as product development. Find Stan's production cost and the dollar markup.

    Cost $224   100% + 275% = 375% $840 − $224 = $616
    Dollar markup $616 $840 ÷ 3.75 = $224

25. Sentry Security Systems sells burglar and fire alarm systems for homes and small businesses. One basic system costs Sentry $720. Sentry marks up the alarm system by $396. Find the selling price, and find the markup percent based on cost.

    Selling price $1,116 $720 + $396 = $1,116
    Markup percent 55% $396 ÷ $720 = 0.55, or 55%

26. After Matt Lord drove his father's car with no oil, the car needed a new engine. A local mechanic charged Matt's father $2,250 for a rebuilt engine that cost him $1,800. All labor was additional. Compute the dollar markup and the markup percent based on cost.

    Dollar markup $450 $2,250 − $1,800 = $450
    Markup percent 25% $450 ÷ $1,800 = 0.25, or 25%

Score for D (24)

Assignment 8.2: Markup Based on Selling Price

Name

Date   Score

Learning Objectives **1** **4** **5**

A (12 points) Calculate the missing terms. (2 points for each correct answer)

	Cost	Dollar Markup	Selling Price		Cost	Dollar Markup	Selling Price
1.	$67.34	$82.15	$149.49	**2.**	$193.19	$265.69	$458.88
	$67.34 + $82.15 = $149.49				$458.88 − $193.19 = $265.69		
3.	$1,819	$840	$2,659	**4.**	$789.25	$476.50	$1,265.75
	$2,659 − $840 = $1,819				$789.25 + $476.50 = $1,265.75		
5.	$62.50	$37.49	$99.99	**6.**	$671.80	$307.15	$978.95
	$99.99 − $62.50 = $37.49				$978.95 − $307.15 = $671.80		

Score for A (12)

B (32 points) In the following problems, the markup percent is based on selling price. Find the missing terms. (2 points for each correct answer)

	Selling Price	Markup Percent	Dollar Markup	Cost		Selling Price	Markup Percent	100% − Markup Percent	Cost
7.	$120	55%	$66	$54	**8.**	$150	25%	75%	$112.50
	0.55 × $120 = $66					100% − 25% = 75%			
	$120 − $66 = $54					0.75 × $150 = $112.50			
9.	$360	40%	$144	$216	**10.**	$1,260	35%	65%	$819
	0.40 × $360 = $144					100% − 35% = 65%			
	$360 − $144 = $216					0.65 × $1,260 = $819			
11.	$1,998	50%	$999	$999	**12.**	$75	70%	30%	$22.50
	0.50 × $1,998 = $999					100% − 70% = 30%			
	$1,998 − $999 = $999					0.30 × $75 = $22.50			
13.	$824	60%	$494.40	$329.60	**14.**	$926	45%	55%	$509.30
	0.60 × $824 = $494.40					100% − 45% = 55%			
	$824 − $494.40 = $329.60					0.55 = $926 = $509.30			

Score for B (32)

Chapter 8 Markup

Assignment 8.2 Continued

C (32 points) In the following problems, the markup percent is based on selling price. Find the missing terms. (2 points for each correct answer)

	Cost	Markup Percent	100% − Markup Percent	Selling Price		Selling Price	Cost	Dollar Markup	Markup Percent
15.	$855	40%	60%	$1,425	**16.**	$220	$143	$77	35%
	\multicolumn{4}{l}{100% − 40% = 60%; $855 ÷ 0.60 = $1425}		\multicolumn{4}{l}{$220 − $143 = $77; $77 ÷ $220 = 0.35}
17.	$143	45%	55%	$260	**18.**	$45	$27	$18	40%
	\multicolumn{4}{l}{100% − 45% = 55%; $143 ÷ 0.55 = $260}		\multicolumn{4}{l}{$45 − $27 = $18; $18 ÷ $45 = 0.40}
19.	$2,520	30%	70%	$3,600	**20.**	$1,400	$924	$476	34%
	\multicolumn{4}{l}{100% − 30% = 70%; $2,520 ÷ 0.70 = $3,600}		\multicolumn{4}{l}{$1,400 − $924 = $476; $476 ÷ $1,400 = 0.34}
21.	$533	35%	65%	$820	**22.**	$840	$462	$378	45%
	\multicolumn{4}{l}{100% − 35% = 65%; $533 ÷ 0.65 = $820}		\multicolumn{4}{l}{$840 − $462 = $378; $378 ÷ $840 = 0.45}

Score for C (32)

D (24 points) Business Applications. In the following problems, the markup percent is based on selling price. Round all percents to the nearest tenth of a percent. (3 points for each correct answer)

23. At the end of summer, Alpine Hardware features garden equipment specials. One rototiller has a selling price of $348. The markup to cover expenses and profit is 50% based on the selling price. Calculate the dollar markup and the cost.

Dollar markup $174   0.50 × $348 = $174
Cost $174 $348 − $174 = $174

24. Parkside Cyclery is a retail bicycle store. For last Christmas season, Parkside purchased one model of mountain bike to use as a Christmas promotion. The bicycles cost $156 each. For this promotion, Parkside's markup was 40% of selling price. Find the selling price and the dollar markup.

Selling price $260   100% − 40% = 60%
Dollar markup $104 $156 ÷ 0.60 = $260; $260 − $156 = $104

25. City TV & Stereo also sells telephones. A two-line cordless telephone set with a speaker phone base, two extra remote handsets, and an answering machine is priced at $182.40. This price includes a markup of $109.44. If this set sells at $182.40, what are the cost and the markup percent based on selling price?

Cost $72.96 $182.40 − $109.44 = $72.96
Markup percent 60% $109.44 ÷ $182.40 = 0.60, or 60%

26. Patio World, a warehouse store, purchased a large volume of teak lounge chairs for $252 each. Upholstered pads were included in the price. To sell the chairs and pads quickly, the store priced the chairs at $360. Compute the dollar markup and the markup percent based on selling price.

Dollar markup $108 $360 − $252 = $108
Markup percent 30% $108 ÷ $360 = 0.30, or 30%

Score for D (24)

Notes

Part 3

Accounting Applications

- **9** Banking
- **10** Payroll
- **11** Taxes
- **12** Insurance

Banking

9

Learning Objectives
By studying this chapter and completing all assignments you will be able to:

Learning Objective 1 — Maintain a checking account.

Learning Objective 2 — Reconcile a bank statement with a checkbook balance.

Using Deposit Slips and Bank Checks

Learning Objective 1

Maintain a checking account.

9.1 Local banks are usually very willing to provide sample forms and have representatives talk about the banking process. Bringing in an outside speaker is an excellent way to introduce this chapter.

Many students will have personal checking accounts. It's a good idea to have students bring in their checkbooks from various local banks.

Bank customers usually make deposits to their checking accounts by using **deposit slips**. Figure 9-1 shows a typical deposit slip, with spaces to list cash and checks being deposited.

In most businesses, each deposit will include a number of checks. Each check is individually listed on each deposit slip. Deposits are also made electronically. Many employees have their pay electronically transmitted directly from their employer's bank accounts to their individual bank accounts.

A bank **check** is a written order directing the bank to pay a certain sum to a designated party, called the **payee**. Banks normally provide checkbooks to their members. Figures 9-2 and 9-3 show typical bank checks, one with the stub on the left and the other with the stub on the top.

Figure 9-1 Deposit Slip

WELLS FARGO BANK
VAN NESS-CALIFORNIA OFFICE 1560 VAN NESS AVENUE SAN FRANCISCO, CA 94109

35-6686 / 3130

DATE _____ 20 ____
DEPOSITS MAY NOT BE AVAILABLE FOR IMMEDIATE WITHDRAWAL

SIGN HERE FOR LESS CASH IN TELLER'S PRESENCE

HART FURNITURE CO.
1039 BROADWAY
SAN FRANCISCO, CA 94103

USE OTHER SIDE FOR ADDITIONAL LISTING. BE SURE EACH ITEM IS PROPERLY ENDORSED.

CASH	CURRENCY	300	00
	COIN	60	49
LIST CHECKS SINGLY	16-30	250	00
	18-21	125	00
	17-17	216	00
TOTAL FROM OTHER SIDE		209	00
TOTAL		1,160	49
LESS CASH RECEIVED		—	
NET DEPOSIT		1,160	49

Back of Deposit Slip:

PLEASE LIST EACH CHECK SEPARATELY BY BANK NUMBER

CHECKS		DOLLARS		CENTS
1	14-36		7 6	75
2	13-22		1 3	25
3	13-22	1	1 9	00
4				
5				
6				
7				
8				
9				
10				
11				
12				
PLEASE FORWARD TOTAL TO REVERSE SIDE		2	0 9	00

Figure 9-2 Check with Check Stub on Left

Today, many bank transactions are completed electronically. Funds that are transmitted electronically, primarily via computers, are called **electronic fund transfers** (**EFTs**). They include **automatic teller machine** (**ATM**) transactions by which customers can check their balances, make deposits, and withdraw funds from their accounts without having to wait for the next available bank teller. Computer programs also initiate many electronic fund transfers. These transactions are processed through the Automated Clearing House Association and include direct deposits of payroll checks and Social Security and other government and pension benefit payments.

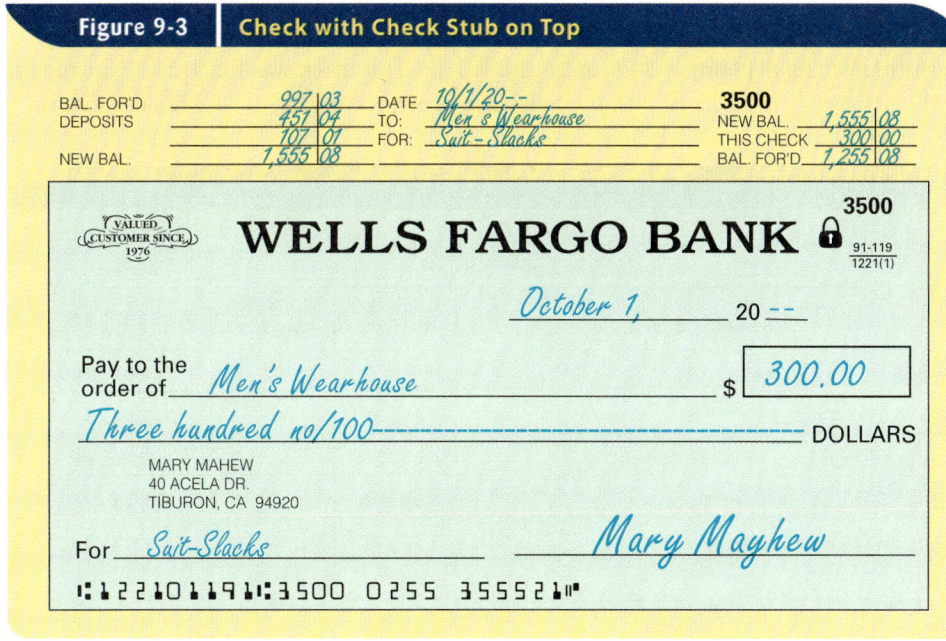

Figure 9-3 Check with Check Stub on Top

Chapter 9 Banking 159

CONCEPT CHECK 9.1

Fill in the total (as necessary) and balance on each check stub. Carry each balance forward to the next stub.

No. 1	$ 65.00
May 1	20--
To Citizens News	
For Advertising	

	$	¢
Balance Bro't Fwd	890	00
Amount Deposited		
Total		
Amount This Check	65	00
Balance Car'd Fwd	825	00

No. 2	$ 79.00
May 4	20--
To District Utilities	
For Gas & electric	

	$	¢
Balance Bro't Fwd	825	00
Amount Deposited		
Total		
Amount This Check	79	00
Balance Car'd Fwd	746	00

No. 3	$ 25.00
May 5	20--
To U.S. Postal Service	
For Stamps	

	$	¢
Balance Bro't Fwd	746	00
Amount Deposited	100	00
Total		
Amount This Check	25	00
Balance Car'd Fwd	821	00

Using Checkbooks and Check Registers

A bank **checkbook** also provides check stubs or a special page on which to record deposits, withdrawals, check numbers, dates, check amounts, other additions and subtractions, and the account balance.

Figure 9-2 shows that check number 2506 was written against the account of Hart Furniture Co. on September 24 to Ace Auto Repair. The check was for $124.35 for repairs to the delivery truck. The stub shows a balance brought forward of $1,332.80, a deposit

Figure 9-4 Check Register

CHECK REGISTER			DEDUCT ALL PER CHECK OR SERVICE CHARGES THAT APPLY			BALANCE
DATE		CHECK NUMBER	CHECKS ISSUED TO OR DEPOSITS RECEIVED FROM	AMOUNT OF CHECK	AMOUNT OF DEPOSIT	$1,332.80
Sept	24		Deposit cash receipts		1,160.49	2,493.29
	24	2506	Ace Auto Repair	124.35		2,368.94
	24	2507	Morton Window Decorators	450.00		1,918.94
	24	2508	Donation to Guide Dogs	100.00		1,818.94
	25	2509	Secure Alarm Systems	150.00		1,668.94
Oct	19	2517	Best Janitorial Service	325.00		855.94
	20		Deposit cash receipts		980.00	1,835.94

160 Part 3 Accounting Applications

on September 24 of $1,160.49, the amount of this check ($124.35), and a balance carried forward of $2,368.94.

Today, most small businesses and many individuals use a **check register**. Like a check stub, a check register provides a place to record information about each bank transaction. Figure 9-4 shows a typical check register. Note that a continuous balance is maintained.

9.2 Today, withdrawals from personal checking accounts are frequently made through ATMs. Thus, it is extremely important to record all withdrawals and to reconcile personal figures with the bank's figures monthly.

✓ CONCEPT CHECK 9.2

In this check register, fill in the cash balance resulting from each transaction.

CHECK REGISTER			DEDUCT ALL PER CHECK OR SERVICE CHARGES THAT APPLY			BALANCE
DATE		CHECK NUMBER	CHECKS ISSUED TO OR DEPOSITS RECEIVED FROM	AMOUNT OF CHECK	AMOUNT OF DEPOSIT	$520.42
Mar	27	123	Replenish petty cash	$ 65.20		455.22
	31	124	Jiffy Janitorial Service	150.00		305.22
Apr	01	125	Sun County Water District	96.72		208.50
	03	–	Deposit weekly receipts		$2,470.80	2,679.30
	03	126	Midtown Mortgage Co.	835.20		1,844.10
	03	127	Sun Gas and Electric Co.	72.18		1,771.92
	04	128	Midtown Weekly Advertiser	32.80		1,739.12
	04	129	Trash Disposal, Inc.	60.00		1,679.12
	04	130	Pacific Plumbing Supplies	906.97		772.15
	10	–	Deposit weekly receipts		2,942.50	3,714.65

Reconciling Bank Statements

Checking account customers receive a printed **bank statement** every month. The bank statement shows an opening balance; deposits and credits, including EFTs; checks paid; withdrawals, including EFTs; service charges; general information about the account; and the balance at the end of the period. In addition, most banks now provide electronic banking through your personal computer. It allows you to view your current bank statement at any time. Figure 9-5 shows a typical bank statement.

The balance shown in the checkbook or check register is usually different from the balance shown on the bank statement. The items that cause this difference are used in reconciling the two balances. These items are as follows:

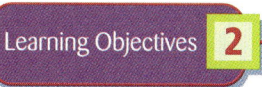

Learning Objectives **2**

Reconcile a bank statement with a checkbook balance.

An **outstanding check** is one that has been written but hasn't yet cleared the bank. Almost always you will have written and recorded some checks that haven't yet been presented to or processed by the bank for payment and charged to the customer's account.

A **bank charge** is a fee for services performed by the bank. At the time the bank statement is made up, your account may have been charged for bank service fees, for printing checks, for bad checks returned, and for EFTs that you haven't yet recorded. These charges would therefore not yet be deducted from your checkbook or check register balance.

Figure 9-5 | Bank Statement

WELLS FARGO BANK

VAN NESS-CALIFORNIA #307
1560 VAN NESS AVE.
SAN FRANCISCO CA 94109

HART FURNITURE CO.
1039 BROADWAY
SAN FRANCISCO, CA 94103

CALL (415) 456-9081
24 HOURS/DAY, 7 DAYS/WEEK
FOR ASSISTANCE WITH
YOUR ACCOUNT.

PAGE 1 OF 1   THIS STATEMENT COVERS: 09/21/– – THROUGH 10/20/– –

WELLS FARGO NEWSLINE: NEW! GET STAMPS AT EXPRESS ATMS WHEN YOU STOP BY FOR CASH. AND, PLEASE NOTE THAT THE COMBINED TOTAL OF CASH WITHDRAWN AND STAMP PURCHASES CANNOT EXCEED YOUR DAILY CASH LIMIT.

REWARD ACCOUNT
31306686

SUMMARY

PREVIOUS BALANCE	$1,332.80
DEPOSITS	1,560.49
WITHDRAWALS	1,081.23
INTEREST	6.30
MONTHLY CHECKING FEE AND OTHER CHARGES	13.00
▶ NEW BALANCE	$1,805.36

MINIMUM BALANCE $980.17
AVERAGE BALANCE $1,336.91

CHECKS AND WITHDRAWALS

CHECK	DATE PAID	AMOUNT
2506	9/26	124.35
2507	9/26	450.00
2508	9/26	100.00
2509	9/27	150.00
2510	10/03	50.00
2511	10/10	132.50
2512	10/20	74.38

DEPOSITS

CUSTOMER DEPOSIT	DATE POSTED	AMOUNT
CUSTOMER DEPOSIT	9/25	1,160.49
EFT CREDIT	9/26	400.00

A **credit** is a deposit or addition to a bank account. In many cases, the bank will have credited your account for an item such as an EFT deposited into the account or interest earned on the account. You the customer don't know the amount of these credits until the bank statement arrives, so the credits haven't yet been entered in your checkbook or check register.

An **outstanding deposit** is a credit that hasn't yet been recorded by the bank. A deposit that you made near the end of the statement period may have been recorded in your checkbook or check register but not recorded by the bank in time to appear on the statement.

Because these items cause a difference between the bank statement balance and your checkbook or check register balance, you should always reconcile the two balances immediately upon receipt of the statement.

To start the reconciliation, compare the check stubs or check register, all deposit slips, and any company records of ATM transactions with the bank statement. Such a comparison is called a **reconciliation of the bank balance.**

When Hart Furniture Company received its monthly bank statement, the bookkeeper noted that the ending balance was $1,805.36 but that the balance in the company checkbook was $1,835.94. To determine the correct balance, the bookkeeper noted the following differences:

1. An EFT credit for $400 had been made to the account and not recorded by Hart.
2. A bank service charge of $13 had been subtracted from Hart's account by the bank.
3. Interest earnings of $6.30 had been added to Hart's account.
4. A deposit on October 20 of $980 had not yet been recorded by the bank.
5. Checks for $27.92, $10, $48.95, $144.25, and $325 had not yet been processed and deducted by the bank.

Most bank statements have printed on the back of the statement a form that can be used to quickly and easily reconcile the customer's checkbook or check register balance with the statement balance. Figure 9-6 shows this form as completed by the Hart Furniture bookkeeper using the information just noted. Note that the adjusted checkbook balance and the adjusted bank balance now agree, showing the correct cash balance of $2,229.24.

Figure 9-6 Reconciliation Form

Balance Your Account

DATE 10/20/--

Checks Outstanding

Check No.	Amount	
2513	27	92
2514	10	00
2515	48	95
2516	144	25
2517	325	00
TOTAL	556	12

1 Check off (✓) checks appearing on your statement. Those checks not checked off (✓) should be recorded in the checks outstanding column.

2

Enter your checkbook balance	$1,835	94
Add any credits made to your account through interest, etc. as shown on this statement. (Be sure to enter these in your checkbook).	6	30
	400	00
SUBTOTAL	2,242	24
Subtract any debits made to your account through bank charges, account fees, etc. as shown on this statement. (Be sure to enter these in your checkbook).	−13	00
Adjusted checkbook balance.	$2,229	24	**A**

3

Bank balance shown on this statement.	$1,805	36
Add deposits shown in your checkbook but not shown on this statement, because they were made and received after date on this statement.	980	00
Subtotal	2,785	36
Subtract checks outstanding	556	12
Adjusted bank balance.	$2,229	24	**B**

Your checkbook is in balance if line **A** agrees with line **B**.

STEPS to Reconcile Bank Balances

1. Reconcile the checkbook (check register) balance. Start with the last balance as recorded in the checkbook.
   a. Add any bank statement credits, such as interest earned or EFT deposits not yet recorded in the checkbook.
   b. Subtract any charges or debits made by the bank, such as service charges, check printing charges, returned check charges, or EFT charges not yet recorded in the checkbook.
   This gives you your **adjusted checkbook balance**.
2. Reconcile the bank balance. Start with the balance as presented on the statement.
   a. Add any deposits or other credits not yet recorded by the bank.
   b. Subtract all outstanding checks.
   This gives you your **adjusted bank balance**.
3. Be sure that the two adjusted balances agree.

✓ CONCEPT CHECK 9.3

At month end, Johnson Hardware received the following bank statement. Use the forms that follow the statement to reconcile the check register used in Concept Check 9.2 and the bank statement.

MIDTOWN BANK
JOHNSON HARDWARE COMPANY
346 POPLAR STREET
MIDTOWN, CA 94872

THIS STATEMENT COVERS: 3/27/-- THROUGH 4/08/--

SUMMARY	
PREVIOUS BALANCE	$ 304.36
DEPOSITS	2,470.80+
WITHDRAWALS	2,416.12-
INTEREST	5.60+
SERVICE CHARGES	7.00-
NEW BALANCE	$ 357.64

CHECKS AND WITHDRAWALS	CHECK	DATE PAID	AMOUNT	CHECK	DATE PAID	AMOUNT
	123	3/29	20.00	130*	4/06	1,743.00
	124*	4/02	100.00			
	126*	4/03	475.00			
	127	4/05	48.32			
	128	4/05	29.80			

DEPOSITS	CUSTOMER DEPOSIT	DATE POSTED	AMOUNT
	CUSTOMER DEPOSIT	4/05	2,470.80

* Indicates checks out of sequence

Enter your checkbook balance	$ 3,109	04
Add any credits made to your account through interest, etc. as shown on this statement. (Be sure to enter these in your checkbook).	5	60
SUBTOTAL	3,114	64
Subtract any debits made to your account through bank charges, account fees, etc. as shown on this statement. (Be sure to enter these in your checkbook).	7	00
Adjusted checkbook balance.	$ 3,107	64

Bank balance shown on this statement.	$ 357	64
Add deposits shown in your checkbook but not shown on this statement, because they were made and received after date on this statement.	2,942	50
Subtotal	3,300	14
Subtract checks outstanding	192	50
Adjusted bank balance.	$ 3,107	64

Your checkbook is in balance if line **A** agrees with line **B**.

Checks Outstanding

Check No.	Amount	
125	$ 132	50
129	60	00
TOTAL	$ 192	50

COMPLETE ASSIGNMENTS 9.1, 9.2, AND 9.3.

Chapter Terms for Review

adjusted bank balance

adjusted checkbook balance

automatic teller machine (ATM)

bank charge

bank statement

check

checkbook

check register

credit

deposit slip

electronic fund transfer (EFT)

outstanding check

outstanding deposit

payee

reconciliation of the bank balance

Try Microsoft® Excel

Try working the following problems using the Microsoft Excel templates found on your student CD. Solutions for the problems are also shown on the CD.

1. Complete the following worksheet by adding formulas in shaded cells to calculate the balance after each transaction in the check register. Formulas should work for either the addition of a deposit or subtraction of a check and be able to be copied down the **Balance** column.

	A	B	C	D	E	F
1	Check Register					Balance
2	Date	Check Number	Checks issued to or deposits received from	Amount of Check	Amount of Deposit	895.42
3	May-04	237	Echo Computer Repair Service	235.00		
4	5		Deposit cash sales		1,569.12	
5	6	238	Glendale Gas Co.	127.90		
6	6	239	Yellow Pages - ad	212.33		
7	8	240	City Stationers - supplies	582.91		
8	10		Deposit cash sales		1,243.32	
9	12	241	Acme Cleaning Service	450.00		
10	13	242	General Telephone	82.57		
11	15		Deposit tax refund		750.00	

2. Jessica Flint's monthly bank statement balance was $1,753.04. Her checkbook balance was $2,590.24. She noted that the following checks were outstanding: #134 for $17.35, #137 for $128.45, and #138 for $52.00. She also noted that a deposit of $974.50 was not yet recorded by the bank. The bank statement lists a service charge of $15 and a bad check of $45.50 returned to Jessica by the bank from a recent deposit.

Enter the data given above in the appropriate cells and complete the worksheet to reconcile the bank statement and checkbook balances by adding formulas in shaded cells.

	A	B	C
1	Checkbook balance		
2	Less bank charges:		
3	Service charge		
4	Bad check		
5	Total subtractions		
6	Adjusted checkbook balance		
7			
8	Bank statement balance		
9	Add unrecorded deposit		
10	Subtotal		
11	Less outstanding checks: #134		
12	#137		
13	#138		
14	Total outstanding check		
15	Adjusted bank balance		

THE BOTTOM LINE

Summary of chapter learning objectives:

Learning Objective	Example
9.1 Maintain a checking account	1. Fill in the New Bal and Bal For'd on each check stub. Carry Bal For'd to the next stub. #1 Bal For'd $100.00 Date 01/17 Deposit 350.50 To AAA New Bal _____ This Ck 175.09 For Ins. Bal For'd _____ #2 Bal For'd _____ Date 01/22 Deposit 375.00 To Longs New Bal _____ This Ck 78.88 For Misc Bal For'd _____
9.2 Maintain a checking account	2. Fill in the cash balance for each date.

CHECK REGISTER

DATE	CHECK NUMBER	CHECK TO—DEPOSIT INFORMATION	DEPOSIT AMOUNT	CHECK AMOUNT	BALANCE
					$453.90
12/11	100	Albertsons		$85.92	
12/12		Monthly Salary Check	$1,580.65		
12/13	101	C.Dobbs-Rent		$850.00	
12/14	102	TJ Max		$ 99.97	
12/15	103	Ace Hardware		$ 107.16	
12/17		Income from Stocks	$212.37		

9.3

Reconcile a bank statement with a checkbook balance

3. Mike Kent's monthly bank statement balance was $1,418. His checkbook balance was $1,620. He noted the following checks outstanding: #119 for $350 and #125 for $197. He noted a deposit of $1,600 as not recorded by the bank. The bank had charged him $17 for checks and $32 for a bad check he had deposited. The bank had credited his account with an electronic transfer for $900. Reconcile the bank and checkbook balances.

Checkbook balance: $1,620
Add electronic transfer: _____
Subtotal   _____
Less bank charges: _____
     _____   _____
Adjusted checkbook balance: _____
Bank balance on statement: $1,418
Add unrecorded deposit: _____
Subtotal   _____
Less outstanding checks: #119 _____
                                   #125 _____   _____
Adjusted bank balance   _____

Answers: 1. $450.50; $275.41; $650.41; $571.53 2. $367.98; $1,948.63; $1,098.63; $998.66; $891.50; $1,103.87 3. $2,471

SELF-CHECK

Review Problems for Chapter 9

1 Each of the following items requires an adjustment to either the bank statement balance or the check register balance. Indicate the correct handling of each item by writing the appropriate letter in the blank.

    A = add to bank statement balance
    B = subtract from bank statement balance
    C = add to checkbook balance
    D = subtract from checkbook balance

_____ (a) Outstanding check written to the landlord for rent

_____ (b) Bank charge for printing checks

_____ (c) A deposit made at the end of the period that was not included on the bank statement

_____ (d) A customer's check that was returned by the bank for insufficient funds (a bounced check)

_____ (e) An error in recording a check in the check register. A check written to Acme Services for $92.20 was recorded in the check register as $95.50

_____ (f) Interest on the checking account

_____ (g) A bank fee of $20 for the bounced check

_____ (h) Bank fees for ATM withdrawals

2 The balance in Ferndale Construction Company's check register May 31 was $12,583.40. The bank statement for Ferndale Construction Company listed the following information:

Previous balance (May 1)	$12,620.10
Deposits	16,265.00
Checks and withdrawals	17,805.95
Interest	52.50
Service charges	20.00
Check returned for insufficient funds	150.00
New balance (May 31)	$10,961.65

By comparing the bank statement and the check register, the company's bookkeeper determined that a deposit of $1,850.15 was not included on the statement and that the following checks were outstanding:

No. 602	$ 35.80
No. 610	212.00
No. 612	95.10

While preparing the reconciliation, the company's bookkeeper noted that check number 585, which had been written for $82.50, had been recorded in the check register as $85.50.

Prepare a bank reconciliation statement for Ferndale Construction Company.

Assignment 9.1: Check Register and Check Stubs

Name

Date   Score

Learning Objective 1

A (20 points) In the following check register, fill in the cash balance resulting from each transaction. (2 points for each correct answer)

1.

CHECK REGISTER			DEDUCT ALL PER CHECK OR SERVICE CHARGES THAT APPLY			BALANCE
DATE		CHECK NUMBER	CHECKS ISSUED TO OR DEPOSITS RECEIVED FROM	AMOUNT OF CHECK	AMOUNT OF DEPOSIT	$1,450.00
Apr	04	842	Alliance Mortgage Company	865.00		585.00
	04	–	Deposit weekly cash receipts		4,197.50	4,782.50
	05	843	U.S. Treasury	1,520.00		3,262.50
	06	844	State Income Tax	990.00		2,272.50
	07	845	General Telephone	65.30		2,207.50
	08	846	Maxwell Office Supply	289.70		1,917.50
	12	–	Deposit weekly cash receipts		3,845.25	5,762.75
	12	847	Eastwood Water Co.	126.42		5,636.33
	12	848	Central Advertising, Inc.	965.00		4,671.33
	12	849	Johnson Tax Services	650.00		4,021.33

Score for A (20)

B (15 points) Fill in the new balance (New Bal) and balance forward (BalFor'd) on each check stub, carrying each balance forward to the next stub. (1½ points for each correct answer)

2.
#101
BalFor'd _920.15_ Date _6-1_   New Bal _1,220.15_
Deposit _300.00_ To _ACE_   This Ck _29.30_
New Bal _1,220.15_ For _REPAIR_ BalFor'd _1,190.85_

3.
#102
BalFor'd _1,190.85_ Date _6-5_   New Bal _1,190.85_
Deposit _____ To _DON_   This Ck _312.80_
New Bal _1,190.85_ For _NOTE_ BalFor'd _878.05_

4.
#103
BalFor'd _878.05_ Date _6-8_   New Bal _1,740.18_
Deposit _862.13_ To _NEC_   This Ck _862.42_
New Bal _1,740.18_ For _COMPUTER_ BalFor'd _877.76_

5.
#104
BalFor'd _877.76_ Date _6-10_ New Bal _3,037.76_
Deposit _2,160.00_ To _CHRON_ This Ck _136.40_
New Bal _3,037.76_ For _AD_   BalFor'd _2,901.36_

6.
#105
BalFor'd _2,901.36_ Date _6-15_ New Bal _3,808.52_
Deposit _907.16_ To _B/A_   This Ck _294.28_
New Bal _3,808.52_ For _CAR PAYMENT_ BalFor'd _3,514.24_

Score for B (15)

Chapter 9 Banking 169

Assignment 9.1 Continued

C (20 points) According to the check register of Kyber Electronics, the cash balance on July 1 was $1,335.60. During the month, deposits of $281.75, $681.10, and $385.60 were made. Checks for $98.99, $307.53, $19.56, $212.40, $287.60, and $88.62 were recorded. (15 points for a correct answer in 7; 5 points for a correct answer in 8)

7. What was the cash balance shown in the check register on July 31? $1,669.35

8. After entering all the items in the check register, the bookkeeper found that the check recorded as $212.40 was actually written as $224.20. What is the correct cash balance? $1,657.55

Balance, July 1		$1,335.60	Correct amount of check	$224.20
Add deposits	$281.75		Amount recorded	212.40
	681.10		Difference	$ 11.80
	385.60	1,348.45		
		$2,684.05	$1,669.35 − 11.80 = $1,657.55	
Subtract checks	$ 98.99			
	307.53			
	19.56			
	212.40			
	287.60			
	88.62	1,014.70		
Balance, July 31		$1,669.35		

Score for C (20) _____

D (45 points) The following problems show the deposits and checks that were recorded on a series of check stubs. In each problem, find the bank balance after each deposit or check. (3 points for each correct answer)

9.			10.			11.		
	Balance	$2,420	80		Balance	$205	55		Balance	$2,670	10
	Check #1	279	10		Check #21	25	10		Deposit	350	00
	Balance	2,141	70		Balance	180	55		Balance	3,020	10
	Check #2	148	20		Deposit	721	45		Check #31	265	72
	Balance	1,993	50		Balance	902	00		Balance	2,754	38
	Deposit	976	80		Check #22	188	14		Check #32	85	70
	Balance	2,970	30		Balance	713	86		Balance	2,668	68
	Check #3	814	00		Check #23	415	92		Deposit	935	62
	Balance	2,156	30		Balance	297	94		Balance	3,604	30
	Check #4	285	17		Check #24	72	38		Check #33	1,230	14
	Balance	$1,871	13		Balance	$225	56		Balance	$2,374	16

Score for D (45) _____

170 Part 3 Accounting Applications

Assignment 9.2: Check Register and Bank Statements

Name

Date   Score

Learning Objectives **1** **2**

A (40 points) Solve the following problems. (10 points for a correct final balance in 1; 30 points for a correct final answer in 2)

1. On October 31, the balance of the account of Hobbies Unlimited at the Citizens Bank was $922.10. This amount was also the balance on the check register at that time. Company checks written and deposits made during November are shown on the check register. Fill in the cash balance for each transaction.

CHECK REGISTER			DEDUCT ALL PER CHECK OR SERVICE CHARGES THAT APPLY			BALANCE
DATE		CHECK NUMBER	CHECKS ISSUED TO OR DEPOSITS RECEIVED FROM	AMOUNT OF CHECK	AMOUNT OF DEPOSIT	$922.10
Nov	01	551	Muni. Water, Inc. (2 mos)	119.60		802.50
	06	552	Fenton Gas Co.	49.60		752.90
	07	553	Olympia Telephone	74.19		678.71
	07	–	Deposit cash receipts		225.50	904.21
	21	554	City Trash Disposal (3 mos)	112.32		791.89
	21	555	Jack's Janitorial Service	33.33		758.56
	24	556	United Fund	12.00		746.56
	24	557	Guide Dogs for the Blind	67.77		678.79
	26	558	Wilson Insurance	212.00		466.79
	28	559	Security Systems, Inc.	138.00		328.79
	28	–	Deposit cash receipts		94.00	422.79

2. On December 3, Hobbies Unlimited, whose check register you completed in problem 1, received the following bank statement. Reconcile the balance on the check register at the end of the month with the final balance on the bank statement. In reconciling the bank statement, you can find which of the checks are outstanding by comparing the list of checks on the statement with the register. Interest and a service charge were recorded on the statement.

C̲B CITIZEN'S BANK

STATEMENT OF ACCOUNT

HOBBIES UNLIMITED
4617 GILMORE ROAD
WHEATLAND, WI 54828-6075

ACCOUNT NUMBER
072 4736

11/30/--
DATE OF STATEMENT

Balance From Previous Statement	Number of Debits	Amount of Checks and Debits	No. of Credits	Amount of Deposits and Credits	Service Charge	Statement Balance
922.10	8	594.81	2	229.70	9.00	547.99

DATE	CHECKS - DEBITS	CHECKS - DEBITS	DEPOSITS - CREDITS	BALANCE
11/03	119.60			802.50
11/05	49.60			752.90
11/09	9.00 SC			743.90
11/09	74.19			669.71
11/09			225.50 ATM	895.21
11/23	112.32	33.33		749.56
11/26	67.77			681.79
11/30	138.00			543.79
11/30			4.20 INT	547.99

PLEASE EXAMINE AND REPORT ANY DISCREPANCIES WITHIN 10 DAYS   DM-Debit Memo   ATM-Automated Teller Machine   CM-Credit Memo
OD-Overdraft   INT-Interest Paid   SC-Service Charge

HOBBIES UNLIMITED
Reconciliation of Bank Statement
November 30

Bank balance on statement	$547.99
Plus deposit not recorded by bank	94.00
	$641.99
Minus outstanding checks:	
#556–12.00	
#558–212.00	224.00
	$417.99
Checkbook balance	$422.79
Plus bank interest	4.20
	$426.99
Minus service charge	9.00
	$417.99

Score for A (40)

Assignment 9.2 Continued

B **(60 points) Solve the following problems. (12 points for each correct answer)**

3. Compute the reconciled balance for each of the problems from the information given.

	Bank Statement Balance	Checkbook Balance	Other Information	Reconciled Balance
a.	$ 769.12	$ 794.47	Outstanding checks: $9.50, $31.15 Automatic transfer to savings: $50.00 Automatic charge, safety deposit box: $16.00	$728.47
b.	$1,559.39	$1,672.00	Outstanding checks: $84.62, $14.20, $55.00 Outstanding deposit: $224.70 Automatic transfer to savings: $50.00 Bank interest credited: $8.27	$1,630.27
c.	$ 893.17	$ 944.73	Outstanding checks: $7.50, $4.18, $62.40 Outstanding deposits: $12.32, $120.00 Bank interest credited: $24.18 Charge for printing new checks: $17.50	$951.41
d.	$ 824.90	$ 739.47	Outstanding checks: $87.50 Deposit of $76.89 shown in check register as $78.96	$737.40
e.	$ 710.00	$1,274.18	Outstanding checks: $150.00, $37.82 Outstanding deposit: $440.00 Deposit of $312.00 shown twice in check register	$962.18

a. Bank statement balance: $769.12 − $40.65 ($9.50 + $31.15) = $728.47
   Checkbook balance: $794.47 − $66.00 (50 + 16) = $728.47
b. Bank statement balance: $1,559.39 − $153.82 ($84.62 + $14.20 + $55.00) + $224.70 = $1,630.27
   Checkbook balance: $1,672.00 − $50.00 + $8.27 = $1,630.27
c. Bank statement balance: $893.17 − $74.08 ($7.50 + $4.18 + $62.40) + $132.32 ($12.32 + $120.00) = $951.41
   Checkbook balance: $944.73 + $6.68 ($24.18 − $17.50) = $951.41
d. Bank statement balance: $824.90 − $87.50 = $737.40
   Checkbook balance: $739.47 − $2.07 ($78.96 − $76.89) = $737.40
e. Bank statement balance: $710.00 − $187.82 ($150.00 + $37.82) + $440.00 = $962.18
   Checkbook balance: $1,274.18 − $312.00 = $962.18

Score for B (60)

Assignment 9.3: Bank Balance Reconciliation Statements

Name

Date   Score

Learning Objective 2

A (50 points) Using the data provided, prepare a bank reconciliation statement in each of the following problems. Space is provided for your solutions. (25 points for each correct reconciliation)

1. The balance shown in the bank statement of Cogswell Cooling, Inc. on November 30 was $1,050.82. The balance shown on the check register was $668.45. The following checks were outstanding:

No. 148	$13.90	No. 161	$96.35
No. 156	235.10	No. 165	34.52

   There was a bank interest credit of $12.00 and a service charge of $9.50 that had not been entered on Cogswell Cooling's check register.

   Cogswell Cooling, Inc.
   Reconciliation of Bank Statement, November 30

Checkbook balance		$ 668.45	Bank balance on statement		$1,050.82
Minus unrecorded bank charges:			Minus outstanding checks:		
Service charge		9.50	No. 148	$ 13.90	
		$ 658.95	No. 156	235.10	
Plus bank interest credit		12.00	No. 161	96.35	
			No. 165	$ 34.52	379.87
Adjusted checkbook balance		$ 670.95	Adjusted bank balance		$ 670.95

2. The June 30 bank statement for Furgison Electric Company shows that a customer's bad check in the amount of $960 was returned and charged against the Furgison Electric Company's account by the bank. This is the first knowledge the company had that one of the checks deposited was not good.

   The balance shown on the Furgison Electric Company's bank statement was $22,367.14. The balance shown on the check register was $24,696.83. The following checks were outstanding:

No. 363	$1,066.20	No. 396	$1,544.14
No. 387	1,972.81	No. 397	772.86

   The following items required adjustment on the bank reconciliation statement:

Outstanding deposit:	$3,001.87
Automatic transfer to note payment:	$4,000.00
Bad check returned and charged to Furgison Electric Company's account by the bank:	$ 960.00
Bank interest credit:	$ 276.17

   Furgison Electric Company
   Reconciliation of Bank Statement, June 30

Checkbook balance		$24,696.83	Bank balance on statement		$22,367.14
Minus unrecorded bank charges:			Plus deposit not recorded by bank		3,001.87
Automatic transfer—note payment	$4,000.00				$25,369.01
Returned check charged to account	960.00	4,960.00	Minus outstanding checks:		
		$19,736.83	No. 363	$1,066.20	
Plus bank interest credit		276.17	No. 387	1,972.81	
			No. 396	1,544.14	
			No. 397	772.86	5,356.01
Adjusted checkbook balance		$20,013.00	Adjusted bank balance		$20,013.00

Score for A (50)

Assignment 9.3 Continued

B (50 points) Using the data provided, prepare a bank reconciliation statement in each of the following problems. Space is provided for your solutions. (25 points for each correct reconciliation)

3. The balance shown on the May 31 bank statement of Linberg Floors was $18,120.16. The balance shown by the check register was $19,512.54. A deposit of $2,004.35 had not been credited by the bank, and the following checks were outstanding:

No. 730	$85.17	No. 753	$462.95	No. 761	$19.75
No. 749	1,216.20	No. 757	512.80	No. 768	982.90

The following items required adjustment on the bank reconciliation statement:

Charge for printing checks	$ 18.00
Automatic insurance payment charged to depositor's account by the bank	$1,765.00
Check deposited by Linberg Floors, returned by bank as bad check	$ 920.00
Interest on bank account credited by the bank	$ 35.20

Linberg Floors
Reconciliation of Bank Statement, May 31

Checkbook balance		$19,512.54	Bank balance on statement		$18,120.16
Plus bank interest credited		35.20	Plus deposit not recorded by bank		2,004.35
		$19,547.74			$20,124.51
Minus unrecorded bank charges:			Minus outstanding checks:		
Service charge	$ 18.00		No. 730	$ 85.17	
Automatic transfer—insurance	1,765.00		No. 749	1,216.20	
Returned check	920.00	2,703.00	No. 753	462.95	
Adjusted checkbook balance		$16,844.74	No. 757	512.80	
			No. 761	19.75	
			No. 768	982.90	3,279.77
			Adjusted bank balance		$16,844.74

4. The balance shown on the June 30 bank statement of Greenwood Stables was $9,527.72. The balance shown on the check register was $7,031.25. The following checks were outstanding:

No. 516	$621.50	No. 521	$93.21	No. 523	$144.80
No. 526	935.11	No. 527	250.00	No. 528	416.35

The following items were listed on the bank statement:

Charge made by the bank for safe deposit box	$ 20.00
Bank error: AA Realty's check charged to Greenwood Stables' account	$ 82.50
Interest on bank account credited by the bank	$ 72.12
Bank charge for printing checks	$ 27.00

Greenwood Stables
Reconciliation of Bank Statement
June 30

Checkbook balance		$7,124.13	Bank statement balance		$9,527.72
Plus bank interest credits		72.12	Add AA Realty's check deducted in error		82.50
		$7,196.25			$9,610.22
Minus unrecorded bank charges:			Minus outstanding checks:		
Safe deposit box charge	$ 20.00		No. 516	$ 621.50	
Check printing charge	27.00	47.00	No. 521	93.21	
Adjusted checkbook balance		$7,149.25	No. 523	144.80	
			No. 526	935.11	
			No. 527	250.00	
			No. 528	416.35	2,460.97
					$7,149.25

Score for B (50)

Payroll Records

10

Learning Objectives
By studying this chapter and completing all assignments you will be able to:

Learning Objective 1 — Prepare a payroll register.

Learning Objective 2 — Compute federal income tax withholding amounts.

Learning Objective 3 — Compute Social Security, Medicare, and other withholdings.

Learning Objective 4 — Complete an employee's earnings record.

Learning Objective 5 — Compute an employer's quarterly federal tax return.

Learning Objective 6 — Compute an employer's federal and state unemployment tax liability.

10.1 Note that on the W-4 Form Kyle Abrum is claiming a total of four allowances (deductions)—most likely himself, his wife, and two children.

Employers must keep payroll records, withhold and pay payroll taxes, and file quarterly and annual reports with state and federal government offices. The payroll records and processes described in this chapter are common to all employers.

Federal taxes paid by all employees include the federal income tax and the two contributions (commonly referred to as taxes) required by the Federal Insurance Contributions Act (FICA): Old-Age, Survivors, and Disability Insurance, commonly called Social Security; and Hospital Insurance, commonly called Medicare.

When hiring new employees, employers must verify each employee's eligibility to work in the United States, get the employee's Social Security number, and have the employee complete a **Form W-4**. The W-4 form shown in Figure 10-1 indicates that Kyle Abrum is married and claims four exemptions, which constitutes his **withholding allowance**.

Preparing a Payroll Register

Learning Objective 1

Prepare a payroll register.

10.2 It's important to note that the payroll register illustrated is generic—today, virtually all payroll processing is done by computer. The forms and processes introduced in this chapter are the same as those used by the many different computer programs.

A **payroll register** is a summary of employee status information, wages earned, payroll deductions, and take-home pay. Whether they do it manually or by computer, all employers maintain some form of payroll register.

A payroll register is prepared for each payroll period. Payroll periods are weekly, biweekly, semimonthly, or monthly. Figure 10-2 shows a payroll register for one weekly period ending March 29. The line for Kyle Abrum shows that he is married, claims four withholding allowances, and is paid on an hourly basis at the rate of $11 per hour ($16.50 for overtime hours). For the current week, he worked 40 regular hours and 6 overtime hours, for gross earnings of $539. From his gross pay he had deductions for Social Security ($33.42), Medicare ($7.82), Federal Income Tax ($14.66), Group Medical Insurance ($39), Group Dental Insurance ($12), and Other ($42), totaling $148.90. His net pay was $390.10.

The Fair Labor Standards Act, commonly called the federal wage and hour law, requires that nonexempt employees be paid 1½ their regular hourly rate for all hours worked in excess of 40 per week. Following the FLSA requirements, the calculations for gross pay are as follows:

STEP 1 Multiply hours worked (up to 40) times the regular rate.

STEP 2 Multiply the regular rate times 1.5 to calculate the overtime rate.

STEP 3 Multiply the hours in excess of 40 times the overtime rate.

STEP 4 Add the results of Steps 1 and 3 to determine gross pay.

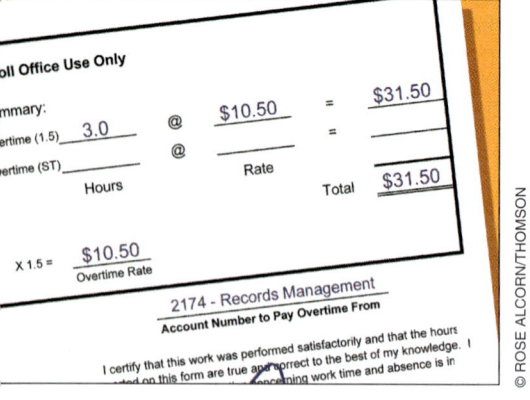

Gross pay calculations for Kyle Abrum:

STEP 1 40 hours × $11 = $440 regular pay

STEP 2 $11 × 1.5 = $16.50 overtime rate

STEP 3 6 hours × $16.50 = $99 overtime pay

STEP 4 $440 + $99 = $539 gross pay

Part 3 Accounting Applications

Figure 10-1 | **Form W-4 (2004)**

Form W-4 (2004)

Purpose. Complete Form W-4 so that your employer can withhold the correct Federal income tax from your pay. Because your tax situation may change, you may want to refigure your withholding each year.

Exemption from withholding. If you are exempt, complete only lines 1, 2, 3, 4, and 7 and sign the form to validate it. Your exemption for 2004 expires February 16, 2005. See **Pub. 505**, Tax Withholding and Estimated Tax.

Note: *You cannot claim exemption from withholding if: (a) your income exceeds $800 and includes more than $250 of unearned income (e.g., interest and dividends) and (b) another person can claim you as a dependent on their tax return.*

Basic instructions. If you are not exempt, complete the **Personal Allowances Worksheet** below. The worksheets on page 2 adjust your withholding allowances based on itemized deductions, certain credits, adjustments to income, or two-earner/two-job situations. Complete all worksheets that apply. However, you may claim fewer (or zero) allowances.

Head of household. Generally, you may claim head of household filing status on your tax return only if you are unmarried and pay more than 50% of the costs of keeping up a home for yourself and your dependent(s) or other qualifying individuals. See line **E** below.

Tax credits. You can take projected tax credits into account in figuring your allowable number of withholding allowances. Credits for child or dependent care expenses and the child tax credit may be claimed using the **Personal Allowances Worksheet** below. See **Pub. 919**, How Do I Adjust My Tax Withholding? for information on converting your other credits into withholding allowances.

Nonwage income. If you have a large amount of nonwage income, such as interest or dividends, consider making estimated tax payments using **Form 1040-ES**, Estimated Tax for Individuals. Otherwise, you may owe additional tax.

Two earners/two jobs. If you have a working spouse or more than one job, figure the total number of allowances you are entitled to claim on all jobs using worksheets from only one Form W-4. Your withholding usually will be most accurate when all allowances are claimed on the Form W-4 for the highest paying job and zero allowances are claimed on the others.

Nonresident alien. If you are a nonresident alien, see the **Instructions for Form 8233** before completing this Form W-4.

Check your withholding. After your Form W-4 takes effect, use Pub. 919 to see how the dollar amount you are having withheld compares to your projected total tax for 2004. See Pub. 919, especially if your earnings exceed $125,000 (Single) or $175,000 (Married).

Recent name change? If your name on line 1 differs from that shown on your social security card, call 1-800-772-1213 to initiate a name change and obtain a social security card showing your correct name.

Personal Allowances Worksheet (Keep for your records.)

- **A** Enter "1" for **yourself** if no one else can claim you as a dependent .... **A** __1__
- **B** Enter "1" if:
  - You are single and have only one job; or
  - You are married, have only one job, and your spouse does not work; or
  - Your wages from a second job or your spouse's wages (or the total of both) are $1,000 or less. ... **B** ____
- **C** Enter "1" for your **spouse**. But, you may choose to enter "-0-" if you are married and have either a working spouse or more than one job. (Entering "-0-" may help you avoid having too little tax withheld.) .... **C** __1__
- **D** Enter number of **dependents** (other than your spouse or yourself) you will claim on your tax return .... **D** __2__
- **E** Enter "1" if you will file as **head of household** on your tax return (see conditions under **Head of household** above) . **E** ____
- **F** Enter "1" if you have at least $1,500 of **child or dependent care expenses** for which you plan to claim a credit .. **F** ____
  (**Note:** *Do not include child support payments. See Pub. 503, Child and Dependent Care Expenses, for details.*)
- **G** **Child Tax Credit** (including additional child tax credit):
  - If your total income will be less than $52,000 ($77,000 if married), enter "2" for each eligible child.
  - If your total income will be between $52,000 and $84,000 ($77,000 and $119,000 if married), enter "1" for each eligible child plus "1" **additional** if you have four or more eligible children. **G** __4__
- **H** Add lines A through G and enter total here. **Note:** *This may be different from the number of exemptions you claim on your tax return.* ▶ **H** __4__

For accuracy, complete all worksheets that apply.
- If you plan to **itemize or claim adjustments to income** and want to reduce your withholding, see the **Deductions and Adjustments Worksheet** on page 2.
- If you have **more than one job** or are **married and you and your spouse both work** and the combined earnings from all jobs exceed $35,000 ($25,000 if married) see the **Two-Earner/Two-Job Worksheet** on page 2 to avoid having too little tax withheld.
- If **neither** of the above situations applies, **stop here** and enter the number from line H on line 5 of Form W-4 below.

Cut here and give Form W-4 to your employer. Keep the top part for your records.

Form W-4 | **Employee's Withholding Allowance Certificate** | OMB No. 1545-0010
Department of the Treasury — Internal Revenue Service | ▶ Your employer must send a copy of this form to the IRS if: (a) you claim more than 10 allowances or (b) you claim "Exempt" and your wages are normally more than $200 per week. | **2004**

1 Type or print your first name and middle initial: **Kyle B.**   Last name: **Abrum**   **2** Your social security number: **123 45 6789**

Home address (number and street or rural route): **4052 Oak Avenue**

3 ☐ Single ☒ Married ☐ Married, but withhold at higher Single rate.
Note: *If married, but legally separated, or spouse is a nonresident alien, check the "Single" box.*

City or town, state, and ZIP code: **Lawton, OK 12345**

4 If your last name differs from that shown on your social security card, check here. You must call 1-800-772-1213 for a new card. ▶ ☐

5 Total number of allowances you are claiming (from line **H** above **or** from the applicable worksheet on page 2) ... **5** __4__
6 Additional amount, if any, you want withheld from each paycheck .... **6** $ ____
7 I claim exemption from withholding for 2004, and I certify that I meet **both** of the following conditions for exemption:
  - Last year I had a right to a refund of **all** Federal income tax withheld because I had **no** tax liability **and**
  - This year I expect a refund of **all** Federal income tax withheld because I expect to have **no** tax liability.
  If you meet both conditions, write "Exempt" here .... ▶ **7** ____

Under penalties of perjury, I certify that I am entitled to the number of withholding allowances claimed on this certificate, or I am entitled to claim exempt status.

Employee's signature (Form is not valid unless you sign it.) ▶ *Kyle B. Abrum*   Date ▶ **5/12/200--**

8 Employer's name and address (Employer: Complete lines 8 and 10 only if sending to the IRS.) | **9** Office code (optional) | **10** Employer identification number (EIN)

For Privacy Act and Paperwork Reduction Act Notice, see page 2.   Cat. No. 10220Q   Form **W-4** (2004)

Figure 10-2 | Weekly Payroll Register

NAME	MARITAL STATUS	WITHHOLDING ALLOWANCES	W = WEEKLY H = HOURLY	RATE	HOURS REG	HOURS O/T	GROSS EARN- INGS	DEDUCTIONS SOCIAL SECURITY	MEDI- CARE	FEDERAL INCOME TAX	GROUP MED. INS.	GROUP DENTAL INS.	OTHER	TOTAL DEDUC- TIONS	NET EARNINGS
Abrum, Kyle	M	4	H	11.00	40	6	539.00	33.42	7.82	14.65	39.00	12.00	42.00	148.89	390.11
Garcia, Fran	S	2	W	680.00	40	—	680.00	42.16	9.86	69.66	18.00	9.00	—	148.68	531.32
Parker, Marie	S	1	H	12.10	32	—	387.20	24.01	5.61	34.69	18.00	—	—	82.31	304.89
Thomas, Robert	M	3	H	9.40	40	4	432.40	26.81	6.27	9.95	39.00	12.00	13.10	107.13	325.27
Weber, James	S	1	H	16.80	40	—	672.00	41.66	9.74	79.45	18.00	9.00	—	157.85	514.15
Totals							2,710.60	168.06	39.30	208.40	132.00	42.00	55.10	644.86	2,065.74

✓ CONCEPT CHECK 10.1

After completion of the payroll register entries, one way to check on the accuracy of computations is to subtract the Total Deductions column from the Gross Pay total; the difference should equal the total of the Net Earnings column. From the payroll register shown in Figure 10-2, check the accuracy of the column totals:

Total of Gross Earnings column	$2,710.60
Less total of Deductions column	644.86
Total of Net Earnings column	$2,065.74

Computing Federal Income Tax Withholding Amounts

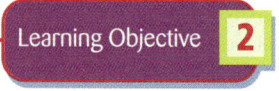

Compute federal income tax withholding amounts.

The federal income tax is a payroll tax that the employer must withhold from the employee's pay and turn over to the Internal Revenue Service (IRS). The amount of the deduction varies with the amount of earnings, the employee's marital status, and the number of withholding allowances claimed.

The *Employer's Tax Guide*, published annually by the Internal Revenue Service, gives employers two primary methods to figure how much income tax to withhold from their employees. These two methods are the **percentage method** and the **wage-bracket method**.

Figure 10-2 shows that Kyle Abrum's federal income tax withholding amount was $14.65, computed by the percentage method. With the percentage method, a deduction is granted for each withholding allowance claimed, based on a chart in the *Employer's Tax Guide*. The amount for each withholding allowance is provided in a table labeled Income Tax Withholding Percentage Method Table. Figure 10-3 illustrates a recent table. It shows that, for weekly pay, a deduction of $59.62 is allowed for each withholding allowance. (For monthly pay, a deduction of $258.33 is allowed for each withholding allowance.)

Figure 10-3 | Percentage Method Amount for One Withholding Allowance

Payroll Period	One Withholding Allowance
Weekly	$59.62
Biweekly	$119.23
Semimonthly	$129.17
Monthly	$258.33

After the total withholding allowance is subtracted from an employee's gross earnings, the amount to be withheld is computed by taking a percentage of the difference. The percentage to be used is given by the IRS in the Tables for Percentage Method of Withholding. Figure 10-4 illustrates a recent table for weekly, biweekly, semimonthly, and monthly payroll periods.

STEPS to Figure the Amount of Federal Income Tax Withholding, Using the Percentage Method

1. Determine the employee's gross earnings.
2. Multiply the appropriate (weekly/monthly) "one withholding allowance" amount (from the withholding table in Figure 10-3) by the number of allowances the employee claims.
3. Subtract that amount from the employee's gross earnings.
4. From the appropriate (weekly/monthly and single/married) percentage method table, subtract the "of excess over" figure to get the amount subject to the tax.
5. Multiply the amount from Step 4 by the appropriate percentage from the percentage method table.
6. If required, add the base tax amount (if any) shown next to the percentage from the percentage method table. (For example, see Table 1, WEEKLY Payroll Period, Married, the second line of the table: $27.50 plus 15% of excess over $429.)

10.3 Note again that the withholding allowance number includes the taxpayer plus the number of dependents the taxpayer claims.

Once the employer withholds federal income tax, the employer *must* promptly deposit the amount withheld with an IRS depository. Today, these deposits are usually made electronically.

Most large companies and payroll processors will use the withholding percentage method, with the formula programmed into a computer.

10.4 The current edition of the *Employer's Tax Guide*, Circular E, should be obtained from the local IRS office. A telephone request will usually suffice to have the current edition mailed quickly. It is also available online at www.IRS.gov.

Chapter 10 Payroll Records

Figure 10-4 | Tables for Percentage Method of Withholding

Tables for Percentage Method of Withholding
(For Wages Paid Through December 2004)

TABLE 1—WEEKLY Payroll Period

(a) SINGLE person (including head of household)—

If the amount of wages (after subtracting withholding allowances) is: The amount of income tax to withhold is:

Not over $51 . . . . . $0

Over—	But not over—		of excess over—
$51	—$187	. . 10%	—$51
$187	—$592	. . $13.60 plus 15%	—$187
$592	—$1,317	. . $74.35 plus 25%	—$592
$1,317	—$2,860	. . $255.60 plus 28%	—$1,317
$2,860	—$6,177	. . $687.64 plus 33%	—$2,860
$6,177	. . . . .	. . $1,782.25 plus 35%	—$6,177

(b) MARRIED person—

If the amount of wages (after subtracting withholding allowances) is: The amount of income tax to withhold is:

Not over $154 . . . . . $0

Over—	But not over—		of excess over—
$154	—$429	. . 10%	—$154
$429	—$1,245	. . $27.50 plus 15%	—$429
$1,245	—$2,270	. . $149.90 plus 25%	—$1,245
$2,270	—$3,568	. . $406.15 plus 28%	—$2,270
$3,568	—$6,271	. . $769.59 plus 33%	—$3,568
$6,271	. . . . .	. . $1,661.58 plus 35%	—$6,271

TABLE 2—BIWEEKLY Payroll Period

(a) SINGLE person (including head of household)—

If the amount of wages (after subtracting withholding allowances) is: The amount of income tax to withhold is:

Not over $102 . . . . . $0

Over—	But not over—		of excess over—
$102	—$373	. . 10%	—$102
$373	—$1,185	. . $27.10 plus 15%	—$373
$1,185	—$2,635	. . $148.90 plus 25%	—$1,185
$2,635	—$5,719	. . $511.40 plus 28%	—$2,635
$5,719	—$12,354	. . $1,374.92 plus 33%	—$5,719
$12,354	. . . . .	. . $3,564.47 plus 35%	—$12,354

(b) MARRIED person—

If the amount of wages (after subtracting withholding allowances) is: The amount of income tax to withhold is:

Not over $308 . . . . . $0

Over—	But not over—		of excess over—
$308	—$858	. . 10%	—$308
$858	—$2,490	. . $55.00 plus 15%	—$858
$2,490	—$4,540	. . $299.80 plus 25%	—$2,490
$4,540	—$7,137	. . $812.30 plus 28%	—$4,540
$7,137	—$12,542	. . $1,539.46 plus 33%	—$7,137
$12,542	. . . . .	. . $3,323.11 plus 35%	—$12,542

TABLE 3—SEMIMONTHLY Payroll Period

(a) SINGLE person (including head of household)—

If the amount of wages (after subtracting withholding allowances) is: The amount of income tax to withhold is:

Not over $110 . . . . . $0

Over—	But not over—		of excess over—
$110	—$404	. . 10%	—$110
$404	—$1,283	. . $29.40 plus 15%	—$404
$1,283	—$2,854	. . $161.25 plus 25%	—$1,283
$2,854	—$6,196	. . $554.00 plus 28%	—$2,854
$6,196	—$13,383	. . $1,489.76 plus 33%	—$6,196
$13,383	. . . . .	. . $3,861.47 plus 35%	—$13,383

(b) MARRIED person—

If the amount of wages (after subtracting withholding allowances) is: The amount of income tax to withhold is:

Not over $333 . . . . . $0

Over—	But not over—		of excess over—
$333	—$929	. . 10%	—$333
$929	—$2,698	. . $59.60 plus 15%	—$929
$2,698	—$4,919	. . $324.95 plus 25%	—$2,698
$4,919	—$7,731	. . $880.20 plus 28%	—$4,919
$7,731	—$13,588	. . $1,667.56 plus 33%	—$7,731
$13,588	. . . . .	. . $3,600.37 plus 35%	—$13,588

TABLE 4—MONTHLY Payroll Period

(a) SINGLE person (including head of household)—

If the amount of wages (after subtracting withholding allowances) is: The amount of income tax to withhold is:

Not over $221 . . . . . $0

Over—	But not over—		of excess over—
$221	—$808	. . 10%	—$221
$808	—$2,567	. . $58.70 plus 15%	—$808
$2,567	—$5,708	. . $322.55 plus 25%	—$2,567
$5,708	—$12,392	. . $1,107.80 plus 28%	—$5,708
$12,392	—$26,767	. . $2,979.32 plus 33%	—$12,392
$26,767	. . . . .	. . $7,723.07 plus 35%	—$26,767

(b) MARRIED person—

If the amount of wages (after subtracting withholding allowances) is: The amount of income tax to withhold is:

Not over $667 . . . . . $0

Over—	But not over—		of excess over—
$667	—$1,858	. . 10%	—$667
$1,858	—$5,396	. . $119.10 plus 15%	—$1,858
$5,396	—$9,838	. . $649.80 plus 25%	—$5,396
$9,838	—$15,463	. . $1,760.30 plus 28%	—$9,838
$15,463	—$27,175	. . $3,335.30 plus 33%	—$15,463
$27,175	. . . . .	. . $7,200.26 plus 35%	—$27,175

EXAMPLE A

Using the six steps given, we compute Kyle Abrum's withholding as follows:

STEP 1 $539.00   (gross earnings from payroll register)

STEP 2
$ 59.62   (one withholding allowance)
× 4   (number of withholding allowances)
$238.48   (total withholding allowance amount)

STEP 3
$539.00   (gross earnings)
238.48   (total withholding allowance amount)
$300.52   (amount subject to withholding)

STEP 4
$300.52   (amount subject to withholding)
− 154.00   (less "excess over" amount in Figure 10-4)
$146.52   (amount subject to percentage computation)

STEP 5
$146.52   (amount subject to percentage computation)
× 0.1   (10% computation)
$14.65   (amount of tax withheld)

STEP 6   The wage range $154–$429 doesn't have a base tax amount and therefore doesn't apply in the case of Kyle Abrum.

The second method of figuring the amount of tax to be withheld from an employee's pay, the wage-bracket method, involves use of a series of wage-bracket tables published in the IRS *Employer's Tax Guide*. Figures 10-5 and 10-6 illustrate the tables for single and married persons who are paid on a weekly basis.

Using the tables from Figure 10-6, we see that a married employee earning a weekly wage of between $530 and $540 and claiming four withholding allowances will have $14 withheld. Note that the amount of federal income tax withheld from Kyle Abrum's pay, using the wage-bracket method, is approximately the same as the amount withheld using the percentage method: $14 versus $14.65. Small differences will frequently result because the wage-bracket method uses tables based on $10 divisions and rounded amounts. Over a period of a year, these differences tend to be relatively insignificant and are accepted by the IRS.

Figure 10-5 Single Persons—Weekly Payroll Period

SINGLE Persons—WEEKLY Payroll Period
(For Wages Paid Through December 2004)

If the wages are—		And the number of withholding allowances claimed is—										
At least	But less than	0	1	2	3	4	5	6	7	8	9	10
		The amount of income tax to be withheld is—										
$0	$55	$0	$0	$0	$0	$0	$0	$0	$0	$0	$0	$0
55	60	1	0	0	0	0	0	0	0	0	0	0
60	65	1	0	0	0	0	0	0	0	0	0	0
65	70	2	0	0	0	0	0	0	0	0	0	0
70	75	2	0	0	0	0	0	0	0	0	0	0
75	80	3	0	0	0	0	0	0	0	0	0	0
80	85	3	0	0	0	0	0	0	0	0	0	0
85	90	4	0	0	0	0	0	0	0	0	0	0
90	95	4	0	0	0	0	0	0	0	0	0	0
95	100	5	0	0	0	0	0	0	0	0	0	0
100	105	5	0	0	0	0	0	0	0	0	0	0
200	210	16	9	3	0	0	0	0	0	0	0	0
210	220	18	10	4	0	0	0	0	0	0	0	0
220	230	19	11	5	0	0	0	0	0	0	0	0
230	240	21	12	6	1	0	0	0	0	0	0	0
240	250	22	13	7	2	0	0	0	0	0	0	0
250	260	24	15	8	3	0	0	0	0	0	0	0
260	270	25	16	9	4	0	0	0	0	0	0	0
270	280	27	18	10	5	0	0	0	0	0	0	0
280	290	28	19	11	6	0	0	0	0	0	0	0
290	300	30	21	12	7	1	0	0	0	0	0	0
300	310	31	22	13	8	2	0	0	0	0	0	0
310	320	33	24	15	9	3	0	0	0	0	0	0
320	330	34	25	16	10	4	0	0	0	0	0	0
330	340	36	27	18	11	5	0	0	0	0	0	0
340	350	37	28	19	12	6	0	0	0	0	0	0
350	360	39	30	21	13	7	1	0	0	0	0	0
360	370	40	31	22	14	8	2	0	0	0	0	0
370	380	42	33	24	15	9	3	0	0	0	0	0
380	390	43	34	25	17	10	4	0	0	0	0	0
390	400	45	36	27	18	11	5	0	0	0	0	0
400	410	46	37	28	20	12	6	0	0	0	0	0
410	420	48	39	30	21	13	7	1	0	0	0	0
420	430	49	40	31	23	14	8	2	0	0	0	0
430	440	51	42	33	24	15	9	3	0	0	0	0
440	450	52	43	34	26	17	10	4	0	0	0	0
450	460	54	45	36	27	18	11	5	0	0	0	0
460	470	55	46	37	29	20	12	6	0	0	0	0
470	480	57	48	39	30	21	13	7	1	0	0	0
480	490	58	49	40	32	23	14	8	2	0	0	0
490	500	60	51	42	33	24	15	9	3	0	0	0
500	510	61	52	43	35	26	17	10	4	0	0	0
510	520	63	54	45	36	27	18	11	5	0	0	0
520	530	64	55	46	38	29	20	12	6	0	0	0
530	540	66	57	48	39	30	21	13	7	1	0	0
540	550	67	58	49	41	32	23	14	8	2	0	0
550	560	69	60	51	42	33	24	15	9	3	0	0
560	570	70	61	52	44	35	26	17	10	4	0	0
570	580	72	63	54	45	36	27	18	11	5	0	0
580	590	73	64	55	47	38	29	20	12	6	0	0
590	600	75	66	57	48	39	30	21	13	7	1	0
600	610	78	67	58	50	41	32	23	14	8	2	0
610	620	80	69	60	51	42	33	24	15	9	3	0
620	630	83	70	61	53	44	35	26	17	10	4	0
630	640	85	72	63	54	45	36	27	18	11	5	0
640	650	88	73	64	56	47	38	29	20	12	6	0
650	660	90	75	66	57	48	39	30	21	13	7	1
660	670	93	78	67	59	50	41	32	23	14	8	2
670	680	95	80	69	60	51	42	33	24	15	9	3
680	690	98	83	70	62	53	44	35	26	17	10	4
690	700	100	85	72	63	54	45	36	27	18	11	5

Figure 10-6 | Married Persons—Weekly Payroll Period

MARRIED Persons—WEEKLY Payroll Period
(For Wages Paid Through December 2004)

If the wages are—		And the number of withholding allowances claimed is—										
At least	But less than	0	1	2	3	4	5	6	7	8	9	10
		The amount of income tax to be withheld is—										
$0	$125	$0	$0	$0	$0	$0	$0	$0	$0	$0	$0	$0
125	130	0	0	0	0	0	0	0	0	0	0	0
130	135	0	0	0	0	0	0	0	0	0	0	0
135	140	0	0	0	0	0	0	0	0	0	0	0
140	145	0	0	0	0	0	0	0	0	0	0	0
145	150	0	0	0	0	0	0	0	0	0	0	0
150	155	0	0	0	0	0	0	0	0	0	0	0
155	160	0	0	0	0	0	0	0	0	0	0	0
160	165	1	0	0	0	0	0	0	0	0	0	0
165	170	1	0	0	0	0	0	0	0	0	0	0
170	175	2	0	0	0	0	0	0	0	0	0	0
175	180	2	0	0	0	0	0	0	0	0	0	0
180	185	3	0	0	0	0	0	0	0	0	0	0
185	190	3	0	0	0	0	0	0	0	0	0	0
190	195	4	0	0	0	0	0	0	0	0	0	0
195	200	4	0	0	0	0	0	0	0	0	0	0
200	210	5	0	0	0	0	0	0	0	0	0	0
210	220	6	0	0	0	0	0	0	0	0	0	0
220	230	7	1	0	0	0	0	0	0	0	0	0
230	240	8	2	0	0	0	0	0	0	0	0	0
240	250	9	3	0	0	0	0	0	0	0	0	0
250	260	10	4	0	0	0	0	0	0	0	0	0
260	270	11	5	0	0	0	0	0	0	0	0	0
270	280	12	6	0	0	0	0	0	0	0	0	0
280	290	13	7	1	0	0	0	0	0	0	0	0
290	300	14	8	2	0	0	0	0	0	0	0	0
300	310	15	9	3	0	0	0	0	0	0	0	0
310	320	16	10	4	0	0	0	0	0	0	0	0
320	330	17	11	5	0	0	0	0	0	0	0	0
330	340	18	12	6	0	0	0	0	0	0	0	0
340	350	19	13	7	1	0	0	0	0	0	0	0
350	360	20	14	8	2	0	0	0	0	0	0	0
360	370	21	15	9	3	0	0	0	0	0	0	0
370	380	22	16	10	4	0	0	0	0	0	0	0
380	390	23	17	11	5	0	0	0	0	0	0	0
390	400	24	18	12	6	0	0	0	0	0	0	0
400	410	25	19	13	7	1	0	0	0	0	0	0
410	420	26	20	14	8	2	0	0	0	0	0	0
420	430	27	21	15	9	3	0	0	0	0	0	0
430	440	28	22	16	10	4	0	0	0	0	0	0
440	450	30	23	17	11	5	0	0	0	0	0	0
450	460	31	24	18	12	6	0	0	0	0	0	0
460	470	33	25	19	13	7	1	0	0	0	0	0
470	480	34	26	20	14	8	2	0	0	0	0	0
480	490	36	27	21	15	9	3	0	0	0	0	0
490	500	37	28	22	16	10	4	0	0	0	0	0
500	510	39	30	23	17	11	5	0	0	0	0	0
510	520	40	31	24	18	12	6	0	0	0	0	0
520	530	42	33	25	19	13	7	1	0	0	0	0
530	540	43	34	26	20	14	8	2	0	0	0	0
540	550	45	36	27	21	15	9	3	0	0	0	0
550	560	46	37	29	22	16	10	4	0	0	0	0
560	570	48	39	30	23	17	11	5	0	0	0	0
570	580	49	40	32	24	18	12	6	0	0	0	0
580	590	51	42	33	25	19	13	7	1	0	0	0
590	600	52	43	35	26	20	14	8	2	0	0	0
600	610	54	45	36	27	21	15	9	3	0	0	0
610	620	55	46	38	29	22	16	10	4	0	0	0
620	630	57	48	39	30	23	17	11	5	0	0	0
630	640	58	49	41	32	24	18	12	6	0	0	0
640	650	60	51	42	33	25	19	13	7	1	0	0
650	660	61	52	44	35	26	20	14	8	2	0	0
660	670	63	54	45	36	27	21	15	9	3	0	0
670	680	64	55	47	38	29	22	16	10	4	0	0
680	690	66	57	48	39	30	23	17	11	5	0	0
690	700	67	58	50	41	32	24	18	12	6	0	0
700	710	69	60	51	42	33	25	19	13	7	1	0
710	720	70	61	53	44	35	26	20	14	8	2	0
720	730	72	63	54	45	36	27	21	15	9	3	0
730	740	73	64	56	47	38	29	22	16	10	4	0

CONCEPT CHECK 10.2

Using the percentage method steps given, verify the federal income tax withholding for Fran Garcia as recorded in the payroll register.

STEP 1 $680.00 (gross earnings from payroll register)

STEP 2 $ 59.62 (one withholding allowance)
  × 2 (number of withholding allowances)
  $119.24 (total withholding allowance amount)

STEP 3 $680.00 (gross earnings)
  119.24 (total withholding allowance amount)
  $560.76 (amount subject to withholding)

STEP 4 $560.76 (amount subject to withholding)
  − 187.00 (less "excess over" amount in Figure 10-4, table 1(a))
  $373.76 (amount subject to percentage computation)

STEP 5 $373.76 (amount subject to percentage computation
  × 0.15 (15% computation)
  $56.06 (amount of tax withheld on percentage computation)

STEP 6 $56.06 (amount of tax withheld on percentage computation)
  13.60 (base tax amount)
  $69.66 (total amount of tax withheld)

Use the wage-bracket method to find the federal income tax withholding for Fran Garcia. Then compute the difference between the percentage method and the wage-bracket method.

Percentage method (Step 6)	$69.66
Wage-bracket method (Figure 10-5 because she is single)	70.00
Difference	$ 0.34

Computing Social Security, Medicare, and Other Withholdings

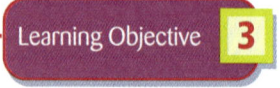

Learning Objective 3

Compute Social Security, Medicare, and other withholdings.

The **Federal Insurance Contributions Act** (**FICA**) provides for a federal system of old-age, survivors, disability, and hospital insurance. The old-age, survivors, and disability insurance part of FICA is financed by the *Social Security tax*. The hospital insurance part of FICA is financed by the *Medicare tax*. These taxes are reported separately and are levied on both the employer and the employee. These taxes have different rates, but only the Social Security tax has a wage base, which is the *maximum* wage that is subject to the tax for the year.

The Social Security tax rate of 6.2% is levied on both the employer and the employee. For 2004, the wage base was $87,900.

The Medicare tax rate of 1.45% is levied on both the employer and the employee. There is no wage-base limit for Medicare; all covered wages are subject to the Medicare tax.

Although both rates are subject to change by legislation, they were current when we compiled the payroll register illustrated in this chapter. All amounts are rounded to the nearest cent. The amounts for Kyle Abrum were $33.42 for Social Security and $7.82 for Medicare.

10.5 The employer is required to deposit these FICA employee deductions and employer contributions in a local depository bank monthly or more often, depending on the amount, and to file reports and complete full payment quarterly.

Depository requirements for FICA taxes are similar to the requirements for federal income tax withholdings.

EXAMPLE B

Social Security deduction:

$539.00	(gross earnings)
× 0.062	(Social Security rate)
$ 33.42	(Social Security amount)

EXAMPLE C

Medicare deduction:

$539.00	(gross earnings)
× .0145	(Medicare rate)
$ 7.82	(Medicare amount)

Many employers today provide some form of group medical insurance for their employees. Frequently, the employee is asked to pay a portion of the premium charged for such insurance, based on the number of dependents the employee has named to be insured. For the payroll register shown in Figure 10-2, we assumed the weekly rates for medical and dental plans shown in Figure 10-7.

Figure 10-7 | Weekly Medical and Dental Plan Rates

	Weekly Medical Plan Premium Paid by Employee	Weekly Dental Plan Premium Paid by Employee
Employee only	$18.00	$9.00
Employee plus one dependent	$22.00	$10.00
Employee plus 2 or more dependents	$39.00	$12.00

The payroll register presented in Figure 10-2 showed that Kyle Abrum subscribed to both the medical and the dental programs. Because of his three dependents, the amounts of his deductions were $39 and $12, respectively.

Frequently, employees will arrange to have special payroll deductions made by the employer to pay union dues, put money into special retirement or savings plans, or make contributions to charitable organizations.

In addition, 42 of the 50 states have some form of state income tax, which normally requires withholding in the same manner as the federal income tax. In such states, state income tax withholding columns are added to the payroll register and withholdings are made according to wage-bracket or percentage charts established by the state, in the same manner as federal income tax withholdings.

The payroll register illustrated in Figure 10-2 reflects a $42 weekly deduction that Kyle Abrum had requested be made for payment of his union dues (other).

CONCEPT CHECK 10.3

Using the format in examples B and C, compute Social Security and Medicare amounts for Fran Garcia, based on her gross weekly earnings of $680.

Social Security deduction:		Medicare deduction:	
$ 680	(gross earnings)	$ 680	(gross earnings)
× 0.062	(Social Security rate)	× 0.0145	(Medicare rate)
$42.16	(Social Security amount)	$9.86	(Medicare amount)

Completing an Employee's Earnings Record

Learning Objective 4

Complete an employee's earnings record.

An employer must submit quarterly and annual reports to the federal government and appropriate state government and pay the amount of taxes withheld from employees' earnings for the period. To obtain the necessary information, most employers keep an **employee's earnings record** for each employee. The employee's earnings record summarizes by quarter the employee's gross earnings, deductions, and net pay.

EXAMPLE D

10.6 When a company processes its payroll by computer, the employee's earnings record is one of the key outputs. It is updated after each payroll period.

Figure 10-8 | Employees Earnings Record

Name: Kyle Abrum
Address: 4052 Oak Ave.
Social Security No.: 234-12-8765
No. of Allowances: 4
Marital Status: Married

| Period Ending | Total Wages | Cumulative Wages | Deductions | | | | | Net Pay |
			Social Security	Medicare	Federal Inc. Tax	Other Deductions	Total	
1/4	$ 550.00	$ 550.00	$ 34.10	$ 7.98	$ 15.75	$ 93.00	$ 150.83	$ 399.17
1/11	550.00	1,100.00	34.10	7.98	15.75	93.00	150.83	399.17
3/29	539.00	7,250.00	33.42	7.82	14.65	93.00	148.89	390.11
Quarter Totals	$7,250.00		$431.20	$ 99.87	$195.79	$908.70	$1,635.56	$5,614.44

The employee's earnings record presented in Figure 10-8 shows that Kyle Abrum is married, claims four allowances, and for the first quarter of the year earned total wages of $7,250. His net pay was $5,614.44 after first-quarter withholdings as follows:

Federal income tax withholding	$ 195.79
Social Security withholding	431.20
Medicare withholding	99.87
Other deductions	908.70
Total deductions	$1,635.56

Part 3 Accounting Applications

CONCEPT CHECK 10.4

Assuming that Fran Garcia's weekly earnings and deductions have remained constant for each of the 13 weeks in the first quarter of the year, compute the following totals, which would appear on her employee's earnings record for the first quarter:

Total wages	$8,840.00	($680.00 × 13)
Federal income tax withholding	905.58	($69.66 × 13)
Social Security withholding	548.08	($42.16 × 13)
Medicare withholding	128.18	($9.86 × 13)
Group medical insurance deductions	234.00	($18.00 × 13)
Group dental insurance deductions	117.00	($9.00 × 13)
Total deductions	$1,932.84	
Net pay	$6,907.16	

Computing an Employer's Quarterly Federal Tax Return

Every employer who withholds federal income tax and FICA taxes (Social Security and Medicare) must file a quarterly return, Form 941—**Employer's Quarterly Federal Tax Return**. Figure 10-9 shows the data that the employer must include on Form 941 (the completed form is slightly abbreviated here). The return must be filed with the IRS within one month after the end of the quarter.

The employer obtains Social Security and Medicare amounts by multiplying the taxable wages paid by 12.4% and 2.9%, respectively. These amounts represent the employees' deductions and matching amounts required to be paid by the employer.

Learning Objective 5

Compute an employer's quarterly federal tax return.

10.7 Form 941 is prepared and submitted quarterly by *all* employers. Copies of the full form are available from the IRS.

EXAMPLE E

For the first quarter of 2004, Yeager Manufacturing paid total wages of $2,132,684.27. The company withheld $372,486.20 for federal income tax. All wages paid were subject to Social Security and Medicare taxes. If during the quarter Yeager had deposited $680,000 toward its taxes due, how much would it be required to send in with its first-quarter Form 941?

Gross wages $2,132,684.27 × 12.4% (Social Security)	$264,452.85
Gross wages $2,132,684.27 × 2.9% (Medicare)	61,847.84
Subtotal	326,300.69
Income taxes withheld	372,486.20
Total	698,786.89
Less deposit	680,000.00
Balance due	$ 18,786.89

Figure 10-9 — Form 941 Employer's Quarterly Federal Tax Return (extract)

Line	Description			Amount
1	Number of employees in the pay period that includes March 12th		1	5
2	Total wages and tips, plus other compensation		2	60,138.12
3	Total income tax withheld from wages, tips, and sick pay		3	4,997.45
4	Adjustment of withheld income tax for preceding quarters of calendar year		4	0.00
5	Adjusted total of income tax withheld (line 3 as adjusted by line 4—see instructions)		5	4,997.45
6	Taxable social security wages	6a 60,138.12 × 12.4% (.124) =	6b	7,457.13
	Taxable social security tips	6c ___ × 12.4% (.124) =	6d	0.00
7	Taxable Medicare wages and tips	7a 60,138.12 × 2.9% (.029) =	7b	1,744.01
8	Total social security and Medicare taxes (add lines 6b, 6d, and 7b). Check here if wages are not subject to social security and/or Medicare tax ▶ ☐		8	9,201.14
9	Adjustment of social security and Medicare taxes (see instructions for required explanation) Sick Pay $ ___ ± Fractions of Cents $ ___ ± Other $ ___ =		9	0.00
10	Adjusted total of social security and Medicare taxes (line 8 as adjusted by line 9—see instructions)		10	9,201.14
11	**Total taxes** (add lines 5 and 10)		11	14,198.59
12	Advance earned income credit (EIC) payments made to employees		12	0.00
13	Net taxes (subtract line 12 from line 11). **If $2,500 or more, this must equal line 17, column (d) below (or line D of Schedule B (Form 941))**		13	14,198.59
14	Total deposits for quarter, including overpayment applied from a prior quarter		14	14,107.58
15	**Balance due** (subtract line 14 from line 13). See instructions		15	91.01
16	**Overpayment.** If line 14 is more than line 13, enter excess here ▶ $ ___ and check if to be: ☐ Applied to next return **or** ☐ Refunded.			

- **All filers:** If line 13 is less than $2,500, you need not complete line 17 or Schedule B (Form 941).
- **Semiweekly schedule depositors:** Complete Schedule B (Form 941) and check here ▶ ☐
- **Monthly schedule depositors:** Complete line 17, columns (a) through (d), and check here ▶ ☐

17 Monthly Summary of Federal Tax Liability. Do not complete if you were a semiweekly schedule depositor.

(a) First month liability	(b) Second month liability	(c) Third month liability	(d) Total liability for quarter

Sign Here — Under penalties of perjury, I declare that I have examined this return, including accompanying schedules and statements, and to the best of my knowledge and belief, it is true, correct, and complete.

Signature ▶   Print Your Name and Title ▶   Date ▶

For Privacy Act and Paperwork Reduction Act Notice, see back of Payment Voucher.   Cat. No. 17001Z   Form **941** (Rev. 1-2001)

✓ CONCEPT CHECK 10.5

As displayed in Figure 10-9, the total taxes due the IRS consist of the $4,997.45 in federal income taxes withheld from employees, plus $7,457.13 and $1,744.01 for Social Security and Medicare taxes, respectively, half of which is withheld from employees and half of which is paid by the employer. Although the employer files Form 941 quarterly, the amount of taxes due is usually deposited in a qualified depository (bank) monthly or more often, and it is only the difference between the monthly deposits and the total taxes due that is sent with the Form 941 report.

Computing an Employer's Federal and State Unemployment Tax Liability

In the preceding section, you learned that the employer must match the employee's contributions to Social Security and Medicare taxes. In addition, employers must pay two payroll taxes for federal and state unemployment programs.

The **Federal Unemployment Tax Act (FUTA)** requires the employer to pay a 6.2% tax on the first $7,000 paid to each employee to fund the federal unemployment compensation program for those who have lost their jobs. Most states have also passed a **State Unemployment Tax Act (SUTA)**, requiring the employer to pay 5.4% tax on the first $7,000 paid to each employee to fund state programs for the unemployed. This 5.4% state tax is *deductible* from the federal tax payment. Thus, in most cases, employers pay the federal government just 0.8% FUTA tax: 6.2% FUTA − 5.4% SUTA = 0.8% requirement.

Learning Objective 6

Compute an employer's federal and state unemployment tax liability.

10.8 Requirements for FUTA and SUTA taxes vary not only from state to state but often from one year to the next in the same state.

EXAMPLE F

During the first quarter, Johnson and Johnson paid wages of $976,550.80. Of this amount, $172,400.60 was paid to employees who had been paid $7,000 earlier in the quarter. What was the employer's liability for FUTA and SUTA taxes, assuming that the state rate was 5.4%?

$976,550.80 − $172,400.60 = $804,150.20 subject to FUTA and SUTA taxes
$804,150.20 × 0.008 = $6,433.20 FUTA tax payment
$804,150.20 × 0.054 = $43,424.11 SUTA tax payment
$6,433.20 + $43,424.11 = $49,857.31

CONCEPT CHECK 10.6

Warner-Lambert Company employed Rojas Perez for 13 weeks during the period January 1 through March 31, 2004. His salary was $1,350 per week. At the end of the quarter, how much in FUTA and SUTA taxes did the company have to pay to the federal and state governments based on Rojas's income?

$1,350 per week × 13 weeks = $17,550 total wage
$7,000 maximum × 0.008 = $56 FUTA tax
$7,000 maximum × 0.054 = $378 SUTA tax
$378 + $56 = $434 total federal and state unemployment taxes

COMPLETE ASSIGNMENTS 10.1 AND 10.2.

Chapter Terms for Review

employee's earnings record

Employer's Quarterly Federal Tax Return

Federal Insurance Contributions Act (FICA)

Federal Unemployment Tax Act (FUTA)

Form W-4

payroll register

percentage method

State Unemployment Tax Act (SUTA)

wage-bracket method

withholding allowance

Try Microsoft® Excel

Try working the following problems using the Microsoft Excel templates found on your Student CD. Solutions for the problems are also shown on the CD.

1. Brighton Company pays its employees at the regular hourly rate for all hours worked up to 40 hours per week. Hours in excess of 40 are paid at $1\frac{1}{2}$ times the regular rate. Set up the following spreadsheet in Excel and add formulas to calculate **Overtime Hours**, **Regular Pay**, **Overtime Pay**, and **Total Gross Pay** for each employee in the shaded cells.

Hint: Use IF function to determine overtime hours.

Employees	Total Hours Worked	Regular Hourly Rate	Overtime Hours	Regular Pay	Overtime Pay	Total Gross Pay
Baker, Jason	42	$ 12.80				
Castro, Jill	38	15.70				
Dobson, Jack	40	12.00				
Ellis, Jennifer	45	14.50				

2. Set up the following worksheet and add formulas in shaded cells to calculate the **Social Security**, **Medicare**, **Total Deductions**, and **Net Pay** for each employee. Assume all wages are taxable and use the following rates: Social Security = 6.2%, Medicare = 1.45%

Employees	Wages	Social Security	Medicare	Income Tax	Total Deductions	Net Pay
Carter, Janes	$460.35			$45.80		
Edison, Alice	289.50			25.00		
Garcia, Joseph	375.00			36.90		
Kilmer, Martha	450.70			52.00		

Summary of chapter learning objectives:

Learning Objective	Example
10.1 Prepare a payroll register	Based on the data presented, complete the following payroll register. Fill out the total wages section and then compute the federal income tax, Social Security, Medicare, and other withholdings. Total all columns and check. Use the percentage method for federal income tax.
10.2 Compute federal income tax withholding amounts	1. G. Lee is paid $14.20 per hour. He works 40 regular hours and 6 overtime hours during the week ending January 7. He is single and claims one withholding allowance. He takes a weekly medical deduction of $7. 2. E. Berg is paid $13 per hour. He worked 40 regular hours and 8 overtime hours during the week of January 7. He is married and claims four withholding allowances. He takes a weekly medical deduction of $15.

10.3

Compute Social Security, Medicare, and other withholdings

Name	Marital Status	W/H Allow	Total Hours	Regular Earnings		Overtime Earnings			Total Wages	Deductions					Net Pay
				Rate per Hour	Amt	Hours Worked	Rate per Hour	Amt		Social Security	Medi-care	Fed. Inc. Tax	Med. Insurance	Total	
Lee, G.															
Berg, E.															

10.4 Complete an employee's earnings record	3. Complete the earnings record for D. Chan. Use 6.2% for Social Security and 1.45% for Medicare taxes. Use the percentage method for federal income tax withholding, on the monthly wages.

Name **D. Chan** Social Security No. **125-11-3296**

Address **7821 Oak Ave.** No. of Allowances **1** Marital Status **Married**

Period Ending	Total Wages	Cumulative Total	Deductions					Net Pay
			Social Security	Medicare	Federal Inc. Tax	Other Deductions	Total	
1/31	$3,100	$3,100				$18.00		
2/28	3,000	6,100				18.00		
3/31	3,450	$9,550				18.00		
Quarter Total	$9,550					$54.00		

Summary of chapter learning objectives:

Learning Objective	Example
10.5 Compute an employer's quarterly federal tax return	4. The Frazer Company had a total payroll of $279,440 for the first quarter of the year. It withheld $29,700 for federal income tax. It made monthly tax deposits of $24,100. Frazer is now filing its quarterly Form 941. Complete the following to determine the amount of the check that Frazer must send to the IRS for undeposited taxes due. a. Social Security tax due for the quarter _____ b. Medicare tax due for the quarter _____ c. Total taxes due for the quarter _____ d. Total deposits for the quarter _____ e. Undeposited taxes due IRS _____
10.6 Compute an employer's federal and state unemployment tax liability	5. Miller Outfitters employed R. Rehnquist for the period from January 1 through March 31, 13 weeks, at a salary of $1,230 per week. At the end of the quarter, how much in FUTA and SUTA taxes are owed to the federal and state governments if the state had a 0.8% FUTA rate and a 5.4% SUTA rate? a. Total wages b. FUTA tax c. SUTA tax d. Total federal and state unemployment taxes paid

Answers: 1 and 2. Lee: Reg Earn $568; O/T Earnings $127.80; Total $695.80; Deductions: SS–$43.14, MC–$10.09; FIT–$85.40; MI–$7; Total Deductions $145.63; Net pay $550.17 Berg: Reg Earn $520; O/T Earnings 156.00; Total $676.00; Deductions: SS–$41.91; MC–$9.80; FIT–$28.78; MI–$15; Total Deductions $95.49; Net Pay $580.51
3. 1/31: $192.20; $44.95; $266.65; $521.80; $2,578.20 2/28: $186.00; $43.50; $251.65; $499.15; $2,500.85 3/31: $213.90; $50.03; $319.15; $601.08; $2,848.92 Totals: $592.10; $138.48; $837.45; $54.00; $1,622.03; $7,929.97
4. a. $34,650.56 b. 8,103.76 c. 72,454.32 d. 72,300.00 e. $154.32 5. a. $15,990.00 b. 56.00 c. 378.00 d. $434.00

SELF-CHECK

Review Problems for Chapter 10

1 Alex Muñoz is paid $15 per hour for the first 40 hours and $1\frac{1}{2}$ times his regular rate for all hours worked over 40 per week.

   a. Determine Alex's gross pay for the week if he works 45 hours.
   b. Calculate the amount to be deducted for Social Security and Medicare taxes for the week.
   c. Determine the amount to be withheld for federal income tax, using the percentage method, if Alex is single and claims one withholding allowance.
   d. What is Alex's net pay for the week, assuming that his only payroll deductions are for Social Security, Medicare, and federal income tax?

2 Determine the amount to be withheld for federal income tax for each of the following, using both the percentage and the wage-bracket methods.

   a. A married employee, claiming two allowances, has weekly gross pay of $650.
   b. A single employee, with one allowance, has weekly gross pay of $525.

3 Calculate the employer's payroll taxes for each of the first three months of the year for three employees who are paid as follows:

   Albertson, K. $3,000 per month
   Becket, W. $4,000 per month
   Jones, C. $2,100 per month

Include FUTA (0.8%), SUTA (5.4%), Social Security (6.2%), and Medicare (1.45%) taxes. Be sure to consider the maximum taxable for unemployment taxes ($7,000) per employee.

4 Determine the taxes to be reported on the quarterly 941 form for an employer who paid total gross wages of $62,000 and withheld $7,800 for federal income tax.

   Social Security   _____
   Medicare   _____
   Federal income tax _____
   Total   _____

5 Determine the amount to be withheld from the current period's gross pay of $6,500 for Social Security and Medicare for an employee whose cumulative wages were $83,200, not including pay for the current period. Use the rates and taxable maximum given in the chapter.

6 Employees of Xper Co. are paid at their regular rate for the first 40 hours, at $1\frac{1}{2}$ times their regular rate for hours worked between 40 and 48, and double their regular rate for all hours worked over 48, per week. Calculate each employee's gross pay for the week.

   John Kowalski, regular rate $12.16, worked 47 hours
   Martha Madison, regular rate $9.50, worked 50 hours
   Joy Weston, regular rate $10.80, worked 42 hours

Notes

Assignment 10.1: Payroll Problems

Name

Date   Score

Learning Objectives **1** **2**

A (52 points) Complete the payroll. (1 point for each correct answer)

1. In this company, employees are paid $1\frac{1}{2}$ times their regular rate for overtime hours between 40 and 48 and 2 times their regular rate for overtime hours over 48, per week.

Name	Total Hours	Regular Rate Per Hour	Regular Earnings		Time and a Half		Double Time		Total Earnings
			Hours	Amount	Hours	Amount	Hours	Amount	
Avila, Susan	49	9.00	40	$ 360.00	8	$108.00	1	$ 18.00	$ 486.00
Carter, Dale	40	8.00	40	320.00	—	—	—	—	320.00
Kula, Mary	50	10.00	40	400.00	8	120.00	2	40.00	560.00
Murphy, Tom	45	9.00	40	360.00	5	67.50	—	—	427.50
Norton, Alice	40	8.80	40	352.00	—	—	—	—	352.00
Payton, Alan	35	8.00	35	280.00	—	—	—	—	280.00
Perry, Lance	47	8.00	40	320.00	7	84.00	—	—	404.00
Polar, Barbara	41	9.00	40	360.00	1	13.50	—	—	373.50
Quinn, Carl	49	8.80	40	352.00	8	105.60	1	17.60	475.20
Reston, Sally	40	8.80	40	352.00	—	—	—	—	352.00
Sacco, Dom	50	9.50	40	380.00	8	114.00	2	38.00	532.00
Warren, Bill	44	10.00	40	400.00	4	60.00	—	—	460.00
TOTALS				$4,236.00		$672.60		$113.60	$5,022.20

Score for A (52)

B (28 points) Solve the following problems. (7 points for each correct answer)

2. Dale LaVine is employed at a monthly salary of $2,700. How much is deducted from his monthly salary for FICA taxes (Social Security and Medicare)? $206.55

Social Security = $2,700 × 0.062 = $167.40
Medicare = $2,700 × 0.0145 = $39.15
Total FICA = $167.40 + $39.15 = $206.55

3. Candace Cooper is employed by a company that pays her $3,600 a month. She is single and claims one withholding allowance. What is her net pay after Social Security, Medicare, and federal income tax withholding? Use the percentage method for federal income tax. $2,808.38

Social Security = $3,600 × 0.062 = $223.20
Medicare = $3,600 × 0.0145 = $52.20
Income tax withholding: From single, monthly percentage table:
   Gross earnings (monthly) $3,600.00   Withhold $322.55 plus 25% of excess over $2,567
   Less one allowance monthly   258.33 $774.67 × .25 = $193.67 $322.55 + $193.67 = $516.22
   Subject to withholding $3,341.67 $3,600 − $223.20 − $52.20 − $516.22 = $2,808.38 net pay

Assignment 10.1 Continued

4. On April 1, the company in problem 3 changed its pay plan from monthly to weekly and began paying Candace $830.77 per week. What is her net weekly pay after Social Security, Medicare, and income tax deductions? Use the percentage method. $648.07

   Social Security: $830.77 × 0.062 = $51.51
   Medicare: $830.77 × 0.0145 = $12.05
   Income tax withholding:

Gross earnings (weekly)	$830.77
Less one allowance weekly	$ 59.62
Subject to withholding	$771.15
From percentage table:	
Withhold $74.35 plus 25% of excess over	$592.00
	$179.15 × 0.25 = $ 44.79
	+ 74.35
	$119.14

   $830.77 − $51.51 − $12.05 − $119.14 = $648.07 net pay

5. William Diggs is married and claims four withholding allowances. His weekly wages are $725. Calculate his Social Security and Medicare deductions and, using the wage-bracket method, his federal income tax withholding. Find his weekly net pay. $633.54

   Social Security: $725 × 0.062 = $44.95
   Medicare: $725 × 0.0145 = $10.51
   Income tax withholding:
   wage-bracket method, married, 4 withholding allowances $36.00
   $725.00 − $44.95 − $10.51 − $36.00 = $633.54 net pay

   Score for B (28)

C (20 points) Compute and compare the federal income tax withholding amounts for each of the following individuals using the percentage method and the wage-bracket method. (Follow the steps in Section 10.2 for the percentage method.) (5 points for each correct difference)

6. Ralph Carson: weekly wages, $320; single; 1 withholding allowance
   - Percentage method: $24.61
   - Wage-bracket method: $25.00
   - Difference: $ 0.39

7. George Wilson: weekly wages, $445; married; 3 withholding allowances
   - Percentage method: $11.21
   - Wage-bracket method: $11.00
   - Difference: $ 0.21

8. Mary Suizo: weekly wages, $292; single; 2 withholding allowances
   - Percentage method: $12.18
   - Wage-bracket method: $12.00
   - Difference: $ 0.18

9. Josephine Creighton: weekly wages, $595; married; 1 withholding allowance
   - Percentage method: $43.46
   - Wage-bracket method: $43.00
   - Difference: $ 0.46

   Score for C (20)

Assignment 10.2: Payroll, Earnings Record, Payroll Tax Returns

Name

Date   Score

Learning Objectives 1 2 3 4 5 6

A (40 points) Solve the following problems. (1 point for each correct answer in the Total Wages column in 1; 2 points for each correct answer in the Net Pay column in 1 and 2)

1. Complete the following weekly payroll register. Workers receive overtime pay for any time worked in excess of 40 hours per week at $1\frac{1}{2}$ the rate of their regular rate per hour. There is a 6.2% deduction for Social Security and 1.45% for Medicare taxes. Use the wage-bracket method for federal income tax withholding. Be sure to use the correct table based on the marital status of each employee.

Name	Marital Status	W/H Allow.	Total Hours	Regular Earnings Rate Per Hour	Regular Earnings Amount	Hours Worked	Overtime Earnings Rate Per Hour	Overtime Earnings Amount	Total Wages	Social Security	Medi-care	Fed. Inc. Tax	Med Ins.	Total	Net Pay
Allen, J.	S	1	40	$12.40	$ 496.00				$ 496.00	$ 30.75	$ 7.19	$ 51.00	$ 15.00	$103.94	$ 392.06
Clark, C.	M	2	43	10.00	400.00	3	15.00	45.00	445.00	27.59	6.45	17.00	12.00	63.04	381.96
Frank, B.	S	0	32	13.50	432.00				432.00	26.78	6.26	51.00	12.00	96.04	335.96
Hanson, K.	M	3	40	15.00	600.00				600.00	37.20	8.70	27.00	18.00	90.90	509.10
Johnson, A.	M	2	48	9.20	368.00	8	13.80	110.40	478.40	29.66	6.94	20.00	18.00	74.60	403.80
Kelly, J.	M	4	44	14.80	592.00	4	22.20	88.80	680.80	42.21	9.87	30.00	18.00	100.08	580.72
Nelson, R.	S	1	40	9.60	384.00				384.00	23.81	5.57	34.00	12.00	75.38	308.62
Olson, B.	M	5	42	14.28	571.20	2	21.42	42.84	614.04	38.07	8.90	16.00	12.00	74.97	539.07
Valdez, M.	S	1	40	12.50	500.00				500.00	31.00	7.25	52.00	15.00	105.25	394.75
TOTALS					$4,343.20			$287.04	$4,630.24	$287.07	$67.13	$298.00	$132.00	$784.20	$3,846.04

2. The total monthly wages of four employees are listed below. Determine the amount of the deductions and the net pay due to each employee. Use 6.2% for Social Security and 1.45% for Medicare tax deductions, and use the percentage method for federal income tax withholding. Determine the deductions and totals.

Name	Marital Status	W/H Allow.	Total Wages	Social Security	Medicare	Federal Income Tax	Total	Net Pay
Ali, Kyber	S	1	$1,750.00	$108.50	$ 25.38	$161.25	$ 295.13	$1,454.87
Dawson, William	M	3	2,100.00	130.20	30.45	65.80	226.45	1,873.55
Garcia, Jessica	S	0	2,580.00	159.96	37.41	325.80	523.17	2,056.83
Lawson, Mary	M	2	2,425.00	150.35	35.16	126.65	312.16	2,112.84
TOTALS			$8,855.00	$549.01	$128.40	$679.50	$1,356.91	$7,498.09

Score for A (40)

Assignment 10.2 Continued

B (20 points) Solve the following problems. (1 point for each correct weekly answer in the Net Pay column and 2 points for the correct quarter total of that column in 3; 1 point for each correct answer in 4)

3. Complete the employee's earnings record for Michelle Lee. Use 6.2% for Social Security and 1.45% for Medicare taxes. Use the percentage method for federal income tax withholding.

Name: Michelle Lee Social Security No. 125-55-1254
Address: 645 Abby Ln No. of Allowances: 2 Marital Status: Married

Period Ending	Total Wages	Cumulative Wages	Deductions					Net Pay
			Social Security	Medicare	Federal Inc. Tax	United Fund	Total	
1/6	$ 450.60	$ 450.60	$ 27.94	$ 6.53	$ 17.74	$ 4.00	$ 56.21	$ 394.39
1/13	412.00	862.60	25.54	5.97	13.88	4.00	49.39	362.61
1/20	412.00	1,274.60	25.54	5.97	13.88	4.00	49.39	362.61
1/27	475.50	1,750.10	29.48	6.89	20.23	4.00	60.60	414.90
2/3	415.20	2,165.30	25.74	6.02	14.20	4.00	49.96	365.24
2/10	490.25	2,655.55	30.40	7.11	21.70	4.00	63.21	427.04
2/17	427.50	3,083.05	26.51	6.20	15.43	4.00	52.14	375.36
2/24	435.90	3,518.95	27.03	6.32	16.27	4.00	53.62	382.28
3/3	510.00	4,028.95	31.62	7.40	23.68	4.00	66.70	443.30
3/10	505.60	4,534.55	31.35	7.33	23.24	4.00	65.92	439.68
3/17	516.00	5,050.55	31.99	7.48	24.28	4.00	67.75	448.25
3/24	498.50	5,549.05	30.91	7.23	22.53	4.00	64.67	433.83
3/31	535.80	6,084.85	33.22	7.77	26.26	4.00	71.25	464.55
Quarter Totals	$6,084.85		$377.27	$88.22	$253.32	$52.00	$770.81	$5,314.04

4. The following is a summary of quarterly earnings of a company's employees. Determine the information requested for the employer's quarterly federal tax return.

Name	Total Wages	Taxes Withheld		
		Social Security	Medicare	Fed. Inc. Tax
Carter, M.	$ 6,084.85	$ 377.27	$ 88.22	$ 451.42
Davis, L.	5,368.00	332.82	77.84	437.50
Gordon, J.	4,266.35	264.51	61.86	398.65
McBride, C.	7,230.00	448.26	104.84	595.80
Taggert, L.	6,240.50	386.91	90.49	465.50
Walton, N.	5,285.92	327.73	76.65	566.00
TOTALS	$34,475.62	$2,137.50	$499.90	$2,914.87

a. Total earnings paid $34,475.62
b. Federal income tax withheld $2,914.87
c. Total Social Security tax paid $ 2,137.50
d. Total Medicare tax paid $ 499.90
e. Total taxes withheld $ 5,552.27

Score for B (20)

Assignment 10.2 Continued

C (40 points) Solve the following problems. (4 points for each correct answer in 5 and 6; 1 point for each correct answer in 7)

5. The quarterly earnings of the employees of the Alpha Company are listed in the following table. Determine the employee information needed for the employer's quarterly federal tax return (Form 941).

Name	Total Wages	Taxes Withheld		
		Social Security	Medicare	Fed. Inc. Tax
Caldwell, Janice	$ 3,420.00	$ 212.04	$ 49.59	$ 423.90
Dorman, J.A.	3,600.00	223.20	52.20	473.67
Eagie, T.W.	4,016.50	249.04	58.24	433.33
Fortune, Mark	3,774.90	234.02	54.74	410.05
Morris, Regina	3,605.40	223.53	52.28	399.83
Tracy, Joseph	4,111.60	254.92	59.62	360.17
TOTALS	$22,528.40	$1,396.75	$326.67	$2,500.95

  a. Total earnings paid $22,528.40
  b. Employee's contribution of Social Security tax $1,396.75
  c. Employee's contribution of Medicare tax $326.67
  d. Federal income tax withheld from wages $2,500.95
  e. Total taxes $5,947.79

6. The Primo Company had a total payroll of $148,600.34 for the first quarter of the current year. It withheld $28,531.27 from the employees for federal income tax during this quarter. The company made the following deposits in a qualified bank depository for the amount of the income and Social Security and Medicare taxes withheld from the employees and for the company's contribution to the FICA tax: $17,050 on February 6; $17,050 on March 4; and $17,050 on April 5. Primo Company's bookkeeper is now filling out Form 941 (quarterly return), which is due by the end of April. Complete the following to determine the amount of the check that the company must send to the IRS for the undeposited taxes due.

  a. Total Social Security and Medicare taxes to be paid for quarter $22,735.85
  b. Total Taxes $51,267.12
  c. Total deposits for quarter (sent to qualified bank depository) $51,150
  d. Undeposited taxes due IRS $117.12

7. Jordan Mills employed Ruth Liebowitz for the period January 1 through March 31 (13 weeks) at a salary of $1,500 per week. At the end of the first quarter of the year, how much in FUTA and SUTA taxes did the company owe to the federal and state governments if the state had an 0.8% FUTA rate and a 5.4% SUTA rate?

  a. Total wages and taxable wages $19,500; $7,000
  b. FUTA tax $56
  c. SUTA tax $378
  d. Total federal and state unemployment taxes paid $434

Score for C (40)

Notes

Taxes

11

Learning Objectives
By studying this chapter and completing all assignments you will be able to:

Learning Objective 1 — Compute sales taxes, using rate tables and percents.

Learning Objective 2 — Compute assessed valuations and property taxes based on assessed valuation.

Learning Objective 3 — Compute tax rates in percents and mills.

Learning Objective 4 — Compute property tax payments involving special assessments, prorations, and exemptions.

Learning Objective 5 — Make basic computations to determine taxable income for taxpayers who use the standard federal income tax Form 1040.

Learning Objective 6 — Make basic computations to determine the tax liability for taxpayers who use the standard federal income tax Form 1040.

Most retail businesses collect a sales tax from customers when a sale occurs. The tax money must be turned over to the government. People and companies owning property usually pay taxes on the property's value. In this chapter we explain calculations involving sales, property, and income taxes.

Computing Sales Taxes

Learning Objective 1

Compute sales taxes, using rate tables and percents.

A **sales tax** is a government **levy**, or charge, on retail sales of certain goods and services. Most states and many cities and other local government entities levy sales taxes. The state **tax rate**—the percent used to compute the amount of sales tax—currently ranges from 3% to 7%, and city and county rates range from 0.925% to 7%.

Retail sales taxes, which usually are a combination of state and local taxes, are calculated as a single percent of taxable sales. For example, a sale is subject to a state sales tax of 5% and a local sales tax of 1%. The combined rate of 6% is applied to all taxable sales in that locality.

11.1 Some goods and services that are frequently exempt from sales taxes are food, pharmacy prescriptions, and charges for installation and repair labor.

SALES TAX AS A PERCENT OF PRICE

Sales taxes generally are rounded to the nearest cent. For example, sales taxes of 4% and 5% on amounts of up to $1 are charged as shown in Figure 11-1.

Figure 11-1 | Sales Taxes

4% on Sales of	Tax Due	5% on Sales of	Tax Due
$0.01 to $0.12	none	$0.01 to $0.09	none
$0.13 to $0.37	$0.01	$0.10 to $0.29	$0.01
$0.38 to $0.62	$0.02	$0.30 to $0.49	$0.02
$0.63 to $0.87	$0.03	$0.50 to $0.69	$0.03
$0.88 to $1.00	$0.04	$0.70 to $0.89	$0.04
		$0.90 to $1.00	$0.05

STEPS to Compute Sales Tax and Total Sales Amount

1. Multiply the taxable sales amount by the tax rate.
2. Add the sales tax amount to the taxable sales amount to get the total sales amount.

EXAMPLE A

If taxable merchandise of $60.39 is sold in a state with a 5% sales tax, what are the amount of tax and the total amount to be paid?

Amount of tax: $60.39 × 0.05 = $3.019, which rounds to $3.02
Total amount to be paid: $60.39 + $3.02 = $63.41

Most retail stores have cash registers that recognize a code such as the Uniform Product Code (UPC) to determine taxable sales and to calculate the sales tax automatically. The sales receipt usually shows the total taxable sales as a subtotal, the sales tax, and the total sales plus tax. Usually, discounts on a sale are subtracted from the sale price before the tax is figured. Shipping and installation labor charges are generally not taxed.

EXAMPLE B

A customer living in a city with a 6% state sales tax and a 1.5% city sales tax purchased a refrigerator regularly priced at $850. He was given a 10% discount. Delivery charges were $45. What were the amount of tax and the total cost to the buyer?

Discount amount: $850 × 10% = $850 × 0.10 = $85
Price after discount: $850 − $85 = $765, or $850 × 0.90 = $765
Sales tax: $765 × (0.06 + 0.015) = $57.38
Cost to buyer: $765 + $57.38 tax + $45 delivery = $867.38

State laws regarding the items subject to sales tax vary. Most states do not tax groceries; however, most do tax meals served in restaurants. Certain nonfood items also sold in grocery stores (such as laundry detergent) are generally taxed. When nontaxable and taxable items are purchased together, the register usually computes the total price of items purchased and automatically adds the correct amount of tax for each taxable item. The taxable items are clearly marked on the register tape along with the total amount of tax charged.

EXAMPLE C

A customer living in a state in which the tax rate is 7% went to a grocery store and purchased a quart of milk for $1.15, a loaf of bread for $2.79, potatoes for $2.25, and two taxable items—laundry detergent for $8.49 and fabric softener for $5.30. What was her total charge at the checkout counter?

Taxable items: $8.49 + $5.30 = $13.79
Tax: $13.79 × 0.07 = $0.9653 = $0.97
Total: $1.15 + $2.79 + $2.25 + $8.49 + $5.30 + $0.97 = $20.95

SALES TAX AS AN AMOUNT PER UNIT

All of the states and the District of Columbia levy special taxes on gasoline and cigarettes, usually stated in cents per unit (gallon or pack). State taxes on gasoline vary widely, from $0.075 in Georgia to $0.285 in Wisconsin; in addition, the federal tax is currently $0.184 per gallon. State taxes on cigarettes currently range from $0.025 to $1.11 per pack; the federal tax is currently $0.71 per pack.

EXCISE TAX AS AN AMOUNT PER UNIT

An **excise tax** is a tax assessed on each unit. In some states both the excise tax and the general sales tax apply to items such as gasoline, cigarettes, and alcoholic beverages. In such instances, the excise tax may be part of the taxable sales price for general sales tax purposes. For example, in a certain locality gasoline costs $1.40 per gallon, plus state and federal excise taxes of $0.40 and is subject to a general sales tax of 6%. The total price per gallon is $1.91 ($1.40 + $0.40 excise tax + $0.11 general sales tax). The general sales tax is calculated as 6% of $1.80.

11.2 Taxes on purchases (sales and excise taxes) fall into two distinct categories: financial and social. Financial sales taxes are those designed to be borne equally by most of the adult population, such as taxes on food and general merchandise. Social sales taxes target segments of the population that use certain items, such as cigarettes and alcohol.

CONCEPT CHECK 11.1

In a state in which the combined state and city sales tax rate is 6%, a customer went to a convenience store and purchased the following items: bread, $1.95; ground meat, $6.79; cheese, $4.79; lightbulbs, $4.25; and motor oil, $1.79. Only the last two items are taxable. Rounding the tax to the nearest cent, compute the total cost of all items and tax.

Nontaxable items: $1.95 + $6.79 + $4.79 = $13.53
Taxable items: $4.25 + $1.79 = $6.04
Total tax: $6.04 × 0.06 tax rate = $0.36

$13.53	Nontaxable items
6.04	Taxable items
0.36	Tax
$19.93	Total

 # Computing Assessed Valuations and Property Taxes

Learning Objective 2

Compute assessed valuations and property taxes based on assessed valuation.

A **property tax** for a business is a tax on real estate or other property, such as machinery, owned by the business. Businesses usually pay property tax bills semiannually. Taxes are based on a value, known as the **assessed valuation**, determined by a representative of the local or state government.

Assessed valuation ordinarily is based on the current **market value** of the property (what the property could be sold for). In many states it is fixed by law at 100%, but it is a fraction of that value in other states. Thus a particular community may use 60% of property values as the basis for tax billing. In most instances, land and buildings are assessed separately.

11.3 The amount of property tax paid as a percentage of the value of real property varies greatly among cities and states. In states that do not have sales taxes or income taxes with special allocations to the cities, property taxes as a percentage of real property value are generally higher. In states where there are significant sales and income tax assessments, some of which are allocated to the cities, property taxes as a percentage of real property value are generally lower.

EXAMPLE D

The Kinsey family lives in a town in which assessed valuation is 60% of market value. The Bailey family lives in a town in which assessed valuation is 75% of market value. Each home has a market value of $260,000. What is the assessed valuation of each home?

Kinsey: $260,000 × 0.60 = $156,000
Bailey: $260,000 × 0.75 = $195,000

Assessed valuation often is increased by improvements to the property, such as the addition of an enclosed porch, a pool, or landscaping. Ordinary maintenance—a new coat of paint, for instance, or repairs to the roof—isn't justification for an increased assessment.

EXAMPLE E

The Lee family and the Kelly family live in a town in which assessed valuation is set by law at 80% of market value. They live in identical houses having a market value of $220,000. The Lee family added an enclosed deck costing $10,500 and a family room costing $23,000. The Kelly family made extensive repairs and repainted the house a new color at a total cost of $15,000. What was the assessed valuation on each home the following year?

Lee: $220,000 + $10,500 + $23,000 = $253,500 × 0.8 = $202,800

Kelly: $220,000 × 0.8 = $176,000 (repairs and painting are not considered improvements)

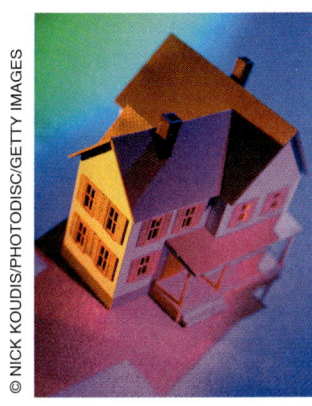

CONCEPT CHECK 11.2

a. The Coles family owns a home with a market value of $300,000 in a community that assesses property at 100% of market value. The Jensen family owns a home with a market value of $400,000 in a community that assesses property at 60% of market value. What is the difference between the actual assessments of the two homes?
Coles: $300,000 × 1 = $300,000
Jensen: $400,000 × 0.6 = $240,000
Difference = $60,000

b. The Bay family home has a present market value of $280,000 in a community that assesses property at 80% of market value. If they add a family room and an additional bathroom at a cost of $42,000, what will be the new assessed valuation?
Revised market value: $280,000 + $42,000 = $322,000
New assessed value: $322,000 × 0.80 = $257,600

Computing Tax Rates in Percents and Mills

PERCENTS

For a city, county, or special district, the tax rate is found by dividing the amount of money the government unit needs to raise by the total assessed valuation of the particular unit.

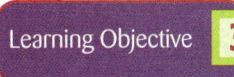

Compute tax rates in percents and mills.

EXAMPLE F

The town of Lakeside has a total assessed valuation of $570,000,000. The amount to be raised by taxation is $9,975,000. What is the tax rate?

The tax rate is

$9,975,000 ÷ $570,000,000 = 0.0175, or 1.75%.

This rate is usually written as 1.75% of value, or $1.75 on each $100 of value.

EXAMPLE G

If a property in Lakeside is assessed for $160,000, what is the tax?
The tax can be found by multiplying the amount by the rate:

$160,000 × 0.0175 = $2,800

MILLS

Tax rates sometimes are expressed in a unit of measure called mills. A **mill** is a tenth of a cent, or $0.001 (one thousandth of a dollar). To convert mills to dollars, divide by 1,000 (move the decimal three places to the left). To convert cents to mills, multiply by 10. Thus a tax rate can be converted from mills to cents or dollars or vice versa by using the following relationships:

mills ÷ 10 = cents	150 mills ÷ 10 = 15¢
mills ÷ 1,000 = dollars	150 mills ÷ 1,000 = $0.15
cents × 10 = mills	15¢ × 10 = 150 mills
dollars × 1,000 = mills	$0.15 × 1,000 = 150 mills

EXAMPLE H

Davis County assesses property at the rate of 182 mills per $100 of assessed value. How much tax would be due on property assessed at $620,000?

$620,000 ÷ 100 = $6,200 to assess millage
182 mills = $0.182
0.182 × $6,200 = $1,128.40 tax

✓ CONCEPT CHECK 11.3

a. A town has a total assessed valuation of $960,000,000. A total of $12,000,000 must be raised by taxation for the operating expenses of the town. What will be the tax rate?
$12,000,000 ÷ $960,000,000 = 0.0125, or 1.25%

b. Convert $0.57 into mills: 57¢ × 10 = 570 mills, or $0.57 × 1,000 = 570 mills

c. If property in a town is assessed at the rate of 140 mills per $100 of assessed value, how much tax will be due on property assessed at $475,000?
$475,000 ÷ 100 = $4,750 to assess millage
140 mills = $0.14
$4,750 × $0.14 = $665 tax due

Computing Special Assessments, Prorations, and Exemptions

Special assessments can be levied for improvements in a community, such as sewers, roads, or sidewalks. Sometimes the cost is spread over a period of years and added to the annual property tax bill of each property owner.

Learning Objective 4

Compute property tax payments involving special assessments, prorations, and exemptions.

EXAMPLE I

The residents of Sonora voted to widen their roads and add sidewalks, at a cost of $480 per residence, with the cost to be spread over a 12-year period. The Walker family had an annual tax bill of $630 before the improvements. If they pay their property taxes semiannually, what will be the amount of their next tax payment?

Annual cost for improvement: $480 ÷ 12 = $40
Annual property tax and improvement payment:
$630 + $40 = $670
Next semiannual tax payment: $670 ÷ 2 = $335

11.4 When real property is sold, the proration of property taxes between the seller and buyer is usually handled through escrow by the escrow agent.

Whenever property is sold, it is customary to *prorate*, or distribute, the taxes between seller and buyer as of the date of the settlement.

EXAMPLE J

A home having an annual tax bill of $720 was sold at the end of the seventh month of the taxable year. The seller had already paid the tax for the full year. How much tax was the seller reimbursed on proration of taxes at the time of the sale?

Months prepaid by seller: 12 − 7 = 5

Tax reimbursed by buyer: $720 × $\frac{5}{12}$ = $300

In almost all states, property used exclusively by nonprofit organizations, such as schools, churches, governments, and charities, is exempt from taxation. Some states also allow partial exemptions for veterans and the elderly.

EXAMPLE K

The town of Hillton assesses property at 75% of market value. The tax rate is 1.2%. A church has a total market value of $560,000. How much does the church save each year by being exempt from property taxes?

$560,000 × 0.75 = $420,000 $420,000 × 0.012 = $5,040 saved

Chapter 11 Taxes **207**

EXAMPLE L

A veteran living in Conton receives a partial exemption of 15% of regular property taxes. The veteran owns property valued at $380,000. If the property is assessed at 80% of value and the current rate is 1.3%, how much tax is due each six months?

Assessed value: $380,000 × 0.80 = $304,000
Regular taxes: $304,000 × 0.013 = $3,952
Taxes due after exemption: $3,952 × 0.85 (100% − 15%) = $3,359.20
Taxes due each six months: $3,359.20 ÷ 2 = $1,679.60

✓ CONCEPT CHECK 11.4

a. The city of Belton voted to build a new library at a cost of $540 per residence, to be spread over a period of 15 years. If the Douglas family presently has a yearly tax bill of $730, paid semiannually, what will be the amount of their next tax payment?
$540 per residence ÷ 15 years = $36 per year
$730 present yearly tax amount + $36 = $766 new yearly tax amount
$766 ÷ 2 = $383 new semiannual tax amount

b. If a home with an annual tax bill of $780 is sold at the end of the third month of the tax year, after taxes have already been paid, how much will the buyer reimburse the seller when taxes are prorated?
12 − 3 = 9 months prepaid by seller

$780 × $\frac{9}{12}$ = $585 reimbursed by buyer

c. A 70-year-old man lives in a state that grants senior citizens a 10% exemption from property taxes. If his home has a market value of $250,000 and the tax rate is 1.3%, how much will be his yearly taxes? The county in which he resides assesses property at 70% of market value.
$250,000 market value × 0.7 = $175,000 assessed valuation
$175,000 assessed valuation × 0.013 = $2,275 regular taxes
$2,275 regular taxes × 0.10 = $227.50 reduction
$2,275 regular taxes − $227.50 reduction = $2,047.50 revised taxes

COMPLETE ASSIGNMENTS 11.1 AND 11.2.

Personal income taxes provide 37% of all income of the federal government. Social Security and Medicare taxes, which you studied in Chapter 10, provide another 33%. Together, these three taxes make up 70% of all federal government income.

Outlays for Social Security, Medicare, and retirement programs constitute 37% of all government expenditures. Payment of interest on government debt represents 7% of all government expenditures.

Figure 11-2 shows the breakdown of federal government income and the allocation of federal government spending.

Determining Taxable Income, Using Standard Form 1040

Form 1040 is the basic form filed by the majority of taxpayers. There are two simplified variations of this form: Form 1040A and Form 1040EZ. The income tax calculation process is illustrated for Form 1040 in Figures 11-3 through 11-8. The label in Figure 11-3 contains spaces for names, address, and Social Security numbers, as well as boxes to check to designate $3 to finance presidential elections.

Learning Objectives 5

Make basic computations to determine taxable income for taxpayers who use the standard federal income tax form 1040.

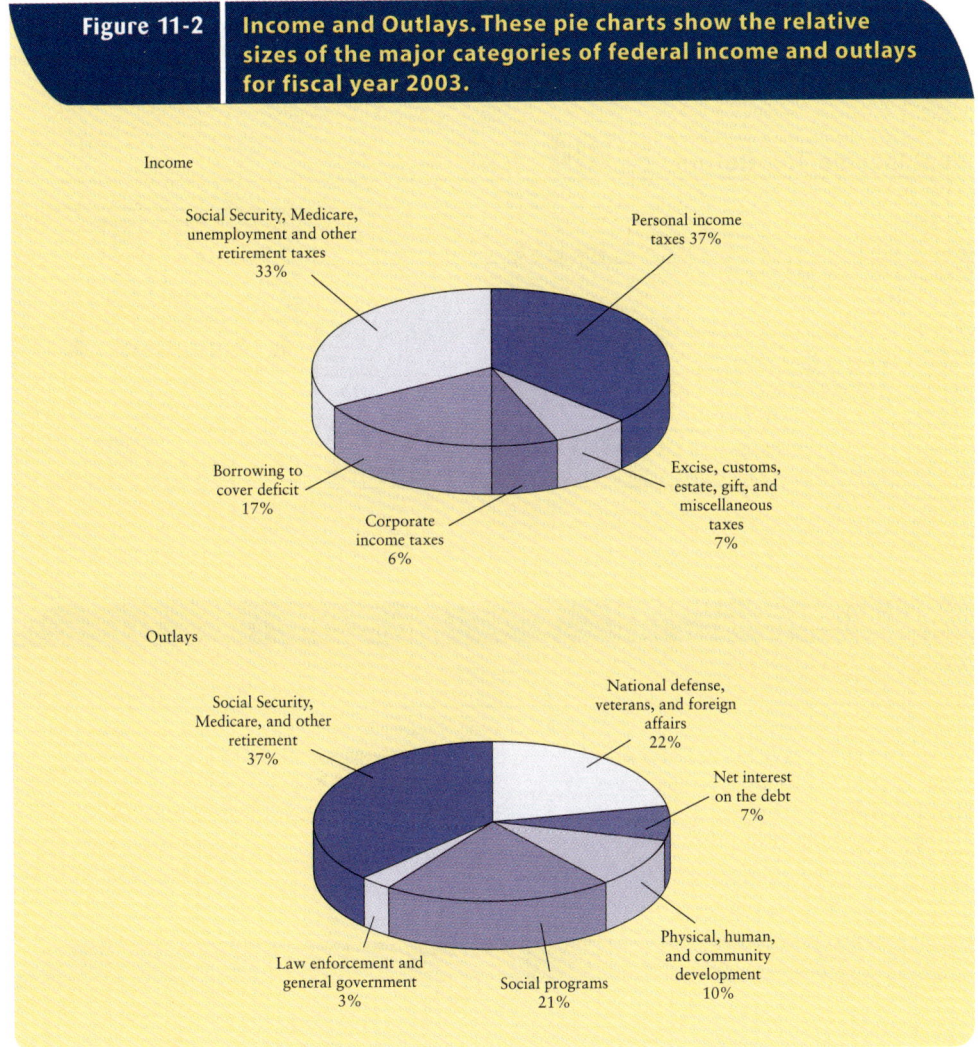

Figure 11-2 Income and Outlays. These pie charts show the relative sizes of the major categories of federal income and outlays for fiscal year 2003.

11.5 In the next few years, both tax rates and tax assessment bases are expected to change often and significantly. The basic concepts and calculation methods used here are based on 2000 laws, regulations, and rates. For an update, consult *Federal Income Taxation*, published by South-Western Publishing Company, and contact your local IRS office to request publications on the current year's tax rates and bases.

A taxpayer's current **filing status** is indicated in the second section of Form 1040, shown in Figure 11-4. Five choices are given. The one selected determines the tax rates the taxpayer uses, as well as many of the taxpayer's deductions.

Personal exemptions, shown in Figure 11-5, are reductions to taxable income for the primary taxpayer and a spouse. One **dependency exemption** is granted for each dependent. Exemptions are phased out for individuals with higher incomes. The amount deducted for each exemption is currently $3,100. This amount is usually adjusted for inflation each year.

11.6 Income from gifts, inheritance, and bequests does not escape taxation; such tax assessments fall under a different category and are paid by the giver instead of the recipient.

Taxable income, shown in Figure 11-6, includes wages, salaries, tips, dividends, interest, commissions, back pay, bonuses and awards, refunds of state and local taxes, alimony received, property received for services, severance pay, accrued leave payments, sick pay, unemployment compensation payments, capital gains, and any other income not specifically exempted by statute. Taxable income may include a portion of Social Security payments, IRA distributions, and pensions and annuities. It also includes income from businesses, professions, farming, partnerships, rents, royalties, estates, trusts, and other sources. It does not include income from gifts, inheritances, bequests, interest on tax-exempt state and local municipal bonds, life insurance proceeds at death, workers' compensation benefits, and certain income items for veterans.

Figure 11-3 **Form 1040 Label Section**

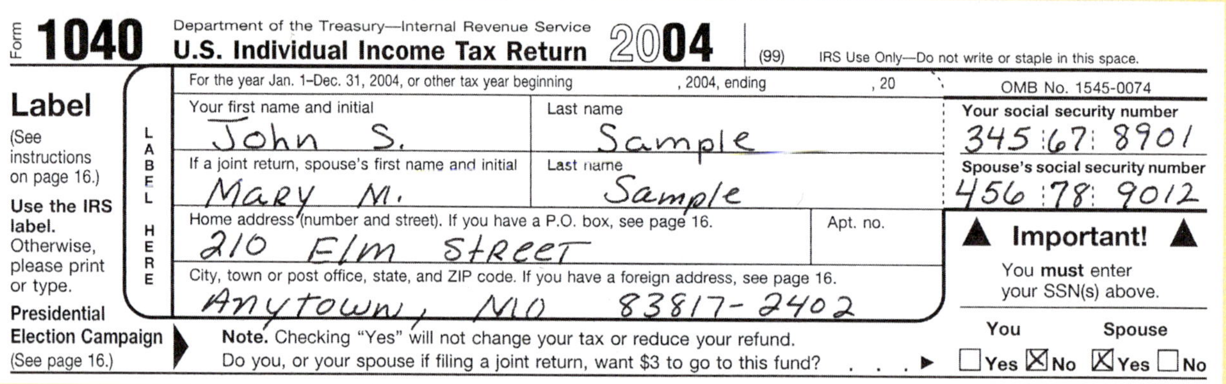
Form 1040 Department of the Treasury—Internal Revenue Service
U.S. Individual Income Tax Return 2004 (99) IRS Use Only—Do not write or staple in this space.
For the year Jan. 1–Dec. 31, 2004, or other tax year beginning , 2004, ending , 20 OMB No. 1545-0074

Label (See instructions on page 16.) Use the IRS label. Otherwise, please print or type.
LABEL HERE
Your first name and initial: John S. Last name: Sample
Your social security number: 345 67 8901
If a joint return, spouse's first name and initial: Mary M. Last name: Sample
Spouse's social security number: 456 78 9012
Home address (number and street). If you have a P.O. box, see page 16.: 210 Elm Street Apt. no.
City, town or post office, state, and ZIP code. If you have a foreign address, see page 16.: Anytown, MO 83817-2402
▲ Important! ▲ You **must** enter your SSN(s) above.
Presidential Election Campaign (See page 16.) **Note.** Checking "Yes" will not change your tax or reduce your refund. Do you, or your spouse if filing a joint return, want $3 to go to this fund? ▶
You ☐ Yes ☒ No Spouse ☒ Yes ☐ No

Figure 11-4 **Form 1040 Filing Status Section**

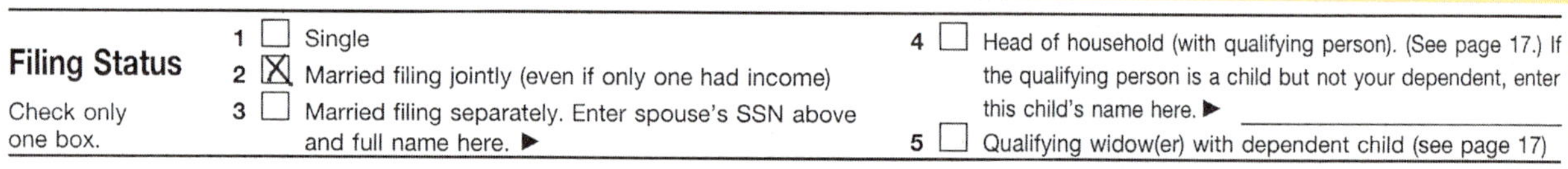
Filing Status
Check only one box.
1 ☐ Single
2 ☒ Married filing jointly (even if only one had income)
3 ☐ Married filing separately. Enter spouse's SSN above and full name here. ▶
4 ☐ Head of household (with qualifying person). (See page 17.) If the qualifying person is a child but not your dependent, enter this child's name here. ▶
5 ☐ Qualifying widow(er) with dependent child (see page 17)

Figure 11-5 **Form 1040 Exemptions Section**

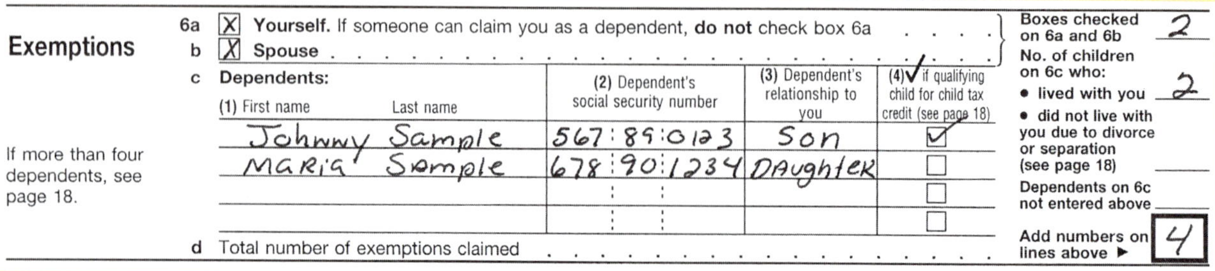
Exemptions
6a ☒ **Yourself.** If someone can claim you as a dependent, **do not** check box 6a
b ☒ **Spouse**
Boxes checked on 6a and 6b: 2
c **Dependents:**

(1) First name Last name	(2) Dependent's social security number	(3) Dependent's relationship to you	(4) ✓ if qualifying child for child tax credit (see page 18)
Johnny Sample	567 89 0123	Son	☑
Maria Sample	678 90 1234	Daughter	☐
			☐
			☐

If more than four dependents, see page 18.
No. of children on 6c who: • lived with you: 2
• did not live with you due to divorce or separation (see page 18)
Dependents on 6c not entered above
d Total number of exemptions claimed
Add numbers on lines above ▶ 4

Figure 11-6 | Form 1040 Income Section

Income	7	Wages, salaries, tips, etc. Attach Form(s) W-2	7	65,000 00
Attach Form(s) W-2 here. Also attach Forms W-2G and 1099-R if tax was withheld.	8a	**Taxable** interest. Attach Schedule B if required	8a	500 00
	b	**Tax-exempt** interest. **Do not** include on line 8a	8b	
	9a	Ordinary dividends. Attach Schedule B if required	9a	
	b	Qualified dividends (see page 20)	9b	
	10	Taxable refunds, credits, or offsets of state and local income taxes (see page 20)	10	
	11	Alimony received	11	
	12	Business income or (loss). Attach Schedule C or C-EZ	12	
	13	Capital gain or (loss). Attach Schedule D if required. If not required, check here ▶ ☐	13	
If you did not get a W-2, see page 19.	14	Other gains or (losses). Attach Form 4797	14	
	15a	IRA distributions . . 15a	b Taxable amount (see page 22) 15b	
	16a	Pensions and annuities 16a	b Taxable amount (see page 22) 16b	
Enclose, but do not attach, any payment. Also, please use Form 1040-V.	17	Rental real estate, royalties, partnerships, S corporations, trusts, etc. Attach Schedule E	17	
	18	Farm income or (loss). Attach Schedule F	18	
	19	Unemployment compensation	19	1,300 00
	20a	Social security benefits . 20a	b Taxable amount (see page 24) 20b	
	21	Other income. List type and amount (see page 24)	21	
	22	Add the amounts in the far right column for lines 7 through 21. This is your **total income** ▶	22	66,800 00

Figure 11-7 | Form 1040 Adjustments to Income Section/Adjusted Gross Income

Adjusted Gross Income	23	Educator expenses (see page 26)	23	
	24	Certain business expenses of reservists, performing artists, and fee-basis government officials. Attach Form 2106 or 2106-EZ	24	
	25	IRA deduction (see page 26)	25	3,000 00
	26	Student loan interest deduction (see page 28)	26	
	27	Tuition and fees deduction (see page 29)	27	
	28	Health savings account deduction. Attach Form 8889	28	
	29	Moving expenses. Attach Form 3903	29	
	30	One-half of self-employment tax. Attach Schedule SE	30	
	31	Self-employed health insurance deduction (see page 30)	31	
	32	Self-employed SEP, SIMPLE, and qualified plans	32	
	33	Penalty on early withdrawal of savings	33	
	34a	Alimony paid b Recipient's SSN ▶	34a	
	35	Add lines 23 through 34a	35	3,000 00
	36	Subtract line 35 from line 22. This is your **adjusted gross income** ▶	36	63,800 00

For Disclosure, Privacy Act, and Paperwork Reduction Act Notice, see page 75.   Cat. No. 11320B   Form **1040** (2004)

The Adjustments to Income section, shown in Figure 11-7, allows the taxpayer to list certain items that are allowed as reductions to the total income. These adjustments include payments by the taxpayer or spouse to an individual retirement account (IRA), student loan interest, payments into a medical savings account, moving expenses, one half of self-employment tax paid, and payments to a retirement plan for the self-employed, penalty on early withdrawal of savings, and alimony paid. **Adjusted gross income (AGI)** is a taxpayer's income after subtraction of adjustments to income from total income. (See lines 36 and 37 of Adjusted Gross Income in Figure 11-7.)

After the adjusted gross income figure is computed, *deductions*—either the standard deduction or itemized deductions—are subtracted in order to figure taxable income (see Figure 11-8). The standard deductions for most taxpayers are shown in Figure 11-9. There are higher standard deductions for individuals who are 65 or over and for individuals who are blind; these are shown in Figure 11-10.

11.7 The standard deduction is based on the assumption that a certain level of medical expenses, charitable gifts, interest payments, and taxes in various forms are part of everyday life, whether or not the individual keeps detailed records of every payment. The amount of the standard deduction is set by congressional action, based on its deliberations about family expenditures and needs.

Figure 11-8 Form 1040 Taxable Income and Income Tax Section

Form 1040 (2004) — Page 2

Tax and Credits

37	Amount from line 36 (adjusted gross income)	63,800.00
38a	Check if: ☐ You were born before January 2, 1940, ☐ Blind. ☐ Spouse was born before January 2, 1940, ☐ Blind. Total boxes checked ▶ 38a	
b	If your spouse itemizes on a separate return or you were a dual-status alien, see page 31 and check here ▶ 38b	
39	Itemized deductions (from Schedule A) or your standard deduction (see left margin)	9,700.00
40	Subtract line 39 from line 37	54,100.00
41	If line 37 is $107,025 or less, multiply $3,100 by the total number of exemptions claimed on line 6d. If line 37 is over $107,025, see the worksheet on page 33	12,400.00
42	Taxable income. Subtract line 41 from line 40. If line 41 is more than line 40, enter -0-	41,700.00
43	Tax (see page 33). Check if any tax is from: a ☐ Form(s) 8814 b ☐ Form 4972	5,540.00

Figure 11-9 Standard Deduction Chart for Most People

Standard Deduction for—

- People who checked any box on line 38a or 38b or who can be claimed as a dependent, see page 31.
- All others:

Single or Married filing separately, $4,850

Married filing jointly or Qualifying widow(er), $9,700

Head of household, $7,150

b If your spouse itemizes on a separate return or you were a dual-status alien, see page 31 and check here ▶ 38b

39 Itemized deductions (from Schedule A) or your standard deduction (see left margin) . . 39

Figure 11-10 Standard Deduction Chart for People Age 65 and Older or Blind (line 39)

Standard Deduction Chart for People Who Were Born Before January 2, 1940, or Were Blind—Line 39

Do not use this chart if someone can claim you, or your spouse if filing jointly, as a dependent. Instead, use the worksheet above.

Enter the number from the box on Form 1040, line 38a . . . . ☐

⚠ CAUTION Do not use the number of exemptions from line 6d.

IF your filing status is...	AND the number in the box above is...	THEN your standard deduction is...
Single	1	$6,050
	2	7,250
Married filing jointly or Qualifying widow(er)	1	$10,650
	2	11,600
	3	12,550
	4	13,500
Married filing separately	1	$5,800
	2	6,750
	3	7,700
	4	8,650
Head of household	1	$8,350
	2	9,550

Some taxpayers choose to itemize deductions rather than use the IRS-approved standard deduction. **Itemized deductions** are deductions allowed for specific payments made by the taxpayer during the tax year. These deductions include charitable contributions, certain interest payments, state and local income (or sales) and property taxes, a portion of medical and dental expenses, casualty and theft losses, tax preparation fees, and other annually identified deductions. Illustrations, examples, and problems in this book are based on the assumption that all state and local taxes and all donations to charity are deductible.

COMPUTING TAXABLE INCOME

Line 42 of Form 1040 shows "taxable income." Taxable income is the amount of income on which the income tax is based. Taxable income for most taxpayers is computed as follows (amounts from the preceding figures):

Total income (income from all sources) (line 22)	$66,800
Less adjustments to income (reductions of Total Income) (line 35)	3,000
Adjusted gross income (line 36)	63,800
Less deductions (from Figure 11-9 or 11-10)	9,700
Less exemptions (line 6d × $3,100, per line 41)	12,400
Taxable income (the amount on which taxes are computed) (line 42)	$41,700

✓ CONCEPT CHECK 11.5

Catherine, a 72-year-old blind widow, had an annual adjusted gross income of $29,000. She filed a return claiming a single exemption and standard deduction. What is her taxable income?

Adjusted gross income	$29,000
Standard deduction: single, over 65, blind	7,250
	21,750
Minus 1 exemption	3,100
Taxable income	$18,650

Determining Taxes Due, Using Standard Form 1040

Taxes are computed from taxable income (line 42). **Tax Rate Schedules** (Figure 11-11) show the tax rate for (1) single, (2) married filing joint return (even if only one had income), (3) married filing separate return, (4) head of household, and (5) qualifying widow or widower. The Tax Rate Schedules shown are used for all illustrations, examples, and problems in this book.

The remaining sections of Form 1040 permit listing of special credits, other taxes, and payments, to arrive at the final refund or amount owed and have spaces for signatures of the taxpayers and of paid preparers.

Learning Objective **6**

Make basic computations to determine the tax liability for taxpayers who use the standard federal income tax Form 1040.

11.8 Note for students that the IRS publishes extensive tax tables annually and that, in most cases for a taxable income of less than $100,000, the amount of tax to be paid is determined directly from these tables after the amount of taxable income is determined.

Figure 11-11 | Tax Rate Schedules

2003 Tax Rate Schedules—Line 16

Schedule X—Use if your **2003** filing status was **Single**

If Schedule J, line 15, is: Over—	But not over—	Enter on Schedule J, line 16		of the amount over—
$0	$7,000	....	10%	$0
7,000	28,400	$700.00 +	15%	7,000
28,400	68,800	3,910.00 +	25%	28,400
68,800	143,500	14,010.00 +	28%	68,800
143,500	311,950	34,926.00 +	33%	143,500
311,950	....	90,514.50 +	35%	311,950

Schedule Y-1—Use if your **2003** filing status was **Married filing jointly** or **Qualifying widow(er)**

If Schedule J, line 15, is: Over—	But not over—	Enter on Schedule J, line 16		of the amount over—
$0	$14,000	....	10%	$0
14,000	56,800	$1,400.00 +	15%	14,000
56,800	114,650	7,820.00 +	25%	56,800
114,650	174,700	22,282.50 +	28%	114,650
174,700	311,950	39,096.50 +	33%	174,700
311,950	....	84,389.00 +	35%	311,950

Schedule Y-2—Use if your **2003** filing status was **Married filing separately**

If Schedule J, line 15, is: Over—	But not over—	Enter on Schedule J, line 16		of the amount over—
$0	$7,000	....	10%	$0
7,000	28,400	$700.00 +	15%	7,000
28,400	57,325	3,910.00 +	25%	28,400
57,325	87,350	11,141.25 +	28%	57,325
87,350	155,975	19,548.25 +	33%	87,350
155,975	....	42,194.50 +	35%	155,975

Schedule Z—Use if your **2003** filing status was **Head of household**

If Schedule J, line 15, is: Over—	But not over—	Enter on Schedule J, line 16		of the amount over—
$0	$10,000	....	10%	$0
10,000	38,050	$1,000.00 +	15%	10,000
38,050	98,250	5,207.50 +	25%	38,050
98,250	159,100	20,257.50 +	28%	98,250
159,100	311,950	37,295.50 +	33%	159,100
311,950	....	87,736.00 +	35%	311,950

● EXAMPLE M

For the Form 1040 illustrated in the text, the tax is computed as follows:

Line 42—Taxable income	$41,700
From Schedule Y-1 (married):	
Tax on $14,300	$1,430
Plus 15% of amount over $14,300	
$41,700 − $14,300 = $27,400 × 0.15	$4,110
Total tax	$5,540

EXAMPLE N

Filing as head of household, Dave has an adjusted gross income of $110,000. He itemizes the following deductions: $700 to Salvation Army, $900 to his church, $8,200 interest on his mortgage, and $3,300 state taxes. He claims two exemptions. Compute his federal tax. Round to the nearest dollar.

Adjusted gross income	$110,000
Minus itemized deductions	13,100
	96,900
Minus 2 exemptions	6,200
Taxable income	$ 90,700
From Schedule Z:	
Tax on $38,900	$5,325
Plus 25% (0.25) of excess over $38,900	
$90,700 − $38,900 = $51,800 × 0.25	12,950
Total tax	$ 18,275

TAX CREDITS AND NET TAX

Credits allowed are subtracted from the tax to calculate the net tax. One of the most common credits is the **Child Tax Credit** (line 51). Taxpayers with dependent children under age 17 can receive a credit of $1,000 per qualifying child. The credit phases out at higher income levels.

Figure 11-12 shows that John and Mary Sample received a Child Tax Credit of $1,000. Look back at Figure 11-5 and note a check mark in the "qualifying child" box for Johnny Sample but not for Maria Sample. This distinction means that the son qualified for the credit because he was under age 17. The daughter qualifies as a dependent for exemption purposes, but no Child Tax Credit is allowed because she is age 17 or older.

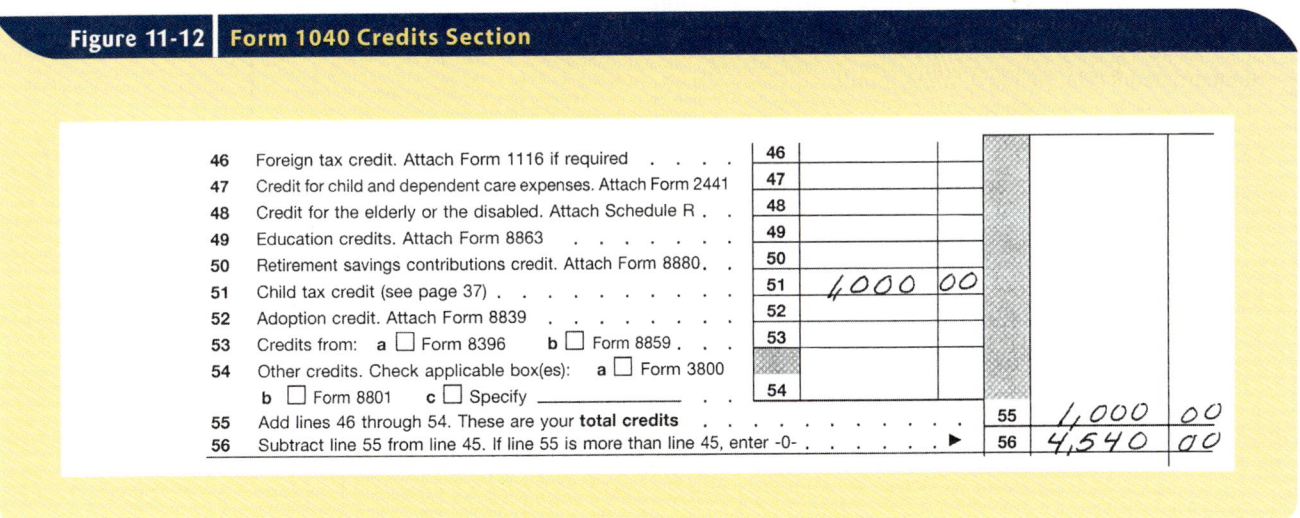

Figure 11-12 Form 1040 Credits Section

EXAMPLE O

Eric and Audrey Vaughn file a joint return. Their adjusted gross income is $48,900, and they take the standard deduction. They have three children, aged 12, 15, and 17, and claim five exemptions. Compute their net federal income tax after credits.

Adjusted gross income	$48,900
Standard deduction (joint)	9,700
	39,200
Minus 5 exemptions × $3,100	15,500
Taxable income	$23,700
From Schedule Y-1:	
Tax on $14,300	$ 1,430
Plus 15% on amount over $14,300	1,410
($23,700 − $14,300) = $9,400 × 0.15	
Total	2,840
Minus child tax credit ($1,000 × 2)	2,000
Net tax after credit	$ 840

✓ CONCEPT CHECK 11.6

Brian and Margaret Lee had wages of $33,200 and interest income of $2,400. They put $3,000 into a deductible IRA. They filed a joint return—claiming three exemptions (Brian, Margaret, and their daughter, aged 5)—and used the standard deduction. During the year, $950 in federal income tax had been withheld from their wages. What was the total tax due with the return?

Total income	$35,600
Adjustments to income: IRA deduction	3,000
Adjusted gross income	32,600
Standard deduction: married, filing jointly	9,700
	22,900
Minus 3 exemptions: 3 × $3,100	9,300
Taxable income	$13,600
From Schedule Y-1:	
$13,600 × 0.10	$ 1,360
Minus child tax credit for one child	1,000
Net tax due after credits	$ 360
Minus federal income tax withheld	950
Refund	$ 590

Chapter Terms for Review

adjusted gross income (AGI)

assessed valuation

Child Tax Credit

dependency exemption

excise tax

filing status

Form 1040

itemized deductions

levy

market value

mill

personal exemptions

property tax

sales tax

tax rate

Tax Rate Schedules

taxable income

Try Microsoft® Excel

Try working the following problems using the Microsoft Excel templates found on your Student CD. Solutions for the problems are also shown on the CD.

1. Set up the following table and complete using Excel formulas to calculate the values in the **Sales Tax Amount** and **Total Sales with Tax** columns using the sales tax rate indicated.

Hint: Use an absolute cell reference for the sales tax rate so that the formula can be copied. Cell references are changed to absolute by adding a $ before both the column letter and the row number. Example: D9

Sales tax rate:	7.25%

Taxable sale	Sales tax amount	Total sale with tax
$12.83		
$81.91		
$20.11		
$111.92		
$0.55		
$7.20		
$328.90		
$1,552.44		
$62.00		

Try Microsoft® Excel is continued on page 226.

THE BOTTOM LINE

Summary of chapter learning objectives:

Learning Objective	Example
11.1 Compute sales taxes, using rate tables and percents	1. The Denver family lives in a state in which the sales tax rate is 6%. When they purchased a dining room table and chairs regularly priced at $990, they were given a discount of 15%. Shipping charges were $50. What was the total cost to the Farleys? 2. Wanda Green lives in a state in which the state tax on gasoline is $0.22 a gallon. Federal tax is $0.19 a gallon. If she purchased an average of 12 gallons per week during the 52-week year, how much did she pay in state and federal taxes combined?
11.2 Compute assessed valuations and property taxes based on assessed valuation	3. The Nguyen family lives in a town in which the assessed valuation on property is 65% of market value. The Parker family lives in a town in which the assessed valuation on property is 80% of market value. Each home has a market value of $162,000. How much is the assessed valuation of each home?

11.3 Compute tax rates in percents and mills	4. The town of Tyler has a total assessed valuation of $850,000,000. For the coming year the city must raise $11,730,000 for operating expenses.
	  a. What will be the tax rate?
	  b. What will the semiannual taxes be on a home with an assessed valuation of $135,000?
	
	5. a. Convert 650 mills to its dollar equivalent.
	  b. Convert $0.12 to mills.

11.4 Compute property tax payments involving special assessments, prorations, and exemptions.	6. A home with annual tax payments of $510 was sold at the end of the tenth month of the taxable year. What was the amount of tax prorated to the buyer? 7. A veteran living in Alameda receives a partial exemption of 10% of regular property taxes. The veteran owns property valued at $312,000. If the property is assessed at 70% of value and the current rate is 1.5%, how much tax is due each six months?

Answers: 1. $941.99 2. $255.84 3. $105,300 (Nguyen), $129,600 (Parker) 4. a. 1.38% b. $931.50 5. a. $0.65 b. 120 mills 6. $85 7. $1,474.20

THE BOTTOM LINE

Summary of chapter learning objectives:

Learning Objective	Example

11.5
 	
Make basic computations to determine taxable income for taxpayers who use the standard federal income Form 1040
	8. Gilbert Black is 28 years old and single. He claimed one exemption. In 200X he earned $47,000 in wages and $675 in taxable interest income. During the year he invested $1,800 in an individual retirement account. Because of a change of jobs, he also had $1,200 in moving expenses, which qualified as an adjustment to income. He had qualifying deductions of $1,000 in deductible medical bills, $300 in church donations, and $9,600 in interest on the condominium he owned. He also paid $150 in state taxes. He had $2,500 in federal income tax withheld during the year. What was the amount of tax due with his return?
	
	Income: $47,000 + $675 = $47,675 $47,675
	Less adjustments to income: $1,800 + $1,200 = $3,000 3,000
	Adjusted gross income $44,675
	Less deductions: $1,000 + $300 + $9,600 + $150 = 11,050
	    $33,625
	Less exemption: $3,100 3,100
	Taxable income $30,525
	Tax computation **Schedule X, Single:**
	   $4,000 on first $29,050 $4,000.00
	   0.25 × ($30,525 − $29,050) $368.75 $ 4,369
	Less tax withheld during the year $4,368.75 2,500
	Tax due with return (rounded to $4,369) $ 1,869

11.6
 	
Make basic computations to determine the tax liability for taxpayers who use the standard federal income Form 1040
	9. Donald and Judy Mason are 72 and 70 years of age, respectively. Judy is blind. They filed as married, filing jointly. Last year they had a total income of $30,000 from investments. They filed a return and claimed the standard deduction. During the year they made quarterly payments of estimated tax in the amount of $1,000. What was the amount of tax due with their return?
	
	Adjusted gross income $30,000
	Less standard deduction
	   From Figure 9-9: Married, filing jointly:
	     both over 65; 1 blind $12,550
	    $17,450
	Less exemptions: $3,100 each $ 6,200
	Taxable income $11,250
	Tax computation Figure 11-11, Schedule Y-1:
	   Married, filing jointly 0.10 × $11,250 $ 1,125
	Less payments made during the year on estimated tax $ 1,000
	Tax due with return $ 125

Answers: 8. $1,869 9. $125

SELF-CHECK

Review Problems for Chapter 11

1. The Dupree Company is considering the purchase of some equipment from two different suppliers. If the sales tax rate is 6%, which of the following offers should Johnson Company accept?

   Company A: Equipment price of $65,000 plus installation and shipping costs of $1,200.

   Company B: Equipment price of $73,500 less 10% discount, no additional charge for installation or shipping.

2. Georgetown needs to raise $7,800,000 in property taxes on property with a total market value of $650,000,000.

   a. What will the tax rate be if property is assessed at 80% of market value?

   b. Determine the amount of semiannual property tax to be paid by each of the following property owners who live in Georgetown.

   Juan Garcia's home in Georgetown has a market value of $350,000.

   Margaret Smith is a senior citizen who receives a 10% exemption from property tax. Her home in Georgetown has a market value of $215,000.

3. The residents of Hunterville voted to add street lights and sidewalks to their city at a cost per residence of $324 to be spread over 12 years.

   a. If Mary Nowitski, a resident of Hunterville, had an annual tax bill of $860 before the special assessment, how much must she now pay semiannually for her property taxes?

   b. If Mary Nowitski sells her home at the end of the eighth month of the tax year and has already paid the property taxes for the full year, including the special assessment, how much of the prepaid property tax should be allocated to the purchaser?

4. Samantha Jones works as a waitress. Last year she earned $15,800 in wages, $8,600 in tips, and $1,500 catering on weekends. She also received $600 interest from her credit union, $800 from a state bond, and an inheritance of $10,000. What was her gross income for federal income tax purposes?

5. Pete and Angel Romero are married and have two children aged 5 and 8. They also support Pete's sister, who lives with them. How much can Pete and Angel subtract from their gross income for exemptions?

6. Jan and Kirsten Bjorg, aged 63 and 66, are married filing a joint tax return. They have itemized deductions totaling $7,900. Should they itemize or use the standard deduction?

7. Eva Jung files as a head of household, has an adjusted gross income of $38,000, claims two exemptions, and uses the standard deduction on her federal return. What is her taxable income?

8. Brad and Justine O'Riley are married, filing a joint return, and have taxable income of $65,000. What is the amount of their income tax?

Assignment 11.1: Sales Tax

Name

Date   Score

Learning Objective **1**

A (50 points) Solve the following problems. (1 point for each correct answer)

1. Jay's fast-food restaurant is in a state with a sales tax rate of 7%. Compute the sales tax, the total sale, and the change given for each transaction.

Amount of Sale	Sales Tax	Total Sale	Cash Paid	Amount of Change
$6.18	$0.43	$6.61	$10.00	$3.39
4.40	0.31	4.71	5.01	0.30
12.89	0.90	13.79	20.00	6.21
19.56	1.37	20.93	25.00	4.07
5.80	0.41	6.21	10.00	3.79
29.41	2.06	31.47	40.00	8.53
18.55	1.30	19.85	20.00	0.15
0.98	0.07	1.05	1.25	0.20
13.99	0.98	14.97	15.00	0.03
15.69	1.10	16.79	20.00	3.21

2. Rosa's Botique is in a city where the state sales tax is 3.5% and the city tax is 2%. Determine the sales tax, the total sale, and the change given for each transaction. Then compute the total sales taxes and total sales.

Amount of Sale	Sales Tax	Total Sale	Cash Paid	Amount of Change
$284.20	$15.63	$299.83	$300.00	$0.17
42.89	2.36	45.25	50.25	5.00
65.98	3.63	69.61	75.00	5.39
227.89	12.53	240.42	250.00	9.58
125.00	6.88	131.88	140.00	8.12
97.72	5.37	103.09	120.00	16.91
Total	$46.40	$890.08		

Score for A (50)

B (30 points) Solve the following problems. Use Figure 11.1 for problems 3 and 4. (points for correct answers as marked)

3. A candy store, operating in a state with a sales tax of 4%, made 758 sales at $0.10; 862 sales at $0.35; 685 sales at $0.49; 950 sales at $0.65; 575 sales at $0.75; and 712 sales at $0.90. How much did the store receive in sales taxes? (8 points) $96.55

    758 sales at $0.10 = no tax = $0.00
    862 sales at $0.35 = 862 × 0.01 = 8.62
    685 sales at $0.49 = 685 × 0.02 = 13.70
    950 sales at $0.65 = 950 × 0.03 = 28.50
    575 sales at $0.75 = 575 × 0.03 = 17.25
    712 sales at $0.90 = 712 × 0.04 = 28.48
                                     $96.55

Assignment 11.1 Continued

4. If the candy store in problem 3 computed the amount of state sales tax submitted to the state based on 4% of gross sales, what would be the difference between the amount of tax the store collected and the amount it submitted to the state? (8 points) $0.44

   758 sales × $0.10 = $75.80
   862         $0.35 =   301.70
   685         $0.49 =   335.65
   950         $0.65 =   617.50
   575         $0.75 =   431.25
   712         $0.90 =   640.80
   Total sales $2,402.70
   $2,402.70 × 0.04 = $96.11 to the state
   $96.55 collected
   _96.11_ to the state
   $0.44 difference, which the store keeps

5. Discount Carpets Company and Oriental Rugs, Inc., each purchased a new delivery van. Discount Carpets is located in a state that has a 5% sales tax and paid the regular price of $21,800 plus tax. Oriental Rugs is located in a state that has a 6% sales tax and received a special discount of $500 off the regular $21,800 price.

   a. Including sales tax, which company paid more for its van? (8 points) Discount Carpets
   b. How much more? (6 points) $312
      Discount Carpets: $21,800 × 0.05 = $1,090 sales tax
         $21,800 + $1,090 = $22,890
      Oriental Rugs: $21,800 − $500 = $21,300 net of discount
         $21,300 × 0.06 = $1,278 sales tax
         $21,300 + $1,278 = $22,578
      $22,890 − $22,578 = $312 more for Discount Carpets

Score for B (30)

C (20 points) Solve the following problems. (points for correct answers as marked)

6. Calico Books has stores in four states. Sales tax rates for the four states are as follows: state A, 8%; state B, 6.2%; state C, $5\frac{1}{2}$%; state D, 3%. Annual sales for the four states last year were as follows: state A, $865,000; state B, $925,000; state C, $539,000; state D, $632,000.

   a. How much did Calico Books collect in sales taxes during the year? (10 points) $175,155
      State A: $865,000 × 8.0% = $69,200
      State B: $925,000 × 6.2% =   57,350
      State C: $539,000 × 5.5% =   29,645
      State D: $632,000 × 3.0% =   _18,960_
                 Total $175,155
   b. If all four states had the same lower sales tax rate of 3%, how much would Calico Books have collected in sales taxes during the year? (5 points) $88,830
      $865,000 + $925,000 + $539,000 + $632,000 = $2,961,000
      $2,961,000 × 0.03 = $88,830
   c. If all four states had the same higher tax rate of 8%, how much would Calico Books have collected in sales taxes during the year? (5 points) $236,880
      $2,961,000 × 0.08 = $236,880

Score for C (20)

Assignment 11.2: Property Taxes

Name

Date   Score

Learning Objectives 2 3 4

A (40 points) Solve the following problems. (4 points for each correct answer)

1. Find the assessed valuation for each of the following towns.

Town	Property Value	Basis for Tax Billing	Assessed Valuation
A	$625,000,000	100%	$625,000,000
B	$862,350,000	85%	$732,997,500
C	$516,800,000	70%	$361,760,000

Town A: $625,000,000 × 1.00 = $625,000,000
Town B: $862,350,000 × 0.85 = $732,997,500
Town C: $516,800,000 × 0.70 = $361,760,000

2. Find the tax rate for each of the following towns. Show your answer as a percent.

Town	Assessed Valuation	Amount to Be Raised	Tax Rate
F	$860,000,000	$13,932,000	1.62%
G	$645,000,000	10,965,000	1.7%
H	$732,000,000	9,150,000	1.25%

Town F: $13,932,000 ÷ $860,000,000 = 0.0162 (1.62%)
Town G: $10,965,000 ÷ $645,000,000 = 0.017 (1.77%)
Town H: $9,150,000 ÷ $732,000,000 = 0.0125 (1.25%)

3. Convert the following percentage tax rates into dollars and cents per $100 of assessed valuation.

Tax Rate	Dollars and cents
1.3%	$1.30
0.98%	$0.98

4. Convert the following percent tax rates into mills per $100 of assessed valuation.

Tax Rate	Mills
1.3%	1,300
0.98%	980

Score for A (40)

Assignment 11.2 Continued

B (24 points) Solve the following problems. (6 points for each correct answer)

5. The Griffin Company is located in a state in which assessed valuation is 100% of market value. The tax rate this year is $1.35 on each $100 of market value. The market value of the company building is $190,000. How much property tax will Griffin pay this year? $2,565
   $190,000 ÷ $100 = 1,900
   1,900 × $1.35 = $2,565

6. The Stockton Corp. is located in an area in which assessed valuation is 80% of market value. The tax rate this year is 1.5%. The market value of Stockton's property is $450,000. How much property tax will Stockton pay this year? $5,400
   $450,000 × 0.80 = $360,000
   $360,000 × 0.015 = $5,400

7. Next year, the assessed valuation in Stockton's area (problem 6) will decrease to 75% of market value and the tax rate will remain the same as this year. How much less tax will Stockton pay next year than it paid this year? $337.50
   $450,000 × 0.75 = $337,500
   $337,500 × 0.015 = $5,062.50
   $5,400 − $5,062.50 = $337.50

8. Perez, Inc., is headquartered in an area in which assessed valuation is 80% of market value. The tax rate this year is $1.40 on each $100 of assessed valuation. Its property has a market value of $320,000. How much property tax will Perez pay this year? $3,584
   $320,000 × 0.80 = $256,000
   $256,000 ÷ $100 = 2,560
   2,560 × $1.40 = $3,584

Score for B (24)

C (24 points) Solve the following problems. Round to the nearest dollar. (3 points for each correct answer)

9a. There are four towns in Hogan county: Lawton, Johnsville, Dover, and Gault. Using the total assessed valuations given and the amount of money the town must raise for operating expenses, compute the necessary tax rate for each town.

Town	Total Assessed Valuation	Money That Must Be Raised	Tax Rate as a Percent
Lawton	$200,000,000	$3,400,000	1.7% (0.017)
Johnsville	$340,000,000	$5,100,000	1.5% (0.015)
Dover	$280,000,000	$3,780,000	1.35% (0.0135)
Gault	$620,000,000	$12,400,000	2.0% (0.02)

b. Convert each of the percentage rates in part a to mills per dollar of assessed valuation.
   Lawton   17 mills
   Johnsville 15 mills
   Dover   13.5 mills
   Gault   20 mills

Score for C (24)

Assignment 11.2 Continued

D **(12 points) Solve the following problems. Round to the nearest dollar. (6 points for each correct answer)**

10. A home with annual tax payments of $624 was sold at the end of the fifth month of the taxable year. The seller had already paid the entire tax for the year. How much tax was the seller reimbursed on proration of taxes at the time of the sale? $364

    $624 ÷ 12 = $52 per month taxes
    12 − 5 = 7 months prepaid
    $52 × 7 = $364 reimbursed to seller

11. A senior citizen lives in a state that grants a 20% exemption on property taxes. Her property is valued at $290,000 and is assessed at 75% of value. The current tax rate is 1.6%. How much tax is due each six months? $1,392

    Assessed valuation: $290,000 × 0.75 = $217,500
    Regular yearly tax payment: $217,500 × 0.016 = $3,480
    Tax after exemption: $3,480 × 0.80 = $2,784
    Due each 6 months: $2,784 ÷ 2 = $1,392

Score for D (12)

Try Microsoft® Excel *(Continued from page 217.)*

2. Adams Company purchased a new copy machine priced at $2,650 less a 10% discount plus delivery and setup charges of $150. Determine the amount of the discount, the sales tax at 6.5%, and the total amount of the sale including delivery and setup costs. Set up the table below on an Excel worksheet and complete by adding formulas for calculations. Hint: Discounts are subtracted before and delivery costs are added after calculating sales tax.

Original price of copy machine	
Discount amount	
Net price after discount	
Sales tax at 6.5%	
Delivery and setup	
Total sale amount	

3. Kingstrom Corporation is located in an area in which assessed valuation is 70% of market value. The current tax rate is 1.35%. Determine Kingstrom's property tax for the year on property with a market value of $652,000. Enter the data below into an Excel worksheet and complete by adding formulas for calculations.

Market value of property	
Assessed valuation at 70%	
Property tax at 1.35%	

Assignment 11.3: Federal Income Tax

Name

Date   Score

Learning Objectives **5** **6**

A (52 points) Complete all problems, using the exemptions, deductions, and tax rates given in the chapter. Round all amounts to the nearest dollar. (Rounding is allowed so long as it is done consistently.) (12 points for correct answers to 2a and 3a; 4 points for other correct answers)

1. Determine the taxable income for each of the following taxpayers.

Adjusted Gross Income	Number of Exemptions	Type of Return	Deductions	Taxable Income
a. $28,700	1	Single	Standard	$20,750
b. $52,450	4	Head of household	Standard	$32,900
c. $23,900	2	Joint	Standard	$8,000
d. $16,452	1	Single	$5,960	$7,392
e. $43,700	6	Joint	$10,212	$14,888

a. $28,700 − $3,100 exemption − $4,850 deduction = $29,750
b. $52,450 − $12,400 exemption − $7,150 deduction = $32,900
c. $23,900 − $6,200 exemption − $9,700 deduction = $8,000
d. $16,452 − $3,100 exemption − $5,960 deduction = $7,392
e. $43,700 − $18,600 exemption − $10,212 deduction = $14,888

2. Sadie Gilford is a 70-year-old single person who lives alone. She takes the standard deduction. Her income during the year was $21,500.

   a. What is Sadie's taxable income? $12,350

Income	$21,500
Less standard deduction	− 6,050
	15,450
Minus exemption	− 3,100
Taxable income	$12,350

   b. What is Sadie's tax? $1,495

   $715 + 15% of amount over $7,150 ($12,350 − 7,150) × 0.15 = $780 + $715 = $1,495

3. George Sampson is 82 years old. His wife Marcia is 83 and is blind. They have $21,000 taxable income. They file a joint return and take the standard deduction.

   a. What is the Sampsons' taxable income? $2,250

Income	$21,000
Less standard deduction	12,550
	8,450
Minus 2 exemptions	6,200
Taxable income	$ 2,250

   b. What is the Sampsons' income tax? $225

   $2,250 × 0.10 = $225

Score for A (52)

Assignment 11.3 Continued

B (48 points) Solve the following problems. (12 points for correct taxable income; 4 points for correct income tax)

4. Alfred Wild is 66 years old; his wife Silvia is 64. They file a joint return. Alfred's salary for the year was $45,000. Silvia's salary was $42,000. They paid mortgage interest of $12,600 and property tax of $1,200 on their home. They paid state income tax of $3,800 during the year. They itemize their deductions.

   a. What is their taxable income? $63,200   b. What is their income tax? $9,275

Income		$87,000	$8,000 + ($63,700 − $58,100) ×
Itemized deductions:	$12,600		0.25 = $9,275
	1,200		
	3,800	17,600	
(standard deduction = $10,650)			
		$69,400	
Minus 2 exemptions ($3,100)		6,200	
Taxable income		$63,200	

5. Michael and Martha Miller are married and have three dependents living with them: their children, aged 17 and 19, and Martha's mother. Michael's salary for the year was $30,000, and Martha's salary was $32,000. They received taxable interest of $1,250 and $500 interest from a state bond. They take the standard deduction and file a joint return.

   a. What is their taxable income? $38,050   b. What is their net tax after credits? $4,993

   Income from state bond is not taxable.   From Schedule Y-1:
   Adjusted gross income $63,250
   Minus standard deduction   9,700   No child tax credit because both children are
                                   53,550   17 or older.
   Minus 5 exemptions   15,500
   Taxable income $38,050 $1,430 × ($38,050 − $14,300) ×
                                                 0.15 = $4,992.50 (rounded to $4,993)

6. Renaldo and Rita Hernandez have three children aged 17, 18, and 12. Renaldo's father lives with them and has no income. Renaldo earned a salary of $46,000 during the year. Rita is not employed. They paid $3,100 property tax and $4,100 mortgage interest on their home. They paid $2,600 principal on their mortgage. They paid state income tax of $2,175. They donated $500 to their church and $500 to the Salvation Army. They spent $5,600 on groceries and $1,100 on utilities. They itemize their deductions.

   a. What is their taxable income? $17,025   b. What is their net income tax after credits? $839

   Principal, groceries, and utilities are not deductible. $1,430 + ($17,025 − $14,300) ×
   Adjusted gross income $46,000   0.15 = $1,838.75 rounded to $1,839
   Itemized deductions: $3,100   Minus child tax credit   1,000
                                4,100   Net tax after credits $839
                                2,175
                                  500
                                  500   10,375
                                          $35,625
   Minus 6 exemptions   18,600
   Taxable income $17,025

Score for B (48)

Insurance

12

Learning Objectives
By studying this chapter and completing all assignments you will be able to:

Learning Objective 1 — Compute costs and savings for auto insurance.

Learning Objective 2 — Compute auto insurance premium rates for high- and low-risk drivers.

Learning Objective 3 — Compute short-rate refunds.

Learning Objective 4 — Compute coinsurance on property losses.

Learning Objective 5 — Compute life insurance premiums.

Learning Objective 6 — Compute cash surrender and loan values.

Learning Objective 7 — Compute medical insurance contributions and reimbursements.

Computing Auto Insurance Costs

Learning Objective 1

Compute costs and savings for auto insurance.

12.1 A building or an auto on which a mortgage or loan is held by a lending institution is generally insured for at least the amount of the loan. The lending institution holds the insurance policy, which states that the lending institution will be paid first in case of damage or loss. Without the insurance industry, the lending industry would have to undergo major changes.

Auto insurance falls into three categories: liability and property damage, comprehensive, and collision. A policy that fully protects the insured will contain all three types.

Auto liability and property damage insurance protects the insured against claims resulting from personal injuries and property damage. Some states require all drivers to carry auto liability and property damage insurance. The amount of protection generally ranges from $50,000 to $1,000,000 per accident.

Auto comprehensive insurance protects the vehicle of the insured against water, theft, vandalism, falling objects, and other damage not caused by collision.

Auto collision insurance protects the vehicle of the insured against collision damage. Such damage may result from a collision with another vehicle or a one-car accident, such as hitting a tree.

The payment for an insurance policy is called a **premium**. Premium rates for auto insurance depend primarily on the coverage included in the policy, the driving record of the insured, and the geographical area where the driver lives.

Auto collision insurance policies usually contain a **deductible clause**, which stipulates that the insured will pay the first portion of collision damage, usually $50 to $500, and that the insurance company will pay the remainder up to the value of the insured vehicle. A deductible clause not only reduces the amount of damages for which the insurance company must pay but also keeps the insurance company from having to get involved in and do paperwork for small repairs costing less than the deductible. Therefore, a deductible clause lowers the premium for collision insurance.

EXAMPLE A

A car was insured for collision damage with a $250 deductible. The premium was $1,750 per year. The insured hit a tree, causing $2,530 damage to his car. How much more did the insured receive than he paid in premiums for that year?

$2,530 damage − $250 deductible = $2,280 paid by insurance
$2,280 received by insured − $1,750 premium paid = $530.

EXAMPLE B

The driver of car A carried auto liability and property damage insurance only. She struck car B, causing $1,400 damage to car B and $700 in injuries to the driver. Car A suffered $940 damage.

a. How much did the insurance company pay for this accident?
   $1,400 for damage to car B + $700 for injuries to driver = $2,100

b. How much did this accident cost the driver of car A?
   $940 in uncovered damage to her own car

No-fault insurance is a term that is used to describe an auto insurance system that requires drivers to carry insurance for their own protection and that limits their ability to sue other drivers for damages. No-fault insurance requires that the driver of each vehicle involved in an injury accident submit a claim to *his or her own insurance company* to cover medical costs for injuries to the driver and passengers in that person's own vehicle. No-fault insurance is mandatory in some states. No-fault insurance doesn't cover damage to either vehicle involved in an accident.

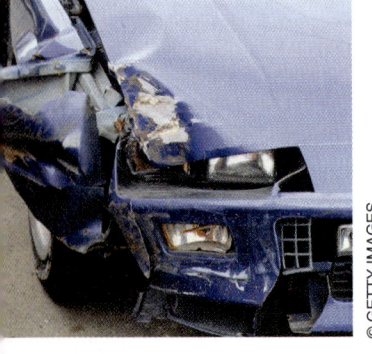

230 Part 3 Accounting Applications

EXAMPLE C

Drivers A and B live in a state in which no-fault insurance is mandatory. Their two cars collided. Driver A and his passengers incurred medical expenses of $3,500. Driver B and her passengers incurred $1,700 in medical expenses. Car A required $1,400 in repairs. Car B required $948 in repairs. How much did the insurance companies pay under the no-fault insurance coverage?

Driver A's insurance company paid $3,500 in medical expenses.
Driver B's insurance company paid $1,700 in medical expenses.
Car repairs are not covered under no-fault insurance.

CONCEPT CHECK 12.1

Driver A lives in a state in which no-fault insurance is mandatory. He carries all three classifications of insurance to be fully protected. His total insurance premium is $2,400, with a collision deductible of $500. Driver A is involved in a major accident when he loses control of his car and hits two parked cars (cars B and C) before colliding with an oncoming car (car D) containing a driver and three passengers. Driver A is alone.

> Damage to Driver A's car is $3,200.
> Damages to cars B, C, and D total $8,600.
> Medical expenses for driver A are $2,800.
> Medical expenses for the driver and passengers of car D are $7,300.
>
> a. How much does driver A's insurance company pay?
>   Damage to car A: $3,200 − $500 deductible = $2,700 covered by collision
>   Damage to cars B, C, and D: $8,600 covered by liability
>   Medical expenses for driver A under no-fault: $2,800
> $2,700 + $8,600 + $2,800 = $14,100 paid by driver A's insurance
> b. How much does driver D's insurance company pay?
>   Medical expenses paid for driver D and passengers (no-fault): $7,300
> c. How much more did driver A's insurance company pay to him and on his behalf for this accident than he paid in insurance expenses for the year? (This is the amount driver A saved this year by being fully insured.)
> $2,400 premium + $500 deductible = $2,900 paid by Driver A
> $14,100 from insurance − $2,900 = $11,200
>   Driver A saved $11,200 this year by being fully insured.

Computing Low-Risk and High-Risk Rates

Auto insurance premium rates reflect the risk involved. Insurance companies study the statistics on automobile accidents relative to driving records. Premium rates are adjusted according to the driving record of the insured. A driver with a clear record of long standing is considered to be a **low-risk driver** and may be rewarded with a discount in the premium rate. Conversely, a driver with a record of numerous citations or accidents is considered to be a **high-risk driver** and may pay double, triple, or even a higher multiple than the normal premium rate.

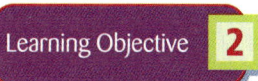

Learning Objective 2

Compute auto insurance premium rates for high- and low-risk drivers.

12.2 Insurance premiums reflect careful and extensive analysis of historical data by insurance companies.

EXAMPLE D

Drivers A and B have identical automobiles and amounts of insurance coverage. The normal premium rate for each is $2,000 per year. Driver A is a low-risk driver and receives a 15% discount on the premium rate. Driver B is a high-risk driver and must pay double the normal rate. How much more does driver B pay for insurance than driver A?

Driver A pays $2,000 × 85% = $1,700 (100% − 15% discount)
Driver B pays $2,000 × 2 = $4,000
Driver B pays $4,000 − $1,700 = $2,300 more

✓ CONCEPT CHECK 12.2

Driver A, a very careful driver, has had the same insurance company for 5 years and has not had a ticket during that 5-year period. Each year, driver A has received a 10% reduction in her premium. Driver B has a record of speeding tickets. He has had one or more every year for 5 years. His premium for year 1 was normal, for years 2 and 3 it was 150%, and for years 4 and 5 it was 200%. The normal annual premium rate for each driver would be $980.

a. How much did driver A pay in premiums over the 5-year period?
$980 × 90% = $882
$882 × 5 = $4,410

b. How much did driver B pay in premiums over the 5-year period?
Year 1: $980
Years 2 and 3: $980 × 1.5 × 2 = $2,940
Years 4 and 5: $980 × 2 × 2 = $3,920
$980 + $2,940 + $3,920 = $7,840

c. How much more did driver B pay during the 5-year period than driver A?
$7,840 − $4,410 = $3,430

Computing Short Rates

Learning Objective 3
Compute short-rate refunds.

Short rates are rates charged for less than a full term of insurance. If an insurance policy is canceled by the **insured** (the person who receives the benefit of the insurance) before the policy's full term is complete, the insured will receive a short-rate return of premium. If a policy is canceled by the insurance company rather than by the insured, the company must refund the entire unused premium.

12.3 Figured into premium rates are expenses incurred for paperwork to issue and service the policy. If the insured cancels the policy before a year's premium has been earned, the premium received will not be commensurate with expenses already incurred.

EXAMPLE E

A driver paid an annual premium of $1,960 for auto insurance. After 3 months, the vehicle was sold and the insurance canceled. The insurance company refunded the remaining portion of the premium at the short rate, based on a penalty of 10% of the full-year premium. What was the refund?

Unused premium: $1,960 × $\frac{3}{4}$ = $1,470   (9 months canceled = $\frac{9}{12}$ = $\frac{3}{4}$ year)

Penalty: $1,960 × 10% = $196
Short-rate refund: $1,470 − $196 = $1,274

✓ CONCEPT CHECK 12.3

A company purchased two cars. Each car was insured at an annual premium of $1,780. At the end of 6 months, the company sold one car and canceled the insurance on that car. At the end of 9 months, the insurance company decided to cancel the insurance on the second car. The insurance company imposes a 10% penalty for short-rate premiums. Compute the refunds the insurance company paid for car 1 and car 2.

Car 1: $1,780 × ½ year = $890 unused premium

$1,780 × 10% = $178 penalty
$890 − $178 = $712 refunded

Car 2: $1,780 × ¼ year = $445 unused and refunded premium

COMPLETE ASSIGNMENT 12.1.

Computing Coinsurance on Property Losses

Property insurance is insurance against loss of or damage to property. A policy can be written to protect the insured against loss from fire, casualty, liability, and theft.

Premium rates, which are quoted in terms of dollars per $1,000 of insurance, depend on the nature of the risk, the location of the property, and the length of time covered by the policy. Short rates and short-rate penalties for less than a full term of insurance apply to property insurance as they do to auto insurance.

Learning Objective 4

Compute coinsurance on property losses.

● EXAMPLE F

A building worth $350,000 is insured for $210,000. The annual premium for the policy is $5,000. A fire causes $80,000 in damage.

a. How much does the insurance company pay?
   $80,000 in damage is less than the $210,000 policy. The insurance company pays the entire $80,000.

b. How much does the property owner pay?
   The property owner pays no damages.

c. How much does the property owner pay that year in damages and insurance?
   $5,000 for the insurance premium only.

12.4 Property insurance is based on the old-world "good neighbor" concept. Before insurance, when one family in a community lost a home or barn to fire or other destruction, neighbors contributed household goods or hours of labor to rebuild the structure. The premium and face value concept of the modern-day insurance policy accomplishes the same purpose.

In an ordinary fire insurance policy, the insured will be paid for the loss up to the amount of the insurance. Policies may be obtained at lower rates if they contain a **coinsurance clause.** This clause specifies that if a property is not insured up to a specified percentage of its value, the owner is responsible for part of the loss and will not be covered for the full amount of damages.

12.5 A precaution commonly taken by insurance companies is to adjust the amount of exposure that one company assumes in a given area through a form of coinsurance. If a major disaster in one locale would jeopardize the ability of an insurance company to continue operations and meet its obligations, the company should acknowledge this fact and limit exposure to a manageable amount through a coinsurance arrangement with another carrier.

It is common practice for a fire insurance policy to have an 80% coinsurance clause. Under this clause, the full amount of the loss will be paid by the insurance company only if the policy amount equals 80% of the property value.

STEPS to Determine the Owner's Share of Property Loss Under Coinsurance

1. Compute the amount of insurance required by multiplying the entire value of the property by the percentage of coinsurance specified.
2. Compute the **recovery amount**, the maximum amount the insurance company will pay, by using the formula
$$\frac{\text{Amount of insurance carried}}{\text{Amount of insurance required}} \times \text{Loss} = \text{Recovery amount}.$$
3. Compare the recovery amount with the amount of the insurance policy.
   a. If the recovery amount is greater than the amount of the policy, the insurance company will limit its payment to the amount of the policy.
   b. If the recovery amount is less than the amount of the policy, the insurance company will pay the recovery amount.
   Note: The insurance company will never pay more than the amount of the loss.
4. Determine the owner's share of the property loss by subtracting the amount the insurance company will pay from the loss amount.

EXAMPLE G

A building valued at $350,000 is insured for $210,000 under a policy with an 80% coinsurance clause. The annual premium is $2,800. A fire causes $200,000 damage to the building.

a. How much will the insurance company pay the insured?

STEP 1 $350,000 \times 80\% = \$280,000$ insurance required

STEPS 2&3 $\dfrac{\$210,000 \text{ amount of insurance carried}}{\$280,000 \text{ amount of insurance required}} \times \$200,000 = \$150,000$ insurance pays

b. How much must the owner pay if the building is repaired for $210,000?

STEP 4 $\$210,000 - \$150,000 = \$60,000$ paid by owner

c. How much does the property owner pay that year for damages and insurance?
$60,000 damages + $2,800 premium = $62,800

d. How much would the insurance company pay if the fire caused $300,000 damage to the building?

$\dfrac{\$210,000}{\$280,000} \times \$300,000 = \$225,000$ recovery amount

The insurance company would limit its payment to $210,000 (the full value of the policy, because the recovery amount exceeds the policy's coverage).

EXAMPLE H

If the amount of insurance carried in example G had been $280,000, how much would the insured have paid for damages and insurance that year?

$2,800 premium only (the 80% coinsurance requirement would have been met)

CONCEPT CHECK 12.4

A building worth $100,000 is insured for $60,000 with an 80% coinsurance clause. A fire causes $70,000 in damage. How much of the repair cost will the insurance company pay, and how much will the insured pay?

$100,000 × 80% = $80,000 insurance required

$\dfrac{\$60,000}{\$80,000} \times \$70,000 = \$52,500$ insurance pays

$70,000 − $52,500 = $17,500 insured pays

COMPLETE ASSIGNMENT 12.2.

Computing Life Insurance Premiums

The policies most commonly issued by life insurance companies are term insurance, straight life (sometimes called ordinary life), limited-payment life, endowment, and annuity.

Term insurance is protection issued for a limited time. A certain premium is paid every year *during the specified time period,* or term. The policy is payable only in case of death of the insured during the term. Otherwise, neither the insured nor the specified beneficiaries receive any payment, and the protection stops at the end of the term.

For **straight (ordinary) life insurance coverage,** a certain premium, or fee, is paid every year *until the death of the insured.* The policy then becomes payable to the **beneficiary.** A policy beneficiary can be a person, a company, or an organization.

Limited-payment life insurance (such as 20-payment life) requires the payment of a specified premium each year for a certain number of years or until the death of the insured, whichever comes first. Should the insured live longer than the specified number of years, the policy is then paid up for the remainder of the insured's life and is payable to the beneficiary on the death of the insured.

Endowment insurance provides insurance payable on the insured's death if it occurs within a specified period. If the insured is alive at the end of the specified period, an endowment of the same amount as the policy is payable.

Annuity insurance pays a certain sum of money to the insured every year after the insured reaches a specified age, until the insured's death.

An **additional death benefit (ADB)**, sometimes referred to as an *accidental death benefit,* accompanies some policies. ADB allows the insured to purchase, at a low rate per thousand dollars of coverage, additional insurance up to the full face value of the policy. In case of death of the insured by accident, both the full value of the policy and the ADB are paid to the beneficiaries. If death occurs other than by accident, the full value of the policy is paid, but no ADB is paid.

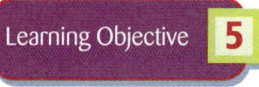

Compute life insurance premiums.

12.6 Businesses frequently use term insurance to insure against losses that would be incurred in the event of the death of one or more of their "key" executives.

Figure 12-1 shows typical annual, semiannual, and quarterly premiums (ages 25–28) for straight life, 20-payment life, and 20-year endowment policies.

Figure 12-1 | Insurance Premium per $1,000

Age	Straight Life			20-Payment Life			20-Year Endowment		
	Annual	Semi-annual	Quarterly	Annual	Semi-annual	Quarterly	Annual	Semi-annual	Quarterly
25	$17.20	$8.94	$4.73	$31.20	$16.26	$8.26	$52.00	$27.04	$14.30
26	17.85	9.28	4.91	31.81	16.52	8.45	52.60	27.35	14.47
27	18.60	9.67	5.11	32.41	16.83	8.64	53.20	27.66	14.63
28	19.30	10.04	5.31	33.06	17.31	8.85	53.86	28.01	14.81

● **EXAMPLE 1**

Using the premiums shown in Figure 12-1, determine the yearly premiums for each of the following $50,000 life insurance policies purchased at age 27.

Type of Insurance	Method of Payment	Premium Computation
Straight Life	Annual	$18.60 × 50 = $930
20-Year Endowment	Quarterly	$14.63 × 4 × 50 = $2,926
20-Payment Life	Semiannual	$16.83 × 2 × 50 = $1,683
20-Year Endowment	Semiannual	$27.66 × 2 × 50 = $2,766
Straight Life	Quarterly	$ 5.11 × 4 × 50 = $1,022

 CONCEPT CHECK 12.5

a. If a person at age 28 purchases a straight life insurance policy having a face value of $150,000 with quarterly premiums, what is the yearly premium?
 $5.31 × 4 × 150 = $3,186

b. If a person at age 25 purchases a 20-payment life insurance policy having a face value of $100,000 with semiannual premiums, what is the yearly premium?
 $16.26 × 2 × 100 = $3,252

c. If a person at age 25 purchases a 20-year endowment insurance policy having a face value of $75,000 with annual premiums, what is the yearly premium?
 $52 × 75 = $3,900

Computing Cash Surrender and Loan Values

Learning Objective 6

Compute cash surrender and loan values.

Except for term insurance, insurance usually has a **cash surrender value,** which is the amount of cash that the company will pay the insured on the surrender, or "cashing in," of the policy. The **loan value** of a policy is the amount that the insured may borrow on the policy from the insurance company. Interest is charged on such loans. The values, often quoted after the third year of the policy, are stated in the policy and increase every year. Figure 12-2 shows typical cash surrender and loan values for policies issued at age 25 per $1,000 of life insurance.

Figure 12-2 | Insurance Values per $1,000

End of Policy Year	Cash Surrender and Loan Values		
	Straight Life	20-Payment Life	20-Year Endowment
3	$ 10	$ 43	$ 88
4	22	68	130
5	35	93	173
10	104	228	411
15	181	380	684
20	264	552	1,000

12.7 Why would an insured take the maximum loan instead of cash surrender? Suppose that a person bought a $20,000 straight life policy 10 years ago. The cash surrender or loan value is $104 × 20 = $2,080. The insured borrows the full amount and dies before repayment. The beneficiary receives $20,000 − $2,080, or $17,920. If the insured had surrendered the policy, no payment would have been made upon death.

EXAMPLE J

Use the figures shown in Figure 12-2 to determine the cash surrender or loan value for each of the following policies.

Policy Year	Type of Policy	Amount of Policy	Cash Surrender or Loan Value
10	Straight Life	$ 75,000	75 × $104 = $ 7,800
5	20-Year Endowment	$ 15,000	15 × $173 = $ 2,595
10	20-Payment Life	$ 50,000	50 × $228 = $11,400
20	Straight Life	$100,000	100 × $264 = $26,400
15	20-Year Endowment	$ 50,000	50 × $684 = $34,200

✔ CONCEPT CHECK 12.6

Use the figures shown in Figure 12-2 to determine the cash surrender or loan value for each of the following policies.

a. Third policy year of a $50,000 20-year endowment policy
 50 × $88 = $4,400
b. Twentieth policy year of a $100,000 straight life policy
 100 × $264 = $26,400
c. Tenth policy year of a $25,000 20-payment life policy
 25 × $228 = $5,700

Computing Medical Insurance Contributions and Reimbursements

Most large employers and many small employers subscribe to a group plan on behalf of their employees. **Group insurance** plans provide medical insurance coverage to large numbers of people at lower premium rates than individuals could obtain separately. Employers generally pay all the premium for employees and a portion of the premium for family members of employees. Many employers now use group plans known as a **health maintenance organization (HMO)** or a **preferred provider organization (PPO)**.

Learning Objective 7

Compute medical insurance contributions and reimbursements.

12.8 Premiums in the examples vary. Explain to students that group medical premiums are adjusted periodically, based on actual medical costs and payments made by the insuring company. A few serious, long-term illnesses by one or more employees can cause premiums to be increased significantly.

EXAMPLE K

Employer A selected a basic health care plan to cover employees who want to participate. Monthly premiums are as follows: employee only, $350; employee with one dependent, $450; and employee with multiple dependents, $530. Employees pay a portion of the premium as follows: employee only, $0; employee with one dependent, $80; and employee with multiple dependents, $120. How much does the employer pay during the year for each category of employee?

Employee only: $350 × 12 = $4,200

Employee with one dependent: ($450 − $80) × 12 = $4,440

Employee with multiple dependents: ($530 − $120) × 12 = $4,920

EXAMPLE L

Employer B selected a total care health plan to cover employees who want to participate. Monthly premiums are as follows: employee only, $300; employee with one dependent, $400; and employee with multiple dependents, $480. The employer pays most of the premium, but employees pay a portion as follows: employee only, $30; employee with one dependent, $80; and employee with multiple dependents, $120. What percent of the premium will be paid by a single employee, an employee with one dependent, and an employee with six dependents?

A single employee: $30 ÷ 300 = 0.10, or 10%

An employee with 1 dependent: $80 ÷ 400 = 0.20, or 20%

An employee with 6 dependents: $120 ÷ 480 = 0.25, or 25%

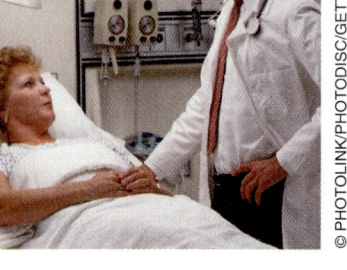

Many group plans include a provision for an annual deductible, which is the cost that must be paid by the employee before any cost is paid by the insurance company. Group medical plans also frequently provide for the payment by the insurance company of a percent of costs over the deductible, usually 70% to 90%, with the remaining 30% to 10% paid by the insured.

EXAMPLE M

Employer C provides group health coverage that includes a $500 annual deductible per family and payment of 70% of the medical charges exceeding the deductible. How much would an employee with three dependents pay if her year's medical bills were $1,500?

$1,500 − $500 deductible = $1,000

$1,000 × 30% paid by employee = $300

$500 deductible + $300 payments = $800 paid by the employee

CONCEPT CHECK 12.7

An employer provides group health coverage that includes a $300 annual deductible per family and payment of 80% of costs over the deductible.

a. How much would an employee with two dependents pay if his year's medical bills were $460?
$460 − $300 deductible = $160
$160 × 20% = $32
$300 deductible + $32 = $332 paid by the employee

b. How much would that employee have paid if total medical bills for the year had been $4,300?
$4,300 medical costs − $300 deductible = $4,000
$4,000 × 20% = $800
$300 deductible + $800 = $1,100

c. How much of the $4,300 in medical bills would that employee have paid if his employer did not provide medical insurance?
$4,300

d. How much did the employer pay if the monthly premium for an employee with multiple dependents was $480?
$480 × 12 = $5,760

COMPLETE ASSIGNMENT 12.3.

Chapter Terms for Review

- additional death benefit (ADB)
- annuity insurance
- auto collision insurance
- auto comprehensive insurance
- auto liability and property damage insurance
- beneficiary
- cash surrender value
- coinsurance clause
- deductible clause
- endowment insurance
- group insurance
- health maintenance organization (HMO)
- high-risk driver
- insured
- limited-payment life insurance
- loan value
- low-risk driver
- no-fault insurance
- preferred provider organization (PPO)
- premium
- property insurance
- recovery amount
- short rates
- straight (ordinary) life insurance
- term insurance

THE BOTTOM LINE

Summary of chapter learning objectives:

Learning Objective	Example
12.1 Compute costs and savings for auto insurance	Drivers A and B live in a state in which no-fault insurance is mandatory. Both drivers carry all three classifications of insurance. Driver A has a deductible of $500; driver B has a deductible of $200. Driver A crashes into driver B. Neither auto has any passengers. Car A has $1,800 in damages; car B has $2,000 in damages. Driver A is not hurt; driver B has $900 in medical bills. 1. How much does driver A's insurance company pay? 2. How much does driver B's insurance company pay?
12.2 Compute auto insurance premium rates for high- and low-risk drivers	3. Juan has an excellent driving record and receives a 10% discount on his annual premium. Dave has a record of numerous tickets and must pay $1\frac{1}{2}$ times the normal annual premium rate. If the normal premium for each driver is $1,500, how much more does Dave pay for his insurance than Juan pays?
12.3 Compute short-rate refunds	4. XYZ company purchased a delivery truck and paid an annual insurance premium of $3,600. XYZ company sold the truck at the end of 8 months and canceled the policy. The insurance company charges a 10% penalty for short-rate refunds. What was the amount of the short-rate refund to XYZ company?
12.4 Compute coinsurance on property losses	5. A building worth $400,000 is insured for $300,000 with an 80% coinsurance clause. Fire causes $200,000 in damage. How much does the insurance company pay?
12.5 Compute life insurance premiums	6. Premiums per $1,000 of straight life insurance at the age of 25 are as follows: annual, $17.20; semiannual, $8.94; and quarterly, $4.73. What will be the total yearly premiums for the following three policies: $50,000, annual; $25,000, semiannual; and $20,000, quarterly?

Answers: 1. $3,300 2. $900 3. $900 4. $840 5. $187,500 6. $1,685.40

Learning Objective	Example
12.6 Compute cash surrender and loan values	7. If cash surrender values for year 15 of a policy are $200 per thousand dollars of coverage for straight life and $380 per thousand dollars of coverage for 20-payment life, what is the total cash surrender value of these two policies: $50,000 straight life and $50,000 20-payment life?
12.7 Compute medical insurance contributions and reimbursements	8a. An employer provides group health coverage that includes a $200 annual deductible per family and payment of 80% of costs over deductible. How much would an employee with four dependents pay if his year's medical bills were as follows: self, $240; dependent 1, $170; dependent 2, $30; dependent 3, $460; and dependent 4, $2,200? b. How much would the employee pay if the annual deductible were $50 per person?

Answers: 7. $29,000 8a. $780 8b. $804

SELF-CHECK

Review Problems for Chapter 12

1. Drivers Jim Olson and Joshua Stein live in a state having no-fault auto insurance. Joshua causes an accident by hitting Jim's car. Joshua isn't hurt. Jim spends 3 days in the hospital at a cost of $5,300. Compute the amount that each driver's insurance company pays toward medical expenses. _____ _____

2. IXP insured an office building for $290,000 for 1 year at a premium rate of $7.20 per thousand. At the end of 9 months, IXP sold the building and canceled the policy. If the insurance company has a short-rate refund policy that includes a 10% penalty, how much refund did IXP receive? _____

3. Driver Devon Cooper has a poor driving record and pays double the usual premium as a high-risk driver. The regular premium would be $490 for a year. If Devon must pay the high-risk premium every year for 5 years, how much more will he pay for insurance premiums than a low-risk driver receiving a 10% discount over the same 5-year period? _____

4. Insurance company A has a standard 90% coinsurance clause for all fire insurance coverage. Insurance company B has a standard 75% coinsurance clause for all fire insurance coverage. A building is valued at $195,000. How much more insurance coverage would insurance company A require than insurance company B for full coinsurance coverage? _____

5. The Morgan Company warehouse was valued at $425,000. The building was insured for $170,000. The policy contained an 80% coinsurance clause. A fire caused $60,000 in damages. Compute the amount of the fire damage the Morgan company had to pay. _____

6. Mike Jankowski, age 27, purchased a $35,000, 20-payment life policy with premiums payable annually. John Jamison, also age 27, purchased a $35,000 straight life policy with premiums payable semiannually. Both Mike and John lived 40 more years. How much more in premiums did John pay the insurance company during his lifetime than Mike paid during his? _____

7. Sally Munson, age 25, purchased a $35,000, 20-payment life policy. Five years later she needed cash. Compute the maximum amount she could borrow on the policy. _____

8. An employer provided group health coverage that includes a $600 annual deductible per family and payment of 80% of costs exceeding the deductible amount. An employee with no dependents incurs $4,800 in medical expenses during the year. How much of the medical costs must the employee pay? _____

Assignment 12.1: Auto Insurance

Name

Date   Score

Learning Objectives **1 2**

A (50 points) Solve the following problems. (5 points for each correct answer)

1. Mary Johnson had full insurance coverage. Her liability and property damage coverage was $100,000 per accident. Her collision insurance had a $500 deductible clause. She struck two cars. Damages to the cars were $640 and $320. Damage to her own car was $470. Her annual insurance premium was $1,180.

   a. What are the total costs to the insurance company for Mary's accident? $960
      $640 + $320 = $960 damage to other cars; her car damage was less than deductible

   b. If this was the only accident that Mary had this year, how much money did the insurance company make on her? $220
      $1,180 premium − $960 cost of accident = $220

   c. What are Mary's total costs this year for insurance and the accident? $1,650
      $1,180 premium + $470 damage to her car = $1,650

   d. What would Mary's total costs for the accident have been without insurance? $1,430
      $640 + $320 + $470 = $1,430 damage to all three cars

2. Renaldo Garcia paid an annual premium of $3,000 for auto collision insurance with a $500 deductible clause. His steering went out, and he hit a tree causing $4,000 damage to his car. How much did he save this year by having insurance? $500
   $3,00 premium + $500 deductible = $3,500 total payments
   $4,000 damages − $3,500 paid = $500 saved

3. Sean O'Day received his driver's license 1 year ago. He has had three citations for speeding, but no accidents. His insurance premium last year was $1,800. This year his premium will be 100% higher because of his driving record.

   a. What will be the amount of his premium this year? $3,600
      $1,800 × 100% = $1,800; $1,800 + $1,800 = $3,600

   b. Four months into the next year, Sean has continued his unsafe driving habits. The insurance company is canceling his policy. What will be the amount of the refund? $2,400
      8/12 × $3,600 = $2,400

   c. Sean O'Day has found an insurance company that will insure him as a high-risk driver at triple the standard annual rate of $1,600. What will be his average monthly insurance premium for the first 28 months of his driving career? (Round your answer to the nearest dollar.) $279
      First 12 months: $1,800
      Next 4 months: $3,600 − $2,400 refund = $1,200
      This 12 months: $1,600 × 3 = $4,800
      $1,800 + $1,200 + $4,800 = $7,800 for 28 months
      $7,800 ÷ 28 = $278.57 = $279.00 rounded average monthly premium

   d. If Sean had been a careful driver and kept the amount of his premium unchanged, how much would he have saved in these first 28 months? (Round your computations to the nearest dollar.) $3,600
      $1,800 original premium × $2\frac{1}{3}$ years (28 months) = $4,200
      $7,800 − $4,200 = $3,600 savings

Assignment 12.1 Continued

4. Drivers A and B have identical insurance coverage. Driver A has an excellent driving record and receives a 15% discount on the standard premium. Driver B has numerous citations and pays 50% above the standard rate. The standard rate in both cases is $1,430. How much more does driver B pay for insurance than driver A? $929.50
   Driver A: $1,430 × 85% = $1,215.50
   Driver B: $1,430 × 150% = $2,145.00
   $2,145 − $1,215.50 = $929.50 more
   Alternative method: $1,430 × 65% = $929.50

Score for A (50)

B (50 points) Solve the following problems. (5 points for each correct answer)

5. Tom Barton carries liability and property damage insurance coverage up to $50,000 per accident, comprehensive insurance, and collision insurance with a $100 deductible clause. He lost control of his car and drove through the display window of a furniture store. Damage to the building was $17,200 and to the inventory was $34,300. Damage to a bike rack on the sidewalk and three bicycles in the rack was $1,840. Damage to his own car was $6,100.
   a. What was the total property damage, excluding damage to Tom's car? $53,340
      $17,200 + $34,300 + $1,840 = $53,340
   b. How much did the insurance company pay for property damage, excluding damage to Tom's car? $50,000
      $50,000, the maximum coverage for liability and property damage
   c. How much did the insurance company pay for damage to Tom Barton's car? $6,000
      $6,100 − $100 deductible = $6,000
   d. How much did the accident cost Tom Barton? $3,440
      $53,340 − $50,000 = $3,340 damage he paid
      $3,340 + $100 deductible = $3,440
   e. If Tom Barton had been in a previous accident this year in which there had been property damage to a parked car of $12,700, how much would the insurance company have paid for damages to everything in the current accident, including Tom Barton's car? $56,000
      $50,000 property damage + $6,000 insured's car = $56,000; insurance coverage is per accident, not cumulative

6. Amy Tan and John Rogers live in a state in which no-fault insurance is mandatory. They have identical full coverage of $50,000 liability and property damage per accident, comprehensive insurance, and collision insurance with a $350 deductible. John lost control of his car on an icy street and struck Amy's car, a parked motorcycle, and a fence. Amy had medical expenses of $780. John had medical expenses of $560. Amy's car had damages of $1,350. John's car had damages of $1,750. Damage to the parked motorcycle was $650 and to the fence was $320.
   a. What did Amy's insurance company pay under the no-fault provision? $780
      $780 was Amy's medical expense
   b. What did John's insurance company pay under the no-fault provision? $560
      $560 was John's medical expense
   c. How much did John's insurance company pay under his liability and property damage coverage? $2,320
      $1,350 for Amy's car + $650 for the motorcycle + $320 for the fence = $2,320
   d. How much did John's insurance company pay under his comprehensive coverage? $1,400
      $1,750 − $350 deductible = $1,400
   e. How much would John's insurance company have paid under his liability and property damage if he had hit Amy's car and five parked cars, with total damage to the six cars of $56,700? $50,000
      $50,000, the maximum coverage per accident

Score for B (50)

244 Part 3 Accounting Applications

Assignment 12.2: Property Insurance

Name

Date   Score

Learning Objectives **3** **4**

A (42 points) Solve the following problems. (6 points for each correct answer)

1. A building valued at $380,000 is insured for its full value. The annual premium is $9.80 per thousand dollars of coverage.

    a. How much does the insured pay to insure his building? $3,724

    $380,000 ÷ $1,000 = 380 thousands

    380 × $9.80 per thousand of coverage = $3,724 premium

    b. If the insurance company cancels the policy at the end of 3 months, how much refund does the insured receive? $2,793

    $\frac{9}{12}$ × $3,724 = $2,793 unused premium

    c. If the insurance company has a 10% penalty clause for short-rate refunds and the insured cancels the policy after 9 months, how much refund does the insured receive? $558.60

    $\frac{3}{12}$ × $3,724 = $931 unused premium

    $3,724 × 10% = $372.40 penalty; $931.00 − $372.40 = $558.60 refund

2. If a company pays an annual premium of $4,800 and the insurance company charges $16 per thousand dollars of insurance, how much insurance does the company carry? $300,000

    $4,800 ÷ $16 per thousand = 300 thousands ($300,000 coverage)

3. A company carries property insurance of $200,000. A fire causes $210,000 in damage. How much does the insurance company pay the insured? $200,000

    $210,000 damage exceeds the coverage; damages of $200,000 are paid.

4. A company carries property insurance of $300,000 with a premium of $13.10 per thousand dollars of coverage. A fire causes $120,000 in damage.

    a. How much does the insurance company pay the insured? $120,000

    Coverage exceed damages; insurance pays full damage of $120,000.

    b. What is the amount of the company's benefits after its annual premium payment? $116,070

    $13.10 × 300 = $3,930

    $120,000 − $3,930 coverage = $116,070

Score for A (42)

Assignment 12.2 Continued

B (58 points) Solve the following problems. (points for correct answers as marked)

5. A building worth $300,000 is insured for $180,000, and the policy carries an 80% coinsurance clause. A fire causes $220,000 in damage.

   a. How much will the insurance company pay? (10 points) $165,000

   $300,000 × 80% = $240,000 insurance required

   $$\frac{\$180,000}{\$240,000} \times \$220,000 \text{ damage} = \$165,000$$

   $165,000 is less than the $180,000 coverage, so the insurance company pays the entire $165,000

   b. How much will the insured pay if the building is repaired for $220,000? (6 points) $55,000

   $220,000 repairs − $165,000 insurance = $55,000

   c. How much would the insurance company pay if damage to the building totaled $300,000? (10 points)

   $180,000

   $$\frac{\$180,000}{\$240,000} \times \$300,000 \text{ damage} = \$225,000$$

   $180,000—the insurance company doesn't pay in excess of coverage

   d. If the damage totaled $300,000, how much would the insured pay if the building were rebuilt for $300,000? (6 points) $120,000

   $300,000 damage − $180,000 insurance = $120,000 paid by the insured

6. A building worth $1,800,000 is insured for $1,200,000, and the policy carries an 80% coinsurance clause. A fire causes $300,000 in damage.

   a. How much does the insurance company pay if the building is repaired for $300,000? (10 points) $280,000

   $1,800,000 × 80% = $1,440,000 required. $1,200,000 ÷ $1,440,000 × 300,000 = $250,000

   b. How much does the insured pay? (6 points) $50,010

   $300,000 damage − $250,000 insurance = $50,000 paid by the insured

7. If an insurance company issues insurance on property valued at $400,000 with a 90% coinsurance clause, what is the amount required to be carried by the insured? (5 points) $360,000

   $400,000 × 90% = $360,000

8. If an insurance company issues insurance on property valued at $200,000 with a 70% coinsurance clause, what is the amount required to be carried by the insured? (5 points) $140,000

   $200,000 × 70% = $140,000

   Score for B (58)

Assignment 12.3: Life and Medical Insurance

Name

Date   Score

Learning Objectives **5 6 7**

A (50 points) Refer to Figures 12-1 and 12-2 in solving the following problems. Assume that every year is a full 12 months long. (points for correct answers as marked)

1. Find the rates per thousand dollars and the premiums for the following policies. (1 point for each correct answer)

Age	Type	Payments Made	Face Value of Policy	Rate per $1,000	Premium Paid Each Year
28	Straight Life	Annually	$200,000	$19.30	$3,860.00
25	20-Payment Life	Quarterly	80,000	$8.26	$2,643.20
25	20-Year Endowment	Semiannually	10,000	$27.04	$540.80
26	Straight Life	Quarterly	120,000	$4.91	$2,356.80
27	20-Payment Life	Semiannually	100,000	$16.83	$3,366.00
28	20-Year Endowment	Annually	85,000	$53.86	$4,578.10

2. Find the cash surrender or loan value for each of the following policies issued at age 25. (1 point for each correct answer)

Policy Year	Type of Policy	Amount of Policy	Cash Surrender or Loan Value
10	Straight Life	$50,000	$5,200
15	20-Payment Life	$25,000	$9,500
10	20-Year Endowment	$50,000	$20,550
3	Straight Life	$20,000	$200
5	20-Payment Life	$75,000	$6,975
4	20-Year Endowment	$60,000	$7,800

3. When Sue Adams was 27 years old, she took out a $75,000, 20-year endowment policy. She paid the premiums annually and survived the endowment period. How much more did she pay in annual premiums than she received from the insurance company at the end of 20 years? (4 points) $3,990

   $53.20 × 75 = $3,990 annual premium
   $3,990 × 20 years = $79,800 paid in
   $79,800 − $75,000 = $4,800

4. Roger Johnson purchased a $50,000 ordinary life policy and an ADB for 50% of the value of the policy. In addition, he purchased a 5-year, $50,000 term policy. He died in an accident 3 years later.
   a. How much money did Roger's beneficiaries receive? (4 points) $125,000
      $50,000 ordinary life + $25,000 ADB + $50,000 term = $125,000
   b. How much money would Roger's beneficiaries have received if he had died in an accident 7 years after purchasing the policies? (4 points) $75,000
      $50,000 ordinary life + $25,000 ADB = $75,000; term insurance would have expired
   c. How much money would Roger's beneficiaries have received if he had died of natural causes 10 years after purchasing the policies? (4 points) $50,000
      $50,000 ordinary life

Chapter 12 Insurance 247

Assignment 12.3 Continued

5. At the age of 25, Carlos Baker purchased a $50,000 straight life policy, with premiums payable annually. He also purchased a $25,000 20-payment life policy, with premiums payable semiannually. At the end of 15 years, he decided to cash in both policies.
   a. How much did Carlos receive for the straight life policy? (4 points) $9,050
      $181 × 50 = $9,050
   b. How much did Carlos receive for the 20-payment life policy? (4 points) $9,500
      $380 ×25 = $9,500
   c. How much more did Carlos pay in premiums than the total amount received for both policies? (8 points) $6,545
      $17.20 × 50 × 15 = $12,900
      $16.26 × 2 × 25 × 15 = $12,195
      $12,900 + $12,195 = $25,09 = total paid
      $9,050 + $9,500 = $18,550 total received
      $25,095 − $18,550 = $6,545

_____
Score for A (50)

B (50 points) Solve the following problems. (10 points for a correct answer to problem 6; 8 points for each other correct answer)

6. An employer provides group health coverage that includes a $250 annual deductible per family and payment of 80% of costs exceeding the deductible. How much would an employee with two dependents pay if her year's medical bills were $550 for herself; $920 for dependent 1; and $230 for dependent 2? $540
   Total medical costs: $550 + $920 + $230 = $1,700
   $1,700 − $250 deductible = $1,450
   $1,450 × 20% paid by employee = $290
   $250 + $290 = $540 total paid by employee

7. An employer provides group health coverage that includes a $400 annual deductible per family and 70% of costs over the deductible.
   a. How much would an employee with no dependents pay if his medical bills were $980 this year? $574
      $980 medical costs − $400 deductible = $580
      $580 × 30% = $174
      $400 + $174 = $574 paid by employee
   b. How much would that employee have paid this year if his medical bills were $7,480? $2,524
      $7,480 medical costs − $400 deductible = $7,080
      $7,080 × 30% = $2,124
      $400 + $2,124 = $2,524 paid by employee

8. An employer provides group health coverage with the following monthly premiums: employee only, $350; employee with one dependent, $450; and employee with multiple dependents, $550.
   a. How much does the employer pay over a 5-year period for an employee with multiple dependents? $33,000
      $550 monthly × 12 = $6,600 annually × 5 = $33,000
   b. If that employee had a dependent with a catastrophic illness that cost $97,000 for hospitalization and treatments during that 5-year period, how much did the insurance company lose on that employee, assuming that she had no other medical claims? $64,000
      $97,000 medical expenses − $33,000 premiums = $64,000
   c. If an employee with no dependents had no illnesses during that same 5-year period, how much did the insurance company make on that employee? $21,000
      $350 monthly × 12 months × 5 years = $21,000

_____
Score for B (50)

Notes

Part 4

Interest Applications

13 Simple Interest
14 Installment Purchases
15 Promissory Notes and Discounting
16 Compound Interest and Present Value

Simple Interest

Learning Objectives

By studying this chapter and completing all assignments you will be able to:

Learning Objective 1 — Compute simple interest with time in years or months.

Learning Objective 2 — Compute ordinary simple interest, using a 360-day year.

Learning Objective 3 — Compute exact simple interest, using a 365-day year.

Learning Objective 4 — Compare ordinary simple interest and exact simple interest.

Learning Objective 5 — Estimate exact simple interest computations.

Learning Objective 6 — Compute the Principal, Rate, and Time from the basic interest formula.

Most businesses and individuals buy at least some assets without making full payment at the time of the purchase. The seller gives immediate possession to the buyer but doesn't require payment until some later date. For example, large retailers such as Macy's Department Store may receive merchandise for the Christmas season but may not be required to pay the seller until January. The seller, who *extends credit* to the buyer, may or may not charge for this privilege. The charge is called **interest,** and it is usually quoted as a percent of the amount of credit extended (the principal). When part of the price is paid at the time of purchase, that part is called a **down payment.**

If the seller charges too much interest or doesn't extend credit, the buyer might borrow money from a third party, such as a bank. A retailer such as Macy's could then buy the merchandise and sell it to repay the bank loan. The amount borrowed is called the **principal,** and the interest charged is a percent of the principal. The bank will charge interest between the loan date and the repayment date. This period of **time** is called the **interest period** or the **term of the loan.**

The promise to repay a loan or pay for merchandise may be oral or written. If it is written, it may be in the form of a letter or it could be one of several special documents known collectively as **commercial paper. Short-term credit** transactions are those for between 1 day and 1 year. **Long-term credit** transactions are those for longer than 1 year. Normally, long-term credit transactions involve major items such as new buildings or equipment rather than supplies or merchandise for sale.

Computing Simple Interest

Learning Objective 1

Compute simple interest with time in years or months.

The easiest type of interest to calculate is called **simple interest.** The calculations are the same for both a loan and a purchase on credit. The interest is a percent of the principal for the period of the loan or credit. The quoted percent usually is an *annual* (yearly) rate. A rate of 10% means that the interest payment for 1 year will be 10% of the principal.

To compute the simple interest on a 1-year loan, simply multiply the Principal by the Rate.

EXAMPLE A

13.1 Most students will have heard the phrase "*I* equals *PRT*." We will write $P \times R \times T$, but remind students that the absence of any sign means to multiply, as in 2*x*, or *PRT*.

Stan McSwain borrowed $1,000 for 1 year at a rate of 8% simple interest. Compute the interest.

The principal is $1,000. The interest for 1 year is 8% of $1,000, or $0.08 \times \$1{,}000 = \$80$.

Most loans, however, are not for a period of exactly 1 year. Loans for longer periods will require the borrower to pay more interest. Likewise, loans for shorter periods will require less interest. To compute the simple interest on loans of any period, multiply the Principal by the Rate and then multiply by the Time, with Time stated in years or in fractions of years. The fundamental formula for simple interest is

> Interest = Principal $\times$ Rate $\times$ Time
> abbreviated as $I = P \times R \times T$ or, even more simply, $I = PRT$.

EXAMPLE B

Find the simple interest on loans of $1,200 when the rate is 6% and the loan periods are $\frac{3}{4}$ year and 4 years.

¾ year

$I = P \times R \times T$

$= \$1,200 \times 0.06 \times \dfrac{3}{4}$

$= \$54$

4 years

$I = P \times R \times T$

$= \$1,200 \times 0.06 \times 4$

$= \$288$

The time period often will be measured in months instead of years. Before computing the interest, change the time into years by dividing the number of months by 12 (the number of months in 1 year).

EXAMPLE C

Compute the interest on credit purchases of $3,000 at 5% for periods of 8 months and 30 months.

8 months

$I = P \times R \times T$

$= \$3,000 \times 0.05 \times \dfrac{8}{12}$

$= \$100$

30 months

$I = P \times R \times T$

$= \$3,000 \times 0.05 \times \dfrac{30}{12}$

$= \$375$

13.2 If students do not use calculators, they should be encouraged to reduce fractions and cancel. Students who are using calculators probably should not reduce fractions because it won't save any time and introduces the possibility of mental errors.

USING CALCULATORS

Today, calculators or computers are used in almost every interest application. The numbers are often large and are always important. The steps are performed on the calculator in the same order as they are written in the formula.

EXAMPLE D

Write the calculator steps for computing the simple interest on $8,000,000 at 9% for 18 months.

$I = P \times R \times T = \$8,000,000 \times 0.09 \times \dfrac{18}{12}$

8 000 000 [×] .09 [×] 18 [÷] 12

[=] 1,080,000, or $1,080,000

With the percent key [%], the steps would be

8 000 000 [×] 9 [%] [×] 18 [÷] 12 [=] 1,080,000, or $1,080,000

✓ CONCEPT CHECK 13.1

The principal is $2,500, the rate is 10%, and interest = Principal × Rate × Time, or $I = P \times R \times T$. Find the interest both for 5 years and for 6 months.

a. If Time is 5 years: $I = P \times R \times T = \$2,500 \times 0.10 \times 5 = \$1,250$

b. If Time is 6 months: $I = P \times R \times T = \$2,500 \times 0.10 \times \dfrac{6}{12} = \$125$

Chapter 13 Simple Interest

Computing Ordinary Interest

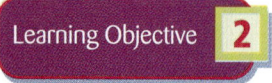

Compute ordinary simple interest, using a 360-day year.

If the term of the loan is stated as a certain number of days, computing interest involves dividing the number of days by the number of days in 1 year—either 360 or 365. Before computers and calculators, interest was easier to compute by assuming that every year had 360 days and that every month had 30 days. The 360-day method, called the **ordinary interest method,** is still used by some businesses and individuals.

EXAMPLE E

Compute the ordinary simple interest on $900 at 9% for 120 days.

$I = P \times R \times T$

$= \$900 \times 0.09 \times \dfrac{120}{360}$

$= \$27$

CONCEPT CHECK 13.2

The Principal is $4,000, the Rate is 7%, and the Time is 180 days. Compute the ordinary simple interest.

Ordinary interest involves use of a 360-day year: $I = P \times R \times T = \$4,000 \times 0.07 \times \dfrac{180}{360} = \$140$

Computing Exact Interest

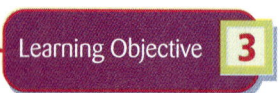

Compute exact simple interest, using a 365-day year.

Banks, savings and loan institutions, credit unions, and the federal government use a 365-day year (366 days for leap years) to compute interest. This method is called the **exact interest method.** The computations are the same as for ordinary simple interest, except that 365 days is used instead of 360 days.

EXAMPLE F

Compute the exact simple interest on $900 at 9% for 120 days.

$I = P \times R \times T$

$= \$900 \times 0.09 \times \dfrac{120}{365}$

$= \$26.6301$, or $26.63

Part 4 Interest Applications

✓ CONCEPT CHECK 13.3

The Principal is $4,000, the Rate is 7%, and the Time is 180 days. Compute the exact simple interest.

> Exact interest involves use of a 365-day year: $I = P \times R \times T = \$4,000 \times 0.07 \times \frac{180}{365} = \$138.08$

Comparing Ordinary Interest and Exact Interest

The 360-day year was very useful before the advent of calculators and computers, so there is a long tradition of using it. However, the 365-day year is more realistic than the 360-day year. Also, the 365-day year is financially better for the borrower because the interest amounts are always smaller. (Why? Because a denominator of 365 gives a smaller quotient than a denominator of 360).

Reexamine examples E and F. The difference between ordinary interest and exact interest is only $27.00 − $26.63, or $0.37. When businesses borrow money, however, the principal may be very large and then the difference will be more significant. Example G is similar to examples E and F, except that the principal is in millions of dollars rather than hundreds.

Learning Objective 4

Compare ordinary simple interest and exact simple interest.

13.3 Before calculators and computers, interest calculations were simplified by using a 360-day year. First, a year could be divided into twelve 30-day months. Second, 360 days made cancellation more likely because 360 has so many divisors. If you discuss cancellation, have students compare the number of divisors of 360 and 365. A good hint is that $360 = 2 \times 2 \times 2 \times 3 \times 3 \times 5$, whereas $365 = 5 \times 73$.

13.4 The difference between ordinary and exact interest is not trivial for large principal amounts.

● EXAMPLE G

Find the difference between ordinary interest and exact interest on $8,000,000 at 9% for 120 days.

Ordinary Interest
$I = P \times R \times T$

$= \$8,000,000 \times 0.09 \times \frac{120}{360}$

$= \$240,000$

Exact Interest
$I = P \times R \times T$

$= \$8,000,000 \times 0.09 \times \frac{120}{365}$

$= \$236,712.3288$ or $\$236,712.33$

The difference is $240,000 − $236,712.33, or $3,287.67.

✓ CONCEPT CHECK 13.4

The principal is $6,000, the rate is 12%, and the time is 120 days. Find the difference between the amounts of simple interest calculated by using the ordinary method (360-day year) and the exact method (365-day year).

> Ordinary interest: $I = P \times R \times T = \$6,000 \times 0.12 \times \frac{120}{360} = \$240.00$
> Exact interest: $I = P \times R \times T = \$6,000 \times 0.12 \times \frac{120}{365} = \$236.71$
> Difference = Ordinary interest − Exact interest = $240.00 − $236.71 = $3.29

Estimating Exact Simple Interest

Learning Objective 5

Estimate exact simple interest computations.

Although calculators are used to compute exact interest, approximation remains very useful. The following calculator solution requires a minimum of 20 key entries.

8 000 000 × .09 × 120 ÷ 365 = 236 712.3288

Pressing any one of the 20 keys incorrectly can result in a large error. By making an estimate of the interest in advance, you may spot a significant calculator error.

13.5 Estimating may be more important now than ever before. When students depend on calculators, they often quit thinking about the "reasonableness" of an answer—they assume that the calculator will be correct.

COMBINATIONS OF TIME AND INTEREST THAT YIELD 1%

To simplify mental approximations, you can round the rate and time to numbers that are easy to compute mentally. Also, use 360 days instead of 365 because it cancels more often. For ordinary interest, several combinations of rate and time are easy to use because their product is 1%. For example, $12\% \times \frac{30}{360} = 12\% \times \frac{1}{12} = 1\%$ and $6\% \times \frac{60}{360} = 6\% \times \frac{1}{6} = 1\%$.

13.6 Again, the reason for using 360 days in estimation is that the various divisors create several possible shortcut combinations.

EXAMPLE H

Approximate the ordinary simple interest on $2,500 at 6.15% for 59 days. Then calculate the actual ordinary simple interest.

Round   6.15% to 6% and 59 days to 60 days.

Estimate: $\$2{,}500 \times 0.06 \times \frac{60}{360} = \$2{,}500 \times 0.01 = \$25.00$

Actual interest: $\$2{,}500 \times 0.0615 \times \frac{59}{360} = \$25.1979$, or $25.20

OTHER RATES AND TIMES

Table 13.1 shows several combinations of rate and time whose products are useful for estimating interest.

Table 13-1: Rate and Time

$4\% \times \frac{90}{360} = 4\% \times \frac{1}{4} = 1\%$		$10\% \times \frac{36}{360} = 10\% \times \frac{1}{10} = 1\%$
$6\% \times \frac{60}{360} = 6\% \times \frac{1}{6} = 1\%$		$12\% \times \frac{30}{360} = 12\% \times \frac{1}{12} = 1\%$
$8\% \times \frac{45}{360} = 8\% \times \frac{1}{8} = 1\%$		$18\% \times \frac{20}{360} = 18\% \times \frac{1}{18} = 1\%$
$9\% \times \frac{40}{360} = 9\% \times \frac{1}{9} = 1\%$		$6\% \times \frac{120}{360} = 6\% \times \frac{1}{3} = 2\%$
$12\% \times \frac{60}{360} = 12\% \times \frac{1}{6} = 2\%$		$12\% \times \frac{90}{360} = 12\% \times \frac{1}{4} = 3\%$
$8\% \times \frac{90}{360} = 8\% \times \frac{1}{4} = 2\%$		$9\% \times \frac{120}{360} = 9\% \times \frac{1}{3} = 3\%$

ESTIMATING EXACT INTEREST

The goal in approximating interest is just to get an estimate. Even though exact interest requires 365 days in a year, you can make a reasonable estimate by assuming that the number of days in a year is 360. This permits the use of all of the shortcut combinations from Table 13.1.

Part 4 Interest Applications

EXAMPLE I

First, compute the actual exact simple interest on $1,200 at 11.8% for 62 days.

Actual interest: $\$1{,}200 \times 0.118 \times \dfrac{62}{365} = \$24.0526$, or $24.05

Second, estimate the amount of interest by using 12% instead of 11.8%, 60 days instead of 62 days, and 360 instead of 365.

Estimate: $\$1{,}200 \times 0.12 \times \dfrac{60}{360} = \$1{,}200 \times 0.02 = \$24$

The difference in $24.05 − $24 = $0.05.

✓ CONCEPT CHECK 13.5

The Principal is $3,750, the Rate is 9.1%, and the Time is 39 days. Calculate the actual exact simple interest. Then make an estimate by using a 360-day year and simpler values for R and T. Compare the results.

Actual interest: $I = P \times R \times T = \$3{,}750 \times 0.09 \times \dfrac{39}{365} = \$36.4623$ or $36.46

Estimate: $I = P \times R \times T = \$3{,}750 \times 0.09 \times \dfrac{40}{360} = \$3{,}750 \times 0.01 = \$37.50$

Difference: Estimate − Actual = $37.50 − $36.46 = $1.04

Computing the Interest Variables

Every simple interest problem has four variables: Interest Amount, Principal, Rate, and Time. Thus far, you have solved for the Interest Amount (I) when the Principal (P), Rate (R), and Time (T) were all given. However, as long as any three variables are given, you can always compute the fourth by just changing the formula $I = P \times R \times T$ into one of its possible variations, as shown in Table 13-2.

Learning Objectives 6

Compute the Principal, Rate, and Time from the basic interest formula.

13.7 If all the students in a class are prepared for algebra, this is a natural place to use it. Students could substitute all the known values into the equation $I = P \times R \times T$ and solve for the missing variable by dividing both sides of the equation.

Table 13-2: PRT formulas

To find	You must know	Use this formula
I	P, R, and T	$I = P \times R \times T$
P	I, R, and T	$P = \dfrac{I}{(R \times T)}$
R	I, P, and T	$R = \dfrac{I}{(P \times T)}$
T	I, P, and R	$T = \dfrac{I}{(P \times R)}$

Assume the use of ordinary interest (a 360-day year) unless the use of exact interest (a 365-day year) is indicated. The stated or computed interest rate is the rate for 1 full year. Also, the length of time used for computing interest dollars must be stated in terms of all or part of a year.

FINDING THE INTEREST AMOUNT, PRINCIPAL, RATE, OR TIME

When any three variables are known, you can solve for the fourth variable, using a formula from Table 13.2. All rates are ordinary simple interest (360-day year).

EXAMPLE J

Find the Principal if the Interest Amount is $75, the Rate is 6%, and the Time is 30 days.

$P = ?; \quad I = \$75; \quad R = 6\%; \quad T = \dfrac{30}{360} \text{ year}$

$P = \dfrac{I}{(R \times T)} = \dfrac{\$75}{\left(0.06 \times \dfrac{30}{360}\right)} = \dfrac{\$75}{0.005} = \$15{,}000$

EXAMPLE K

Find the Rate if the Interest Amount is $22, the Principal is $2,000, and the Time is 30 days.

$R = ?; \quad I = \$22; \quad P = \$2{,}000; \quad T = \dfrac{30}{360} \text{ year}$

$R = \dfrac{I}{(P \times T)} = \dfrac{\$22}{\left(\$2{,}000 \times \dfrac{30}{360}\right)} = \dfrac{\$22}{\$166.67} = 0.132, \text{ or } 13.2\%$

EXAMPLE L

Find the Time if the Interest Amount is $324, the Principal is $4,800, and the Rate is 9%. Express Time in days, based on a 360-day year.

$T = ?; \quad I = \$324; \quad P = \$4{,}800; \quad R = 9\%$

$T = \dfrac{I}{(P \times R)} = \dfrac{\$324}{(\$4{,}800 \times 0.09)} = \dfrac{\$324}{\$432} = 0.75 \text{ year}$

Based on a 360-day year, 0.75 year = 0.75 × 360 days = 270 days.

CONCEPT CHECK 13.6

Each of the following problems gives three of the four variables. Find the missing variable. All rates are ordinary simple interest (360-day year). Round P and I to the nearest cent; round R to the nearest $\frac{1}{10}$%; round T to the nearest whole day, assuming that 1 year has 360 days. Use one of the four formulas:

$$I = P \times R \times T, \quad P = \frac{I}{(R \times T)}, \quad R = \frac{I}{(P \times T)}, \quad \text{and} \quad T = \frac{I}{(P \times R)}$$

a. Principal = $1,240; Rate = 6%; Time = 270 days
   Find Interest Amount:
   $$I = P \times R \times T = \$1,240 \times 0.06 \times \frac{270}{360} = \$55.80$$

b. Principal = $8,000; Interest Amount = $50; Time = 45 days
   Find Rate:
   $$R = \frac{I}{(P \times T)} = \frac{\$50}{\left(\$8,000 \times \frac{45}{360}\right)} = 0.05, \text{ or } 5\%$$

c. Principal = $1,280; Interest Amount = $64; Rate = 10%
   Find Time:
   $$T = \frac{I}{(P \times R)} = \frac{\$64}{(\$1,280 \times 0.10)} = 0.5 \text{ year}$$
   In a 360-day year, T = 0.5 year = 0.5 × 360 days = 180 days.

d. Interest Amount = $90; Rate = 9%; Time = 60 days
   Find Principal:
   $$T = \frac{I}{(R \times T)} = \frac{\$90}{\left(0.09 \times \frac{60}{360}\right)} = \$6,000$$

COMPLETE ASSIGNMENTS 13.1 AND 13.2.

Chapter Terms for Review

commercial paper
down payment
exact interest method
interest
interest period
long-term credit

ordinary interest method
principal
short-term credit
simple interest
term of the loan
time

Chapter 13 Simple Interest

THE BOTTOM LINE

Summary of chapter learning objectives:

Learning Objective	Example
13.1 Compute simple interest with time in years or months	Find the simple interest using the basic formula: **Interest = Principal × Rate × Time**, or $I = P \times R \times T$ 1. Principal = \$3,500; Rate = 9%; Time = 2.5 years 2. Principal = \$975; Rate = 8%; Time = 9 months
13.2 Compute ordinary simple interest, using a 360-day year	3. Find the ordinary simple interest for a 360-day year: Principal = \$3,000; Rate = 10%; Time = 240 days
13.3 Compute exact simple interest, using a 365-day year	4. Find the exact simple interest for a 365-day year: Principal = \$2,800; Rate = 7%; Time = 75 days
13.4 Compare ordinary simple interest and exact simple interest	5. Find the difference between ordinary simple interest and exact simple interest: Principal = \$5,000; Rate = 6%; Time = 75 days
13.5 Estimate simple interest computations	6. Estimate the exact interest by using a 360-day year and simpler values for Rate and Time: Principal = \$2,100; Rate = 5.8%; Time = 62 days
13.6 Compute the Principal, Rate, and Time from the basic interest formula	Solve for Principal, Rate, and Time using a 360-day year and the formulas $P = \dfrac{I}{(R \times T)}$, $R = \dfrac{I}{(P \times T)}$, and $T = \dfrac{I}{(P \times R)}$ 7. Interest Amount = \$42; Rate = 6% Time = 105 days 8. Principal = \$1,600; Interest Amount = \$30; Time = 75 days 9. Principal = \$7,200; Interest Amount = \$135; Rate = 15%

Answers: 1. \$787.50 2. \$58.50 3. \$200.00 4. \$40.27 5. \$62.50 − \$61.64 = \$0.86 6. \$2,100 × 0.06 × $\frac{60}{360}$ = \$21 7. 2,400 8. 0.09, or 9% 9. 45 days

SELF-CHECK

Review Problems for Chapter 13

In problems 1 and 2, compute the amount of (a) ordinary simple interest and (b) the amount of exact simple interest. Then compute (c) the difference between the two interest amounts

	Principal	Rate	Time	Ordinary Interest	Exact Interest	Difference
1	$1,680	6%	270 Days	a. _____	b. _____	c. _____
2	$10,500	8%	60 Days	a. _____	b. _____	c. _____

In problems 3 and 4, first compute (a) the actual exact simple interest. Then, change each rate and time to the closest numbers that permit use of the shortcuts shown in Table 13.1 and compute (b) the *estimated* amount of exact interest. Finally, compute (c) the difference between the actual and estimated exact interest.

	Principal	Rate	Time	Actual Exact Interest	Estimated Exact Interest	Difference
3	$12,000	3.8%	92 Days	a. _____	b. _____	c. _____
4	$2,000	9.2%	117 Days	a. _____	b. _____	c. _____

5. Dick Liebelt borrowed money for 240 days at a rate of 9% ordinary simple interest. How much did Dick borrow if he paid $90 in interest? _____

6. Linda Rojas loaned $1,000 to one of her employees for 90 days. If the employee's interest amount was $12.50, what was the ordinary simple interest rate? _____

7. Tessa O'Leary loaned $10,000 to a machine shop owner who was buying a piece of used equipment. The interest rate was 6% exact simple interest, and the interest amount was $360. Compute the number of days of the loan. _____

8. Kaye Mushalik loaned $2,500 to Fay Merritt, a good friend since childhood. Because of their friendship, Kaye charged only 3% ordinary simple interest. Two months later, when Fay received her annual bonus, she repaid the entire loan and all the interest. What was the total amount that Fay paid? _____

9. Katherine Wu and her sister Madeline have a home decorating and design business. Often, they buy antiques and fine art objects and then resell the items to their clients. They have a line of credit at their bank to provide short-term financing, if necessary, for these purchases. The bank always charges exact simple interest, but the rate varies depending on the economy. Katherine and Madeline need to borrow $22,400 for 90 days to buy a large collection of antique furniture at an estate sale. If the bank charges 5.25%, how much interest would they pay? _____

Notes

Assignment 13.1: Simple Interest

Name

Date   Score

Learning Objectives **1 2 3 4 5**

A (20 points) Compute the simple interest. If the time is given in months, let one month be $\frac{1}{12}$ of a year. If the time is in days, let one year be 360 days. (2 points for each correct answer)

	Principal	Rate	Time	Interest		Principal	Rate	Time	Interest
1.	$500	6.0%	1 year	$30	**2.**	$4,000	8%	3 years	$960
	$500 × 0.06 × 1					$4,000 × 0.08 × 3			
3.	$1,800	8%	4 months	$48	**4.**	$960	5%	21 months	$84
	$1,800 × 0.08 × $\frac{4}{12}$					$960 × 0.05 × $\frac{21}{12}$			
5.	$7,500	5%	180 days	$187.50	**6.**	$3,600	12%	30 months	$1,080
	$7,500 × 0.05 × $\frac{180}{360}$					$3,600 × 0.12 × $\frac{30}{12}$			
7.	$12,800	7%	2.5 years	$2,240	**8.**	$450	5%	$3\frac{1}{2}$ years	$78.75
	$12,800 × 0.07 × 2.5					$450 × 0.05 × 3.5			
9.	$5,200	10%	90 days	$130	**10.**	$20,000	7.5%	8 months	$1,000
	$5,200 × 0.10 × $\frac{90}{360}$					$20,000 × 0.075 × $\frac{8}{12}$			

Score for A (20)

Assignment 13.1 Continued

B (30 points) Compute the ordinary interest, the exact interest, and their difference. Round answers to the nearest cent. (2 points for each correct interest; 1 point for each correct difference)

	Principal	Rate	Time	Ordinary Interest	Exact Interest	Difference
11.	$2,400	4%	180 days	$48.00	$47.34	$0.66
				$2,400 \times 0.04 \times \frac{180}{360}$	$2,400 \times 0.04 \times \frac{180}{365}$	$48.00 - $47.34
12.	$4,800	5%	75 days	$50.00	$49.32	$0.68
				$4,800 \times 0.05 \times \frac{75}{360}$	$4,800 \times 0.05 \times \frac{75}{365}$	$50 - $49.32
13.	$12,000	6%	240 days	$480.00	$473.42	$6.58
				$12,000 \times 0.06 \times \frac{240}{360}$	$12,000 \times 0.06 \times \frac{240}{365}$	$480.00 - $473.42
14.	$1,400	15%	60 days	$35.00	$34.52	$0.48
				$1,400 \times 0.15 \times \frac{60}{360}$	$1,400 \times 0.15 \times \frac{60}{365}$	$35 - $34.52
15.	$7,500	8%	225 days	$375.00	$369.86	$5.14
				$7,500 \times 0.08 \times \frac{225}{360}$	$7,500 \times 0.08 \times \frac{225}{365}$	$375 - $369.86
16.	$365	4%	30 days	$1.22	$1.20	$0.02
				$365 \times 0.04 \times \frac{30}{360}$	$365 \times 0.04 \times \frac{30}{365}$	$1.22 - $1.20

Score for B (30)

Assignment 13.1 Continued

C (20 points) In each problem, first find the actual exact simple interest. Then, estimate the interest by assuming a 360-day year and round each rate and time to the nearest numbers that will permit the shortcuts in Table 13-1. Finally, find the difference. Round answers to the nearest cent. (2 points for each correct estimate and actual interest; 1 point for each correct difference)

	Principal	Rate	Time	Actual Exact Interest	Estimate	Difference
17.	$625	8.1%	46 days	$6.38	$6.25	$0.13
	$625 \times 0.081 \times \frac{46}{365}$	$8\% \times \frac{45}{360} = 1\%$		625×0.01		$6.38 - $6.25
18.	$5,600	3.99%	92 days	$56.32	$56	$0.32
	$5,600 \times 0.0399 \times \frac{92}{365}$	$4\% \times \frac{90}{360} = 1\%$		$5,600 \times 0.01$		$56.32 - $56
19.	$2,000	8.95%	123 days	$60.32	$60	$0.32
	$2,000 \times 0.0895 \times \frac{123}{365}$	$9\% \times \frac{120}{360} = 3\%$		$2,000 \times 0.03$		$60.32 - $60
20.	$10,000	6%	61 days	$100.27	$100	$0.27
	$10,000 \times 0.06 \times \frac{61}{365}$	$6\% \times \frac{60}{360} = 1\%$		$10,000 \times 0.01$		$100.27 - $100.00

Score for C (20)

D (30 points) Determine the missing variable by using one of the formulas

$$I = P \times R \times T, \quad P = \frac{I}{(R \times T)}, \quad R = \frac{I}{(P \times T)}, \quad \text{or} \quad T = \frac{I}{(P \times R)}.$$

For problems 21–25, use a 360-day year. For problems 26–30, use a 365-day year. Round dollar amounts to the nearest cent. Round interest rates to the nearest $\frac{1}{10}$ of a percent. Find the time in days, rounded to the nearest whole day. (3 points for each correct answer)

	Principal	Rate	Time	Interest
21.	$4,800	11%	240 days	$352.00
	$P = \frac{I}{R \times T} = \frac{\$352}{\left(0.11 \times \frac{240}{360}\right)} = \$4,800$			
22.	$12,000	5%	30 days	$50.00
	$T = \frac{I}{P \times R} = \frac{\$50}{(\$12,000 \times 0.05)} = 0.083$ year		$0.083 \times 360 = 29.88$, or 30 days	

Chapter 13 Simple Interest

Assignment 13.1 Continued

	Principal	Rate	Time	Interest
23.	$600	8%	45 days	$6.00

$$R = \frac{I}{P \times T} = \frac{\$6}{\left(\$600 \times \frac{45}{360}\right)} = 0.08$$

24.	$2,480	6%	75 days	$31.00

$$I = P \times R \times T = \$2,480 \times 0.06 \times \frac{75}{360} = \$31$$

25.	$25,000	4%	225 days	$625.00

$$T = \frac{I}{(P \times R)} = \frac{\$625}{(\$25,000 \times 0.04)} = 0.625 \text{ year} \quad 0.625 \times 360 = 225 \text{ days}$$

26.	$8,618.06	8%	270 days	$510.00

$$P = \frac{I}{(R \times T)} = \frac{\$510}{\left(0.08 \times \frac{270}{365}\right)} = \$8,618.06$$

27.	$1,350	7.6%	120 days	$33.73

$$I = P \times R \times T = \$1,350 \times 0.076 \times \frac{120}{365} = \$33.73$$

28.	$34,950	5.5%	75 days	$395.00

$$T = \frac{I}{(P \times R)} = \frac{\$395}{(\$34,950 \times 0.055)} = 0.2055 \text{ year} \quad 0.2055 \times 365 = 75 \text{ days}$$

29.	$16,000	7.5%	90 days	$296.00

$$R = \frac{I}{(P \times T)} = \frac{\$296}{\left(\$16,000 \times \frac{90}{365}\right)} = 0.0750$$

30.	$2,758.88	4.9%	135 days	$50.00

$$P = \frac{I}{(R \times T)} = \frac{\$50}{\left(0.049 \times \frac{135}{365}\right)} = \$2,758.88$$

Score for B (30)

Assignment 13.2: Simple Interest Applications

Name

Date   Score

Learning Objectives **1** **2** **6**

A (50 points) Solve each of the following ordinary simple interest problems by using a 360-day year. Find both the interest dollars and the total amount (i.e., principal plus interest) of the loan. (7 points for each correct interest; 3 points for each correct amount)

1. Tom Titus plans to lend $850 to his friend Bill White so that Bill can fly with him to Canada for vacation. Tom is charging Bill only 3% ordinary simple interest. Bill repays everything, interest plus principal, to Tom 180 days later. How much does Bill pay?

    $850 \times 0.03 \times \frac{180}{360}$ Interest $12.75

    $850 + $12.75   Amount $862.75

2. Tony Woo and Helen Lee are planning to start a business that will export American food to China. They estimate that they will need $75,000 to pay for organizational costs, get product samples, and make three trips to Shanghai. They can borrow the money from their relatives for 4 years. Tony and Helen are willing to pay their relatives 9% ordinary simple interest. Compute the total amount that Tony and Helen will owe their relatives in 4 years.

    $75,000 \times 0.09 \times 4$ Interest $27,000

    $75,000 + $27,000 = $102,000   Amount $102,000

3. Carolyn Wilfert owns a temporary services employment agency. Businesses call her when they need to hire various types of workers for a short period of time. The businesses pay a fee to Carolyn, who pays the salaries and benefits to the employees. One benefit is that Carolyn will make small, short-term loans to her employees. After a flood, employee Judy Hillstrom needed to borrow $3,600 to have her house cleaned and repainted. Judy repaid the loan in 6 months. If Carolyn charged 5% ordinary simple interest, how much did Judy repay?

    $3,600 \times 0.05 \times \frac{6}{12}$ Interest $90

    $3,600 + $90   Amount $3,690

4. Several years ago, Dick Shanley and Karl Coke formed a partnership to rent musical instruments to school districts that do not want to own and maintain the instruments. In the spring, they investigate borrowing $80,000 to buy trumpets and trombones. Because they collect their rental fees in advance, they anticipate being able to repay the loan in 135 days. How much will they need to repay if the ordinary simple interest rate is 6.5%?

    $80,000 \times 0.065 \times \frac{135}{360}$ Interest $1,950

    $80,000 + $1,950   Amount $81,950

5. With her husband, Ruby Williams owns and manages a video game arcade. A manufacturer developed a new line of games and offered very low interest financing to encourage arcade operators such as Ruby to install the new games. Ruby was able to finance $75,000 worth of games for 8 months for 3.2% ordinary simple interest. Calculate how much Ruby will repay.

    $75,000 \times 0.032 \times \frac{8}{12}$ Interest $1,600

    $75,000 + $1,600   Amount $76,600

Score for A (50)

Assignment 13.2 Continued

B (50 points) Solve each of the following exact simple interest problems by using a 365-day year. Find both the interest dollars and the total amount (i.e., principal plus interest) of the loan. (7 points for each correct interest; 3 points for each correct amount)

6. Robert Burke, managing partner of a local transportation company, thinks that the company should borrow money to upgrade its truck repair facility. After investigating several sources of short-term loans, Robert determines that the company can borrow $400,000 for 200 days at 5.5% exact simple interest. If the company agrees to take out this loan, how much will it need to repay at the end of the 200 days?

    $400,000 \times 0.055 \times \frac{200}{365}$ Interest $12,054.79

    $400,000 + $12,054.79   Amount $412,054.79

7. Dave Engle, a former teacher, now has a business selling supplemetary educational materials such as books and computer software to parents and schools. In June, he borrowed $45,000 from his bank to buy some new educational computer games that he hopes to sell during August and September. The bank's rate is 6.25% exact simple interest as long as the time does not exceed half a year. If Dave repays everything in 120 days, how much will he pay?

    $45,000 \times 0.0625 \times \frac{120}{365}$ Interest $924.66

    $45,000 + $924.66   Amount $45,924.66

8. After working in construction for 5 years, Jerry Weekly had saved almost enough money to buy a fishing boat and move to Alaska to become a commercial fisherman. He still needed $9,500, which his wife could borrow from her parents until the end of the first fishing season. The parents charged 5% exact simple interest, and Jerry repaid them after 95 days. How much interest did he pay, and what was the total amount?

    $9,500 \times 0.05 \times \frac{95}{365}$ Interest $123.63

    $9,500 + $123.63   Amount $9,623.63

9. Bill and Carol Campbell need to purchase two new saws for their retail lumber yard. The company that sells the saws offers them some short-term financing at the relatively high rate of 11% exact simple interest. They decide to accept the financing offer, but only for $5,000 and only for 45 days. How much will Bill and Carol repay at the end of the 45 days?

    $5,000 \times 0.11 \times \frac{45}{365}$ Interest $67. 81

    $5,000 + $67.81   Amount $5,067.81

10. After working for a large accounting firm for 10 years, Bette Ryan, C.P.A., decided to open her own office. She borrowed $60,000 at 7.2% exact simple interest. She made enough during the first income tax season to repay the loan in 190 days. How much did Bette repay?

    $60,000 \times 0.072 \times \frac{190}{365}$ Interest $2, 284.77

    $60,000 + $2,248.77   Amount $62,248.77

Score for B (50)

Installment Purchases

Learning Objectives
By studying this chapter and completing all assignments you will be able to:

Learning Objective 1 — Convert between annual and monthly interest rates.

Learning Objective 2 — Compute simple interest on a monthly basis.

Learning Objective 3 — Compute finance charges for credit account purchases.

Learning Objective 4 — Compute costs of installment purchases.

Learning Objective 5 — Compute effective rates.

Learning Objective 6 — Amortize a loan.

Learning Objective 7 — Compute the monthly payment on a home mortgage.

Most individuals today can purchase goods or services on credit if they choose. The buyer gets immediate possession or immediate service but delays payment. Either the seller extends the credit or the buyer uses a **credit card,** or loan, from a third party.

Credit is usually offered for an interest charge, which is usually computed each month. A summary of the purchases, payments, and interest charges is sent to the borrower (credit purchaser) each month. It may not be simple to compare the methods used to compute interest by competing lenders. Some lenders may charge interest on the **average daily balance**. Although it is a simple concept, and easy for a computer to calculate, it may be difficult for the purchaser to reconcile when he or she makes many purchases and/or merchandise returns in a single month.

In addition to interest, a lender may charge additional fees to extend credit or loan money. These might include items such as loan origination fees, membership fees, credit check fees, administrative fees, and insurance premiums. All of the fees together are called **finance charges**. These additional fees, whether one-time, annual, or monthly, also make it difficult to compare lenders because each lender could be slightly different. It is of some help to consumers that there are laws that mandate that lenders must explain their various fees and rates.

Converting Interest Rates

Learning Objective 1

Convert between annual and monthly interest rates.

The general concept behind charging for credit purchases is to compute finance charges on the unpaid balance each month. The formula is still $I = P \times R \times T$, where P is the unpaid balance. However, T is not years or a fraction of a year (as in Chapter 13)—T is in months, and R, the rate, is a monthly rate. For example, the rate might be 1.5% *per month.*

Understanding the relationship between monthly and annual rates is important.

Rule: To convert an annual rate to a monthly rate, divide the annual rate by 12; to convert a monthly rate to an annual rate, multiply the monthly rate by 12.

EXAMPLE A

a. Convert 9% per year to the equivalent monthly rate.
   9% annually ÷ 12 = 0.75% monthly

b. Convert 0.5% per month to the equivalent annual rate.
   0.5% monthly × 12 = 6% annually

CONCEPT CHECK 14.1

a. Convert an 18% annual rate to the equivalent monthly rate.
   Divide the annual rate by 12 to get the monthly rate: 18% ÷ 12 = 1.5% per month

b. Convert a 1.25% monthly rate to the equivalent annual rate.
   Multiply the monthly rate by 12 to get the annual rate: 1.25% × 12 = 15% per year

Computing Simple Interest on a Monthly Basis

In terms of single-payment simple interest, 1.5% *per month* is identical to 18% *per year*.

Rule: If the rate is annual, the time must be in years; if the rate is monthly, the time must be in months.

Learning Objective 2

Compute simple interest on a monthly basis.

EXAMPLE B

Compute the simple interest on $1,000 for 2 months at 18% per year, on an annual basis and on a monthly basis.

Annual: $I = P \times R \times T = \$1{,}000 \times 0.18 \text{ per year} \times \frac{2}{12} \text{ year} = \$30$

Monthly: 18% per year = 18% ÷ 12 = 1.5% per month

$I = P \times R \times T = \$1{,}000 \times 0.015 \text{ per month} \times 2 \text{ months}$
$= \$30$

Reminder: Both computations differ from those in Chapter 13, where you counted the exact number of days and divided by either 360 or 365.

✓ CONCEPT CHECK 14.2

Compute the simple interest on $800 for 3 months at 0.5% per month.

$I = P \times R \times T = \$800 \times 0.5\% \text{ per month} \times 3 \text{ months} = \$800 \times 0.005 \times 3 = \$12$

Computing Finance Charges

To enable consumers to compute the total cost of credit, Congress has passed several laws, beginning with the Consumer Credit Protection Act of 1968 (CCPA). Title I of the CCPA is known as the **Truth in Lending Act (TILA)**. TILA is administered by the Federal Reserve Board. Among other major legislation, Congress also passed the Consumer Leasing Act of 1976, administered by the Federal Trade Commission, and the Home Ownership and Equity Protection Act of 1994, administered by the Department of Housing and Urban Development. All of these require lenders to make certain disclosures to consumers.

Among several mandates, TILA requires creditors to tell consumers these three things:

1. The total of all finance charges, including interest, carrying charges, insurance, and special fees
2. The annual percentage rate (APR) of the total finance charge
3. The method by which they compute the finance charge

As noted in the previous section, an annual interest rate is a monthly interest rate multiplied by 12. However, as the term is used in TILA, the **annual percentage rate (APR)** is a specific, defined term that must include all finance charges, not just interest.

Learning Objective 3

Compute finance charges for credit account purchases.

14.1 There is much consumer protection information on different Websites. For example, you can direct students to go to www.federalreserve.gov, click on "Consumer Information", and click on "Consumer Handbook to Credit Protection Laws." Or go to www.ftc.gov, click on "For Consumers," click on "Credit", and click on "Rules and Acts." These Websites do change, so please try each Website yourself before advising students to try.

14.2 The terminology and calculation rules of the Truth in Lending Act are summarized in the Comptroller's Handbook on Truth in Lending published in 1996 by the Office of the Comptroller of the Currency. You can find it, or direct students to it, at www.occ.treas.gov/handbook/til.pdf.

14.3 The term *annual percentage rate* has more than one meaning in business. It may refer to compound interest rather than simple interest. In TILA, however, it has a specific meaning and may include more than just interest.

14.4 Credit card interest calculations can be very complex—sometimes almost impossible for the cardholder to reconcile. Taking real terms from an actual credit card statement, yours or a student's, may be interesting. It may be useful or necessary to write to the issuing company for an example of how it calculates charges. You may be able to find the information on the company's Website. And even with that information, the method of calculation may be difficult to duplicate without a computer.

Furthermore, under TILA, lenders are permitted to use more than one method to compute the APR. Lenders may even use either a 360-day year or a 365-day year. TILA does not set limits on rates.

As mentioned, TILA does require that total finance charges be stated clearly, that the finance charges also be stated as an annual percentage rate, and the method of computation be given. Although the method that is mentioned may be stated clearly, it may not always be simple for a consumer to calculate. One difficulty might be to determine the account balance that is to be used in the calculation. A wide variety of methods may be applied. For example:

1. The finance charge may be based on the amount owed at the beginning of the current month, ignoring payments and purchases.
2. The finance charge may be based on the amount owed at the beginning of the month, after subtracting any payments during the month and ignoring purchases.
3. The finance charge may be based on the average daily balance. (Add the unpaid balance each day; divide the total by the number of days in the month.) Payments are usually included; new purchases may or may not be included.
4. A variation of the average daily balance method is to compute the interest charge each day, on a daily basis, and then add all the daily interest charges for the month.

Although the total finance charges, and the annual percentage rate, and the method of calculation may all be clearly stated, some consumers will have difficulty reconstructing the interest and finance charges on their bills. A consumer who wants to understand more can write to the creditor for a more detailed explanation and even an example of how to do the calculations.

Figure 14-1 is the lower portion of a typical statement of a retail store. Examples C and D illustrate two simple methods used to compute finance charges.

Figure 14-1 Retail Statement of Account

PREVIOUS BALANCE	FINANCE CHARGE	PAYMENTS	CREDITS	PURCHASES	NEW BALANCE	MINIMUM PAYMENT	CLOSING DATE
624.00	9.36	500.00	62.95	364.57	434.98	45.00	10-16-99

IF WE RECEIVE PAYMENT OF THE FULL AMOUNT OF THE NEW BALANCE BEFORE THE NEXT CYCLE CLOSING DATE, SHOWN ABOVE, YOU WILL AVOID A FINANCE CHARGE NEXT MONTH. THE FINANCE CHARGE, IF ANY, IS CALCULATED ON THE PREVIOUS BALANCE BEFORE DEDUCTING ANY PAYMENTS OR CREDITS SHOWN ABOVE. THE PERIODIC RATES USED ARE 1.5% OF THE BALANCE ON AMOUNTS UNDER $1,000 AND 1% OF AMOUNTS IN EXCESS OF $1,000, WHICH ARE ANNUAL PERCENTAGE RATES OF 18% AND 12% RESPECTIVELY.

EXAMPLE C

Compute the finance charge and the new balance for the statement shown in Figure 14-1 based on the previous balance, $624, ignoring all payments, credits, and purchases.

Finance charge = $624 × 1.5% × 1 month = $9.36
New balance = $624.00 + $9.36 − $500.00 − $62.95 + $364.57 = $434.98

EXAMPLE D

Assume that the finance charge in Figure 14-1 is based on the previous balance, less any payments or credits, but ignores subsequent purchases. Compute the finance charge and the new balance.

The finance charge is based on $624.00 − $500.00 − $62.95 = $61.05.
Finance charge = $61.05 × 1.5% × 1 month = $0.91575, or $0.92
New balance = $624.00 + $0.92 − $500.00 − $62.95 + $364.57 = $426.54

✓ CONCEPT CHECK 14.3

The finance terms given in the charge account statement of Figure 14-1 indicate that the finance charge, if any, is charged on the previous balance, before deducting payments or credits or adding purchases. Calculate the finance charge and the unpaid balance if the previous balance was $2,425.90, the payment was $1,200, there were no credits, and there were $572.50 in new purchases.

An interest rate of 1.5% applies to the first $1,000 and 1% applies to the excess:
$2,425.90 − $1,000 = $1,425.90.

0.015 × $1,000 = $15.00
0.01 × $1,425.90 = $14.26
Finance charge = $15.00 + $14.26 = $29.26
New balance = $2,425.90 − $1,200 + $29.26 + $572.50 = $1,827.66

COMPLETE ASSIGNMENT 14.1.

Computing Costs of Installment Purchases

In a credit sale, the buyer pays the purchase price plus credit charges. Usually, the buyer makes monthly payments called **installments**. Just as you saw in the previous section, the method of computing the interest is just as important as the interest rate. Most often, the interest is based on the unpaid balance and is calculated each month using a monthly interest rate. Sometimes, the interest may be calculated only once at the beginning using an annual interest rate, but the interest might be paid in equal installments along with the principal installments.

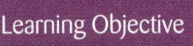

Learning Objective 4

Compute costs of installment purchases.

EXAMPLE E

Nancy Bjonerud purchases $4,000 worth of merchandise. She will repay the principal in four equal monthly payments of $1,000 each. She will also pay interest each month on the unpaid balance for that month, which is calculated at an annual rate of 12%. First, calculate each of the monthly interest payments. Then, display the results in a table.

Given the annual interest of 12%, the monthly rate is 12% ÷ 12 = 1% per month.

Month 1: $4,000 × 1% = $40 Month 3: $2,000 × 1% = $20
Month 2: $3,000 × 1% = $30 Month 4: $1,000 × 1% = $10
Total interest = $40 + $30 + $20 + $10 = $100

Month	Unpaid Balance	Monthly Interest	Principal Payment	Total Payment	New Balance
1	$ 4,000	$ 40	$1,000	$1,040	$3,000
2	3,000	30	1,000	1,030	2,000
3	2,000	20	1,000	1,020	1,000
4	1,000	10	1,000	1,010	0
	$10,000	$100	$4,000	$4,100	

14.5 Although the interest is calculated on the unpaid balance, example E is not "amortization." Usually amortized means that all of the total payments are equal and that each month, principal payments increase and the interest payments decrease. Amortization is covered in examples J, K, and L. After studying example L, compare its table to the tables in examples E and F.

14.6 The method of computing and paying interest in example F may seem strange when compared to example E. However, the method in example F is somewhat similar to *bank discounting*, which will be covered in Chapter 15. In both, the interest/discount amount is calculated on a total amount at the beginning. In example F, this amount is paid over time. In bank discounting, it is all paid at the very beginning. In fact, it is deducted from the face value, leaving the proceeds.

● **EXAMPLE F**

Carmel Dufault purchases $4,000 worth of merchandise. She will pay interest of 12% on $4,000 for four months. First, calculate the total amount of interest. Carmel will repay one-fourth of the interest amount each month. In addition, she will repay the $4,000 in four equal monthly amounts of $1,000 each. Display the results in a table.

$4,000 × 12% × $\frac{4}{12}$ = $160

$160 ÷ 4 = $40 per month for interest

Month	Unpaid Balance	Monthly Interest	Principal Payment	Total Payment	New Balance
1	$ 4,000	$ 40	$1,000	$1,040	$3,000
2	3,000	40	1,000	1,040	2,000
3	2,000	40	1,000	1,040	1,000
4	1,000	40	1,000	1,040	0
	$10,000	$160	$4,000	$4,160	

✓ CONCEPT CHECK 14.4

A kitchen stove is priced at $600 and is purchased with a $100 down payment. The $500 remaining balance is paid in two successive monthly payments of $250 each. Compute interest using the following methods:

a. Interest of 1.5% is calculated on the unpaid balance each month (18% annual rate).
   Month 1: $500 × 0.015 = $7.50
   Month 2: New balance is $250. $250 × 0.015 = $3.75
   Total interest = $7.50 + $3.75 = $11.25

b. Simple interest is calculated on the entire $500 for 2 months at 1.5% per month (18% annual rate).
   $500 × 0.015 per month × 2 months = $15.00

Computing Effective Interest Rates

Learning Objective 5

Compute effective rates.

Examples E and F are very similar, but not quite identical. The numbers are the same: Both purchases are for $4,000; both repay the $4,000 principal in four equal monthly payments; both use a 12% annual interest rate. The only difference is the method of calculating the interest. In example E, the total amount of interest is $100; in example F, it is $160. In example F, it is more expensive to borrow the same money than in example E. In example F, interest is calculated as if the entire $4,000 were borrowed for 4 months ($4,000 × 0.12 × 4/12). But Carmel repays $1,000 of the money after only 1 month.

The true interest rate, or the **effective interest rate**, cannot be the same in each example because it costs more in example F to borrow the same amount of money for the same length of time. To calculate the effective interest rate, we use the familiar formula from Chapter 13, $R = \dfrac{I}{P \times T}$, where I is the amount of interest in dollars, T is the time of the loan in years, and P is the **average unpaid balance** (or the *average principal*) over the period of the loan. The average unpaid balance is the sum of all of the unpaid monthly balances divided by the number of months. (*Note:* The term *effective interest rate* is also used in other contexts where a different formula is used to find the effective rate.)

EXAMPLE G

Use the formula $R = \dfrac{I}{P \times T}$ to compute the effective interest rates for (a) example E and (b) example F. In both examples, the time of the loan is $T = \dfrac{4}{12}$ of a year. Using the preceding tables, for each example, the average unpaid balance is

$$P = \dfrac{\$4{,}000 + \$3{,}000 + \$2{,}000 + \$1{,}000}{4} = \dfrac{\$10{,}000}{4} = \$2{,}500.$$ But in example E, $I = \$100$ and in example F, $I = \$160$.

a. Example F: $T = \dfrac{4}{12}$; $P = \$2{,}500$; $I = \$100$; so that

$$R = \dfrac{I}{P \times T} = \dfrac{\$100}{\$2{,}500 \times \dfrac{4}{12}} = \dfrac{\$100}{\$833.33} = 0.120000, \text{ or } 12\%$$

b. Example F: $T = \dfrac{4}{12}$; $P = \$2{,}500$; $I = \$160$; so that

$$R = \dfrac{I}{P \times T} = \dfrac{\$160}{\$2{,}500 \times \dfrac{4}{12}} = \dfrac{\$160}{\$833.33} = 0.1920008 \text{ or } 19.2\%$$

14.7 One example of a different use of the term *effective rate* is in compound interest (Chapter 16). Consider the interest rate 9% compounded monthly. The "effective rate" is the *annual rate* that would yield the same amount of monetary return over one year as 9% compounded monthly does. The formula is $R = (1 + 0.09/4)^4 - 1$, so that $R = 0.09308$, or 9.308%, compounded annually.

14.8 Students may be surprised at the large difference. Compare the loans in examples E and F with a simple loan of $4,000 for 4 months at 12%, where the entire $4,000 and the $160 interest are repaid at the end. In examples E and F, the borrowers do not have the entire $4,000 for the entire 4 months. In example E, the interest amount is calculated as if the borrower had the entire $4,000, but on average the principal is only $2,500 for the 4 months.

Rule: When the interest is calculated on the unpaid balance each month, the quoted rate and the effective rate will always be the same. When interest is computed only once on the original principal, but the principal is repaid in installments, then the effective interest rate will always be higher than the quoted rate.

The preceding rule is true even when the principal is not repaid in equal installments each month.

EXAMPLE H

Look back at example E where Nancy Bjonerud made four equal principal payments of $1,000 each. Suppose instead that she repays the principal in four monthly payments of $900, $1,200, $1,100, and $800. As in example E, she will also pay interest each month on the unpaid balance for that month, which is calculated at an annual rate of 12%. Compute the interest amount for each month and display the results in a table. Then, compute the average unpaid balance and the effective interest rate using the formula $R = \dfrac{I}{P \times T}$

Given annual interest of 12%, the monthly rate is 12% ÷ 12 = 1% per month.

Month 1: $4,000 × 1% = $40   Month 3: $1,900 × 1% = $19
Month 2: $3,100 × 1% = $31   Month 4: $800 × 1% = $8
Total interest = $40 + $31 + $19 + $8 = $98

Month	Unpaid Balance	Monthly Interest	Principal Payment	Total Payment	New Balance
1	$4,000	$40	$ 900	$ 940	$3,100
2	3,100	31	1,200	1,231	1,900
3	1,900	19	1,100	1,119	800
4	800	8	800	808	0
	$9,800	$98	$4,000	$4,098	

$$P = \dfrac{\$4{,}000 + \$3{,}100 + \$1{,}900 + \$800}{4} = \dfrac{\$9{,}800}{4} = \$2{,}450$$

$$R = \dfrac{I}{P \times T} = \dfrac{\$98}{\$2{,}450 \times \dfrac{4}{12}} = \dfrac{\$98}{\$816.67} = 0.11999951, \text{ or } 12\%$$

INCREASING THE EFFECTIVE RATE

Example F shows how the effective rate in an installment sale can be increased by using a different method to calculate interest. Of course, a reputable lender should indicate the true interest rate in the terms of the agreement. But in installment sales, the interest rate may be only one of several variables in the total cost of purchasing. Any additional fees to make the installment purchase increase the actual cost of borrowing.

Naturally, some businesses will attempt to attract buyers by offering very low purchase prices, even "guaranteeing to match all competitors' advertised prices for 30 days." Others may offer installment purchases at low or even 0% interest rates and no additional fees—but they will charge a higher base price. Different consumers are attracted by different things—some by low prices; some by favorable terms of purchase. For many consumers, buying is simply an emotional response with very little actual thought given to actual costs.

Lender and sellers "effectively" increase the cost of borrowing money or buying in installments by charging or suggesting additional fees. If it is a purchase of merchandise, the lender could require that the merchandise be insured for the term of purchase. Or the lender could charge a credit application fee.

Consider the following modification to example E, which had an effective rate of 12% in example G, part a.

EXAMPLE I

Look back at example G, part a, where we used $R = \dfrac{I}{P \times T}$ to calculate the effective rate for example E, with I equal to the total interest charge of $100. Suppose instead that the lender had charged Nancy the interest of $100, AND a loan origination fee of 1% of the purchase price, AND an insurance premium of $1 per month for the term of the loan. Use the formula $R = \dfrac{I}{P \times T}$ to compute the effective interest rate, but let I be the total finance charge.

The average unpaid balance is still $P = \dfrac{\$4{,}000 + \$3{,}000 + \$2{,}000 + \$1{,}000}{4} = \dfrac{\$10{,}000}{4} = \$2{,}500.$

I = Total finance charge = Interest + Loan origination fee + Insurance
Interest only = $40 + $30 + $20 + $10 = $100
Loan origination fee = 1% of $4,000 = 0.01 × $4,000 = $40
Insurance = $1 × 4 months = $4
Therefore, I = $100 + $40 + $4 = $144

$$R = \dfrac{I}{P \times T} = \dfrac{\$144}{\$2{,}500 \times \dfrac{4}{12}} = \dfrac{\$144}{\$833.33} = 0.17280069,\ \text{or } 17.3\%$$

14.9 The effective rate of 17.3% in example I is still not as high as the 19.2% of example G, part b, because the total finance charge in example I was $144 and in example G, part b, it was $160. If the $40 loan origination fee and the $4 insurance fee were included in example G part b, the total finance charge would be $204 and the effective rate of borrowing would be 24.48%.

Because the interest in example E was paid on the unpaid balance, the effective rate was 12%, the same as the quoted interest rate. If these same additional finance charges from example I were applied to example F, the results would be even more dramatic.

✓ CONCEPT CHECK 14.5

From Concept Check 14.4, a kitchen stove priced at $600 is purchased with a $100 down payment. The remaining balance of $500 may be financed over 2 months with either of the following installment payment plans.
Plan 1: Two monthly principal payments of $250 each and a total interest amount of $11.25
Plan 2: Two monthly principal payments of $250 each and a total interest amount of $15.00
Calculate the effective annual rate of each plan, using $R = \dfrac{I}{(P \times T)}$, where P is the average unpaid monthly balance and T is $\dfrac{2}{12}$ year. In each plan, the monthly unpaid balances are $500 in month 1 and $250 in month 2.
The average unpaid balance is $\dfrac{(\$500 + \$250)}{2} = \dfrac{\$750}{2} = \$375$, so $P = \$375$.

Plan 1: $R = \dfrac{I}{(P \times T)} = \dfrac{\$11.25}{(\$375 \times \frac{2}{12})} = \dfrac{\$11.25}{\$62.50} = 0.18$, or 18% effective annual rate

Plan 2: $R = \dfrac{I}{(P \times T)} = \dfrac{\$15.00}{(\$375 \times \frac{2}{12})} = \dfrac{\$15.00}{\$62.50} = 0.24$, or 24% effective annual rate

COMPLETE ASSIGNMENT 14.2.

Chapter 14 Installment Purchases

Amortizing a Loan

Learning Objective 6

Amortize a loan.

14.10 Instead of all equal payments, some amortized loans have equal payments with one larger "balloon" payment at the end. Part C of assignment 14.3 has a balloon payment.

In example E, interest was calculated on the unpaid balance, but the total payment was different each month: $1,040, $1,030, $1,020, and $1,010. Equal monthly payments are usually simpler, especially for the borrower. In example F, the total payments were the same each month, always $1,040. However, the interest was not calculated on the unpaid balance. In example E, the effective interest rate was equal to the quoted interest rate of 12%. But in example F, the effective rate was much higher, 19.2%.

Taking the best features of each example, consider a loan where the total payments are equal each month AND the interest is calculated on the unpaid balance each month. Such a loan is said to be *amortized*; the method is called **amortization**. (The word *amortize* is also used in different contexts and there is more than one way to amortize a loan.) Although possible for any time purchase, amortization is especially relevant for larger purchases made over longer periods of time. Loans to pay for homes and automobiles are usually amortized. There may, or may not, be a down payment.

COMPUTING THE MONTHLY PAYMENT

The basic concept to amortize a loan is to multiply the loan amount by a **amortization payment factor**. The product is the amount of the monthly payment. This factor may be derived from a calculator or computer or from a book of financial tables. When lenders amortize loans today, they use computers to do the final calculations. Initial calculations, however, are often made using calculators or tables. Chapter 23 will describe how to use a calculator to make amortization calculations. In Chapter 14, we will use tables. Both methods are still used, and both lead to the same results. (You can also go to the Internet, search on "amortization calculations," and find Websites that help you to do the calculations.)

Table 14-1 illustrates the concept of tables for amortization payment factors. Actual tables would have many pages and would be much more detailed. If you study other courses in business mathematics, accounting or finance, you may use tables that are slightly different than Table 14.1. In Chapter 23, we will encounter one such table. Regardless of the exact format of the table, the concepts are the same. And, to repeat, financial calculators and computers will eventually completely eliminate the need for any of these tables.

Notice that the title of Table 14-1 is "Amount of Monthly Payment per $1,000 Borrowed." Therfore, you must first determine the amount of the loan in "thousands of dollars," not the number of dollars. The annual interest rates in Table 14.1 were selected because they are evenly divisible by 12. This will eliminate the necessity to round off interest rates when you convert an annual rate into a monthly rate.

> **STEPS** to Find the Monthly Payment of an Amortized Loan Using Table 14-1
>
> 1. Divide the loan amount by $1,000 to get the number of thousands of dollars.
> 2. Locate the amortization payment factor in Table 14-1.
> 3. Multiply the quotient in Step 1 by the amortization payment factor. The product is the amount of the monthly payment.

Table 14-1: Amortization Payment Factors—Amount of Monthly Payment per $1,000 Borrowed

Term of Loan		Annual Interest Rate					
		4.5%	6%	7.5%	9%	10.5%	12%
1	month	1003.75000	1005.00000	1006.25000	1007.50000	1008.75000	1010.00000
2	months	502.81425	503.75312	504.69237	505.63200	506.57203	507.51244
3	months	335.83645	336.67221	337.50865	338.34579	339.18361	340.02211
4	months	252.34814	253.13279	253.91842	254.70501	255.49257	256.28109
5	months	202.25561	203.00997	203.76558	204.52242	205.28049	206.03980
6	months	168.86099	169.59546	170.33143	171.06891	171.80789	172.54837
1	year	85.37852	86.06643	86.75742	87.45148	88.14860	88.84879
2	years	43.64781	44.32061	44.99959	45.68474	46.37604	47.07347
3	years	29.74692	30.42194	31.10622	31.79973	32.50244	33.21431
4	years	22.80349	23.48503	24.17890	24.88504	25.60338	26.33384
5	years	18.64302	19.33280	20.03795	20.75836	21.49390	22.24445
10	years	10.36384	11.10205	11.87018	12.66758	13.49350	14.34709
15	years	7.64993	8.43857	9.27012	10.14267	11.05399	12.00168
20	years	6.32649	7.16431	8.05593	8.99726	9.98380	11.01086
25	years	5.55832	6.44301	7.38991	8.39196	9.44182	10.53224
30	years	5.06685	5.99551	6.99215	8.04623	9.14739	10.28613

EXAMPLE J

Find the monthly payment required to amortize a $4,000 loan over 4 months at 12% (1% per month).

STEP 1 Divide $4,000 by $1,000; $4,000 ÷ $1,000 = 4 thousands

STEP 2 Find the intersection of the 12% column and the 4-month row in Table 14-1. The amortization payment factor is $256.28109 per each one thousand dollars.

STEP 3 Multiply the 4 (from step 1) by the amortization payment factor.
4 × $256.28109 = $1,025.12436, or $1,025.12 monthly.

EXAMPLE K

Judith Kranz agrees to purchase an automobile for $18,300. Judith will make a $2,000 down payment and amortize the balance with monthly payments over 4 years at 9% (0.75% per month). Determine Judith's monthly payment.

$18,300 − $2,000 = $16,300 amount financed

STEP 1 $16,300 ÷ $1,000 = 16.3 thousands

STEP 2 Find the intersection of the 9% column and the 4-year row in Table 14-1. The amortization payment factor is $24.88504 per thousand.

STEP 3 Multiply the 16.3 (from step 1) by the amortization payment factor.
16.3 × $24.88504 = $405.62615, or $405.63 monthly.

LOAN PAYMENT SCHEDULE

After determining the amount of the monthly payments, a lender can prepare a schedule of loan payments called an **amortization schedule**. The payment for the last month is determined in the schedule, and it may be slightly different from the payment in the other months.

> **STEPS to Create an Amortization Schedule**
>
> For each row except the last:
> 1. Interest payment = Unpaid balance × Monthly interest rate
> 2. Principal payment = Monthly payment − Interest payment
> 3. New unpaid balance = Old unpaid balance − Principal payment
>
> For the last row (i.e., for the final payment):
> 1. Interest payment = Unpaid balance × Monthly interest rate
> 2. Monthly payment = Unpaid balance + Interest payment
> 3. Principal payment = Unpaid balance

EXAMPLE L

Create an amortization schedule for the loan in example J, a $4,000 loan amortized at 12% over 4 months. The interest rate is 1% per month.

Month	Unpaid Balance	Interest Payment	Principal Payment	Total Payment	New Balance
1	$ 4,000.00	$ 40.00	$ 985.12	$1,025.12	$3,014.88
2	3,014.88	30.15	994.97	1,025.12	2,019.91
3	2,019.91	20.20	1,004.92	1,025.12	1,014.99
4	1,014.99	10.15	1,014.99	1,025.14	0
Totals	$10,049.78	$100.50	$4,000.00	$4,100.50	

Note: In example L, the last monthly payment is 2 cents larger than the others. Because the interest payments need to be rounded, the final payment usually will be slightly different from the previous payments.

Since amortization implies that interest is paid on the unpaid balance, the formula $R = \dfrac{I}{P \times T}$ should show that the effective rate is the same as the quoted rate of 12%. Looking at the table for example L, the average unpaid balance is

$$P = \frac{\$4,000.00 + \$3,014.88 + \$2,019.91 + \$1,1014.99}{4} = \frac{\$10,049.78}{4}$$
$$= \$2,512.45$$

The total interest paid is $I = \$40.00 + \$30.15 + \$20.20 + \$10.15 = \$100.50$. Therefore,

$$R = \frac{I}{P \times T} = \frac{\$100.50}{\$2,512.45 \times \dfrac{4}{12}} = \frac{\$100.50}{\$837.48} = 0.1200029, \text{ or } 12\%$$

The reason that the result was 12.00029% instead of 12%, is that all of the payments were rounded to the nearest cent. You can easily verify that if you round all payments to five decimal places, $R = 12.0000007$. However, also be sure to calculate the monthly payment to five places, or $1,025.12436.

✓ CONCEPT CHECK 14.6

A $2,000 purchase is amortized over 2 months at an annual rate of 9%. First use Table 14-1 to calculate the monthly payment for month 1. Then show the calculations to construct a 2-month amortization schedule.

$2,000 ÷ $1,000 = 2 thousands
Amortization payment factor from Table 14-1 is $505.63200.
2 × $505.63200 = $1,011.264, or $1,011.26 for month 1

Month	1		2	
Unpaid balance	Original principal:	$2,000.00	From end of month 1:	$1,003.74
Monthly rate	0.09 ÷ 12 = 0.0075			
Interest payment	$2,000.00 × 0.0075 =	$ 15.00	$1,003.74 × 0.0075 =	$ 7.53
Total payment	From above:	$1,011.26	$1,003.74 + $7.53 =	$1,011.27
Principal payment	$1,011.26 − $15.00 =	$ 996.26		$1,003.74
New balance	$2,000.00 − $996.26 =	$1,003.74	$1,003.74 − $1,003.74 =	$ 0.00

COMPLETE ASSIGNMENT 14.3

Finding the Monthly Payment of a Home Mortgage

Persons who decide to purchase a home usually borrow the majority of the money. The amount that is borrowed is usually amortized, and usually for a long time, such as 15, 20, or 30 years. Such a home loan is called a **mortgage**. The interest rate may be **fixed**, which means that it stays the same for the entire length of the loan. Also popular are **variable-rate loans**, which permit the lender to periodically adjust the interest rate depending on current financial market conditions. Whether a borrower decides on a fixed or variable rate loan depends on several factors, such as how long he or she plans to remain in that home.

A mortgage loan is still a loan. And amortizing a mortgage is the same as amortizing any other loan: Look up the amortization payment factor in Table 14-1 and multiply by the number of thousands of dollars that are borrowed.

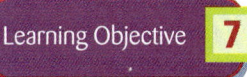

Learning Objective 7

Compute the monthly payment on a home mortgage.

EXAMPLE M

George and Kathy Jarvis bought a home priced at $190,000. They made a $20,000 down payment and took out a 30-year, 6% mortgage on the balance. Find the size of their monthly payment.

$19,000 − $20,000 = $17,000 amount borrowed

STEP 1 Divide $170,000 by $1,000 to get 170.

STEP 2 Find the amortization factor in the 6% column and 30-year row of Table 14-1. It is $5.99551.

STEP 3 Multiply the 170 from Step 1 by $5.99551 to get $1,019.23670. The monthly payment will be $1,019.24.

AMORTIZATION SCHEDULE FOR A MORTGAGE

An amortization schedule for a mortgage is computed line-by-line just as the amortization schedules for other loans such as the one in example L. However, a 30-year loan will have 360 lines, one for each month of the loan. This could be about six or seven pages of paper with three calculations per line, or 1,080 calculations. Today, these tables are always produced with a computer. You can create an amortization schedule using EXCEL or you can find several sources on the Internet to do the calculations for you. However, to review the concept, examine example N.

EXAMPLE N

Construct the first three lines of an amortization schedule for the Jarvis's home mortgage loan in example M.

The Jarvis's $170,000 mortgage has a monthly payment of $1,019.24.
For a 6% annual interest rate, the monthly rate is 6% ÷ 12 = 0.5%.
For each row, 1. Monthly interest = Unpaid balance × 0.005
2. Principal payment = Total payment − Monthly interest
3. New balance = Unpaid balance − Principal payment

Month	Unpaid Balance	Monthly Interest	Principal Payment	Total Payment	New Balance
1	$170,000.00	$850.00	$169.24	$1,019.24	$169,830.76
2	169,830.76	849.15	170.09	1,019.24	169,660.67
3	169,660.67	848.30	170.94	1,019.24	169,489.73

✓ Concept Check 14.7

A home cost $180,000. The buyers made a down payment of $30,000. Compute the monthly payment on a 25-year mortgage with an annual interest rate of 7.5%. Use Table 14-1.

> The amount borrowed is $180,000 − $30,000 = $150,000.
> The amortization payment factor from Table 14-1 is 7.38991.
> The amount of the loan in thousands is $150,000 ÷ $1,000 = 150.
> The monthly mortgage payment is 150 × $7.38991 = $1,108.49.

Chapter Terms for Review

- amortization
- amortization payment factor
- amortization schedule
- annual percentage rate (APR)
- average daily balance
- average principal
- average unpaid balance
- credit card
- effective interest rate
- finance charge
- fixed interest rate
- installments
- mortgage
- Truth in Lending Act (TILA)
- variable-rate loans

THE BOTTOM LINE

Summary of chapter learning objectives:

Learning Objective	Example
14.1 Convert between annual and monthly interest rates	1. Convert 0.75% per month to an annual rate. 2. Convert 15% per year to a monthly rate
14.2 Compute simple interest on a monthly basis	3. Compute the simple interest on $1,500 for 7 months at 0.5% per month (6% per year).
14.3 Compute finance charges for credit account purchases	4. Charge account terms apply a 1.25% finance charge to the previous balance, less any payments and credits, ignoring purchases. Find the finance charge and new balance when the previous balance is $1,683.43, payments plus credits total $942.77, and purchases are $411.48.
14.4 Compute costs of installment purchases	5. Furniture worth $2,500 is paid for with a $400 down payment and three payments of $700, plus monthly interest of 1% on the unpaid balance. Find the total interest paid. The monthly balances are $2,100, $1,400, and $700.
14.5 Compute effective rates	6. A $2,400 purchase is to be repaid in 3 equal monthly principal payments of $800 each. There will be one interest payment of $60 (10% of $2400 for three months) and insurance premiums of $1 each month. Calculate the effective rate of interest. The monthly balances are $2,400, $1,600, and $800.
14.6 Amortize a loan	7. A $2,000 loan will be amortized over 6 months at an annual rate of 9%. Find the payment, using Table 14-1, and calculate the unpaid balance after the first month.
14.7 Compute the monthly payment on a home mortgage.	8. A $130,000 home mortgage is for 20 years at 4.5% annual interest. Find the monthly payment.

Answers: 1. 9% per year 2. 1.25% per month 3. $52.50 4. Finance charge, $9.26; new balance, $1,161.40 5. $42 6. 15.75% 7. Payment, $342.14; unpaid balance, $1,672.86 8. $822.44

SELF-CHECK

Review Problems for Chapter 14

1 Change the monthly rates to annual rates.

  a. 0.75% = _____   b. 0.6% = _____   c. 1.2% = _____   d. $\frac{2}{5}$% = _____

2 Change the annual rates to monthly rates.

  a. 6% = _____   b. 15% = _____   c. 13.2% = _____   d. 9.6% = _____

3 A store offers the following credit terms: "There will be no finance charge if the full amount of the new balance is received on or before the due date. Unpaid balances after the due date will be charged interest based upon the previous balance, less any payments and credits before the due date. The rates are 1.75% on the first $1,000 of the unpaid balance and 1.25% on the part of the unpaid balance that exceeds $1,000."

Calculate (a) the finance charge and (b) the new balance on an account that had a previous balance of $2,752.88; a payment of $800; credits of $215; and purchases of $622.75.

4 Neta Prefontaine buys $3,000 worth of merchandise. She agrees to pay $1,000 per month on the principal. In addition, she will pay interest of 1% per month (12% annually) on the unpaid balance. Complete the following table.

Month	Unpaid Balance	Interest Payment	Principal Payment	Total Payment	New Balance
1	$3,000.00	a. ____	$1,000.00	b. ____	c. ____
2	d. ____	e. ____	$1,000.00	f. ____	g. ____
3	h. ____	i. ____	$1,000.00	j. ____	$0.00

5 Use the results of problem 4 and compute the effective annual interest rate using the formula $R = \frac{I}{P \times T}$, where P is the average unpaid balance, I is the total interest paid, and T is the period of the loan in years.

6 Use Table 14-1 to find the monthly payment of a $125,000 mortgage that is amortized over 15 years at 7.5%.

7 A $3,000 loan is amortized over 3 months at 12%. The first two monthly payments are $1,020.07; the final payment may differ. Complete the following table.

Month	Unpaid Balance	Interest Payment	Total Payment	Principal Payment	New Balance
1	$3,000.00	a. ____	$1,020.07	b. ____	c. ____
2	d. ____	e. ____	$1,020.07	f. ____	g. ____
3	h. ____	i. ____	j. ____	k. ____	$0.00

Notes

Assignment 14.1 Monthly Finance Charges

Name

Date   Score

Learning Objectives 1 2 3

A (19 points) Problem 1: Change the rates from annual to monthly. Problem 2: Change the rates from monthly to annual. (1 point for each correct answer)

1a. 18% = __1.5%__   **b.** 15% __1.25%__   **c.** 16.8% __1.4%__   **d.** 7.2% __0.6%__

   18% ÷ 12   15% ÷ 12   16.8% ÷ 12   7.2% ÷ 12

e. 6% __0.5%__   **f.** 19.2% __1.6%__   **g.** 14.4% __1.2%__   **h.** 8.4% __0.7%__

   6% ÷ 12   19.2% ÷ 12   14.4% ÷ 12   8.4% ÷ 12

i. 9% __0.75%__   **j.** 9.6% __0.8%__

   9% ÷ 12   9.6% ÷ 12

2a. 0.5% = __6%__   **b.** 0.7% = __8.4%__   **c.** 1.3% = __15.6%__   **d.** 1.25% = __15%__

   0.5% × 12   0.7% × 12   1.3% × 12   1.25% × 12

e. 1.1% = __13.2%__   **f.** 0.75% = __9%__   **g.** 0.9% = __10.8%__   **h.** 1.15% = __13.8%__

   1.1% × 12   0.75% × 12   0.9% × 12   1.15% × 12

i. 0.4% = __4.8%__

   0.4% × 12

Score for A (19)

Chapter 14 Installment Purchases 287

Assignment 14.1 Continued

B (33 points) Lakeside Furniture Store offers the credit terms shown to its retail customers. In problems 3–5 compute the finance charge, if any, and the new balance. Assume that all payments are made within the current billing cycle. (3 points for each correct answer)

TERMS: There will be no finance charge if the full amount of the new balance is received within 25 days after the cycle-closing date. The finance charge, if any, is based upon the entire previous balance *before* any payments or credits are deducted. The rates are 1.5% per month on amounts up to $1,000 and 1.25% on amounts in excess of $1,000. These are annual percentage rates of 18% and 15%, respectively.

Cycle Closing	Previous Balance	Payment Amount	Credits	Finance Charge	Purchases	New Balance
3. 3/20/200–	$2,147.12	$900.00	$175.50	$29.34	$647.72	$1,748.68
	$0.015 \times \$1,000 = \$15.00$	$0.0125 \times \$1,147.12 = \$14.34$			$\$15.00 + \$14.34 = \$29.34$	

$\$2,147.12 - \$900 - \$175.50 + \$29.34 + \$647.72 = \$1748.68$

4. 6/20/200–	$743.72	$0.00	$15.00	$11.16	$609.88	$1,349.76
	$0.015 \times \$743.72 = \$11.16$	$\$743.72 - \$15.00 + \$11.16 + \$609.88 = \$1,349.76$				

5. 9/20/200–	$3,412.27	$3,000.00	$212.98	$45.15	$907.51	$1,151.95
	$0.015 \times \$1,000 = \$15.00$	$0.0125 \times \$2,412.27 = \$30.15$			$\$15.00 + \$30.15 = \$45.15$	

$\$3,412.27 - \$3,000.00 - \$212.98 + \$45.15 + \$907.51 = \$1,151.95$

Assignment 14.1 Continued

In problems 6 and 7, Lelia McDaniel has an account at Lakeside Furniture Store. Compute the missing values in Lelia's account summary for the months of August and September. The previous balance in September is the same as the new balance in August.

Cycle Closing	Previous Balance	Payment Amount	Credits	Finance Charge	Purchases	New Balance
6. 8/20/200–	$1,636.55	$900.00	$36.00	$22.96	$966.75	$1,690.26
	0.015 × $1,000 = $15.00		0.0125 × $636.55 = $7.96		$15.00 + $7.96 = $22.96	

$1,636.55 − $900.00 − $36.00 + $22.96 + $966.75 = $1,690.26

7. 9/20/200–	$1,690.26	$1,200.00	$109.75	$23.63	$589.41	$993.55
	0.015 × $1,000 = $15		0.0125 × $690.26 = $8.63		$15.00 + $8.63 = $23.63	

$1,690.26 − $1,200.00 − $109.75 + $23.63 + $589.41 = $993.55

Score for B (33)

C (48 points) Devlin's Feed & Fuel offers the credit terms shown to its retail customers. In problems 8-12 compute the missing values in the charge accounts shown. Assume that all payments are made within 30 days of the billing date. (3 points for each correct answer)

TERMS: Finance Charge is based on the Net Balance, if payment is received within 30 days of the billing date. If payment is made after 30 days, then the Finance Charge is based on the Previous Balance. Net Balance equals Previous Balance less Payments and Credits. In either case, the monthly rate is 1.25% on the first $500 and 1% on any amount over $500. These are annual percentage rates of 15% and 12%, respectively.

Billing Date	Previous Balance	Payment Amount	Credit	Net Balance	Finance Charge	New Purchases	New Balance
8. 4/25/200–	$2,621.05	$1,700.00	$0.00	$921.05	$10.46	$751.16	$1,682.67
	$2,621.05 − $1,700.00 = $921.05			0.0125 × $500 = $6.25		0.01 × $421.05 = $4.21	

$6.25 + $4.21 = $10.46 $921.05 + 10.46 + $751.16 = $1,682.67

Assignment 14.1 Continued

Billing Date	Previous Balance	Payment Amount	Credit	Net Balance	Finance Charge	New Purchases	New Balance
9. 3/25/200–	$1,827.15	$700.00	$28.75	$1,098.40	$12.23	$672.39	$1,783.02

$1,827.15 − $700.00 − $28.75 = $1,098.40   0.0125 × $500 = $6.25 $0.01 × $598.40 = $5.98

$6.25 + $5.98 = $12.23 $1,098.40 + $12.23 + $672.39 = $1,783.02

10. 11/25/200–	$1,241.88	$250.00	$84.09	$907.79	$10.33	$351.94	$1,270.06

$1,241.88 − $250.00 − $84.09 = $907.79   0.0125 × $500 = $6.25   0.01 × $407.79 = $4.08

6.25 + $4.08 = $10.33 $907.79 + $10.33 + $351.94 = $1,270.06

In problems 11 and 12 compute the missing values in Jimmy Petrasek's charge account summary at Devlin's for the months of June and July. The previous balance in July is the same as the new balance in June.

11. 6/25/200–	$1,571.62	$500.00	$62.00	$790.12	$9.15	$772.35	$1,571.62

$1,352.12 − $500.00 − $62 = $790.12   0.0125 × $500 = $6.25   0.01 × $290.12 = $2.90

$6.25 + $2.90 = $9.15 $790.12 + 9.15 + $772.35 = $1,571.62

12. 7/25/200–	$1,571.62	$600.00	$67.77	$903.85	$10.29	$743.95	$1,658.09

$1,571.62 − $600.00 − $67.77 = $903.85   0.0125 × $500 = $6.25   0.01 × $403.85 = $4.04

$6.25 + $4.04 = $10.29 $903.85 + $10.29 + $743.95 = $1,658.09

Score for C (48)

Assignment 14.2 Installment Sales and Effective Rates

Name

Date   Score

Learning Objectives **4** **5**

A (60 points) Bob Wallis needed to purchase office equipment costing $4,800. He was able to finance his purchase over 3 months at a 9% annual interest rate. Following are three different payment options under these conditions. Complete the installment purchase table for each payment option. (2 points for each correct answer)

1. Bob pays the $1,600 per month on the principal and pays interest of 0.75% of the unpaid balance each month (9% annual rate).

Month	Unpaid Balance	Monthly Interest	Principal Payment	Total Payment	New Balance
1	$4,800.00	$36.00	$1,600.00	$1,636.00	$3,200.00
2	3,200.00	24.00	1,600.00	1,624.00	1,600.00
3	1,600.00	12.00	1,600.00	1,612.00	0.00
			4,800.00		

$4,800 × 0.0075 = $36 $1,600 + $36 = $1,636 $4,800 − $1,600 = $3,200
$3,200 × 0.0075 = $24 $1,600 + $24 = $1,624 $3,200 − $1,600 = $1,600
$1,600 × 0.0075 = $12 $1,600 + $12 = $1,612 $1,600 − $1,600 = $0

2. Bob makes monthly payments of $1,400, $1,400, and $2,000 on the principal and pays interest of 0.75% of the unpaid balance each month (9% annual rate).

Month	Unpaid Balance	Monthly Interest	Principal Payment	Total Payment	New Balance
1	$4,800.00	$36.00	$1,400.00	$1,436.00	$3,400.00
2	3,400.00	25.50	1,400.00	1,425.50	2,000.00
3	2,000.00	15.00	2,000.00	2,015.00	0.00
			4,800.00		

$4,800 × 0.0075 = $36.00 $1,400 + $36 = $1,436 $4,800 − $1,400 = $3,400
$3,400 × 0.0075 = $25.50 $1,400 + $25.50 = $1,425.50 $3,400 − $1,400 = $2,000
$2,000 × 0.0075 = $15 $2,000 + $15 = $2,015 $2,000 − $2,000 = $0

3. Bob pays $1,600 principal on the principal. The total interest charge is 9% of the original principal for 3 months. Bob pays $\frac{1}{3}$ of the interest each month.

Month	Unpaid Balance	Monthly Interest	Principal Payment	Total Payment	New Balance
1	$4,800.00	$36.00	$1,600.00	$1,636.00	$3,200.00
2	3,200.00	36.00	1,600.00	1,636.00	1,600.00
3	1,600.00	36.00	1,600.00	1,636.00	0.00
			4,800.00		

$I = P \times R \times T = \$4,800 \times 0.09 \times \frac{3}{12} = \$108$ $108 ÷ 3 = $36 per month

$1,600 + $36 = $1,636 per month, every month
$4,800 − $1,600 = $3,200 after 1 month
$3,200 − $1,600 = $1,600 after 2 months
$1,600 − $1,600 = $0 after 3 months

Score for A (60)

Assignment 14.2 Continued

B (40 points) For each of the following problems calculate the effective rate using the formula $R = \dfrac{I}{P \times T}$.
(Points for each correct answer as shown)

4. Compute R = effective rate for the table in problem 1 in Part A, with P = average unpaid balance and I = total interest charge.
   a. P = Average unpaid balance $3,200 (3 pts)
   b. I = Total interest charge $72 (3 pts)
   c. R = Effective interest rate 9% (4 pts)

$$P = \frac{\$4{,}800 + \$3{,}200 + \$1{,}600}{3} = \frac{\$9{,}600}{3} = \$3{,}200$$

$I = \$36 + \$24 + \$12 = \$72$

$$R = \frac{I}{P \times T} = \frac{\$72}{\$3{,}200 \times \frac{3}{12}} = \frac{\$72}{\$800} = 0.09, \text{ or } 9\%$$

5. Compute R = effective rate for the table in problem 1 in Part A, with P = average unpaid balance and I = total **finance** charge. The finance charge is the total interest, plus a loan origination fee of $\frac{1}{2}$% of the original principal, plus $6 of insurance premiums ($2 per month).
   a. P = Average unpaid balance $3,200 (3 pts)
   b. I = Total **finance** charge $102 (3 pts)
   c. R = Effective interest rate 12.75% (4 pts)

$P = \$3{,}200$ Loan fee $= 0.005 \times \$4{,}800 = \$24$

I = Interest + Loan fee + Insurance = $72 + $24 + 6 = $102

$$R = \frac{I}{P \times T} = \frac{\$102}{\$3{,}200 \times \frac{3}{12}} = \frac{\$102}{\$800} = 0.1275, \text{ or } 12.75\%$$

6. Compute R = effective rate for the table in problem 2 in Part A, with P = average unpaid balance and I = total interest charge.
   a. P = Average unpaid balance $3,400 (3 pts)
   b. I = Total interest charge $76.50 (3 pts)
   c. R = Effective interest rate 9% (4 pts)

$$P = \frac{\$4{,}800 + \$3{,}400 + \$2{,}000}{3} = \frac{\$10{,}200}{3} = \$3{,}400$$

$I = \$36 + \$25.50 + \$15 = \$76.50$

$$R = \frac{I}{P \times T} = \frac{\$76.50}{\$3{,}400 \times \frac{3}{12}} = \frac{\$76.50}{\$850} = 0.09, \text{ or } 9\%$$

7. Compute R = effective rate for the table in problem 3 in Part A, with P = average unpaid balance and I = total interest charge.
   a. P = Average unpaid balance $3,200 (3 pts)
   b. I = Total interest charge $108 (3 pts)
   c. R = Effective interest rate 13.5% (4 pts)

$$P = \frac{\$4{,}800 + \$3{,}200 + \$1{,}600}{3} = \frac{\$9{,}600}{3} = \$3{,}200$$

$I = \$36 + \$36 + \$36 = \$108$, or $I = \$4{,}800 \times 0.09 \times \dfrac{3}{12} = \$108$

$$R = \frac{I}{P \times T} = \frac{\$108}{\$3{,}200 \times \frac{3}{12}} = \frac{\$108}{\$800} = 0.135, \text{ or } 13.5\%$$

Score for B (40)

Assignment 14.3 Amortization and Mortgages

Name

Date   Score

Learning Objectives 6 7

A (16 points) Lincoln Lending Corp. amortizes all of mortgage loans and many of its personal loans on a monthly basis. The total monthly payments are equal each month and include both interest and principal. Use Table 14-1 to find the amortization payment factor for each loan. Then compute the monthly payment. (2 points for each correct answer)

Loan and Terms of Amortization	Amortization Payment Factor	Monthly Payment
1. $5,000 over 6 months at 7.5%	$170.33143	$ 851.66
$5,000 ÷ $1,000 = 5		
5 × $170.33143 = $851.65715, or $851.66		
2. $16,000 over 2 years at 10.5%	$ 46.37604	$ 742.02
$16,000 ÷ $1,000 = 16		
16 × $46.37604 = $742.01664, or $742.02		
3. $175,000 over 25 years at 6%	$ 6.44301	$1,127.53
$175,000 ÷ $1,000 = 175		
175 × $6.44301 = $1,127.52675, or $1,127.53		
4. $230,000 over 30 years at 7.5%	$ 6.99215	$1,608.19
$230,000 ÷ $1,000 = 230		
230 × $6.99215 = $1,608.19450, or $1,608.19		

Score for A (16)

B (32 points) On April 13, Braunda Johannesen borrowed $6,000 from her bank to help her pay her federal income taxes for the previous year. The bank amortized her loan over 4 months at an annual rate of 9%. Braunda paid interest of 0.75% of the unpaid balance each month. Find the amortization payment factor in Table 14-1. This factor makes a total payment of $1,528.23 each month except the last. For the last month, the total payment is the interest payment plus the unpaid balance. Complete the following amortization schedule. (2 points for each correct answer.)

5. Amortization factor from Table 14-1: $254.70501
Multiply the amortization factor by 6 to get the total payment shown for months 1, 2, and 3.

	Month	Unpaid Balance	Interest Payment	Total Payment	Principal Payment	New Balance
6.	1	$6,000.00	$45.00	$1,528.23	$1,483.23	$4,516.77
7.	2	4,516.77	33.88	1,528.23	1,494.35	3,022.42
8.	3	3,022.42	22.67	1,528.23	1,505.56	1,516.86
9.	4	1,516.86	11.38	1,528.24	1,516.86	0.00

$6,000.00 × 0.0075 = $45.00; $1,528.23 − $45.00 = $1,483.23; $6,000.00 − $1,483.23 = $4,516.77
$4,516.77 × 0.0075 = $33.88; $1,528.23 − $33.88 = $1,494.35; $4,516.77 − $1,494.35 = $3,022.42
$3,022.42 × 0.0075 = $22.67; $1,528.23 − $22.67 = $1,505.56; $3,022.42 − $1,505.56 = $1,516.86
$1,516.86 × 0.0075 = $11.38; $1,516.86 + $11.38 = $1,528.24; $1,516.86 − $1,516.86 =   0.00

Score for B (32)

Assignment 14.3 Continued

C (30 points) Refer to Part B, in which Braunda Johannesen borrowed $6,000 to help pay her federal income taxes. Now suppose that Braunda agreed to make payments of $1,200 in months 1, 2, and 3. The bank will compute the interest on the unpaid balance at a rate of 0.75% (9%/12) each month and deduct the interest from the $1,200. In the last (fourth) month, Braunda will pay all of the remaining unpaid balance plus the interest for the last month. Complete the table, using the same procedure as in Part B. (2 points for each correct answer)

	Month	Unpaid Balance	Interest Payment	Total Payment	Principal Payment	New Balance
10.	1	$6,000.00	$45.00	$1,200.00	$1,155.00	$4,845.00
11.	2	4,845.00	36.34	1,200.00	1,163.66	3,681.34
12.	3	3,681.34	27.61	1,200.00	1,172.39	2,508.95
13.	4	2,508.95	18.82	2,527.77	2,508.95	0.00

$6,000 \times 0.0075 = \$45$; $\$1,200 - \$45 = \$1,155$; $\$6,000 - \$1,155 = \$4,845$
$\$4,845 \times 0.0075 = \$36.34$; $\$1,200 - \$36.34 = \$1,163.66$; $\$4,845 - \$1,163.66 = \$3,681.34$
$\$3,681.34 \times 0.0075 = \$27.61$; $\$1,200 - \$27.61 = \$1,172.39$; $\$3,681.34 - \$1,172.39 = \$2,508.95$
$\$2,508.95 \times 0.0075 = \$18.82$; $\$2,508.95 + \$18.82 = \$2,527.77$

Score for C (30)

D (22 points) Mr. and Mrs. Paul Yeiter sold their previous home and used the profits as a down payment to buy a new home. They took out a $160,000, 25-year mortgage from Colonial Home Finance. The mortgage had an annual interest rate of 6%. From Table 14-1, the amortization payment factor is $6.44301 and the monthly payment is $1,030.88. Complete the first three rows of the amortization schedule for the Yeiters' mortgage. (2 points for each correct answer)

Amortization Schedule for Mortgage

Month	Unpaid Balance	Monthly Interest	Principal Payment	Total Payment	New Balance
1	$160,000.00	$800.00	$230.88	$1,030.88	$159,769.12
2	$159,769.12	$798.85	$232.03	1,030.88	$159,537.09
3	$159,537.09	$797.69	$233.19	1,030.88	$159,303.90

Month 1
Interest payment: $160,000 \times 0.005 = \$800.00$
Principal payment: $\$1,030.88 - \$800.00 = \$230.88$
New balance: $\$160,000 - \$230.88 = \$159,769.12$

Month 2
Interest payment: $\$159,769.12 \times 0.005 = \$798.85$
Principal payment: $\$1,030.88 - \$798.85 = \$232.03$
New balance: $\$159,769.12 - \$232.03 = \$159,537.09$

Month 3
Interest payment: $\$159,537.09 \times 0.005 = \$797.69$
Principal payment: $\$1,030.88 - \$797.69 = \$233.19$
New balance: $\$159,537.09 - \$233.19 = \$159,303.90$

Score for D (22)

Promissory Notes and Discounting

15

Learning Objectives
By studying this chapter and completing all assignments you will be able to:

Learning Objective 1 — Compute the number of interest days of a promissory note.

Learning Objective 2 — Determine the due date of a promissory note.

Learning Objective 3 — Compute the maturity value of a promissory note.

Learning Objective 4 — Discount a promissory note.

Learning Objective 5 — Compute the proceeds and actual interest rate on a bank discount loan.

Learning Objective 6 — Compute the savings from borrowing money to take a cash discount.

Business and individuals both use long-term loans (more than 1 year) to purchase large items such as equipment or buildings. Likewise, businesses and individuals also use short-term loans when they are convenient. Long-term and short-term loans are written in the form of various financial documents, one of which is called a **promissory note**. It is a promise by a borrower to repay a certain amount of money on a certain date. Sometimes the promissory note can be sold to a third party, in which case the note is called a **negotiable promissory note**. Because the buyer of the note is assuming some risk that the borrower will not repay, he or she will not likely pay the entire value of the note. Such a note is said to have been **discounted**. Similarly, an individual may go to a bank to borrow money, and the bank may deduct the entire amount of the interest in advance. This is called **bank discounting**.

Unlike individuals, however, businesses may borrow large amounts of money for only a few days. For example, a retail business buys merchandise from manufacturers and wholesalers. But the retailer may know immediately that it cannot sell the merchandise in time to pay the supplier's invoice. Perhaps the supplier also offers a cash discount if the buyer pays the invoice within a few days (see Chapter 7). The retailer can usually save money by borrowing enough cash to pay the invoice and take advantage of the cash discount. If the amounts are large, the savings can be significant.

Promissory Notes

A promissory note is an unconditional promise by the **maker** of the note (the borrower) to repay money to the **bearer** of the note (the lender) at some time in the future. This date is called the **due date** or the **maturity date**. The dollar amount written on the note is called the **face value** of the note. It is the same as the principal (P in Chapter 13). Most promissory notes are **interest-bearing,** especially if one or both parties is a business. This means that the maker must also pay interest to the bearer on the maturity date. The sum of the face value and the **interest dollars** (I in Chapter 13) is the **maturity value** (**MV**) of the note. Figure 15-1 illustrates a simple promissory note.

Figure 15-1 | **Promissory Note**

$ _2,000 00_ ATLANTA, GEORGIA _March 15_ 20 _--_
— Sixty days — AFTER DATE _I, Sylvia Cometta,_ PROMISE TO PAY TO
THE ORDER OF _William Dale Crist_
PAYABLE AT Bank of the South
Two thousand and 00/100 DOLLARS
VALUE RECEIVED WITH EXACT INTEREST AT _10%_ PER ANNUM
NO. _47_ DUE _May 14, 20--_

Sylvia Cometta

Computing the Number of Interest Days of a Note

To define the interest period, or term, of a promissory note, the lender either specifies the due date of the note or states the number of interest days. When the due date is given, the number of interest days must be computed before the interest charge can be computed.

To do so you need the number of days in each month, as shown in Table 15-1. February has 29 days in leap years. A leap year is any year that is evenly divisible by 4, except for certain years ending in 00 (e.g., 1900 and 2000). In order to be leap years, years ending in 00 must be evenly divisible by 400; thus 2000 was a leap year, but 1900 wasn't.

Learning Objective 1

Compute the number of interest days of a promissory note.

15.1 Some students may be interested in knowing that the reason 2000 was a leap year and 1900 was not is that a year is not exactly 365.25 days. A year actually is a bit less—365.242199 days. By adding one leap year every four years, eventually too many days are added (0.7801 day in one century). Therefore, every 100 years a leap year is skipped. But this correction also isn't exact, so every fourth century a leap year isn't skipped.

STEPS to Compute the Number of Interest Days Between Two Dates

1. Determine the number of interest days in the beginning month.
2. Determine the number of interest days in the middle months.
3. Add the numbers from Steps 1 and 2 to the number of interest days in the final month. (For the final month, the number of interest days is equal to the number of the due date.)

EXAMPLE A

A promissory note is made on July 25. The due date is October 8. Use Table 15-1 to help you determine the number of interest days between July 25 and October 8.

15.2 Remind students of the mnemonic: 30 days has September, April, June and November; all the rest have 31 except February, which has 28.

Table 15-1 Days in Each Month (non-leap years)

Month	Number of Days	Month	Number of Days	Month	Number of Days
January	31 days	May	31 days	September	30 days
February	28 days	June	30 days	October	31 days
March	31 days	July	31 days	November	30 days
April	30 days	August	31 days	December	31 days

STEP 1	STEP 2	STEP 3
31 days in July	August has 31 days	6 days in July
− 25 days of note	September has 30 days	31 days in August
6 days of interest in July		30 days in September
		+ 8 days in October
		75 total interest days in the promissory note

✓ CONCEPT CHECK 15.1

A promissory note is dated October 20. The maturity date (due date) is February 20. Determine the number of interest days.

As October has 31 days and the note is dated October 20, there are $31 - 20 = 11$ days of interest in October. Since the note is due on February 20, there are 20 interest days in February. The total can be expressed as

October		November		December		January		February		Total Interest Days
11	+	30	+	31	+	31	+	20	=	123

JAVIER PIERIN/PHOTODISC/GETTY IMAGES

Determining the Due Date of a Note

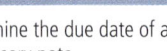 Learning Objective **2**

Determine the due date of a promissory note.

15.3 If the due date isn't a business day, the due date will be advanced to the next business day. The same principle applies to income taxes: The filing deadline is April 15 unless that is a Saturday or Sunday, in which case the deadline is the following Monday.

When the promissory note explicitly states the number of interest days, then you must determine the due date. The procedure is somewhat the reverse of finding the number of interest days.

STEPS to Determine the Due Date

1. Determine the number of interest days in the beginning month.
2. Determine the number of interest days that remain after the first month.
3. Determine the number of interest days remaining at the end of each succeeding month by subtracting. Continue subtracting until less than 1 month remains. The due date is the number of interest days remaining in the final month.

298 Part 4 Interest Applications

● **EXAMPLE B**

A promissory note is made on July 25. The note is for 75 days. Determine the due date.

STEP 1

```
  31  days in July
− 25  days of note
   6  days of interest in July
```

STEP 2

```
  75  days of interest in the note
−  6  days of interest in July
  69  days left in term after end of July
```

STEP 3

```
  69  days of interest left after July
− 31  days in August
  38  days of interest left after August
− 30  days in September
   8  days of interest left after end of September,
      or 8 days of interest in October
```

The due date is October 8.

Although the procedure looks somewhat cumbersome on paper, it goes very quickly on a calculator. You can subtract repeatedly to deduct the days of each month, and after each subtraction, the calculator will display the number of interest days remaining. You don't need to write down all the intermediate results.

When the length of the interest period is expressed in months, the date is advanced by the number of months given. The due date is the same date of the month as the date of the note. For example, a 3-month note dated July 3 will be due on September 3. The exact number of interest days must then be computed, as shown previously. If the note is dated the 31st of a month and the month of maturity is April, June, September, or November, the due date is the 30th. If the month of maturity is February, the due date is the 28th (or 29th in a leap year).

● **EXAMPLE C**

Find the due date of a 3-month note dated January 31 (the last day of the month).

Maturity month: April (count "February, March, April")
Last day: 30 (last day of April)
                  Therefore the due date is April 30.

✓ CONCEPT CHECK 15.2

a. A 90-day promissory note is dated February 5 in a non-leap year. Determine the due date.
   Since February has 28 days, the note has 28 − 5 = 23 days of interest in February.

Total Interest Days	February	March	April
90	− 23	− 31	− 30
=	67	= 36	= 6 days remaining after April

   The due date is May 6.

b. A 4-month promissory note is dated April 30. Determine the due date.
   Four months after April 30 is August 30. The due date is August 30.

Chapter 15 Promissory Notes and Discounting

Computing the Maturity Value of a Note

Learning Objective 3

Compute the maturity value of a promissory note.

The maturity value (*MV*) of a promissory note is the sum of the face value (principal) of the note and the interest:

Maturity value = Principal + Interest   or $MV = P + I$

15.4 Sometimes the sum $P + I$ is also called the *amount* of the investment or the future value of the investment. We use these terms in Chapter 16 on compound interest.

EXAMPLE D

Compute the maturity value of the interest-bearing promissory note illustrated in Figure 15-1.

The face value (*P*) of the note is $2,000. The interest rate (*R*) is 10% exact interest per year. The loan period of the note is 60 days, so the time in years (*T*) is $\frac{60}{365}$.

$$I = P \times R \times T = \$2,000 + 0.10 \times \frac{60}{365} = \$32.88$$

$$MV = P + I = \$2,000 + \$32.88 = \$2,032.88$$

✓ CONCEPT CHECK 15.3

A 90-day promissory note has a face value of $2,800 and an exact simple interest rate of 7.5%. Compute the maturity value.

$I = P \times R \times T = \$2,800 \times 0.075 \times \frac{90}{365} = \$51.78$ $MV = P + I = \$2,800 + \$51.78 = \$2,851.78$

COMPLETE ASSIGNMENT 15.1.

Discounting Promissory Notes

Learning Objective 4

Discount a promissory note.

Often, when a lender holds a promissory note as security for a loan to a borrower, the lender may need cash before the maturity date of the note. One option is for the lender to "sell" the note to a third party. Such a note is said to be *negotiable*.

However, now the third party is assuming the risk that the original borrower might not pay everything on the maturity date. Therefore, to acquire the note, the third party will pay the original lender less money than the maturity value. The note is said to "sell at a discount."

There is are several new vocabulary terms involved in discounting promissory notes. The calculations, however, are straightforward and very similar to simple interest calculations. This can be explained by using examples.

EXAMPLE E

On August 19, Telescan Medical Instruments borrows $75,000 from a private investor, Margaret Wegner. In return, Telescan gives Margaret Wegner a 120-day promissory note at an ordinary simple interest rate of 8% (360-day year). Compute the due date and the maturity value of the promissory note.

Due date: August 19 + 120 days = December 17

Interest: $I = P \times R \times T = \$75{,}000 \times 0.08 \times \dfrac{120}{360} = \$2{,}000$

Maturity value: $MV = \text{Principal} + \text{Interest} = \$75{,}000 + \$2{,}000 = \$77{,}000$

In the example, Telescan Medical must pay $77,000 to Margaret Wegner until December 17. During the 120 days, Margaret has only the promissory note—no cash. If the note is negotiable, Margaret can sell the note to a third party at any time before December 17. Suppose that Margaret sells the note on October 5 to Auburn Financial Corporation. October 5 is called the **discount date**. The time between October 1 and December 17 is the **discount period**. The length of the discount period is the number of days between October 5 and December 17. Since the original 8% interest rate was ordinary simple interest (360-day year), we will also use a 360-day year in the discount calculation.

Auburn Financial agrees to buy the note at a discount of 12% of the maturity value. 12% is the **discount rate**. The **discount amount** is calculated using a formula similar to ordinary simple interest:

Discount Amount = Maturity value × Discount rate × Time (Discount period)

Maturity value: $77,000
Discount rate: 12%
Discount period: October 5 to December 17 = (31 − 5) + 30 + 17 = 73 days

Discount Amount = $\$77{,}000 \times 0.12 \times \dfrac{73}{360} = \$1{,}873.67$

The difference between the maturity value and the discount amount is called the **proceeds**. It is the amount that Auburn Financial will pay to Margaret Wegner for her promissory note from Telescan Medical Systems.

Proceeds = Maturity value − Discount amount
          = $77,000 − $1,873.67 = $75,126.33

To summarize

15.5 In all the exercises in this book, the discount rate is higher than the original interest rate. You may want to discuss why this seems reasonable and ask students if they can imagine circumstances under which a person discounting a note might be willing to take a lower rate than the original lender did.

STEPS to Discount a Promissory Note

1. Compute the interest amount and maturity value (MV) of the promissory note.
2. Determine the maturity (due) date of the note.
3. Compute the number of days in the discount period. The time, T, is the number of days in the discount period divided by 360 (or by 365).
4. Compute the discount amount, using $D = MV \times R \times T$, where R is the discount rate.
5. Compute the proceeds by subtracting the discount amount from the maturity value.

NON-INTEREST-BEARING PROMISSORY NOTES

Sometimes the original lender may not charge any interest at all. In this situation, the maturity value of the note is equal to the face value. Similarly, the original lender may require that all of the interest must be completely paid in advance. Therefore, this is another type of promissory note that does not have any interest dollars in the maturity value, so the maturity value is equal to the face value. To find the proceeds of a **non-interest-bearing promissory note,** follow the same steps that were listed above. But, in Step 1, the amount of interest is $0 and the maturity value is the face value.

EXAMPLE F

Willie Smith, owner of a True-Value Hardware Store, is holding a 75-day, non-interest-bearing note for $3,500. The note is dated June 21. On August 10, Willie sells the note to the Marshfield Lending Company, which discounts the note at 11%. Find the discount amount and the proceeds using a 365-day year.

STEP 1 Interest amount = $0; Maturity value = Face value = $3,500

STEP 2 Due date: June 21 + 75 days = September 4

STEP 3 Discount period: August 10 to September 4 = 25 days

STEP 4 Discount amount: Maturity value × Discount rate × time

$$= \$3{,}500 \times 0.11 \times \frac{25}{365}$$

$$= \$26.37$$

STEP 5 Proceeds: Maturity value − Discount amount
= $3,500 − $26.37
= $3,473.63

✓ CONCEPT CHECK 15.4

A 75-day promissory note, bearing interest at 10%, is dated December 11 and has a face value of $5,000. On January 24, the note is discounted at 14%. Find the discount amount and the proceeds. Note: The interest amount, the maturity value, the maturity date, and the days of discount must first be determined. Use a 365-day year for all interest and discount calculations.

Interest amount: $\$5{,}000 \times 0.10 \times \frac{75}{365} = \$102.74$

Maturity value: $5,000 + $102.74 = $5,102.74

Maturity date: Dec. 11 + 75 days = Feb. 24

Days of discount: Jan. 24 to Feb. 24 = 31 days

Discount amount: $\$5{,}102.74 \times 0.14 \times \frac{31}{365} = \$60.67$

Proceeds: $5,102.74 − $60.67 = $5,042.07

COMPLETE ASSIGNMENT 15.2.

Bank Discounting

In Chapter 13 and at the beginning of this chapter, we studied the simple procedure to borrow and repay money: Determine the Principal, Rate, and Time; compute the interest amount; the maturity value (amount due) is the principal plus the interest.

Learning Objective 5

Compute the proceeds and actual interest rate on a bank discount loan.

EXAMPLE G

Rueben Cortez, owner/operator of a fast-food restaurant, borrows $50,000 from his bank for 60 days at 9% ordinary simple interest. Using a 360-day year, compute the interest and the maturity value.

$P = \$50,000; R = 9\%; T = \dfrac{60}{360}$

Interest $(I) = P \times R \times T = \$50,000 \times 0.09 \times \dfrac{60}{360} = \$750$

Maturity value $(MV) = P + I = \$50,000 + \$750 = \$50,750$

Please observe: Rueben will keep the entire $50,000 for the entire 60 days and then repay a total of $50,750 on the due date.

In the previous section, we studied promissory notes that were discounted at some date between the date of the loan and the due date. Similarly, sometimes banks will discount loans immediately, at the time they are written. The steps to discount a loan are the same as discounting promissory notes, but even simpler because (a) the face value is equal to the maturity value, (b) the discount date is the same as the loan date, and (3) the number of discount days is the same as the period of the loan.

STEPS to Discount a Bank Loan

1. Compute the discount amount, using $D = FV \times R \times T$, where R is the discount rate.
2. Compute the proceeds by subtracting the discount amount from the face value.

EXAMPLE H

Rueben Cortez, owner/operator of a fast-food restaurant, goes to his bank to borrow money. Rueben signs a 60-day note with a $50,000 face value at a 9% discount rate. Using a 360-day year, compute the discount amount and the proceeds of the loan.

$FV = \$50,000; R = 9\%; T = \dfrac{60}{360}$

STEP 1 Discount amount $(D) = FV \times R \times T = \$50,000 \times 0.09 \times \dfrac{60}{360} = \$750$

STEP 2 Proceeds = Face value − Discount amount = $50,000 − $750 = $49,250

Please observe: In Example H, Rueben will keep $49,250 for the entire 60 days and then repay a total of $50,000 on the due date.

As mentioned earlier, some persons refer to this type of discounted loan as "non-interest-bearing" because the amount to be repaid is the "face value." However, the term *non-interest-bearing* is misleading because the loan is NOT "interest-free." There is a charge of $750 to borrow $49,250 for 60 days.

COMPARING A DISCOUNT RATE TO AN INTEREST RATE

Discount rates are less familiar to those consumers who have encountered only interest rates. There is the possibility of misunderstanding or confusion. In Example G, Rueben Cortez borrowed $50,000 for 60 days and paid $750. The ordinary simple interest rate was 9%. In Example H, Rueben borrowed $49,250 for 60 days and paid $750. Although a discount rate (9%) was given, a simple interest rate was not given. To compute the actual simple interest rate, use the formula from Chapter 14:

$$R = \frac{I}{P \times T}, \text{ letting } I = \$750, P = \$49,250, \text{ and } T = \frac{60}{360}$$

$$R = \frac{I}{P \times T} = \frac{\$750}{\$49,250 \times \frac{60}{360}} = \frac{\$750}{\$8,208.33} = 0.09137, \text{ or } 9.14\%$$

The interest rate in Example H is actually 9.14%; the discount rate is 9%. They are different rates, but both lead to a $750 fee to borrow $49,250 for 60 days. A borrower must understand the difference between interest rates and discount rates and how each is used in loan calculations.

✓ CONCEPT CHECK 15.5

A bank made a 90-day loan on a discount basis. The face value was $64,000, and the discount rate was 11%. Compute the discount amount and the proceeds. Then compute the actual interest rate, using the proceeds as the principal of the loan instead of the face value. Use a 360-day year in all calculations.

$$\text{Discount amount} = FV \times R \times T = \$64,000 \times 0.11 \times \frac{90}{360} = \$1,760$$

$$\text{Proceeds} = \text{Face value} - \text{Discount amount} = \$64,000 - \$1,760 = \$62,240$$

$$\text{Actual Interest Rate} = \frac{I}{(P \times T)} = \frac{\$1,760}{\left(\$62,240 \times \frac{90}{360}\right)} = 0.1131, \text{ or } 11.31\%$$

Borrowing Money to Take a Cash Discount

In Chapter 7, we described how manufacturers and wholesalers use cash discounts to encourage their customers to pay their invoices early. Recall that the terms "2/10, net 30" mean that the buyer will receive a 2% discount by paying the invoice within 10 days and that the entire invoice is due within 30 days. However, it would be normal that a buyer would not have the immediate cash to pay the invoice early. The buyer may need to sell the merchandise to get the cash to pay the invoice. Normally, a buyer can save money by borrowing money to pay the invoice early and earn the cash discount.

Learning Objective 6

Compute the savings from borrowing money to take a cash discount.

EXAMPLE I

DVD Central purchased $100,000 worth of CDs and DVDs. The invoice was dated October 4 with terms of 2/10, net 30. Compute the due date, the discount date, the cash discount, and the total remittance required to get the cash discount. (Review Chapter 12 if necessary.)

Due date = October 4 + 30 days = November 3
Discount date = October 4 + 10 days = October 14
If paid by October 14:
   Cash discount = $100,000 × 0.02 = $2,000
   Total remittance: $100,000 − $2,000 = $98,000

Regardless of whether it takes the discount, DVD Central needs to pay $100,000 by November 3. The company may want to save the $2,000, but perhaps it doesn't have the $98,000 now. Or maybe it has the money but wants to spend it on something else. In either situation, DVD Central might be able to borrow the money from October 14 until November 3. Before borrowing, DVD Central should compare the savings from the cash discount with the interest on a loan.

EXAMPLE J

DVD Central can borrow $98,000 for 20 days (October 14 to November 3) by paying 10% exact simple interest (365-day year). Compute the interest on the loan and the savings for DVD Central if it borrows to take the discount.

Interest = $P \times R \times T$ = $98,000 × 0.10 × $\dfrac{20}{365}$ = $536.99

Savings = $2,000 discount − $536.99 interest = $1,463.01

The reason for borrowing only between the discount date and the due date is to delay making payments as long as possible, whether to get discounts or to avoid penalties. The discount date is the latest possible date to pay and get the discount; the due date is the latest possible date to pay and avoid a penalty.

Although borrowing and taking the discount is almost always cheaper, the actual dollar amount may determine what DVD Central decides. If the original purchase were only $1,000, the savings would be only $14.63. Such an amount may not be worth the effort of getting a loan. However, for borrowing small amounts regularly, businesses often have "revolving lines of credit." These allow them to borrow and repay frequently, without always making a new loan application.

15.6 As a review of Chapter 14, you or the students could calculate what interest rate on $98,000 for 20 days is required to earn $2,000 interest. The answer is 37.24%. This is the rate for any purchase with terms of 2/10, net 30. By not taking the discount, the buyer is missing the opportunity to make a 37.24% investment.

15.7 The decision as to whether to pay an invoice early or prolong payment for as long as possible will also depend on how strictly the seller enforces the late payment penalty. When the buyer is very large compared to the seller, the seller may permit late payments just to get and keep the business. In such cases, the buyer may prefer to delay payment for a long time at no penalty instead of taking the discount.

CONCEPT CHECK 15.6

A retailer purchases merchandise under the terms 1.5/20, net 45. The invoice is for $45,000 and is dated July 22. For the cash discount, calculate the due date, the discount date, the amount of the cash discount, and the total remittance required. The retailer borrows enough money to pay the entire remittance. The interest rate is 12% exact simple interest, and the loan is for the length of time between the last date to take advantage of the cash discount and the due date. Calculate the amount of the interest and the savings gained by borrowing the remittance to take the discount.

Discount:
- Due date: July 22 + 45 days = September 5
- Discount date: July 22 + 20 days = August 11
- Cash discount: $45,000 × 0.015 = $675
- Remittance: $45,000 − $675 = $44,325

Loan:
- Interest days: August 11 to September 5 = 25 days

$$\text{Interest} = P \times R \times T = \$44{,}325 \times 0.12 \times \frac{25}{365} = \$364.32$$

Savings: $675 cash discount − $364.32 interest = $310.68

COMPLETE ASSIGNMENT 15.3.

Chapter Terms for Review

- bank discount
- bearer
- discount a note
- discount amount (D)
- discount date
- discount period
- discount rate
- due date
- face value
- interest dollars (I)
- interest-bearing note
- maker
- maturity date
- maturity value (MV)
- negotiable promissory note
- non-interest-bearing note
- proceeds
- promissory note

THE BOTTOM LINE

Summary of chapter learning objectives:

Learning Objective	Example
15.1 Compute the number of interest days of a promissory note	1. Find the number of days between December 15 and February 27.
15.2 Determine the due date of a promissory note	2. Find the due date of a 60-day note written on April 20.
15.3 Compute the maturity value of a promissory note	3. Find the maturity value of a 90-day promissory note with a face value of $6,500 and an exact interest rate of 8%.
15.4 Discount a promissory note	4a. A 30-day note, bearing an interest rate of 9%, is dated November 6 and has a face value of $8,000. On November 15, the note is discounted at 12%. Use a 365-day year to find the interest amount, the discount amount, and the proceeds. 4b. A 60-day non-interest-bearing note has a face value of $2,500 and is dated May 13. On June 3, the note is discounted at 11%. Use a 365-day year to find the discount amount and the proceeds.
15.5 Compute the proceeds and actual interest rate on a bank discount loan	5. A 60-day bank loan with a face value of $3,900 is made on a discount basis at a discount rate of 12%. Use the 360-day year to compute the discount amount and the proceeds. Then find the actual interest rate, based on the proceeds rather than on the face value.
15.6 Compute the savings from borrowing money to take a cash discount	6. A $20,000 invoice dated March 15 has terms of 2/5, net 25. Find the due date, discount date, cash discount, and required remittance. Next, calculate the interest amount of borrowing the remittance at 9% exact interest for the time between the last date to take advantage of the cash discount and the due date. Finally, calculate the savings.

Answers: 1. 74 days 2. June 19 3. $6,628.22 4a. Interest, $59.18; discount, $55.64; proceeds, $8,003.54; 4b. Discount, $29.38; proceeds, $2,470.62 5. Discount, $78; proceeds, $3,822; interest rate, 12.24% 6. Due date, April 9; discount date, March 20; cash discount, $400; remittance, $19,600; interest, $96.66; savings, $303.34

SELF-CHECK

Review Problems for Chapter 15

1. A 75-day promissory note for $3,500 is dated November 24, 2006. Find (a) the due date and (b) the maturity value, if the rate is 7% ordinary simple interest.

2. A promissory note for $4,400 is dated December 12, 2005 and has a due date of May 12, 2006. Find (a) the number of interest days and (b) the maturity value, if the rate is 6% ordinary simple interest.

3. A 135-day promissory note for $15,000 is dated August 24, 2007. Find (a) the due date and (b) the maturity value, if the rate is 4.6% exact simple interest.

4. A promissory note for $2,980 is dated May 20, 2008 and has a due date of September 20, 2008. Find (a) the number of interest days and (b) the maturity value, if the rate is 6.5% exact simple interest.

5. Vernon Lee holds a 120-day interest-bearing note for $2,960 that is dated May 15 and has a rate of 8% exact simple interest. On July 15, Vernon sells it at a discount rate of 15%. Using a 365-day year, calculate (a) the interest amount, (b) the maturity value, (c) the maturity date, (d) the days of discount, (e) the discount amount, and (f) the proceeds.

6. Contractor Allen Kimmel is holding a 90-day non-interest-bearing note for $3,100 dated November 10. On December 10, Mr. Kimmel sells the note to Thrift's Financing, Inc. at a discount rate of 12%. Using a 365-day year, calculate (a) the maturity value, (b) the maturity date, (c) the days of discount, (d) the discount amount, and (e) the proceeds.

7. Eastside Bank & Trust Co. made a 120-day loan for $4,500 on a discount basis, using a discount rate of 9%. Using a 360-day year, calculate (a) the discount amount, (b) the proceeds, and (c) the actual interest rate (to two decimal places).

8. Jankowski Corporation just received an invoice for $1,600 that has cash discount terms of 2/10, net 30. Jankowski borrows enough money from Eastside Bank & Trust Co. at 10% exact simple interest (365-day year) to take advantage of the cash discount. It borrows the money only for the time period between the due date and the last day to take advantage of the discount. Calculate (a) the amount of the cash discount, (b) the number of interest days, (c) the amount of interest on the loan, and (d) the amount of its savings.

Assignment 15.1: Dates, Times, and Maturity Value

Name

Date   Score

Learning Objectives 1 2 3

A (36 points) Problems 1–6: Find the number of interest days. Problems 7–12: Find the due date. Be sure to check for leap years. (3 points for each correct answer)

	Date of Note	Due Date	Days of Interest
1.	April 6, 2006	October 11, 2006	188
	$(30 - 6) + 31 + 30 + 31 + 31 + 30 + 11 = 188$		
2.	June 30, 2008	October 6, 2008	98
	$(30 - 30) + 31 + 31 + 30 + 6 = 98$		
3.	February 9, 2007	June 11, 2007	122
	$(28 - 9) + 31 + 30 + 31 + 11 = 122$		
4.	June 14, 2005	September 13, 2005	91
	$(30 - 14) + 31 + 31 + 13 = 91$		
5.	November 8, 2006	March 9, 2007	121
	$(30 - 8) + 31 + 31 + 28 + 9 = 121$		
6.	July 14, 2008	October 1, 2008	79
	$(31 - 14) + 31 + 30 + 1 = 79$		

	Date of Note	Interest Days	Due Date
7.	November 1, 2005	90 days	January 30, 2006
	$90 - (30 - 1) - 31 = 30$		
8.	August 17, 2006	180 days	February 13, 2007
	$180 - (31 - 17) - 30 - 31 - 30 - 31 - 31 = 13$		
9.	September 24, 2008	75 days	December 8, 2008
	$75 - (30 - 24) - 31 - 30 = 8$		
10.	April 28, 2006	60 days	June 27, 2006
	$60 - (30 - 28) - 31 = 27$		
11.	November 7, 2005	120 days	March 7, 2006
	$120 - (30 - 7) - 31 - 31 - 28 = 7$		
12.	March 25, 2007	3 months	June 25, 2007
	Mar. 25 + 3 months = June 25		

Score for A (36)

Chapter 15 Promissory Notes and Discounting

Assignment 15.1 Continued

B (64 points) For each of the following promissory notes, find the missing entry for days of interest or maturity date (due date). Then compute the amount of interest due at maturity and the maturity value. For problems 13–16, use a 360-day year; for problems 17–20, use a 365-day year. (Points indicated at the top of each column.)

	Face Value	Date of Note	Days of Interest (3 pts)	Maturity Date (3 pts)	Rate	Interest Amount (3 pts)	Maturity Value (2 pts)
13.	$26,000	Oct. 11, 2006	90	Jan. 9, 2007	6.2%	$403	$26,403

$90 - (31 - 11) - 30 - 31 = 9$ $\$26{,}000 \times 0.062 \times \dfrac{90}{360} = \$403$ $\$26{,}000 + \$403$

14.	$12,500	Mar. 28, 2006	101	July 7, 2006	8.5%	$298.09	$12,798.09

$(31 - 28) + 30 + 31 + 30 + 7 = 101$ $\$12{,}500 \times 0.085 \times \dfrac{101}{360} = \$298.09$ $\$12{,}500 + \$298.09$

15.	$35,750	July 15, 2005	105	Oct. 28, 2005	5.6%	$583.92	$36,333.92

$105 - (31 - 15) - 31 - 30 = 28$ $\$35{,}750 \times 0.056 \times \dfrac{105}{360} = \$583.92$ $\$35{,}750 + \$583.92$

16.	$950	Jan. 26, 2007	66	April 2, 2007	7.2%	$12.54	$962.54

$(31 - 26) + 28 + 31 + 2 = 66$ $\$950 \times 0.072 \times \dfrac{66}{360} = \$12.54$ $\$950 + \$12.54$

17.	$11,800	Nov. 23, 2005	125	Mar. 28, 2006	4.9%	$198.01	$11,998.01

$(30 - 23) + 31 + 31 + 28 + 28 = 125$ $\$11{,}800 \times 0.049 \times \dfrac{125}{365} = \$198.01$ $\$11{,}800 + \$198.01$

18.	$18,420	May 7, 2007	136	Sept. 20, 2007	6.75%	$463.28	$18,883.28

$(31 - 7) + 30 + 31 + 31 + 20 = 136$ $\$18{,}420 \times 0.0675 \times \dfrac{136}{365} = \$463.28$ $\$18{,}420 + \$463.28$

19.	$52,000	Feb. 10, 2005	180	Aug. 9, 2005	8.25%	$2,115.62	$54,115.62

$180 - (28 - 10) - 31 - 30 - 31 - 30 - 31 = 9$ $\$52{,}000 \times 0.0825 \times \dfrac{180}{365} = \$2{,}115.62$ $\$52{,}000 + \$2{,}115.62$

20.	$31,860	June 2, 2008	105	Sept. 15, 2008	7.5%	$687.39	$32,547.39

$105 - (30 - 2) - 31 - 31 = 15$ $\$31{,}860 \times 0.075 \times \dfrac{105}{365} = \$687.39$ $\$31{,}860 + \$687.39$

Score for B (64)

Part 4 Interest Applications

Assignment 15.2: Discounting Promissory Notes

Name

Date   Score

Learning Objective 4

A (50 points) Compute the missing information to discount the following interest-bearing and non-interest-bearing promissory notes. Use a 360-day year for all interest and discount calculations. (Points for each correct answer are shown in parentheses.)

1. Sharon Wilder had been holding a 75-day note for $2,500. The note had a 6% interest rate and had been written on March 1. To pay income taxes, Sharon sold the note on April 13 to a loan company. The loan company discounted the note at 11%.

   Interest amount (3 pts) $31.25
   Maturity value (2 pts) $2,531.25
   Maturity date (2 pts) May 15
   Days of discount (2 pts) 32
   Discount amount (3 pts) $24.75
   Proceeds (2 pts) $2,506.50

   $I = \$2,500 \times 0.06 \times \dfrac{75}{360} = \$31.25$

   $MV = \$2,500 + \$31.25 = \$2,531.25$

   Date: $75 - (31 - 1) - 30 = 15$

   Days: $(30 - 13) + 15 = 32$

   $D = \$2,531.25 \times 0.11 \times \dfrac{32}{360} = \$24.75$

   Proceeds $= \$2,531.25 - \$24.75 = \$2,506.50$

2. On September 7, Carol Swift Financial Services bought a $12,500 promissory note. The note had been written on July 7, was for 150 days, and had an interest rate of 9%. Carol's company discounted the note at 12%.

   Interest amount (3 pts) $468.75
   Maturity value (2 pts) $12,968.75
   Maturity date (2 pts) Dec. 4
   Days of discount (2 pts) 88
   Discount amount (3 pts) $380.42
   Proceeds (2 pts) $12,588.33

   $I = \$12,500 \times 0.09 \times \dfrac{150}{360} = \$468.75$

   $MV = \$12,500 + \$468.75$

   Date: $150 - (31 - 7) - 31 - 30 - 31 - 30 = 4$

   Days: $(30 - 7) + 31 + 30 + 4 = 88$

   $D = \$12,968.75 \times 0.12 \times \dfrac{88}{360} = \$380.42$

   Proceeds $= \$12,968.75 - \$380.42 = \$12,588.33$

3. Jim Walter was holding a 105-day non-interest-bearing note for $4,500. The note was dated October 10. To raise Christmas cash, Jim sold the note to a local finance company on December 15. The company discounted the note at 10%.

   Interest amount (1 pt) $0
   Maturity value (1 pt) $4,500
   Maturity date (2 pts) Jan. 23
   Days of discount (2 pts) 39
   Discount amount (3 pts) $48.75
   Proceeds (2 pts) $4,451.25

   $I = \$0; MV = FV = \$4,500$

   Date: $105 - (31 - 10) - 30 - 31 = 23$

   Days: $(31 - 15) + 23 = 39$

   $D = \$4,500 \times 0.10 \times \dfrac{39}{360} = \$48.75$

   Proceeds $= \$4,500 - \$48.75 = \$4,451.25$

4. Barbara Finell owned a finance company. On July 19 she purchased a 180-day non-interest-bearing promissory note for $6,000. The note had been written on May 23. Because of the high financial risk involved, Barbara discounted the note at 15%.

   Interest amount (1 pt) $0
   Maturity value (1 pt) $6,000
   Maturity date (2 pts) Nov. 19
   Days of discount (2 pts) 123
   Discount amount (3 pts) $307.50
   Proceeds (2 pts) $5,692.50

   $I = \$0; MV = FV = \$6,000$

   Date: $180 - (31 - 23) - 30 - 31 - 31 - 30 - 31 = 19$

   Days: $(31 - 19) + 31 + 30 + 31 + 19 = 123$

   $D = \$6,000 \times 0.15 \times \dfrac{123}{360} = \$307.50$

   Proceeds $= \$6,000 - \$307.50 = \$5,692.50$

Score for A (50)

Assignment 15.2 Continued

B (50 points) Compute the missing information to discount the following interest-bearing and non-interest-bearing promissory notes. Use a 365-day year for all interest and discount calculations. (Points for each correct answer are shown in parentheses.)

5. As payment for services, Pat Chard held a 90-day, 8% note for $3,600 that was dated April 20. On June 5, Pat took the note to a financial services company, which bought the note at a 13% discount rate.

   Interest amount (3 pts) $71.01
   Maturity value (2 pts) $3,671.01
   Maturity date (2 pts) July 19
   Days of discount (2 pts) 44
   Discount amount (3 pts) $57.53
   Proceeds (2 pts) $3,613.48

   $I = 3{,}600 \times 0.08 \times \dfrac{90}{365} = \$71.01$

   $MV = \$3{,}600 + \$71.01 = \$3{,}671.01$

   Date: $90 - (30 - 20) - 31 - 30 = 19$
   Days: $(30 - 5) + 19 = 44$

   $D = \$3{,}671.01 \times 0.13 \times \dfrac{44}{365} = \$57.53$

   Proceeds $= \$3{,}671.01 - \$57.53 = \$3{,}613.48$

6. Joslin Builders received a 135-day, 7% note dated October 11. The face value was $12,450, which was for remodeling a client's garage. On December 20, Joslin sold the note to McGraw Lending Corp., which discounted the note at 12%.

   Interest amount (3 pts) $322.34
   Maturity value (2 pts) $12,772.34
   Maturity date (2 pts) Feb. 23
   Days of discount (2 pts) 65
   Discount amount (3 pts) $272.94
   Proceeds (2 pts) $12,499.40

   $I = \$12{,}450 \times 0.07 \times \dfrac{135}{365} = \$322.34$

   $MV = \$12{,}450 + \$322.34 = \$12{,}772.34$

   Date: $135 - (31 - 11) - 30 - 31 - 31 = 23$
   Days: $(31 - 20) + 31 + 23 = 65$

   $D = \$12{,}772.34 \times 0.12 \times \dfrac{65}{365} = \$272.94$

   Proceeds $= \$12{,}772.34 - \$272.94 = \$12{,}499.40$

7. Teri Chung loaned $4,000 to a client who gave Teri a non-interest-bearing note dated August 4. The note was for 75 days. On September 3, Teri sold the note to her finance company, which discounted it at 10%.

   Interest amount (1 pt) $0
   Maturity value (1 pt) $4,000
   Maturity date (2 pts) Oct. 18
   Days of discount (2 pts) 45
   Discount amount (3 pts) $49.32
   Proceeds (2 pts) $3,950.68

   $I = \$0;\ MV = FV = \$4{,}000$
   Date: $75 - (31 - 4) - 30 = 18$
   Days: $(30 - 3) + 18 = 45$

   $D = \$4{,}000 \times 0.10 \times \dfrac{45}{365} = \$49.32$

   Proceeds $= \$4{,}000 - \$49.32 = \$3{,}950.68$

8. Patti Gentry was holding a 60-day non-interest-bearing note for $6,200. The note was dated June 22. On July 16, Patti sold the note to a lender who discounted the note at 14%.

   Interest amount (1 pt) $0
   Maturity value (1 pt) $6,200
   Maturity date (2 pts) Aug. 21
   Days of discount (2 pts) 36
   Discount amount (3 pts) $85.61
   Proceeds (2 pts) $6,114.39

   $I = \$0;\ MV = FV = \$6{,}200$
   Date: $60 - (30 - 22) - 31 = 21$
   Days: $(31 - 16) + 21 = 36$

   $D = \$6{,}200 \times 0.14 \times \dfrac{36}{365} = \$85.61$

   Proceeds $= \$6200 - \$85.61 = \$6{,}114.39$

Score for B (50)

Assignment 15.3: Bank Discounting and Cash Discounts

Name

Date   Score

Learning Objectives 5 6

A (36 points) The Citizens' Bank of New England made six new loans on a discount basis. Compute the discount amount and the proceeds. Then compute the actual interest rate based on the proceeds rather than the face value. Use a 360-day year for problems 1–3 and use a 365-day year for problems 4–6. Round the actual interest rates to the nearest 1/100 of a percent. (2 points for each correct answer.)

	Face Value	Discount Rate	Time	Discount Amount	Proceeds	Actual Interest Rate
1.	$7,500	10%	120 days	$250	$7,250	10.34%

$$\$7{,}500 \times 0.10 \times \frac{120}{360} = \$250 \qquad R = \frac{I}{P \times T} = \frac{\$250}{\$7{,}250 \times \frac{120}{360}} = 0.103448$$

$$\$7{,}500 - \$250 = \$7{,}250$$

2.	$4,450	6%	90 days	$66.75	$4,383.25	6.09%

$$\$4{,}450 \times 0.06 \times \frac{90}{360} = \$66.75 \qquad R = \frac{\$66.75}{\$4{,}383.25 \times \frac{90}{360}} = 0.060914$$

$$\$4{,}450 - \$66.75 = \$4{,}383.25$$

3.	$16,500	12%	150 days	$825	$15,675	12.63%

$$\$16{,}500 \times 0.12 \times \frac{150}{360} = \$825 \qquad R = \frac{\$825}{\$15{,}675 \times \frac{150}{360}} = 0.126316$$

$$\$16{,}500 - \$825 = \$15{,}675$$

4.	$6,750	8.2%	75 days	$113.73	$6,636.27	8.34%

$$\$6{,}750 \times 0.082 \times \frac{75}{365} = \$113.73 \qquad R = \frac{\$113.73}{\$6{,}636.27 \times \frac{75}{365}} = 0.083403$$

$$\$6{,}750 - \$113.73 = \$6{,}636.27$$

5.	$980	7.5%	135 days	$27.18	$952.82	7.71%

$$\$980 \times 0.075 \times \frac{135}{365} = \$27.18 \qquad R = \frac{\$27.18}{\$952.82 \times \frac{135}{365}} = 0.077125$$

$$\$980 - \$27.18 = \$952.82$$

6.	$18,250	9.6%	105 days	$504	$17,746	9.87%

$$\$18{,}250 \times 0.096 \times \frac{105}{365} = \$504 \qquad R = \frac{\$504}{\$17{,}746 \times \frac{105}{365}} = 0.098726$$

$$\$18{,}250 - \$504 = \$17{,}746$$

Score for A (36)

Assignment 15.3 Continued

B (64 points) William Bros. Home Builders made several purchases from vendors who offered various terms of payment. How much can William Bros. save on each invoice if it borrows the money to pay the invoice early and receive the cash discount? The loan interest rates are all exact simple interest (365-day year). Assume that the number of interest days is the time between the due date and the last day to take advantage of the cash discount. (2 points for each correct answer)

	Invoice	Terms	Cash Discount	Interest Rate on Loan	Interest Days	Interest Amount	Savings
7.	$5,000	2/10, n/30	$100.00	10%	20	$26.85	$73.15

$5,000 × 0.02 = $100 $4,900 × 0.10 × $\frac{20}{365}$ = $26.849

$5,000 − $100 = $4,900 $100 − $26.85 = $73.15

8.	$8,500	1.5/15, n/30	$127.50	6.25%	15	$21.50	$106

$8,500 × 0.015 = $127.50 $8,372.50 × 0.0625 × $\frac{15}{365}$ = $21.5047

$8,500 − $127.50 = $8,372.50 $127.50 − $21.50 = $106

9.	$17,500	3/5, n/25	$525.00	8%	20	$74.41	$450.59

$17,500 × 0.03 = $525 $16,975 × 0.08 × $\frac{20}{365}$ = $74.411

$17,500 − $525 = $16,975 $525 − $74.41 = $450.59

10.	$18,600	1/15, n/45	$186.00	9%	30	$136.21	$49.79

$18,600 × 0.01 = $186 $18,414 × 0.09 × $\frac{30}{365}$ = $136.213

$18,600 − $186 = $18,414 $186 − $136.21 = $49.79

11.	$9,200	1/30, n/60	$92.00	9.6%	30	$71.87	$20.13

$9,200 × 0.01 = $92 $9,108 × 0.096 × $\frac{30}{365}$ = $71.866

$9,200 − $92 = $9,108 $92 − $71.87 = $20.13

12.	$12,500	2/10, n/45	$250	8%	35	$93.97	$156.03

$12,500 × 0.02 = $250 $12,250 × 0.08 × $\frac{35}{365}$ = $93.973

$12,500 − $250 = $12,250 $250 − $93.97 = $156.03

13.	$26,000	2.5/5, n/25	$650.00	8.5%	20	$118.07	$531.93

$26,000 × 0.025 = $650 $25,350 × 0.085 × $\frac{20}{365}$ = $118.068

$26,000 − $650 = $25,350 $650 − $118.07 = $531.93

14.	$65,400	3/10, n/25	$1,962.00	7.5%	15	$195.53	$1,766.47

$65,400 × 0.03 = $1,962 $63,438 × 0.075 × $\frac{15}{365}$ = $195.528

$65,400 − $1,962 = $63,438 $1,962 − $195.53 = $1,766.47

Score for B (64)

Compound Interest and Present Value

16

Learning Objectives
By studying this chapter and completing all assignments you will be able to:

Learning Objective 1 — Compute future values and compound interest.

Learning Objective 2 — Compute present values.

Learning Objective 3 — Use present value tables and/or formulas.

16.1 Daily compounding is not required in any of the problems of this chapter because each student would need a calculator with an exponent key. An example is given in example G.

Most Americans will buy at least one product that is financed over 1 or more years. The product will probably be large, such as a car or a home. The interest on the loan for the car or home is not the simple interest you studied in Chapter 13; it is *compound* interest. Interest on car loans or home loans is normally compounded monthly. Most banks offer savings accounts and certificates of deposit (CDs) for which interest is compounded daily. Credit unions may pay interest that is compounded quarterly (four times a year). To evaluate the value of corporate bonds, an investor bases calculations on interest compounded semiannually (twice a year).

To understand even the simplest financial decisions in today's world, you need to understand the fundamentals of compound interest, future values, and present values.

Computing Future Values and Compound Interest

Learning Objective 1

Compute future values and compound interest.

Simple interest is computed with the formula $I = P \times R \times T$, which you learned in Chapter 13. For example, the simple interest on $2,000 invested at 6% for 2 years is $I = P \times R \times T = \$2,000 \times 0.06 \times 2 = \$240$. The amount, or future value, of the investment is $A = P + I = \$2,000 + \$240 = \$2,240$.

Compound interest means that the computations of the simple interest formula are performed periodically during the term of the investment. The money from the previous interest computation is added to the principal before the next interest computation is performed. If an investment is *compounded annually for 2 years,* the simple interest is computed once at the end of each year. The simple interest earned in year 1 is added to the principal for the beginning of year 2. The total value of an investment is the principal plus all the compound interest, called the **future value** or the **compound amount**. In finance, the principal is usually called the present value.

16.2 As the chapter progresses, the term *principal* is used less and the term *present value* is used more. See examples H, I, and J.

EXAMPLE A

Don Robertson invests $2,000 for 2 years in an account that pays 6% compounded annually. Compute the total compound interest and future value (compound amount).

	$2,000.00	Original principal
×	0.06	Interest rate
	$120.0000	First-year interest
+	2,000.00	First-year principal
	$2,120.00	Second-year principal
×	0.06	Interest rate
	$127.2000	Second-year interest
	2,120.00	Second-year principal
	$2,247.20	Final compound amount (future value)
−	2,000.00	Original principal
	$247.20	Total compound interest

On the $2,000 investment in example A, the total amount of compound interest paid is $247.20, compared to $240 simple interest over the same 2 years.

The computations in example A are time-consuming and become more tedious with each compounding. Twice as many computations would be required for a 4-year

Part 4 Interest Applications

investment. In actual practice, compound interest is computed using **compound interest tables**, calculators or computers.

Table 16-1, on pages 338 and 339, is part of a future value table. The numbers in the table are called **future value factors** or **compound amount factors**. The columns (vertical) represent interest rates, and the rows (horizontal) represent the number of times that interest is compounded. The following steps explain how to use Table 16-1 to find future values (compound amounts) and compound interest.

> **STEPS** **to Use the Future Value Table**
>
> 1. Locate the factor in the proper row and column of Table 16-1.
> 2. Multiply the principal (present value) by the factor. The product is the future value.
> 3. Subtract the principal (present value) from the future value. The difference is the total amount of compound interest.

EXAMPLE B

Use Table 16-1 to compute the future value and total amount of compound interest of a 2-year $2,000 investment at 6% compounded annually.

STEP 1 The interest rate is 6%. Interest is compounded twice—once each year for 2 years. Locate the intersection of the 6.00% column and row 2. The future value factor is 1.12360.

STEP 2 Future value = $2,000 × 1.12360 = $2,247.20

STEP 3 Compound interest = $2,247.20 − $2,000 = $247.20

These results are identical to the results in example A.

EXAMPLE C

Mary Simmons loans $5,000 to her son for 6 years at 4% compounded annually. Compute the future value and total compound interest. Use Table 16-1.

STEP 1 The interest rate is 4%. Interest is computed six times, once each year for 6 years. The future value factor in the 4.00% column and row 6 is 1.26532.

STEP 2 Future value = $5,000 × 1.26532 = $6,326.60

STEP 3 Compound interest = $6,326.60 − $5,000 = $1,326.60

FUTURE VALUE FORMULA

If you prefer, Step 2 may be summarized as a formula in words or symbols:

Future value = Principal (Present value) × Future value factor (from Table 16-1)
or, $FV = PV \times FVF$

VARIOUS COMPOUNDING PERIODS

In examples A, B, and C, the compounding was annual (i.e., done once each year). Compounding is also done daily (every day), monthly (every month), quarterly (every quarter), or semiannually (every half-year). The word **period** is the unit of time of the compounding. The period will be a day, a month, a quarter, a half-year, or a year. You can use Table 16-1 with some interest rates for all these compounding periods except 1 day. Daily compounding requires the use of a calculator with an exponent key.

To do monthly, quarterly, or semiannual compounding using Table 16-1, follow the same steps you used to do annual compounding. The only differences are that the column will be the **periodic interest rate** (i) and that the row will be the **number of compounding periods** (n). Sometimes the periodic rate and the number of periods will be stated clearly. More often perhaps, the interest rate will be given as an annual rate (r) and the time will be stated in years (t). When that happens, find the row and column as described in the steps below. The letter m is the number of compounding periods in one year.

$[m = 1, 2, 4, 12, 365]$
$[i = \frac{r}{m}]$
$[n = m \cdot t]$

> **STEPS** to Determine the Periodic Rate and the Number of Compounding Periods
>
> **i.** Determine the number of compounding periods in 1 year ($m = 1$ for annually, $m = 2$ for semiannually, $m = 4$ for quarterly, $m = 12$ for monthly, $m = 365$ for daily).
>
> **ii.** Divide the stated annual rate (r) by the number of periods in 1 year (m). The quotient is the periodic rate (i), the correct column.
>
> **iii.** Multiply the number of periods in 1 year (m) by the number of years (t). The product is the total number of compounding periods (n), the correct row.

16.3 12% may be a high interest rate, but it is convenient. It is divisible by 2, 4, and 12.

EXAMPLE D

Find the periodic interest rate and the number of compounding periods in 2 years when 12% is compounded (a) semiannually ($m = 2$ times per year), (b) quarterly ($m = 4$ times per year), and (c) monthly ($m = 12$ times per year). Then find the future value factors in Table 16-1.

Each term is for 2 years; each rate is 12%, but compounded differently:

$[i = \frac{0.12}{2} = 0.06;$
$r = 2 \times 2 = 4]$

$[i = \frac{0.12}{4} = 0.03;$
$n = 4 \times 2 = 8]$

$[i = \frac{0.12}{12} = 0.01;$
$n = 12 \times 2 = 24]$

	STEP i	STEP ii	STEP iii
	Periods per Year	**Periodic Interest Rate**	**Compounding Periods**
a.	2	12% ÷ 2 = 6%	2 × 2 years = 4 periods
b.	4	12% ÷ 4 = 3%	4 × 2 years = 8 periods
c.	12	12% ÷ 12 = 1%	12 × 2 years = 24 periods

Future value factors from Table 16-1 are as follows.

a.	Semiannually	6.00% column and row 4	Factor = 1.26248
b.	Quarterly	3.00% column and row 8	Factor = 1.26677
c.	Monthly	1.00% column and row 24	Factor = 1.26973

To compute the future value and the compound interest, first determine the periodic rate and the number of compounding periods using Steps i, ii, and iii. Then do Steps 1, 2, and 3, as illustrated in examples B and C previously.

Calculators and Exponents

Note: This is an optional section that describes how to use a calculator to compute future value factors. It requires some knowledge of exponents and exponential notation and it requires a calculator that has an exponent key. Some persons will prefer to use Table 16-1, but others may prefer to use a calculator or to use a calculator just to check their work.

The expression 2^3 means $2 \times 2 \times 2 = 8$. The 3 is called an **exponent**, or we can say that "2 is raised to the 3rd **power**." Many calculators have a key labeled $\boxed{y^x}$ that is used for exponents. To compute 2^3, enter the following keystrokes: 2 $\boxed{y^x}$ 3 $\boxed{=}$. The answer on the calculator display is 8.

The future value factors in Table 16-1 can be calculated directly by anyone who has a calculator that will compute exponents. For the periodic interest rate of i (decimal) and for the number of compounding periods equal to n, the future value factor is $FVF = (1 + i)^n$. Note: The interest rate must be entered as a decimal, not a percent.

● **EXAMPLE E**

Use a calculator with an exponent key to compute the future value factor for each of the following:

a. 12% compounded semiannually for 2 years: $i = 0.12 \div 2 = 0.06$; $n = 2 \times 2 = 4$ periods

$FVF = (1 + i)^n = (1 + 0.06)^4 = 1.26247696$

The calculator keystrokes might be   1.06 $\boxed{y^x}$ 4 $\boxed{=}$

The exact calculator strokes will depend upon your own calculator. Refer to your calculator's manual. It is usually faster to mentally add the 1 and 0.06 because the sum is just 1.06, but many calculators also have keys for "parenthesis."

b. 12% compounded quarterly for 2 years: $i = 0.12 \div 4 = 0.03$; $n = 2 \times 4 = 8$ periods

$FVF = (1 + i)^n = (1 + 0.03)^8 = 1.26677008$ 1.03 $\boxed{y^x}$ 8 $\boxed{=}$

c. 12% compounded monthly for 2 years: $i = 0.12 \div 12 = 0.01$; $n = 2 \times 12 = 24$ periods

$FVF = (1 + i)^n = (1 + 0.01)^{24} = 1.26973465$ 1.01 $\boxed{y^x}$ 24 $\boxed{=}$

In the example, we wrote each future value factor with eight decimal places. The factors in Table 16-1 have only five decimal places. Throughout this chapter, factors that have five decimal places come from the tables and factors with eight decimal places come from the formula using a calculator. If you use a calculator and more than five decimal places, usually you will get a slightly different answer than if you use only five decimal places. The more decimal places you use, the more accurate the answers will be. In this book, all of the solutions assume the use of the tables and only five decimal places.

Throughout the chapter, any calculator solutions will be shown in brackets in the margins.

EXAMPLE F

Barbara Scoble and her husband deposit $20,000 in her credit union, which pays interest of 8% compounded quarterly. Find the future value and the total compound interest after 2 years. (Use Table 16-1, or a calculator.)

$[m = 4]$

$[i = \frac{0.08}{4} = 0.02]$

$[n = 4 \times 2 = 8]$

$[FVF = (1 + i)^n = (1.02)^8$
$\quad = 1.17165938]$

$[FV = PV \times (1 + i)^n$
$\quad = \$20,000 \times 1.17165938$
$\quad = \$23,433.1876$
$\quad \text{or } \$23,433.19]$

STEP i   There are $m = 4$ compounding periods in 1 year.

STEP ii   Periodic interest rate $= 8\% \div 4 = 2\%$ per period

STEP iii   Number of periods $= 4 \times 2$ years $= 8$ periods

STEP 1   Using Table 16-1, the 2.00% column and row 8:
Future value factor $(FVF) = 1.17166$.

STEP 2   Future value $= \$20,000.00 \times 1.17166 = \$23,433.20$

STEP 3   Total compound interest $= \$23,433.20 - \$20,000.00 = \$3,433.20$

Thus, using Table 16-1 to find the *FVF*, $20,000 invested at 8% compounded quarterly will be worth $23,433.20 in 2 years. If you use a calculator to find the *FVF*, the future value is $23,433.19. You should use whichever method seems more clear to you.

Effective Rates

In Chapter 14, we said that the term "effective rate" is used in more than one context. In Chapter 14, "effective rate" was related to the interest rate paid on the "average unpaid balance" in an installment purchase. Here in Chapter 16, "effective rate" refers to the true annual yield an investor earns when her/his money is compounded more than once per year.

In example F, Barbara Scoble and her husband earned 8% compounded quarterly. Their $20,000 deposit was worth $23,433.20 after 2 years. 8% is an *annual* rate, not a *quarterly* rate. But 8% was not really used in the compounding; the rate that was actually compounded was 2% per quarter. Thus, 8% is not the true annual rate, or the effective rate. The 8% in example F is called a "nominal" rate because the *name* of the rate is 8% and the 8% is compounded quarterly.

The effective rate is the rate that the Scobles would earn if their money had been compounded annually instead of quarterly. You can either use Table 16.1 or use a formula with a calculator to find the effective rate.

To use Table 16.1, find the future value factor of 2% for 4 quarters (1 year). It is 1.08243. Subtract 1 to get 0.08243, or 8.243%. The effective rate is 8.243% per year. What this means is that the Scobles are actually earning 8.243% per year on an investment that has been quoted as earning "8% compounded quarterly."

The formula for the effective rate R is $R = (1 + \frac{0.08}{4})^4 - 1$. In example F, $R = (1 + \frac{0.08}{4})^4 - 1 = (1 + 0.02)^4 - 1 = 1.08243216 - 1 = 0.08243216$, or 8.243216%. Rounded to four decimal places, the effective rate is $R = 8.2432\%$.

DAILY COMPOUNDING

Most banks offer daily compounding on several different savings accounts and certificates of deposit. Tables to do daily compounding would be cumbersome and impractical. However, using a calculator with an exponent, the computation is just as simple as other compounding. Assume that there are 365 days in a year.

EXAMPLE G

Use a calculator to find the future value of $20,000 invested for 2 years at 8% compounded daily. First find the periodic interest rate (i) as a decimal, and find the number of days (n) in two years. Then find the future value factor to eight decimal places.

STEP i   There are $m = 365$ compounding periods in 1 year.

STEP ii i = periodic interest rate = $0.08 \div 365 = 0.00021918$

STEP iii n = number of periods = $365 \times 2 = 730$ periods

STEP 1 $FVF = (1 + i)^n = (1 + 0.00021918)^{730} = 1.17349194$

STEP 2   Future value = $20,000 \times 1.17349194 = $23,469.84$

Compare the two future values from Examples F and G. The future value using quarterly compounding is $23,433.19 (using a calculator to find the FVF). With daily compounding, the future value is $23,469.84, a difference of $36.65.

16.4 Because calculators may not be permitted in all classes, and/or because calculators that do exponents may not be available to all students, there are no more examples or problems that involve daily compounding in this book.

✓ CONCEPT CHECK 16.1

a. If $2,600 is invested for 5 years at 6% compounded semiannually, compute the future value of the investment. (Use Table 16-1 or a calculator.)

Semiannually means $m = 2$ periods per year.
Periodic rate = $6\% \div 2 = 3\%$ per half-year
Number of periods = 2×5 years = 10 periods
The FVF from row 10 of the 3.00% column in Table 16-1 is 1.34392.
Future value = $2,600 \times 1.34392 = $3,494.192$, or $3,494.19

$[m = 2]$
$[i = \frac{0.06}{2} = 0.03]$
$[n = 2 \times 5 = 10]$
$[FVF = (1 + 0.03)^{10}$
$\quad = 1.34391638]$
$[FV = PV \times (1 + 0.03)^{10}$
$\quad = 2,600 \times 1.34391638$
$\quad = \$3,494.18]$

b. If $3,200 is invested for 1 year at 9% compounded monthly, what is the compound interest on the investment?

Monthly means $m = 12$ periods per year.
Periodic rate = $9\% \div 12 = 0.75\%$ per month.
Number of periods is 12×1 year = 12 periods.
The FVF from row 12 of the 0.75% column in Table 16-1 is 1.09381.
Future value = $3,200 \times 1.09381 = $3,500.192$, or $3,500.19
Compound interest = Future value − Present value (Principal)
$\qquad\qquad\qquad\qquad = \$3,500.19 − \$3,200 = \$300.19$

$[m = 12]$
$[i = \frac{0.09}{12} = 0.0075]$
$[n = 12 \times 1 = 12]$
$[FVF = (1 + 0.0075)^{12}$
$\quad = 1.09380690]$
$[FV = PV \times (1 + 0.0075)^{12}$
$\quad = \$3,200 \times 1.09380690$
$\quad = \$3,500.18]$

COMPLETE ASSIGNMENT 16.1.

Computing Present Values

Learning Objective 2

Compute present values.

The basic investment problem is to compute what a given sum of money invested today will be worth in the future. Example F was such a future value problem. There we found that $20,000 original principal (or present value) invested today at 8% compounded quarterly will have a future value of $23,433.20 in 2 years.

Some savers and investors want to compute future values; others want to compute present values. Consider the following present value problem.

● EXAMPLE H

16.5 For most families, a more realistic situation would be one in which they were saving money each month for the children's education. If the deposits were equal, this would be a problem involving an annuity. See examples in Chapter 23.

Polly Layer has a 12-year-old son and a 10-year-old daughter. Polly inherits $100,000. Friends tell Polly that she should plan to have $60,000 cash available for her son's education when he turns 18. She should also have $70,000 cash available for her daughter's education when she turns 18. Polly wants to put enough money in an investment for each child so that in 6 and 8 years the two accounts will be worth $60,000 and $70,000, respectively. If Polly can earn 5% compounded annually, how much money should she put into each investment today?

Polly knows the future value of the investments—$60,000 and $70,000. What she wants to compute is the **present value**—the amounts that she needs to invest today for each child. We will solve this problem later, in example L.

Businesses make investments in the present to provide future revenues. Sometimes a business will estimate its future revenues and costs (future values). Then the business might use these numbers to compute the required amounts to invest initially (present values).

As given earlier, the formula for future value is

$$[FV = PV \times (1 + i)^n]$$

Future value = Present value × Future value factor (from Table 16-1 or a calculator)

Rewriting the formula to solve for present value gives

$$\left[PV = \frac{FV}{(1 + i)^n}\right]$$

Present value = Future value ÷ Future value factor (from Table 16-1 or a calculator)

● EXAMPLE I

How much money must be invested today to end up with $6,326.60 in 3 years? The interest rate is 8% compounded semiannually. (Use Table 16-1 or a calculator.)

$[m = 2]$
$\left[i = \frac{0.08}{2} = 0.04\right]$
$[n = 2 \times 3 = 6]$
$[FVF = (1 + 0.04)^6$
$\quad = 1.26531902]$

The $6,326.60 is the future value for which we want to find the present value. Interest is computed six times—twice each year for 3 years. The future value factor in Table 16-1 in the 4.00% column and row 6 is 1.26532. Substitute these values into the formula to solve for present value.

Present value = Future value ÷ Future value factor (from Table 16-1)
= $6,326.60 ÷ 1.26532 = $5,000

Compare this result to that of example C, in which $5,000 was invested for 6 years at 4% compounded annually. The future value was $6,326.60.

EXAMPLE J

Edison Motors estimates that in 2 years it will cost $20,000 to repair a diagnostic machine. How much must Edison invest today to have $20,000 in 2 years, if the interest rate is 6% compounded monthly? How much interest will Edison Motors earn on its investment?

$20,000 is the future value for which Edison wants to know the present value.

STEP i   There are 12 compounding periods in 1 year (monthly). $[m = 12]$

STEP ii   Periodic rate = 6% ÷ 12 = 0.5% $[i = \frac{0.06}{12} = 0.005]$

STEP iii   Number of compounding periods = 12 × 2 years = 24 $[n = 12 \times 2 = 24]$
The future value factor in the 0.5% column and row 24 of Table 16-1 is 1.12716. $[FVF = (1 + 0.005)^{24} = 1.12715978]$

Substitute these values into the formula to solve for present value:

Present value = Future value ÷ Future value factor (from Table 16-1)
= $20,000 ÷ 1.12716 = $17,743.71 to the nearest cent

If Edison Motors invests $17,743.71 today at 6% compounded monthly, it will have $20,000 at the end of 2 years.

The $20,000 is the sum of the amount invested plus the total compound interest earned. To find the interest, subtract the amount invested from $20,000.

Interest = Future value − Present value = $20,000 − $17,743.71 = $2,256.29

✓ CONCEPT CHECK 16.2

What present value (principal) invested for 3 years at 10% compounded semiannually will result in a total future value of $4,000? (Use Table 16-1 or a calculator.)

Semiannually means 2 periods per year.
Periodic rate = 10% ÷ 2 = 5% per half-year
Number of periods = 3 years × 2 = 6 periods
The future value factor from row 6 of the 5.00% column in Table 16-1 is 1.34010.

Present value = Future value ÷ Future value factor
= $4,000 ÷ 1.34010 = $2,984.852, or $2,984.85

$[m = 2]$
$[i = \frac{0.10}{2} = 0.05]$
$[n = 2 \times 3 = 6]$

$[FVF = (1 + 0.05)^6 = 1.34009564]$

Using Present Value Tables and/or Formulas

You may prefer to solve for present values by using **present value factors** (**PVF**) rather than future value factors, as in the preceding formula. Table 16-2, on pages 340 and 341, is a table of present value factors. Use exactly the same procedure (Steps i, ii, and iii) to find present value factors as you used to find future value factors.

Learning Objective 3

Use present value tables and/or formulas.

PRESENT VALUE FORMULA

If you use the present value factors (Table 16-2 or a calculator), you use a different formula, the present value formula,

With a calculator,
$PVF = \frac{1}{FVF} = \frac{1}{(1+i)^n}$

> Present value = Future value × Present value factor (from Table 16-2 or a calculator)
> or, $PV = FV \times PVF$

EXAMPLE K

Rework example J using Table 16-2 and the present value formula. How much must Edison Motors invest today to have $20,000 in 2 years if the interest rate is 6% compounded monthly?

$20,000 is the future value, for which Edison wants to know the present value.

$[m = 12]$

STEP i   There are 12 compounding periods in 1 year (monthly).

$[i = \frac{0.06}{12} = 0.005]$

STEP ii   Periodic rate = 6% ÷ 12 = 0.5%

$[n = 12 \times 2 = 24]$

STEP iii   Number of compounding periods = 12 × 2 years = 24

$[PVF = \frac{1}{(1 + 0.005)^{24}}$
$= 0.88718567]$

The present value factor in the 0.5% column and row 24 of Table 16-2 is 0.88719.

Substitute these values into the present value formula.

16.6 Some students who use calculators to find PVFs might be interested to see that they can calculate $(1 + i)^n$ and then press the $^1/_x$ key, or even that $(1 + i)^{-n}$ is the same as $1 \div (1 + i)^n$.

Present value = Future value × Present value factor (from Table 16-2)
$= \$20,000 \times 0.88719 = \$17,743.80$

The answer to example J was $17,743.71. The discrepancy between that result and $17,743.80 in example K is due to rounding. If the two tables had more decimal places instead of just five, this discrepancy would disappear. In fact, using the calculator PVF from the margin, we get $PV = \$20,000 \times 0.88718567 = \$17,743.7134$, which is identical to the nearest cent.

EXAMPLE L

Solve the present value problem from example H. If Polly can earn 5% compounded annually, how much should she deposit today in investments for her son and daughter so that the investments will be worth $60,000 and $70,000 in 6 and 8 years, respectively?

Son
$[m = 1]$
$[i = \frac{0.05}{1} = 0.05]$
$[n = 1 \times 6 = 6]$
$[PVF = \frac{1}{(1 + 0.05)^6}$
$= 0.74621540]$

Daughter
$[m = 1]$
$[i = \frac{0.05}{1} = 0.05]$
$[n = 1 \times 8 = 8]$
$[PVF = \frac{1}{(1 + 0.05)^8}$
$= 0.67683936]$

		Son	Daughter
	Future value:	$60,000	$70,000
	Term:	6 years	8 years
	Rate:	5% compounded annually	5% compounded annually
STEP i	Periods per year:	1 (annual)	1 (annual)
STEP ii	Periodic rate:	5% ÷ 1 = 5%	5% ÷ 1 = 5%
STEP iii	Compounding periods:	1 × 6 years = 6	1 × 8 years = 8
	PV factor (Table 16-2):	0.74622	0.67684
	Present value:	$60,000 × 0.74622 = $44,773.20	$70,000 × 0.67684 = $47,378.80

Part 4   Interest Applications

The reason for two formulas and two tables is historical, predating handheld calculators. Without a calculator, a multiplication problem is typically easier than a division problem with the same two numbers.

Theoretically, we need only one formula and one table. The second present value formula and the table of present value factors permit us to solve present value problems by using multiplication instead of division. Look at example J. To solve the problem requires that we divide $20,000 by 1.12716, which is extremely time-consuming to do without a calculator. (The answer is $17,743.71.) Using Table 16-2, we can solve the same problem by multiplying $20,000 by 0.88719, a relatively easy calculation without a calculator. (The answer is $17,743.80; the difference is due to rounding in the creation of the table.)

16.7 Finance courses and financial decision-making normally involve more work with present values than with future values. Before the age of calculators, a traditional finance textbook might not have used a future value table very often.

NOTES ABOUT THE FUTURE VALUE AND PRESENT VALUE TABLES

The numbers in the future value table (Table 16-1) are actually just the future value of $1.00 at a specific interest rate and for a specific period of time. For example, suppose that you invest $1.00 for 2 years at 6% compounded annually. This is the same problem as example A, except that the principal is only $1.00 instead of $2,000.00.

The calculations shown at the right have not been rounded off. The answer, which is $1.1236, is the future value of the $1.00 investment. Now, find row 2 and the 6.00% column of Table 16-1. The future value factor is 1.12360—exactly the same as $1.1236, without the dollar sign and with five decimal places.

$1.00	Original principal
× 0.06	Interest rate
$0.0600	First-year interest
+ 1.00	First-year principal
$1.0600	Second-year principal
× 0.06	Interest rate
$0.0636	Second-year interest
+ 1.06	Second-year principal
$1.1236	Final compound amount

16.8 Emphasize again that the present value formula and table were developed to make an "easier" multiplication problem compared to a very tedious long division problem in "precalculator days"—less than 35 years ago.

Each number in the present value table (Table 16-2) can be calculated directly from the corresponding number in the future value table. The corresponding numbers are *reciprocals* of each other. Recall that the reciprocal of a number is found by dividing the number into 1.

Look back at examples J and K, which showed two different ways to solve the same problem. In example J we used a future value factor, which was 1.12716. In example K we used a present value factor, which was 0.88719. Each factor is in row 24 and the 0.50% column of its respective table. With your calculator, divide 1 by 1.12716 to get 0.88718549, which rounded to five places is 0.88719. And dividing 1 by 0.88719 gives 1.12715427.

1 ÷ 1.12716 = 0.88718549, or 0.88719
1 ÷ 0.88719 = 1.12715427, or 1.12716

Examine your calculator. You may have a reciprocal key, labeled "1/x." If you have such a key, enter 1.12716 and press the 1/x key. The calculator will display 0.88718549. Press the 1/x key again and the calculator will display 1.12716, or perhaps 1.12716000.

CONCEPT CHECK 16.3

a. What present value (principal) invested for 3 years at 10% compounded semiannually will result in a total future value of $4,000? (Use Table 16-2 or a calculator.)

Semiannually means 2 periods per year.
Periodic rate = 10% ÷ 2 = 5% per half-year
Number of periods = 2 × 3 years = 6 periods
The present value factor from row 6 of the 5.00% column in Table 16-2 is 0.74622.
Present value = Future value × Present value factor
= $4,000 × 0.74622 = $2,984.88

$[m = 2]$
$[i = \frac{0.10}{2} = 0.05]$
$[n = 2 \times 3 = 6]$
$[PVF = \frac{1}{(1 + 0.05)^6}$
$= 0.74621540]$

Note: The answers to Concept Checks 16.2 and 16.3a are essentially the same. If the future value table and the present value table had values with six decimals, the answers would both be $2,984.86.

b. Seven years ago, a woman invested money at 9% compounded annually. If the investment is now worth $6,000, how much compound interest did she earn in the 7 years? (Use Table 16-2 or a calculator.)

Annually means 1 period per year.
Periodic rate = 9% ÷ 1 = 9% per year
Number of periods is 1 × 7 years = 7 periods
The present value factor from row 7 of the 9.00% column in Table 16-2 is 0.54703.
Present value = Future value × Present value factor
= $6,000 × 0.54703 = $3,282.18
Compound interest = Future value − Present value
= $6,000 − $3,282.18 = $2,717.82

$[m = 1]$
$[i = \frac{0.09}{1} = 0.09]$
$[n = 1 \times 7 = 7]$
$[PVF = \frac{1}{(1 + 0.09)^7}$
$= 0.54703424]$

COMPLETE ASSIGNMENT 16.2.

Chapter Terms for Review

compound amount
compound amount factors
compound interest
compound interest table
exponent
future value
future value factors

number of compounding periods
period (compounding period)
periodic interest rate
power
present value
present value factors

THE BOTTOM LINE

Summary of chapter learning objectives:

Learning Objective	Example
16.1 Compute future values and compound interest	1. Compute the future value of $9,000 invested at 6% compounded monthly for 2 years. 2. Compute the compound interest earned on $5,000 invested at 6% compounded quarterly for 5 years.
16.2 Compute present values	3. Compute the present value that has to be invested at 10% compounded semiannually for 6 years to result in $8,000.
16.3 Use present value tables and/or formulas	4. If $6,000 is the future value after 13 years at 9% compounded annually, compute the principal (present value). 5. An investment made 16 months ago is worth $5,634.95 today. If the interest rate was 9% compounded monthly, what was the amount of compound interest?

Answers: 1. $10,144.44 2. $1,734.30 3. $4,454.69 or $4,454.72 4. $1,957.08 5. $634.95

SELF-CHECK

Review Problems for Chapter 16

1 Calculate the future value (compound amount) and compound interest. (Use Table 16-1 or a calculator.)

Principal	Rate	Time	Future Value	Interest
$ 4,000	6% compounded monthly	3 yr	a. _____	b. _____
$12,000	8% compounded quarterly	7 yr	c. _____	d. _____
$20,000	9% compounded annually	11 yr	e. _____	f. _____
$ 8,000	10% compounded semiannually	10 yr	g. _____	h. _____

2 Calculate the present value (principal) and compound interest. (Use Table 16-2 or a calculator.)

Future Value	Rate	Time	Present Value	Interest
$30,000	5% compounded annually	7 yr	a. _____	b. _____
$ 6,000	8% compounded semiannually	12 yr	c. _____	d. _____
$15,000	9% compounded monthly	4 yr	e. _____	f. _____
$40,000	6% compounded quarterly	5 yr	g. _____	h. _____

3 Vernon Lee received a $6,000 bonus from his employer. He can invest it safely in his credit union at 4% compounded quarterly. What will be the value of the investment in 7 years?

4 Donna Takeuchi inherited $6,200. She invested it immediately in an investment fund paying 6% compounded semiannually. How much interest would Donna earn if she left principal and interest invested for 10 years?

5 Sandy Hopkins was planning to buy a new car in 3 years. She has some money today that she can invest for 3 years in an account that will pay 6% compounded quarterly. How much of it would she need to deposit today so that she will have $8,000 in her account in 3 years?

6 Doug Jurgensen will need to buy a $25,000 wood lathe in 2 years. He can deposit excess profits from this year in an investment that should pay 9% compounded monthly. If Doug earns the $25,000 in 2 years, how much will he earn in interest?

Assignment 16.1: Future Value (Compound Amount)

Name

Date   Score

Learning Objective 1

A (28 points) Find the future value (compound amount) and the compound interest, as indicated, for each of the following investments. Round answers to the nearest cent. Use Table 16-1 or a calculator. (2 points for each correct answer)

	Principal	Rate	Term	Future Value	Compound Interest
1.	$6,000	6% compounded monthly	4 years	$7,622.94	$1,622.94

6% ÷ 12 = 0.5% $6,000 × 1.27049 = $7,622.94
12 × 4 = 48 periods $7,622.94 − $6,000 = $1,622.94

2.	$750	8% compounded semiannually	13 years	$2,079.35	$1,329.35

8% ÷ 2 = 4% $750 × 2.77247 = $2,079.35
2 × 13 = 26 periods $2,079.35 − $750 = $1,329.35

3.	$20,000	8% compounded quarterly	8 years	$37,690.80	$17,690.80

8% ÷ 4 = 2% $20,000 × 1.88454 = $37,690.80
4 × 8 = 32 periods $37,690.80 − $20,000 = $17,690.80

4.	$8,400	10% compounded annually	20 years	$56,511.00	$48,111.00

10% ÷ 1 = 10% $8,400 × 6.72750 = $56,511.00
1 × 20 = 20 periods $56,511 − $8,400 = $48,111.00

5.	$5,000	9% compounded monthly	18 months	$5,719.80	$719.80

9% ÷ 12 = 0.75% $5,000 × 1.14396 = $5,719.80
18 months = 18 periods $5,719.80 − $5,000 = $719.80

6.	$14,450	6% compounded quarterly	4 years	$18,336.91	$3,886.91

6% ÷ 4 = 1.5% $14,450 × 1.26899 = $18,336.91
4 × 4 = 16 periods $18,336.91 − $14,450 = $3,886.91

7.	$4,000	4% compounded semiannually	9 years	$5,713.00	$1,713.00

4% ÷ 2 = 2% $4,000 × 1.42825 = $5,713
2 × 9 = 18 periods $5,713 − $4,000 = $1,713

Score for A (28)

Note to teachers: Answers were computed using Table 16-1. Calculator answers may vary slightly.

Assignment 16.1 Continued

B (32 points) Find the future value (compound amount) or the compound interest, as indicated, for each of the following investments or loans. Round answers to the nearest cent. Use Table 16-1 or a calculator. (4 points for each correct answer)

8. Compute the future value (compound amount) of $4,500 invested for 10 years at 5% compounded quarterly.

   $7,396.29

   5% ÷ 4 = 1.25% $4,500 × 1.64362 = $7,396.29
   4 × 10 = 40 periods

9. How much compound interest will you pay if you borrow $25,000 for 13 months at 15% compounded monthly?

   $4,381.50

   15% ÷ 12 = 1.25% $25,000 × 1.17526 = $29,381.50
   13 months = 13 periods $29,381.50 − $25,000 = $4,381.50

10. Calculate the future value (compound amount) on a loan of $6,500 at 10% compounded annually for 5 years.

    $10,468.32

    10% ÷ 1 = 10% $6,500 × 1.61051 = $10,468.32
    1 × 5 = 5 periods

11. How much compound interest will you earn if you loan $7,900 for 16.5 years at 12% compounded semiannually?

    $46,140.66

    12% ÷ 2 = 6% $7,900 × 6.84059 = $54,040.66
    2 × 16.5 = 33 periods $54,040.66 − $7,900 = $46,140.66

12. What total amount (principal and interest) must be repaid in $2\frac{1}{2}$ years on a loan of $15,000 at 9% compounded monthly?

    $18,769.05

    9% ÷ 12 = 0.75% $15,000 × 1.25127 = $18,769.05
    12 × $2\frac{1}{2}$ = 30 periods

13. Determine the total compound interest that you will have to pay if you borrow $845 at 10% compounded semiannually and don't pay it back for 11 years.

    $1,626.84

    10% ÷ 2 = 5% $845 × 2.92526 = $2,471.84
    2 × 11 = 22 periods $2,471.84 − $845 = $1,626.84

14. How much compound interest will you earn if you invest $10,000 for 13 years at 8% compounded annually?

    $17,196.20

    8% ÷ 1 = 8% $10,000 × 2.71962 = $27,196.20
    1 × 13 = 13 periods $27,196.20 − $10,000 = $17,196.20

15. Compute the future value (compound amount) of $18,000 invested for 4.5 years at 5% compounded quarterly.

    $22,510.44

    5% ÷ 4 = 1.25% $18,000 × 1.25058 = $22,510.44
    4 × 4.5 = 18 periods

Score for B (32)

Assignment 16.1 Continued

C **(40 points) Business Applications.** Find the future value (compound amount) or the compound interest, as indicated. Round answers to the nearest cent. Use Table 16-1 or a calculator. (4 points for each correct answer)

16. Kathy Shutter thinks that she needs to borrow $7,500 for 2 years. She doesn't have a very good credit rating, so most finance companies want to charge her a high interest rate. She finally finds a lender that will loan her the money at 12% compounded monthly. How much interest will Kathy have to pay to this particular lender?

    $2,022.98

    12% ÷ 12 = 1% $7,500 × 1.26973 = $9,522.98
    12 × 2 = 24 periods $9,522.98 − $7,500 = $2,022.98

17. Mary Sousa receives a telephone call from a salesperson who describes "an incredible investment opportunity." The investment promises a return of 16% compounded semiannually for investments of $5,000 or more. One disadvantage is that no money will be paid out for a long time. Another disadvantage is that the investment is very risky. Mary doesn't think that she will need the money for 6 years, so she decides to invest $5,000. If the investment pays what it promises, how much interest will Mary earn in the 6 years?

    $7,590.85

    16% ÷ 2 = 8% $5,000 × 2.51817 = $12,590.85
    2 × 6 = 12 periods $12,590.85 − $5,000 = $7,590.85

18. William Wang wants to borrow money from his father to buy a car. William's father is trying to teach him how to manage money, so he agrees to loan him the money, but at 5% compounded quarterly. William borrows $11,200 and repays everything—principal plus all of the interest—in $3\frac{1}{2}$ years. How much does William pay back to his father?

    $13,327.44

    5% ÷ 4 = 1.25% $11,200 × 1.18995 = $13,327.44
    4 × $3\frac{1}{2}$ = 14 periods

19. Don Hildebrand is trying to decide whether to invest money in a bank or in something a little riskier that will pay a higher return. One very simple investment promises to pay a minimum of 9% compounded annually, but he must leave all of money and interest invested for 6 years. How much will Don earn during the 6 years if he invests $4,500 and the investment pays the minimum?

    $3,046.95

    9% ÷ 1 = 9% $4,500 × 1.67710 = $7,546.95
    1 × 6 = 6 periods $7,546.95 − $4,500 = $3,046.95

20. Marcia Juarez and her brother-in-law have a successful business with several employees. They decide to borrow $15,000 to pay their quarterly payments for payroll taxes and federal income tax. They get the money at 9% compounded monthly and repay all interest and principal after 9 months. How much do they repay?

    $16,043.40

    9% ÷ 12 = 0.75% $15,000 × 1.06956 = $16,043.40
    9 months = 9 periods

Assignment 16.1 Continued

21. Sammie Crass inherited $16,780. She wants to invest it in something relatively safe so that she can transfer all the money to her children's college fund in about 8 years. One investment brochure (called a prospectus) states that it will pay a return of 8% compounded quarterly. How much will Sammie have total, principal plus interest, after 8 years?

    $31,622.58

    8% ÷ 4 = 2% $16,780 × 1.88454 = $31,622.58
    4 × 8 = 32 periods

22. To help his daughter and son-in-law purchase their first new car, Robert Chow loans them $15,000. They agree on an interest rate of 3% compounded annually, and Mr. Chow tells them that they can pay it all back, the $15,000 plus the interest, in 5 years. How much interest will Mr. Chow receive from them?

    $2,389.05

    3% ÷ 1 = 3% $15,000 × 1.15927 = $17,389.05
    1 × 5 = 5 periods $17,389.05 − $15,000 = $2,389.05

23. Sandee Millet owns and operates an art supply store in a suburban shopping center. Sandee learns about an investment that claims to pay a return of 8% compounded semiannually for 4 years. Sandee decides to invest $4,750. Compute the amount of interest that she will earn in the 4 years.

    $1,750.71

    8% ÷ 2 = 4% $4,750 × 1.36857 = $6,500.71
    2 × 4 = 8 periods $6,500.71 − $4,750 = $1,750.71

24. Ken Ortman is a student at medical school. He borrowed $32,000 for 26 months at the rate of 6% compounded monthly. How much total, principal plus compound interest, must Ken repay at the end of the 26 months?

    $36,430.72

    6% ÷ 12 = 0.5% $32,000 × 1.13846 = $36,430.72
    26 months = 26 periods

25. The County Employees Credit Union pays an interest rate of 8% compounded quarterly on savings accounts of $1,000 or more, with the requirement that the money be deposited for at least 6 months. How much interest will Marilyn Bunnell earn if she deposits $1,800 and leaves it in the credit union for 2 years?

    $308.99

    8% ÷ 4 = 2% $1,800 × 1.17166 = $2,108.99
    4 × 2 = 8 periods $2,108.99 − $1,800 = $308.99

Score for C (40)

Assignment 16.2: Present Value

Name

Date   Score

Learning Objectives **2** **3**

A (28 points) Find the present value (principal) and the compound interest, as indicated, for each of the following investments. (*Hint:* Subtract the present value from the future value to find the compound interest.) Use Table 16-1, Table 16-2, or a calculator. Round answers to the nearest cent. (2 points for each correct answer)

	Future Value	Rate	Term	Present Value	Compound Interest
1.	$3,900	6% compounded semiannually	3 years	$3,266.17	$633.83

    6% ÷ 2 = 3% $3,900 ÷ 1.19405 = $3,266.19   or $3,900 × 0.83748 = $3,266.17
    2 × 3 = 6 periods $3,900 − $3,266.17 = $633.83

2.	$15,000	8% compounded quarterly	7 years	$8,615.55	$6,384.45

    8% ÷ 4 = 2% $15,000 ÷ 1.74102 = $8,615.64   or $15,000 × 0.57437 = $8,615.55
    4 × 7 = 28 periods $15,000 − $8,615.55 = $6,384.45

3.	$35,000	5% compounded annually	9 years	$22,561.35	$12,438.65

    5% ÷ 1 = 5% $35,000 ÷ 1.55133 = $22,561.29   or $35,000 × 0.64461 = $22,561.35
    1 × 9 = 9 periods $35,000 − $22,561.35 = $12,438.65

4.	$6,800	9% compounded monthly	4 years	$4,750.55	$2,049.45

    9% ÷ 12 = 0.75% $6,800 ÷ 1.43141 = $4,750.56   or $6,800 × 0.69861 = $4,750.55
    12 × 4 = 48 periods $6,800 − $4,750.55 = $2,049.45

5.	$10,000	6% compounded quarterly	10 years	$5,512.60	$4,487.40

    6% ÷ 4 = 1.5% $10,000 ÷ 1.81402 = $5,512.62   or $10,000 × 0.55126 = $5,512.60
    4 × 10 = 40 periods $10,000 − $5,512.60 = $4,487.40

6.	$50,000	8% compounded semiannually	6 years	$31,230.00	$18,770.00

    8% ÷ 2 = 4% $50,000 ÷ 1.60103 = $31,229.90   or $50,000 × 0.62460 = $31,230.00
    2 × 6 = 12 periods $50,000 − $31,230 = $18,770

7.	$2,500	6% compounded monthly	18 months	$2,285.35	$214.65

    6% ÷ 12 = 0.5% $2,500 ÷ 1.09393 = $2,285.34   or $2,500 × 0.91414 = $2,285.35
    18 months = 18 periods $2,500 − $2,285.35 = $214.65

Score for A (28)

Note to teachers: Problems were solved using both Table 16-1 and Table 16-2. The given answers come from the Table 16-2 solutions. Table 16-1 and calculator solutions may vary slightly.

Assignment 16.2 Continued

B (32 points) Find the present value (principal) or the compound interest, as indicated, for each of the following investments or loans. Use Table 16-1, Table 16-2, or a calculator. Round answers to the nearest cent. (4 points for each correct answer)

8. Compute the present value (principal) if the future value 20 years from now is $25,000 and if the interest rate is 8% compounded semiannually.

   $\underline{\$5,207.25}$

   8% ÷ 2 = 4% $25,000 ÷ 4.80102 = $5,207.23   or $25,000 × 0.20829 = $5,207.25
   2 × 20 = 40 periods

9. How much compound interest would you pay if you repay a total of $8,425 1 year and 6 months after borrowing the principal at 9% compounded monthly?

   $\underline{\$1,060.20}$

   9% ÷ 12 = 0.75% $8,425 ÷ 1.14396 = $7,364.77   or $8,425 × 0.87416 = $7,364.80
   12 × 1½ = 18 periods $8,425 − $7,364.80 = $1,060.20

10. Calculate the present value (principal) of a loan made 3 years ago at 8% compounded quarterly if the borrower repays a total of $6,250.

    $\underline{\$4,928.06}$

    8% ÷ 4 = 2% $6,250 ÷ 1.26824 = $4,928.09   or $6250 × 0.78849 = $4,928.06
    4 × 3 = 12 periods

11. Compute the amount that a company must invest (the present value) at 10% compounded annually if it wants to have $100,000 available (the future value) in 25 years.

    $\underline{\$9,230.00}$

    10% ÷ 1 = 10%
    1 × 25 = 25 periods $100,000 ÷ 10.83471 = $9,229.60   or $100,000 × 0.09230 = $9,230.00

12. How much compound interest is earned on a 6.5-year investment that has a rate of return of 6% compounded quarterly and repays a total compound amount (future value) of $9,600?

    $\underline{\$3,081.41}$

    6% ÷ 4 = 1.5% $9,600 ÷ 1.47271 = $6,518.59   or $9,600 × 0.67902 = $6,518.59
    4 × 6.5 = 26 periods $9,600 − $6,518.59 = $3,081.41

13. Determine the present value (principal) of a single deposit that is worth exactly $4,750 after 15 months at 6% compounded monthly.

    $\underline{\$4,407.62}$

    6% ÷ 12 = 0.5% $4,750 ÷ 1.07768 = $4,407.62   or $4,750 × 0.92792 = $4,407.62
    15 months = 15 periods

14. Calculate the amount of compound interest that has accrued on an investment that is now worth $15,000 after 14 years at 10% compounded semiannually.

    $\underline{\$11,173.65}$

    10% ÷ 2 = 5% $15,000 ÷ 3.92013 = $3,826.40   or $15,000 × 0.25509 = $3,826.35
    2 × 14 = 28 periods $15,000 − $3,826.35 = $11,173.65

Assignment 16.2 Continued

15. Compute the present value (principal) if the future value is $50,000 after 50 years at 6% compounded annually.

$\underline{\$2,714.50}$

6% ÷ 1 = 6%
1 × 50 = 50 periods $50,000 ÷ 18.42015 = $2,714.42   or $50,000 × 0.05429 = $2,714.50

$\underline{\hspace{3cm}}$
Score for B (32)

C **(40 points) Business Applications. Find the present value (principal) or the compound interest, as indicated. Use either Table 16-1, Table 16-2, or a calculator. Round answers to the nearest cent. (4 points for each correct answer)**

16. Ben Mahaffy needs to buy another used logging truck. His mother will loan him part of the money at only 4% compounded quarterly. If Ben estimates that he will be able to repay his mother a total of $27,500 in $1\frac{1}{2}$ years, how much can he borrow from her today?

$\underline{\$25,906.38}$

4% ÷ 4 = 1% $27,500 ÷ 1.06152 = $25,906.25   or $27,500 × 0.94205 = $25,906.38
4 × $1\frac{1}{2}$ = 6 periods

17. Six years ago, Eleanor Baker invested money at 8% compounded annually. Today she received a check for $6,000 that represented her total payment of principal and interest. Compute the amount of the interest that she earned.

$\underline{\$2,218.97}$

8% ÷ 1 = 8% $6,000 ÷ 1.58687 = $3,781.03   or $6,000 × 0.63017 = $3,781.02
1 × 6 = 6 periods $6,000 − $3,781.02 = $2,218.98

18. Lee Oman wants to have $30,000 available at the end of 3 years to help purchase a computerized metal lathe for his machine stop. If he can invest money at 6% compounded semiannually, how much should he invest?

$\underline{\$25,124.40}$

6% ÷ 2 = 3% $30,000 ÷ 1.19405 = $25,124.58   or $30,000 × 0.83748 = $25,124.40
2 × 3 = 6 periods

19. As part of their financial planning, Janice Garcia's grandparents made monetary gifts to each of their grandchildren. In addition, Janice's grandfather told her that, if she would save part of her gift for at least a year, he would pay her interest of 9% compounded monthly. Janice decided to save just enough so that she would have $5,000 at the end of 21 months, when she will be 16 years old. How much should she save?

$\underline{\$4,273.90}$

9% ÷ 12 = 0.75% $5,000 ÷ 1.16989 = $4,273.91   or $5,000 × 0.85478 = $4,273.90
21 months = 21 periods

20. Marilyn Whitehorse estimated that she would need $12,600 in $5\frac{1}{2}$ years to buy new equipment for her pottery shop. Having extra cash, she invested money in an extremely safe investment that advertised a return of 6% compounded semiannually. Marilyn invested just enough money to end up with the $12,600. How much of the $12,600 did Marilyn earn on her investment?

$\underline{\$3,497.51}$

6% ÷ 2 = 3% $12,600 ÷ 1.38423 = $9,102.53   or $12,600 × 0.72242 = $9,102.49
2 × $5\frac{1}{2}$ = 11 periods $12,600 − $9,102.49 = $3,497.51

Assignment 16.2 Continued

21. Keith Smith is a financial advisor. A client would like to have $25,000 in 5 years for possible weddings for her twin daughters who are now 18 years old. After comparing the projected returns with the risk, Keith recommends an investment that will pay 6% compounded quarterly. To end up with the $25,000, how much must the client invest today?

$\underline{\$18,561.75}$

6% ÷ 4 = 1.5% $25,000 ÷ 1.34686 = $18,561.69   or $25,000 × 0.74247 = $18,561.75
4 × 5 = 20 periods

22. A small company estimated that a modest investment today would realize a return of 10% compounded annually. The company wants a total sum of $20,000 in 5 years. If the company invests the appropriate amount to reach the $20,000 objective, how much of the $20,000 will be earned by the investment?

$\underline{\$7,581.60}$

10% ÷ 1 = 10% $20,000 ÷ 1.61051 = $12,418.43   or $20,000 × 0.62092 = $12,418.40
1 × 5 = 5 periods $20,000 − $12,418.40 = $7,581.60

23. Linda Anderson inherited $10,000. She knew that she would need $8,000 in 3 years to pay additional tuition for her children's education. Linda wanted to save enough to have the $8,000 3 years from now. She found an incredible, relatively safe investment that would pay 15% compounded monthly for the entire 3 years—if she agreed to leave the money untouched for 3 years. If Linda invests enough of the inheritance to guarantee the $8,000, how much will she have left over from the $10,000 inheritance?

$\underline{\$4,884.72}$

15% ÷ 12 = 1.25% $8,000 ÷ 1.56394 = $5,115.29   or $8,000 × 0.63941 = $5,115.28
12 × 3 = 36 periods $10,000 − $5,115.28 = $4,884.72

24. Charles Peterson owns an antique store in New England. He is planning a buying trip to France for next spring—in 9 months. Charles estimates the cost of the trip will be $8,000 in 9 months. How much should Charles set aside today to have $8,000 in 9 months? He can earn 8% compounded quarterly.

$\underline{\$7,538.56}$

8% ÷ 4 = 2% $8,000 ÷ 1.06121 = $7,538.56   or $8,000 × 0.94232 = $7,538.56
9 months ÷ 3 = 3 periods

25. Technology advances so rapidly that printers for higher-end computer systems are obsolete almost before they come onto the market. Frances Leung thinks that it would be reasonable to budget $500 next year for an up-to-date printer. Frances can make a safe investment paying 9% compounded monthly for a year. If she invests the necessary amount of her money, how much of the $500 will be paid by the investment?

$\underline{\$42.88}$

9% ÷ 12 = 0.75% $500 ÷ 1.09381 = $457.12   or $500 × 0.91424 = $457.12
12 × 1 = 12 periods $500 − $457.12 = $42.88

Score for C (40)

Notes

Table 16-1: Future Value (Compound Amount) Factors

Period	0.50%	0.75%	1.00%	1.25%	1.50%	2.00%	3.00%	4.00%	5.00%	6.00%	8.00%	9.00%	10.00%	12.00%
1	1.00500	1.00750	1.01000	1.01250	1.01500	1.02000	1.03000	1.04000	1.05000	1.06000	1.08000	1.09000	1.10000	1.12000
2	1.01003	1.01506	1.02010	1.02516	1.03023	1.04040	1.06090	1.08160	1.10250	1.12360	1.16640	1.18810	1.21000	1.25440
3	1.01508	1.02267	1.03030	1.03797	1.04568	1.06121	1.09273	1.12486	1.15763	1.19102	1.25971	1.29503	1.33100	1.40493
4	1.02015	1.03034	1.04060	1.05095	1.06136	1.08243	1.12551	1.16986	1.21551	1.26248	1.36049	1.41158	1.46410	1.57352
5	1.02525	1.03807	1.05101	1.06408	1.07728	1.10408	1.15927	1.21665	1.27628	1.33823	1.46933	1.53862	1.61051	1.76234
6	1.03038	1.04585	1.06152	1.07738	1.09344	1.12616	1.19405	1.26532	1.34010	1.41852	1.58687	1.67710	1.77156	1.97382
7	1.03553	1.05370	1.07214	1.09085	1.10984	1.14869	1.22987	1.31593	1.40710	1.50363	1.71382	1.82804	1.94872	2.21068
8	1.04071	1.06160	1.08286	1.10449	1.12649	1.17166	1.26677	1.36857	1.47746	1.59385	1.85093	1.99256	2.14359	2.47596
9	1.04591	1.06956	1.09369	1.11829	1.14339	1.19509	1.30477	1.42331	1.55133	1.68948	1.99900	2.17189	2.35795	2.77308
10	1.05114	1.07758	1.10462	1.13227	1.16054	1.21899	1.34392	1.48024	1.62889	1.79085	2.15892	2.36736	2.59374	3.10585
11	1.05640	1.08566	1.11567	1.14642	1.17795	1.24337	1.38423	1.53945	1.71034	1.89830	2.33164	2.58043	2.85312	3.47855
12	1.06168	1.09381	1.12683	1.16075	1.19562	1.26824	1.42576	1.60103	1.79586	2.01220	2.51817	2.81266	3.13843	3.89598
13	1.06699	1.10201	1.13809	1.17526	1.21355	1.29361	1.46853	1.66507	1.88565	2.13293	2.71962	3.06580	3.45227	4.36349
14	1.07232	1.11028	1.14947	1.18995	1.23176	1.31948	1.51259	1.73168	1.97993	2.26090	2.93719	3.34173	3.79750	4.88711
15	1.07768	1.11860	1.16097	1.20483	1.25023	1.34587	1.55797	1.80094	2.07893	2.39656	3.17217	3.64248	4.17725	5.47357
16	1.08307	1.12699	1.17258	1.21989	1.26899	1.37279	1.60471	1.87298	2.18287	2.54035	3.42594	3.97031	4.59497	6.13039
17	1.08849	1.13544	1.18430	1.23514	1.28802	1.40024	1.65285	1.94790	2.29202	2.69277	3.70002	4.32763	5.05447	6.86604
18	1.09393	1.14396	1.19615	1.25058	1.30734	1.42825	1.70243	2.02582	2.40662	2.85434	3.99602	4.71712	5.55992	7.68997
19	1.09940	1.15254	1.20811	1.26621	1.32695	1.45681	1.75351	2.10685	2.52695	3.02560	4.31570	5.14166	6.11591	8.61276
20	1.10490	1.16118	1.22019	1.28204	1.34686	1.48595	1.80611	2.19112	2.65330	3.20714	4.66096	5.60441	6.72750	9.64629
21	1.11042	1.16989	1.23239	1.29806	1.36706	1.51567	1.86029	2.27877	2.78596	3.39956	5.03383	6.10881	7.40025	10.80385
22	1.11597	1.17867	1.24472	1.31429	1.38756	1.54598	1.91610	2.36992	2.92526	3.60354	5.43654	6.65860	8.14027	12.10031
23	1.12155	1.18751	1.25716	1.33072	1.40838	1.57690	1.97359	2.46472	3.07152	3.81975	5.87146	7.25787	8.95430	13.55235
24	1.12716	1.19641	1.26973	1.34735	1.42950	1.60844	2.03279	2.56330	3.22510	4.04893	6.34118	7.91108	9.84973	15.17863
25	1.13280	1.20539	1.28243	1.36419	1.45095	1.64061	2.09378	2.66584	3.38635	4.29187	6.84848	8.62308	10.83471	17.00006

Table 16-1: Future Value (Compound Amount) Factors *(continued)*

Period	0.50%	0.75%	1.00%	1.25%	1.50%	2.00%	3.00%	4.00%	5.00%	6.00%	8.00%	9.00%	10.00%	12.00%
26	1.13846	1.21443	1.29526	1.38125	1.47271	1.67342	2.15659	2.77247	3.55567	4.54938	7.39635	9.39916	11.91818	19.04007
27	1.14415	1.22354	1.30821	1.39851	1.49480	1.70689	2.22129	2.88337	3.73346	4.82235	7.98806	10.24508	13.10999	21.32488
28	1.14987	1.23271	1.32129	1.41599	1.51722	1.74102	2.28793	2.99870	3.92013	5.11169	8.62711	11.16714	14.42099	23.88387
29	1.15562	1.24196	1.33450	1.43369	1.53998	1.77584	2.35657	3.11865	4.11614	5.41839	9.31727	12.17218	15.86309	26.74993
30	1.16140	1.25127	1.34785	1.45161	1.56308	1.81136	2.42726	3.24340	4.32194	5.74349	10.06266	13.26768	17.44940	29.95992
31	1.16721	1.26066	1.36133	1.46976	1.58653	1.84759	2.50008	3.37313	4.53804	6.08810	10.86767	14.46177	19.19434	33.55511
32	1.17304	1.27011	1.37494	1.48813	1.61032	1.88454	2.57508	3.50806	4.76494	6.45339	11.73708	15.76333	21.11378	37.58173
33	1.17891	1.27964	1.38869	1.50673	1.63448	1.92223	2.65234	3.64838	5.00319	6.84059	12.67605	17.18203	23.22515	42.09153
34	1.18480	1.28923	1.40258	1.52557	1.65900	1.96068	2.73191	3.79432	5.25335	7.25103	13.69013	18.72841	25.54767	47.14252
35	1.19073	1.29890	1.41660	1.54464	1.68388	1.99989	2.81386	3.94609	5.51602	7.68609	14.78534	20.41397	28.10244	52.79962
36	1.19668	1.30865	1.43077	1.56394	1.70914	2.03989	2.89828	4.10393	5.79182	8.14725	15.96817	22.25123	30.91268	59.13557
37	1.20266	1.31846	1.44508	1.58349	1.73478	2.08069	2.98523	4.26809	6.08141	8.63609	17.24563	24.25384	34.00395	66.23184
38	1.20868	1.32835	1.45953	1.60329	1.76080	2.12230	3.07478	4.43881	6.38548	9.15425	18.62528	26.43668	37.40434	74.17966
39	1.21472	1.33831	1.47412	1.62333	1.78721	2.16474	3.16703	4.61637	6.70475	9.70351	20.11530	28.81598	41.14478	83.08122
40	1.22079	1.34835	1.48886	1.64362	1.81402	2.20804	3.26204	4.80102	7.03999	10.28572	21.72452	31.40942	45.25926	93.05097
41	1.22690	1.35846	1.50375	1.66416	1.84123	2.25220	3.35990	4.99306	7.39199	10.90286	23.46248	34.23627	49.78518	104.21709
42	1.23303	1.36865	1.51879	1.68497	1.86885	2.29724	3.46070	5.19278	7.76159	11.55703	25.33948	37.31753	54.76370	116.72314
43	1.23920	1.37891	1.53398	1.70603	1.89688	2.34319	3.56452	5.40050	8.14967	12.25045	27.36664	40.67611	60.24007	130.72991
44	1.24539	1.38926	1.54932	1.72735	1.92533	2.39005	3.67145	5.61652	8.55715	12.98548	29.55597	44.33696	66.26408	146.41750
45	1.25162	1.39968	1.56481	1.74895	1.95421	2.43785	3.78160	5.84118	8.98501	13.76461	31.92045	48.32729	72.89048	163.98760
46	1.25788	1.41017	1.58046	1.77081	1.98353	2.48661	3.89504	6.07482	9.43426	14.59049	34.47409	52.67674	80.17953	183.66612
47	1.26417	1.42075	1.59626	1.79294	2.01328	2.53634	4.01190	6.31782	9.90597	15.46592	37.23201	57.41765	88.19749	205.70605
48	1.27049	1.43141	1.61223	1.81535	2.04348	2.58707	4.13225	6.57053	10.40127	16.39387	40.21057	62.58524	97.01723	230.39078
49	1.27684	1.44214	1.62835	1.83805	2.07413	2.63881	4.25622	6.83335	10.92133	17.37750	43.42742	68.21791	106.71896	258.03767
50	1.28323	1.45296	1.64463	1.86102	2.10524	2.69159	4.38391	7.10668	11.46740	18.42015	46.90161	74.35752	117.39085	289.00219

Table 16-2: Present Value Factors

Period	0.50%	0.75%	1.00%	1.25%	1.50%	2.00%	3.00%	4.00%	5.00%	6.00%	8.00%	9.00%	10.00%	12.00%
1	0.99502	0.99256	0.99010	0.98765	0.98522	0.98039	0.97087	0.96154	0.95238	0.94340	0.92593	0.91743	0.90909	0.89286
2	0.99007	0.98517	0.98030	0.97546	0.97066	0.96117	0.94260	0.92456	0.90703	0.89000	0.85734	0.84168	0.82645	0.79719
3	0.98515	0.97783	0.97059	0.96342	0.95632	0.94232	0.91514	0.88900	0.86384	0.83962	0.79383	0.77218	0.75131	0.71178
4	0.98025	0.97055	0.96098	0.95152	0.94218	0.92385	0.88849	0.85480	0.82270	0.79209	0.73503	0.70843	0.68301	0.63552
5	0.97537	0.96333	0.95147	0.93978	0.92826	0.90573	0.86261	0.82193	0.78353	0.74726	0.68058	0.64993	0.62092	0.56743
6	0.97052	0.95616	0.94205	0.92817	0.91454	0.88797	0.83748	0.79031	0.74622	0.70496	0.63017	0.59627	0.56447	0.50663
7	0.96569	0.94904	0.93272	0.91672	0.90103	0.87056	0.81309	0.75992	0.71068	0.66506	0.58349	0.54703	0.51316	0.45235
8	0.96089	0.94198	0.92348	0.90540	0.88771	0.85349	0.78941	0.73069	0.67684	0.62741	0.54027	0.50187	0.46651	0.40388
9	0.95610	0.93496	0.91434	0.89422	0.87459	0.83676	0.76642	0.70259	0.64461	0.59190	0.50025	0.46043	0.42410	0.36061
10	0.95135	0.92800	0.90529	0.88318	0.86167	0.82035	0.74409	0.67556	0.61391	0.55839	0.46319	0.42241	0.38554	0.32197
11	0.94661	0.92109	0.89632	0.87228	0.84893	0.80426	0.72242	0.64958	0.58468	0.52679	0.42888	0.38753	0.35049	0.28748
12	0.94191	0.91424	0.88745	0.86151	0.83639	0.78849	0.70138	0.62460	0.55684	0.49697	0.39711	0.35553	0.31863	0.25668
13	0.93722	0.90743	0.87866	0.85087	0.82403	0.77303	0.68095	0.60057	0.53032	0.46884	0.36770	0.32618	0.28966	0.22917
14	0.93256	0.90068	0.86996	0.84037	0.81185	0.75788	0.66112	0.57748	0.50507	0.44230	0.34046	0.29925	0.26333	0.20462
15	0.92792	0.89397	0.86135	0.82999	0.79985	0.74301	0.64186	0.55526	0.48102	0.41727	0.31524	0.27454	0.23939	0.18270
16	0.92330	0.88732	0.85282	0.81975	0.78803	0.72845	0.62317	0.53391	0.45811	0.39365	0.29189	0.25187	0.21763	0.16312
17	0.91871	0.88071	0.84438	0.80963	0.77639	0.71416	0.60502	0.51337	0.43630	0.37136	0.27027	0.23107	0.19784	0.14564
18	0.91414	0.87416	0.83602	0.79963	0.76491	0.70016	0.58739	0.49363	0.41552	0.35034	0.25025	0.21199	0.17986	0.13004
19	0.90959	0.86765	0.82774	0.78976	0.75361	0.68643	0.57029	0.47464	0.39573	0.33051	0.23171	0.19449	0.16351	0.11611
20	0.90506	0.86119	0.81954	0.78001	0.74247	0.67297	0.55368	0.45639	0.37689	0.31180	0.21455	0.17843	0.14864	0.10367
21	0.90056	0.85478	0.81143	0.77038	0.73150	0.65978	0.53755	0.43883	0.35894	0.29416	0.19866	0.16370	0.13513	0.09256
22	0.89608	0.84842	0.80340	0.76087	0.72069	0.64684	0.52189	0.42196	0.34185	0.27751	0.18394	0.15018	0.12285	0.08264
23	0.89162	0.84210	0.79544	0.75147	0.71004	0.63416	0.50669	0.40573	0.32557	0.26180	0.17032	0.13778	0.11168	0.07379
24	0.88719	0.83583	0.78757	0.74220	0.69954	0.62172	0.49193	0.39012	0.31007	0.24698	0.15770	0.12640	0.10153	0.06588
25	0.88277	0.82961	0.77977	0.73303	0.68921	0.60953	0.47761	0.37512	0.29530	0.23300	0.14602	0.11597	0.09230	0.05882

Table 16-2: Present Value Factors *(continued)*

Period	0.50%	0.75%	1.00%	1.25%	1.50%	2.00%	3.00%	4.00%	5.00%	6.00%	8.00%	9.00%	10.00%	12.00%
26	0.87838	0.82343	0.77205	0.72398	0.67902	0.59758	0.46369	0.36069	0.28124	0.21981	0.13520	0.10639	0.08391	0.05252
27	0.87401	0.81730	0.76440	0.71505	0.66899	0.58586	0.45019	0.34682	0.26785	0.20737	0.12519	0.09761	0.07628	0.04689
28	0.86966	0.81122	0.75684	0.70622	0.65910	0.57437	0.43708	0.33348	0.25509	0.19563	0.11591	0.08955	0.06934	0.04187
29	0.86533	0.80518	0.74934	0.69750	0.64936	0.56311	0.42435	0.32065	0.24295	0.18456	0.10733	0.08215	0.06304	0.03738
30	0.86103	0.79919	0.74192	0.68889	0.63976	0.55207	0.41199	0.30832	0.23138	0.17411	0.09938	0.07537	0.05731	0.03338
31	0.85675	0.79324	0.73458	0.68038	0.63031	0.54125	0.39999	0.29646	0.22036	0.16425	0.09202	0.06915	0.05210	0.02980
32	0.85248	0.78733	0.72730	0.67198	0.62099	0.53063	0.38834	0.28506	0.20987	0.15496	0.08520	0.06344	0.04736	0.02661
33	0.84824	0.78147	0.72010	0.66369	0.61182	0.52023	0.37703	0.27409	0.19987	0.14619	0.07889	0.05820	0.04306	0.02376
34	0.84402	0.77565	0.71297	0.65549	0.60277	0.51003	0.36604	0.26355	0.19035	0.13791	0.07305	0.05339	0.03914	0.02121
35	0.83982	0.76988	0.70591	0.64740	0.59387	0.50003	0.35538	0.25342	0.18129	0.13011	0.06763	0.04899	0.03558	0.01894
36	0.83564	0.76415	0.69892	0.63941	0.58509	0.49022	0.34503	0.24367	0.17266	0.12274	0.06262	0.04494	0.03235	0.01691
37	0.83149	0.75846	0.69200	0.63152	0.57644	0.48061	0.33498	0.23430	0.16444	0.11579	0.05799	0.04123	0.02941	0.01510
38	0.82735	0.75281	0.68515	0.62372	0.56792	0.47119	0.32523	0.22529	0.15661	0.10924	0.05369	0.03783	0.02673	0.01348
39	0.82323	0.74721	0.67837	0.61602	0.55953	0.46195	0.31575	0.21662	0.14915	0.10306	0.04971	0.03470	0.02430	0.01204
40	0.81914	0.74165	0.67165	0.60841	0.55126	0.45289	0.30656	0.20829	0.14205	0.09722	0.04603	0.03184	0.02209	0.01075
41	0.81506	0.73613	0.66500	0.60090	0.54312	0.44401	0.29763	0.20028	0.13528	0.09172	0.04262	0.02921	0.02009	0.00960
42	0.81101	0.73065	0.65842	0.59348	0.53509	0.43530	0.28896	0.19257	0.12884	0.08653	0.03946	0.02680	0.01826	0.00857
43	0.80697	0.72521	0.65190	0.58616	0.52718	0.42677	0.28054	0.18517	0.12270	0.08163	0.03654	0.02458	0.01660	0.00765
44	0.80296	0.71981	0.64545	0.57892	0.51939	0.41840	0.27237	0.17805	0.11686	0.07701	0.03383	0.02255	0.01509	0.00683
45	0.79896	0.71445	0.63905	0.57177	0.51171	0.41020	0.26444	0.17120	0.11130	0.07265	0.03133	0.02069	0.01372	0.00610
46	0.79499	0.70913	0.63273	0.56471	0.50415	0.40215	0.25674	0.16461	0.10600	0.06854	0.02901	0.01898	0.01247	0.00544
47	0.79103	0.70385	0.62646	0.55774	0.49670	0.39427	0.24926	0.15828	0.10095	0.06466	0.02686	0.01742	0.01134	0.00486
48	0.78710	0.69861	0.62026	0.55086	0.48936	0.38654	0.24200	0.15219	0.09614	0.06100	0.02487	0.01598	0.01031	0.00434
49	0.78318	0.69341	0.61412	0.54406	0.48213	0.37896	0.23495	0.14634	0.09156	0.05755	0.02303	0.01466	0.00937	0.00388
50	0.77929	0.68825	0.60804	0.53734	0.47500	0.37153	0.22811	0.14071	0.08720	0.05429	0.02132	0.01345	0.00852	0.00346

Part 5

Business Applications

17 Inventory and Turnover
18 Depreciation
19 Financial Statements
20 International Business

Inventory and Turnover

17

Learning Objectives
By studying this chapter and completing all assignments you will be able to:

Learning Objective 1 — Account for inventory by inventory sheets and reports from a perpetual inventory system.

Learning Objective 2 — Compute inventory value by the average cost, LIFO, and FIFO methods.

Learning Objective 3 — Compute inventory by using the lower of cost or market value.

Learning Objective 4 — Estimate inventory by using cost of goods sold.

Learning Objective 5 — Compute inventory turnover.

A company's inventory is the amount of goods it has on hand at any particular time. Retailers and wholesalers have only one kind of inventory—*merchandise,* which are the goods they sell.

Accounting for Inventory

INVENTORY SHEETS

Learning Objective 1

Account for inventory by inventory sheets and reports from a perpetual inventory system.

At least once each year, businesses undertake a **physical inventory**—an actual counting of the merchandise on hand. Some stores that require close control take a physical inventory every six months, quarterly, or even monthly. Sometimes retail stores use outside firms that specialize in taking inventory.

When inventory is counted, a description of each item, the quantity, the unit cost or retail price, and the **extension** (quantity × price) are recorded on an **inventory sheet**, as shown in Figure 17-1. The inventory value is then compared with accounting records, and any needed adjustments are made.

17.1 A restaurant must keep supplies of meat, potatoes, and other foodstuffs to prepare its menu items. These are counted and constitute the physical inventory for such establishments.

Figure 17-1 | Inventory Sheet

WARREN'S AUTO PARTS
Inventory Sheet
April 30, 20—

Description	Quantity	Unit Price (Average Cost)	Extension
Ignition terminals—#746083	318	$36.14	$11,492.52
Odometer cables—#007614	73	9.97	727.81
Wiper blades, compact—#417654	38	4.71	178.98
Spark plugs, 0.14—#772034	354	2.34	828.36
Hood/truck latches—#476508	58	13.42	778.36
Total			$14,006.03

PERPETUAL INVENTORY SYSTEMS

Some firms keep a **perpetual inventory**—a running count of all inventory items, based on tracking each item as it comes into and goes out of inventory. In businesses that handle high-cost items, such as cars or large appliances, the perpetual system keeps track of each item by serial number and price.

Businesses that handle small items, such as candy bars or shoes, have difficulty identifying each specific item. Their perpetual inventory systems keep a count of the number of units on hand, not individual prices and serial numbers.

Data for a perpetual inventory system are usually kept on a computer. Figure 17-2 illustrates a computer printout of an inventory record sheet. The last item in the Balance on Hand column shows how many units are on hand on the 4/30 recording date—354 Quickstart spark plugs: 0.14, part number 772034.

Figure 17-2 Inventory Sheet

WARREN'S AUTO PARTS
Inventory Record Sheet

ITEM: QUICKSTART SPARK PLUG: 0.14

PART NUMBER: #772034

LOCATION: Aisle 72, Bin 4, Box C

MINIMUM STOCK: 200   MAXIMUM STOCK: 800

ORDER FROM:
Northwest Distributors
2337 Colfax Avenue
Milbrae, CA 93233
Phone—(415) 345-7654

ORDER: 100–800

Purchase Orders (PO)			Inventory Control					
Date	PO No.	Quantity	Date	Source Code	Units In	Unit Cost	Units Out	Balance on hand
2/03	F0129	400	1/01	—		$2.10		350
3/15	M1678	300	1/31	SJ01			120	230
3/22	M2076	200	2/28	SJ02			58	172
4/26	A3210	400	3/02	F0129	400	2.36		572
			3/31	SJ03			315	257
			4/03	M1678	300	2.40		557
			4/20	M2076	200	2.64		757
			4/30	SJ04			403	354

Note: The 400 units ordered 4/26 have not yet been received.

✓ CONCEPT CHECK 17.1

The CompuParts wholesale computer store maintains a perpetual inventory of computer parts received and removed or shipped out. The following inventory record sheet shows the data for May. Compute the balance on hand after each transaction.

COMPUPARTS WHOLESALE
Inventory Record Sheet

ITEM: MONITOR CORD #A718
Location: BIN #C7
Minimum Stock: 250   Maximum Stock: 1,000

Order From:
Myers Distributors
1422 Oak Drive
Stockton, CA 97777
Fax: 209-775-7823

	Units In	Unit Cost	Units Out	Balance on Hand
5/01		$15.40		390
5/11	470	$15.80		860
5/15			260	600
5/28	320	$15.90		920
5/31			410	510

Computing Inventory, Using the Average Cost, FIFO, and LIFO Methods

Learning Objective 2

Compute inventory value by the average cost, LIFO, and FIFO methods.

In all inventory systems, the cost of the inventory on hand at the end of the period is called **ending inventory (EI)**. The ending inventory must be computed before financial statements can be prepared.

To compute ending inventory, a business usually adopts one of three cost methods: average cost, first-in, first-out (FIFO); or last-in, first-out (LIFO). Once selected, the method must be followed consistently. We use the cost data from Figure 17-2 to illustrate computations for the three cost methods.

THE AVERAGE COST METHOD

The **average cost method** is based on the assumption that the cost for each item on hand is the average cost for items from the opening inventory and items purchased during the period.

EXAMPLE A

The average cost of the units on the inventory record sheet for stock part #772034 (Quickstart spark plugs: 0.14) is computed as follows:

Date	Units Purchased	Cost	Extension
1/01	350	$2.10	$ 735.00
3/02	400	2.36	944.00
4/03	300	2.40	720.00
4/20	200	2.64	528.00
	1,250		$2,927.00

Average cost per unit: $2,927 ÷ 1,250 = $2.34
Ending inventory (EI) at average cost: 354 units × $2.34 = $828.36

THE FIFO METHOD

17.2 In periods of rising costs, the FIFO method will result in *higher* gross profit; in periods of decreasing costs, the FIFO method will result in *lower* gross profit. Just the opposite is true for the LIFO method. In periods of rising costs, the LIFO method will result in *lower* gross profit; in periods of decreasing costs, the LIFO method will result in *higher* gross profit.

The **first-in, first-out (FIFO) costing method** is based on the assumption that the cost for units sold is determined in the order in which the units were purchased. Thus the cost of the inventory remaining is assumed to be based on the price of the units received most recently.

EXAMPLE B

Under the FIFO method, the inventory of 354 units would consist of the 200 units last purchased plus 154 units from the preceding purchase.

Date		
4/20	200 units × $2.64 = $528.00	
4/03	154 units × $2.40 = $369.60	
	354	$897.60 Ending inventory at FIFO cost

346 Part 5 Business Applications

THE LIFO METHOD

The **last-in, first-out (LIFO) costing method** is based on the assumption that the cost of the inventory remaining is determined by the cost of the units purchased the earliest.

EXAMPLE C

Under the LIFO method, the 354 units would consist of the 350 units on hand on 1/01 plus 4 units from the first purchase on 3/02.

Date		
1/01	350 units × $2.10 = $735.00	
3/02	4 units × $2.36 = $ 9.44	
	354	$744.44 Ending inventory at LIFO cost

✓ CONCEPT CHECK 17.2

The inventory record sheets for Hairbrushes at Debbie's Beauty Supply show 5,000 units purchased (or on hand) at a total cost of $10,240. The inventory at year's end was 1,500 units. Compute the value of the ending inventory by each of the three methods: average cost, FIFO, and LIFO.

Date	Units Purchased	Cost	Extension	
1/01	2,000	$2.00	$ 4,000	Average Cost: $10,240 ÷ 5,000 = $2.048
1/30	200	2.10	420	1,500 × $2.05 (rounded) = $3,075
2/20	700	2.10	1,470	
3/17	1,100	2.00	2,200	FIFO: (500 × $2.10) + (500 × $2.20) +
10/30	500	2.20	1,100	(500 × $2.00) = $3,150
11/17	500	2.10	1,050	
	5,000		$10,240	LIFO: (1,500 × $2.00) = $3,000

Computing Inventory at the Lower of Cost or Market Value

Financial statements usually present the ending inventory at its cost value, computed by using the average, FIFO, or LIFO costing method. However, in some cases the market value (current replacement cost) of goods is lower than their original or average cost. Most companies prefer to show the **lower of cost or market value** in their inventories. When market value exceeds the cost, the cost is used; when the cost exceeds market value, market value is used.

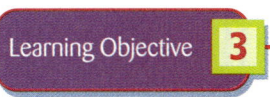

Learning Objective 3

Compute inventory by using the lower of cost or market value.

17.3 Examples of inventory with a current market value lower than cost include fashion merchandise carried in inventory so long that its value has fallen below what the merchant originally paid and electronic equipment that was state-of-the-art when originally purchased but was soon superseded by more advanced and cheaper equipment.

> **STEPS to Determine the Lower of Cost or Market (LCM) Inventory Value**
>
> 1. Compute the unit or total cost for each type of inventory item, using the average, FIFO, or LIFO costing method.
> 2. Determine the market value for each inventory item.
> 3. Compare the cost value from Step 1 with the market value from Step 2 and choose the lower of the two.
> 4. Compute the extension amount for each item based on the lower amount.
> 5. Sum the amounts in Step 4 to determine the total inventory value under LCM.

EXAMPLE D

Under LCM, using the average cost method illustrated in example A, the total inventory shown in Figure 17-1 would be valued at $13,802.13.

		STEP 1	STEP 2	STEP 3	STEP 4	
	(A)	(B) Unit Price	(C) Unit Price	(D) Lower of		
Description	Quantity	(Average Cost)	at Market	(B) or (C)	Extension (A × D)	
Ignition terminals—#746083	318	$36.14	$35.50	$35.50	$11,289.00	Market
Odometer cables—#007614	73	9.97	11.00	9.97	727.81	Cost
Wiper blades, compact—#417654	38	4.71	4.70	4.70	178.60	Market
Spark plugs, 0.14—#772034	354	2.34	2.64	2.34	828.36	Cost
Hood/trunk latches—#476508	58	13.42	14.10	13.42	778.36	Cost
Total					$13,802.13	STEP 5

EXAMPLE E

Under LCM, using the FIFO cost method illustrated in example B, the FIFO cost for the inventory for Quickstart spark plugs would be $897.60. Combining LCM with FIFO for the Quickstart spark plugs illustrated in example B, the ending inventory for this one item would be valued at $897.60 because the market value ($934.56) is higher than the FIFO cost.

			Market Value			
	(A)	(B)	(C) Unit Price	(D) Total	Lower of	
Description	Quantity	FIFO Cost	at Market	(A × C)	(B) or (D)	
Spark plugs	354	$897.60	$2.64	$934.56	$897.60	Cost

CONCEPT CHECK 17.3

L & L Records' inventory shows the following. Compute the inventory value at the lower of cost or market.

Description	Quantity	Cost	Market	Extension	
Classical #3	300	$ 7.07	$10.10	$2,121.00	Cost
Western #8	180	9.10	8.07	1,452.60	Market
Modern—light #11	410	11.17	12.08	4,579.70	Cost
Rock—new #4	89	12.10	12.10	$1,076.90	Cost/market
Total				$9,230.20	

Estimating Inventory Value

For monthly financial statements, inventory frequently is estimated without a physical count or a perpetual inventory system. The method usually used to estimate month-end inventory is called the **gross profit method**. This method involves estimating the cost of goods sold and subtracting this amount from the sum of the opening inventory and purchases made during the month. Note that **beginning inventory (BI)** is the ending inventory from the month before and **purchases (P)** are those goods for sale that have been purchased during the current month. The gross profit method is based on the formula

Learning Objective 4

Estimate inventory by using cost of goods sold.

```
  Beginning inventory (BI)
+ Purchases (P)
  Cost of goods available for sale
− Cost of goods sold (CGS) (estimated)
  Ending inventory (EI) (estimate)
```

Without a physical inventory, a precise cost of goods sold can't be determined. In this case, it is estimated by applying a markup percentage rate to **net sales** (total sales less sales returned and adjustments for the period). The net sales (100%) less this markup rate (percent) equals the cost of goods sold (percent). For instance, if the markup rate were 30%, the cost of goods sold would be 100% − 30% = 70%. If the rate of markup were 40%, the cost of goods sold would be 100% − 40% = 60%.

17.4 Usually, the percent of markup used to estimate cost of goods sold will be slightly lower than the standard or average percent of markup used by the retailer. A lower rate is used to account for special discounts, theft, loss through breakage, and the like.

EXAMPLE F

Assume that Warren's Auto Parts had a beginning inventory of $80,000. During the month, the company purchased and received $50,000 in goods and had net sales of $90,000. Throughout the month, Warren's maintained a 40% markup on all sales. Its cost of goods sold would be computed as follows.

Net sales for the month	$90,000
Cost of goods sold (estimated)	$54,000 [$90,000 × (100% − 40%) = $90,000 × 0.60]

Warren's Auto Parts would then determine its ending inventory (estimated) as follows:

Inventory, beginning of month	$ 80,000
Purchases for month	+ 50,000
Goods available for sale	$130,000
Cost of goods sold (estimated)	− 54,000
Ending inventory (estimated)	$ 76,000

© ROSE ALCORN/THOMSON

Sometimes a company's markup rate is based on cost rather than selling price. In this case, if the markup on cost were 30%, the cost of goods sold would be net sales divided by 130%. If the markup on cost were 40%, the cost of goods sold would be net sales divided by 140%.

EXAMPLE G

Assume that Warren's Auto Parts had a beginning inventory of $80,000. During the month, it had purchases of $50,000 and net sales of $90,000. Throughout the month, Warren's maintained a markup of 50% based on cost. What were Warren's cost of goods sold and ending inventory?

Beginning inventory	$80,000	
Purchases	+ 50,000	
Cost of goods available for sale	$130,000	
Cost of goods sold (estimated)	− 60,000	($90,000 ÷ 150%)
Ending inventory (estimated)	$70,000	

✓ CONCEPT CHECK 17.4

C & S Electronics records show the following. Compute the estimated ending inventory at cost.

Beginning inventory	$24,000	Net sales for period	$60,000
Purchases for period	$33,000	Markup based on retail	40%

$24,000 + $33,000 = $57,000 cost of goods available
$60,000 × 60% = $36,000 cost of goods sold
$57,000 − $36,000 = $21,000 ending inventory

Computing Inventory Turnover

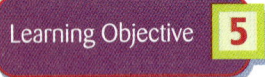

Compute inventory turnover.

Inventory turnover is the number of times the average inventory is converted into sales during the year. Inventory turnover is very high for a grocery store or ice cream parlor; it is very low for a specialty jewelry store or an antique shop. Standard turnover rates for corporate businesses are published. Some standard rates are 3.5 for hardware stores, 12.7 for grocery stores, 3.3 for nurseries, and 39.3 for stations and mini-marts.

Before turnover can be determined, average inventory must be calculated. **Average inventory** is the average of the inventories taken over a specific period of time—annually, semiannually, quarterly, or monthly.

17.5 Inventory turnover is becoming extremely important to large retail outlets such as Target, Kmart, Office Depot, and Sears. Many stock market analysts use inventory turnover in evaluating retail stores' relative strength.

Inventory is taken	Average inventory (at retail or cost)
Annually (once a year)	(BI + EI) ÷ 2
Semiannually (every six months)	(BI + end of 6 months + EI) ÷ 3
Quarterly (every three months)	(BI + 3 quarterlys + EI) ÷ 5
Monthly (every month)	(BI + 11 monthlys + EI) ÷ 13

Computation of inventory turnover can be based on either retail (selling) price or cost. **Inventory turnover at retail** is net sales divided by average inventory.

STEPS to Compute Inventory Turnover at Retail

1. Determine net sales.
2. Compute average inventory using retail price.
3. Compute inventory turnover at retail: Net sales ÷ Average inventory at retail.

EXAMPLE H

Assume that inventories for the year, based on selling price, are as follows: beginning, $90,000; end of month 3, $80,000; end of month 6, $100,000; end of month 9, $70,000; and end of month 12 (ending), $60,000. Net sales for the year are $520,000. Compute the inventory turnover at retail.

STEP 2
Average inventory = ($90,000 + $80,000 + $100,000 + $70,000 + $60,000) ÷ 5
= $400,000 ÷ 5 = $80,000

STEP 3
Inventory turnover at retail = $520,000 net sales ÷ $80,000 average inventory = 6.5 times

Note: The value of net sales and average inventory must both be figured at retail.

Some retailers prefer to express their rate of inventory turnover in terms of cost. **Inventory turnover at cost** is obtained by dividing the cost of goods sold (CGS) during a period by the average inventory for the same period computed at cost prices. (CGS is simply net sales at cost.)

STEPS to Compute Inventory Turnover at Cost

1. Compute the cost of goods sold using the formula BI + P − EI = CGS.
2. Compute the average inventory at cost.
3. Compute inventory turnover at cost: Cost of goods sold ÷ Average inventory at cost.

EXAMPLE 1

Assume that beginning inventory cost $60,000, purchases cost $300,000, and ending inventory cost $80,000. Compute the inventory turnover at cost.

STEP 1

Cost of goods sold:	Inventory at beginning of year	$ 60,000
	Purchases during year	+ 300,000
	Goods available for sale	$ 360,000
	Inventory at end of year	− 80,000
	Cost of goods sold	$ 280,000

STEP 2

Average inventory = ($60,000 BI + $80,00 EI) ÷ 2
= $140,000 ÷ 2 = $70,000

STEP 3

Inventory turnover at cost = $280,000 cost of goods sold
÷ $70,000 average inventory = 4.0 times

Note: The value of goods sold and average inventory must both be figured at cost.

✓ CONCEPT CHECK 17.5

Brinkman Scooter Shop has two branches (A and B), each using a markup of 50% of retail. Compute ending inventory, average inventory, and inventory turnover based on retail for each branch from the following data.

	Branch A	Branch B
Net sales	$1,400,000	$1,200,000
Beginning inventory	220,000	300,000
Inventory (March 31)	190,000	400,000
Inventory (June 30)	280,000	350,000
Inventory (September 30)	280,000	360,000
Inventory (December 31)	200,000	300,000

Average inventory: A—($220,000 + 190,000 + 280,000 + 280,000 + 200,000) ÷ 5 = $234,000
B—($300,000 + 400,000 + 350,000 + 360,000 + 300,000) ÷ 5 = $342,000

Retail turnover: A—$1,400,000 ÷ $234,000 = 5.98
B—$1,200,000 ÷ $342,000 = 3.51

COMPLETE ASSIGNMENTS 17.1 and 17.2.

Chapter Terms for Review

- average cost method
- average inventory
- beginning inventory (BI)
- cost of goods sold (CGS)
- ending inventory (EI)
- extension
- first-in, first-out (FIFO) costing method
- gross profit method
- inventory sheet
- inventory turnover
- inventory turnover at cost
- inventory turnover at retail
- last-in, first-out (LIFO) costing method
- lower of cost or market value (LCM)
- market value
- net sales
- perpetual inventory
- physical inventory
- purchases (P)

THE BOTTOM LINE

Summary of chapter learning objectives:

Learning Objective	Example
17.1 Account for inventory by inventory sheets and reports from a perpetual inventory system	1. Compute the Balance on Hand after each transaction: <table><tr><th>Date</th><th>Units In</th><th>Units Out</th><th>Balance on Hand</th></tr><tr><td>12/01</td><td></td><td></td><td>34,768</td></tr><tr><td>12/17</td><td>7,789</td><td></td><td>_____</td></tr><tr><td>12/19</td><td></td><td>17,072</td><td>_____</td></tr><tr><td>12/20</td><td>11,789</td><td></td><td>_____</td></tr><tr><td>12/31</td><td></td><td>14,490</td><td>_____</td></tr></table>
17.2 Compute inventory value by the average cost, LIFO, and FIFO methods	2. From the data shown, compute the ending inventory by the average cost, FIFO, and LIFO methods for Redwood Stove Company's stove part #717. The ending inventory, by physical count, was 300. **Stove Part #717** <table><tr><th>Date</th><th>Units In</th><th>Cost</th><th>Extension</th><th>Ending Inventory Value:</th></tr><tr><td>1/12</td><td>200</td><td>$3.00</td><td>$600</td><td></td></tr><tr><td>1/14</td><td>300</td><td>3.20</td><td>960</td><td>Average cost: _____</td></tr><tr><td>1/15</td><td>500</td><td>3.00</td><td>1,500</td><td>FIFO: _____</td></tr><tr><td>1/17</td><td>200</td><td>3.10</td><td>620</td><td>LIFO: _____</td></tr><tr><td>1/18</td><td>400</td><td>3.00</td><td>1,200</td><td></td></tr><tr><td>Total</td><td>1,600</td><td></td><td>$4,880</td><td></td></tr></table>
17.3 Compute inventory by using the lower of cost or market value	3. Compute Redwood Stove Company's inventory value at the lower of cost or market value. <table><tr><th>Description</th><th>Quantity</th><th>Cost</th><th>Market</th><th>Extension</th></tr><tr><td>Stoves</td><td>24</td><td>$277.50</td><td>$350.50</td><td>_____</td></tr><tr><td>Piping</td><td>90</td><td>34.50</td><td>27.00</td><td>_____</td></tr><tr><td>Hearths</td><td>75</td><td>78.00</td><td>78.00</td><td>_____</td></tr><tr><td>Screens</td><td>50</td><td>105.00</td><td>125.00</td><td>_____</td></tr><tr><td>Tool Sets</td><td>28</td><td>65.50</td><td>55.00</td><td>_____</td></tr><tr><td>Total</td><td></td><td></td><td></td><td>_____</td></tr></table>

Answers: 1. 42,557; 25,485; 37,274; 22,784 2. Average cost, $915; FIFO, $900; LIFO, $920 3. $6,660; $2,430; $5,850; $5,250; $1,540; $21,730

THE BOTTOM LINE

Summary of chapter learning objectives:

Learning Objective	Example
17.4 Estimate inventory by computing an estimated cost of goods sold	4. Redwood Stove Company has a markup of 50% of retail. Last year it had total sales of $400,000. It had a beginning inventory of $150,000 based on cost. It purchased merchandise for $180,000 during the year. Compute the ending inventory at cost.
17.5 Compute inventory turnover	5. Two years ago Redwood Stove Company used a markup of 65% of cost. That year's data are shown. Compute ending inventory, average inventory, and inventory turnover at retail. Net sales $900,000 Purchases (cost) 600,000 Beginning inventory—retail 300,000 Inventory—retail (June 30) 450,000

Answers: 4. Inventory $130,000 5. Ending inventory at retail, $390,000; average inventory, $380,000; turnover, 2.37

SELF-CHECK

Review Problems for Chapter 17

1. The D&D Company has 45 units on hand January 1. During the month, units in total 320 and units out total 285. What is the balance on hand January 31? _____

2. According to physical count, Dawson Lumber had 3,250 units in inventory March 31. Dawson Lumber's beginning inventory and purchases for the first quarter were as follows:

Jan. 1	Beginning Inventory	2,500 units @ $25.00
Jan. 15	Purchased	5,000 units @ $27.50
Feb. 5	Purchased	6,000 units @ $26.25
Mar. 10	Purchased	3,000 units @ $27.00

Calculate the value of the inventory March 31 and cost of goods sold for the quarter based on the average, FIFO, and LIFO costing methods.

	Inventory Value	Cost of Goods Sold
a. Average cost:	_____	_____
b. FIFO cost:	_____	_____
c. LIFO cost:	_____	_____

3. Lansky Company's inventory January 1 was valued at $41,000. During the first quarter, $365,000 of goods were purchased and sales totaled $550,000. Estimate the inventory March 31 if Baxter's markup is 40% based on selling price. _____

4. Compute the average inventory cost of goods sold, and turnover based on cost using the following data. Kelly Pet Supplies takes inventory every 6 months and had inventory of $35,000 on January 1, $42,600 on June 30, and $38,200 on December 31. Kelly's purchased goods totaling $275,000 during the year and had sales of $390,000.

   a. Average inventory: _____

   b. Cost of goods sold: _____

   c. Turnover: _____

Assignment 17.1: Inventory Cost

Name

Date   Score

Learning Objectives 1 2 3

A **(40 points) Compute the extensions and totals. (1 point for each correct answer)**

1. The inventory of Michelle's Clock Shop shows the following items, at both costs and market prices. Determine the total value of the inventory at the lower of cost or market price for each item.

Description	Quantity	Unit Cost Price	Unit Market Price	Extension at Lower of Cost or Market
Quartz clock and pen set	22	$36.00	$34.80	$765.60
Travel alarm clock	42	15.60	19.20	$655.20
Ultrasonic travel clock	16	23.00	23.70	$368.00
Digital alarm clock	40	19.80	18.60	$744.00
AM/FM clock radio	85	21.00	21.00	$1,785.00
Digital clock radio	9	54.00	57.50	$486.00
Total				$4,803.80

2. A retail furniture dealer counted the following goods in inventory on December 31. An accountant recommended that the inventory items be valued at the lower of cost or market price. Compute the total value of the inventory based on the lower of cost or market price.

Article	Quantity	Unit Cost Price	Extension at Cost	Unit Market Price	Extension at Market	Inventory Value at Lower of Cost or Market
Armchairs, wood	24	$ 40.00	$960.00	$ 68.50	$1,644.00	$960.00
Armchairs, tapestry	6	75.00	$450.00	105.00	$630.00	$450.00
Armchairs, Windsor	12	115.00	$1,380.00	85.00	$1,020.00	$1,020.00
Beds, bunk	8	85.50	$684.00	75.00	$600.00	$600.00
Bedroom suites	3	297.50	$892.50	410.00	$1,230.00	$892.50
Tables, coffee	30	63.00	$1,890.00	62.00	$1,860.00	$1,860.00
Chairs, kitchen	24	23.00	$552.00	32.00	$768.00	$552.00
Dining tables	8	117.40	$939.20	95.70	$765.60	$765.60
Dining suites	5	288.80	$1,444.00	395.00	$1,975.00	$1,444.00
Sofa sets	9	479.60	$4,316.40	325.00	$2,925.00	$2,925.00
Total			$13,508.10		$13,417.60	$11,469.10

Score for A (40)

Assignment 17.1 Continued

B (60 points) Compute the value of ending inventory. (10 points for each correct answer)

3. Garcia Manufacturing Company made purchases of a material as shown in the following listing. The inventory at the end of the year was 3,500 units. Compute the value of the inventory by each of the three methods: (a) average cost; (b) first-in, first-out; and (c) last-in, first-out.

Date	Units	Unit Cost	Total Cost
Jan 5	3,600	$6.20	$ 22,320
Mar. 11	3,000	5.80	17,400
May 14	5,300	6.00	31,800
July 8	1,600	6.30	10,080
Sept. 7	4,000	6.20	24,800
Nov. 10	2,500	6.40	16,000
Total	20,000		$122,400

a. Average cost: $22,950
$122,400 \div 20,000 = \$6.12$
$3,500 \times \$6.12 = \$21,420$

b. First-in, first-out: $22,200
$2,500 \times \$6.40 = \$16,000$
$1,000 \times \$6.20 = \ \ 6,200$
3,500 \ \ \ \ \ \ \ \ \ \ \ \ \ $22,200

c. Last-in, first-out: $21,700
$3,500 \times \$6.20 = \$21,700$

4. The Willand Company had 320 units on hand at the beginning of the year, with a unit cost of $4.20. The number and unit cost of units purchased and the number of units sold during the year are shown. What would be the value of the ending inventory of 380 units based on the (a) average cost; (b) first-in, first-out; and (c) last-in, first-out costing methods?

Date	Units Purchased	Unit Cost	Units Sold	Units on Hand
Jan. 1		$4.20		320
Feb. 2			190	130
Apr. 16	200	$4.32		330
June 10	300	$4.40		630
Aug. 5			280	350
Oct. 12	250	$4.48		600
Nov. 27			220	380

a. Average cost: $1,650.72
b. First-in, first-out: $1,692.00
c. Last-in, first-out: $1,603.20

a. Average cost: Beginning inventory   320 units × $4.20 = $1,344
April 16 purchase   200 units × 4.32 =   864
June 10 purchase   300 units × 4.40 = 1,320
Oct. 12 purchase   250 units × 4.48 = 1,120
                   1,070 units   = $4,648

$4,648 ÷ 1,070 units = $4.344
$4.344 × 380 units = $1,650.72 or 4.34 × 380 = 1,649.20 (rounded)

b. FIFO: Oct. 12 purchase   250 units × $4.48 = $1,120
June 10 purchase   130 units × 4.40 =   572
Ending inventory   380 units   = $1,692

c. LIFO: Beginning inventory   320 units × $4.20 = $1,344.00
April 16 purchase   60 units × 4.32 =   259.20
Ending inventory   380 units   = $1,603.20

Score for B (60)

Assignment 17.2: Inventory Estimating and Turnover

Name

Date   Score

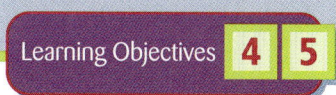

A (50 points) Solve the following problems. (2 points for each correct answer)

1. Fill in the blanks in each of the following calculations with the correct amount. Use the formulas
   Beginning inventory + Purchases = Goods available for sale
   Goods available for sale − Cost of goods sold = Ending inventory

	Store A	Store B	Store C	Store D	Store E
Beginning inventory	$ 80,000	$120,000	$ 37,000	$96,000	$42,000
Purchases	120,000	90,000	72,000	21,000	93,000
Goods available for sale	200,000	210,000	109,000	117,000	135,000
Less cost of goods sold	125,000	128,000	86,000	30,000	74,000
Ending inventory	$ 75,000	82,000	$ 23,000	87,000	$61,000

A: 200,000 GAFS − 80,000 BI = 120,000 P   200,000 GAFS − 125,000 CGS = 75,000 EI

B: 210,000 GAFS − 90,000 P = 120,000 BI   210,000 GAFS − 128,000 CGS = 82,000 EI

C: 109,000 GAFS − 37,000 BI = 72,000 P   109,000 GAFS − 23,000 EI = 86,000 CGS

D: 117,000 GAFS − 21,000 P = 96,000 BI   117,000 GAFS − 30,000 CGS = 87,000 EI

E: 135,000 GAFS − 42,000 BI = 93,000 P   135,000 GAFS − 74,000 CGS = 61,000 EI

2. Each of the five stores in problem 1 had the net sales shown. What was the average percent of markup, based on cost, for each of the five stores? What was the average percent of markup, based on selling price, for each of the five stores?

	Store A	Store B	Store C	Store D	Store E
Net sales	$200,000	$150,000	$172,000	$40,000	$100,000
Markup—cost	60%	17.19%	100%	33.33%	35.14%
Markup—selling price	37.5%	14.67%	50%	25.0%	26%

A: 200,000 NS − 125,000 CGS = 75,000 MU   75,000 MU ÷ 125,000 CGS = 60%
   75,000 MU ÷ 200,000 = 37.5%

B: 150,000 NS − 128,000 CGS = 22,000 MU   22,000 MU ÷ 128,000 CGS = 17.19%
   22,000 MU ÷ 150,000 = 14.67%

C: 172,000 NS − 86,000 CGS = 86,000 MU   86,000 MU ÷ 86,000 CGS = 100%
   86,000 MU ÷ 172,000 = 50%

D: 40,000 NS − 30,000 CGS = 10,000 MU   10,000 MU ÷ 30,000 CGS = 33.33%
   10,000 MU ÷ 40,000 = 25.0%

E: 100,000 NS − 74,000 CGS = 26,000 MU   26,000 MU ÷ 74,000 CGS = 35.14%
   26,000 MU ÷ 100,000 = 26%

Assignment 17.2 Continued

3. The Country Kitchen takes inventory at retail sales price every 3 months. Its inventory at the beginning of last year was $40,500; at 3 months, $45,000; at 6 months, $52,500; at 9 months, $49,500; and at the end of the year, $44,000. Net sales for the year were $296,800.

   a. What was the average inventory? $46,300

   b. What was the turnover? 6.41

   a. $40,500 + $45,000 + $52,500 + $49,500 + $44,000 = $231,500
      $231,500 ÷ 5 = $46,300 average inventory
   b. $296,800 ÷ $46,300 = 6.41

4. Steve's Auto Shop began the year with an inventory of $33,500. Purchases during the year totaled $194,200. The inventory at the end of the year was $36,400.

   a. What was the cost of goods sold? $191,300

   b. What was the average inventory? $34,950

   c. What was the turnover? 5.47

   a. Beginning inventory, January 1 $33,500
      Purchases   + 194,200
      Goods available for sale $227,700
      Inventory, December 31   − 36,400
      Cost of goods sold $191,300
   b. ($33,500 + $36,400) ÷ 2 = $34,950 average inventory
   c. $191,300 ÷ $34,950 = 5.47

Score for A (50)

B (50 points) Solve the following problems. (Points for each correct answer as marked)

5. Jackson Wholesalers' records showed these figures.

	Cost	Retail Price		
Beginning inventory	$19,793	$32,990	Net sales for the year	$61,450
Purchases for the year	$47,200	$78,665	Markup based on sales	40%

   Compute the ending inventory:

   a. At cost: $30,123 $\left(2\frac{1}{2} \text{ points}\right)$

   b. At retail price: $50,205 $\left(2\frac{1}{2} \text{ points}\right)$

   a. Beginning inventory (cost price) $19,793
      Purchases (cost price)   + 47,200
      Goods available for sale $66,993
      Cost of goods sold   − 36,870   ($61,450 × 0.6)
      Ending inventory (at cost) $30,123
   b. Beginning inventory (selling price) $ 32,990
      Purchases (selling price)   + 78,665
      Goods available for sale $111,655
      Sales   − 61,450
      Ending inventory (selling price) $50,205

360 Part 5 Business Applications

Assignment 17.2 Continued

6. The JM Clothing store kept all merchandise records in terms of selling price. On July 1, the JM books showed the following information.

Beginning inventory, January 1:	$23,500
6-month purchases:	99,000
6-month net sales:	87,800

   What was the estimated ending inventory on July 1? (5 points) $34,700

Beginning inventory (selling price)	$ 23,500
6-month purchases	+ 99,000
Goods available for sale	$122,500
Net sales	− 87,800
Ending inventory, July 1 (selling price)	$34,700

7. The Kid's Land Clothing Store kept all purchase and inventory records on a cost basis. The owner marked up all goods at 40.0% of the cost price. On July 1, the Kid's Land books showed the following information.

Beginning inventory, January 1:	$1,126,000
6-month purchases:	2,221,400
6-month net sales:	2,508,200

   What was the estimated inventory, at cost, on July 1? (5 points) $1,555,829

Beginning inventory (cost price)	$1,126,000	
6-month purchases (cost price)	+ 2,221,400	
Goods available for sale	$3,347,400	
Net sales	− 1,791,571	($2,508,200 ÷ 1.40)
Ending inventory	$1,555,829	

8. Amy's Art Shop kept all inventory and sales records on the basis of retail prices. It recorded purchases at cost and marked up its merchandise at 120% of cost. On January 1, its inventory of art was $260,000. During the year, its purchases were $300,000 and net sales were $730,000. What was its ending inventory? (5 points) $190,000

Beginning inventory	$260,000	
Purchases	+ 660,000	($300,000 + 120% markup)
Goods available for sale	$920,000	
Net sales	− 730,000	
Ending inventory	$190,000	

Assignment 17.2 Continued

9. From the information given, calculate the estimated cost of goods sold and ending inventory. Round to the nearest dollar. (1 point for each correct answer)

	Cost of Goods Available for Sale	Net Sales	Markup Based on Cost	Markup Based on Sales	Estimated Cost of Goods Sold	Estimated Ending Inventory
a.	$204,000	$260,000	30%		$200,000	$4,000
b.	268,000	260,000		30%	$182,000	$86,000
c.	444,000	350,000		27%	$255,500	$188,500
d.	444,000	350,000	27%		$275,591	$168,409
e.	37,500	36,000	50%		$24,000	$13,500
f.	368,000	400,000		60%	$160,000	$208,000
g.	420,000	600,000		40%	$360,000	$60,000
h.	440,000	360,000	15%		$313,043	$126,957
i.	125,000	180,000	60%		$112,500	$12,500
j.	130,000	200,000	100%		$100,000	$30,000

a. $260,000 ÷ 130% = $200,000 $204,000 − $200,000 = $4,000
b. $260,000 × 70% = $182,000 $268,000 − $182,000 = $86,000
c. $350,000 × 73% = $255,500 $444,000 − $255,500 = $188,500
d. $350,000 ÷ 127% = $275,591 $444,000 − $275,591 = $168,409
e. $36,000 ÷ 150% = $24,000 $37,500 − $24,000 = $13,500
f. $400,000 × 40% = $160,000 $368,000 − $160,000 = $208,000
g. $600,000 × 60% = $360,000 $420,000 − $360,000 = $60,000
h. $360,000 ÷ 115% = $313,043 $440,000 − $313,043 = $126,957
i. $180,000 ÷ 160% = $112,500 $125,000 − $112,500 = $12,500
j. $200,000 ÷ 200% = $100,000 $130,000 − $100,000 = $30,000

10. Maurice Company sells hair products. From the following inventory record sheets for Baby Soft Shampoo, determine the total units in, total amount, and the value of the ending inventory of 300 bottles based on average cost, FIFO, and LIFO. (2 points for each correct answer)

Date	Units In	Cost	Amount
1/11	400	$3.40	$1,360
1/23	50	3.00	150
2/10	100	3.20	320
2/20	200	3.30	660
2/25	80	3.50	280
	830		$2,770

Average cost: $1,002

FIFO: $1,004

LIFO: $1,020

Average cost: $2,770 ÷ 830 = $3.34; 300 × $3.34 = $1,002
FIFO: (80 × $3.50) + (200 × $3.30) + (20 × $3.20) = $1,004
LIFO: 300 × $3.40 = $1,020

Score for B (50)

Depreciation 18

Learning Objectives
By studying this chapter and completing all assignments you will be able to:

Learning Objective 1 — Compute depreciation using the straight-line method.

Learning Objective 2 — Compute depreciation using the units of production method.

Learning Objective 3 — Compute depreciation using the declining-balance method.

Learning Objective 4 — Compute depreciation using the sum-of-the-years-digits method.

Learning Objective 5 — Compute depreciation for income tax purposes using the Modified Accelerated Cost Recovery System (MACRS).

Learning Objective 6 — Compute partial-year depreciation using the five primary different depreciation methods covered.

Depreciation is the decrease in the value of assets owned by a business, such as automobiles, buildings, and computers. Depreciation is caused by wear or by **obsolescence** (becoming out-of-date). In the toy manufacturing industry, some dies and tools last only 1 or 2 years because of changing fads. An automobile will wear out after a number of years or miles of use. Buildings lose value as wood, electrical wiring, and fixtures deteriorate and as design characteristics and owners' needs change. A business computer frequently becomes obsolete in 3 to 5 years.

In business, depreciation is figured on almost all physical assets owned and in use. Depreciation is deducted from gross profits as an expense. In this chapter, we present five common methods of calculating depreciation: the straight-line, units-of-production, declining-balance, sum-of-the-years-digits, and Modified Accelerated Cost Recovery System methods.

Computing Depreciation with the Straight-Line Method

Learning Objective 1

Compute depreciation using the straight-line method.

The **straight-line (SL) method** of determining depreciation is the easiest method. It distributes depreciation evenly over the useful life of an asset, assigning equal amounts to designated units (miles, number of items made, etc.) or periods (usually months or years). It is based on the assumption that wear and obsolescence occur evenly over the life of the property. The three factors used to compute depreciation by the straight-line method are

1. The **original cost**, which includes the price paid for an item and any freight charges and expenses for installation. Cost includes anything necessary to get the asset to where it is to be used and in a condition to be used.
2. The **estimated service life**, which is the length of time the buyer expects to be able to use an asset. The estimated service life may be stated in terms of years or months that normally may be expected during the life of the asset.
3. The estimated **scrap value (SV)**, which is the amount the owner of an asset expects to receive upon disposing of it at the end of its estimated service life.

18.1 When an asset is purchased, some sales and other business taxes are subject to interpretation. When possible, some companies will expense these costs immediately. In other cases, all taxes paid will become part of the original cost of the depreciable asset.

The basic formula for computing the amount of depreciation under the straight-line method is

(Original cost − Scrap value) ÷ Estimated service life in periods of time
= Depreciation amount for 1 unit or period

EXAMPLE A

An office computer costing $12,500 has an estimated life of 5 years and an estimated scrap value of $900. What is the annual depreciation amount?

$12,500 cost − $900 SV = $11,600 estimated total depreciation
$11,600 ÷ 5 estimated total years = $2,320 annual depreciation

Computing Depreciation with the Units-of-Production Method

The **units-of-production method** of determining depreciation distributes depreciation based on how much the asset is used. It is usually expressed in miles driven, hours used, tons hauled, or units produced. Calculation is like that used in the straight-line method except that miles, hours, tons, or units are used rather than months or years. The basic formula for computing the amount of depreciation under the units-of-production method is

> (Original cost − Scrap value) ÷ Estimated life in service units
> = Depreciation amount for 1 unit

Learning Objective 2

Compute depreciation using the units-of-production method.

Example B shows depreciation of an asset based on the number of hours it is used. First you must find the hourly depreciation and then multiply it by the number of hours operated during a particular month or year.

EXAMPLE B

A machine costing $10,000 has an estimated life of 60,000 hours of operation and an estimated scrap value of $400. If it was operated for 2,800 hours during the first year, how much depreciation expense will be shown for the first year?

$10,000 cost − $400 SV = $9,600 estimated total depreciation
$9,600 ÷ 60,000 estimated total hours = $0.16 hourly depreciation
2,800 hours operated × $0.16 = $448 first year's depreciation.

Example C shows depreciation in terms of the number of units that it will produce during its lifetime: Divide the number of units into the estimated total depreciation amount to get the depreciation per unit.

EXAMPLE C

A press that costs $145,000 will produce an estimated 3,500,000 units in its life and has an estimated scrap value of $5,000. If it produced 235,000 units this year, how much depreciation will be shown for the year?

$145,000 cost − $5,000 SV = $140,000 estimated total depreciation
$140,000 ÷ 3,500,000 estimated total units = $0.04 depreciation per unit
235,000 units produced × $0.04 = $9,400 first year's depreciation

BOOK VALUE

The **book value** of an asset is the original cost minus the **accumulated depreciation**, or the total of all depreciation to that time.

EXAMPLE D

At the end of the first year, the book value of the press in example C would be

$145,000 cost − $9,400 accumulated depreciation = $135,600

The book value can be determined at any time in the life of an asset.

EXAMPLE E

At the end of the third year, the book value of the computer in example A would be computed as follows:

$2,320 annual depreciation × 3 years = $6,960 accumulated depreciation
$12,500 cost − $6,960 = $5,540.

✓ CONCEPT CHECK 18.1

On January 1, Oakdale Appliances bought a new delivery truck for $48,000. Oakdale's accountant estimated a truck life of 200,000 miles and a scrap (trade-in) value of $4,000. In the first year, the truck was driven 38,000 miles; in the second year, it was driven 46,000 miles. Compute the depreciation and book value for the first 2 years.

$48,000 cost − $4,000 SV = $44,000 estimated total depreciation
$44,000 ÷ 200,000 miles = $0.22 depreciation per mile

Year 1: 38,000 miles × $0.22 per mile = $8,360 depreciation
         $48,000 cost − $8,360 = $39,640 book value

Year 2: 46,000 miles × $0.22 per mile = $10,120 depreciation
         $39,640 year 1 book value − $10,120 = $29,520 new book value
         or
         $48,000 cost − ($8,360 + $10,120) accumulated depreciation = $29,520 book value

Computing Depreciation with the Declining-Balance Method

Learning Objective 3

Compute depreciation using the declining-balance method.

The **declining-balance (DB) method** is based on the theory that depreciation is greatest in the first year and less in each succeeding year.

STEPS to Compute Depreciation, Using the DB Method

1. Divide 100% by the estimated years of useful life to determine the **basic depreciation rate**.
2. Multiply the basic depreciation rate by 2 (**double-declining-balance**) or by 1.5 (**150%-declining-balance**) to determine the **declining-balance depreciation rate**.
3. Multiply the declining-balance depreciation rate by the book value of the asset at the beginning of the year to determine the depreciation amount for that year. (For the first year, the book value at the beginning of the year equals the asset cost. Do not subtract the scrap value.)

Step 3 is repeated each year, using the new (declined) book value (last year's beginning book value minus last year's depreciation amount). The same rate is used each year. The declining-balance rate continues to apply until the scrap value is reached. The item may not be depreciated below its scrap value.

EXAMPLE F

Use the declining-balance method with an annual double-declining balance to depreciate the office computer in example A.

STEP 1 $100\% \div 5$ years $= 20\%$

STEP 2 $20\% \times 2 = 40\%$ annual double-declining-balance rate.

Year		Beginning Book Value	Rate	STEP 3 Depreciation
1		$12,500	×40% =	$5,000
2	$12,500 − $5,000 =	7,500	×40% =	3,000
3	7,500 − 3,000 =	4,500	×40% =	1,800
4	4,500 − 1,800 =	2,700	×40% =	1,080
5	2,700 − 1,080 =	1,620	×40% =	648
6	1,620 − 648 =	972	×40% =	388.80 $72*

*As book value ($972) is larger than estimated scrap value ($900), there is some depreciation in the sixth year. However, the calculated depreciation ($388.80) is greater than book value minus scrap value ($972 − $900 = $72). Thus depreciation is limited to the smaller amount, $72.

CONCEPT CHECK 18.2

On January 1, Oakdale Appliances bought a new delivery truck for $48,000. Oakdale's accountant estimated a truck life of 4 years and a scrap value of $4,000. Compute the depreciation for the first 2 years using the 150%-declining-balance method.

$48,000 cost
$100\% \div 4$ years $= 25\%$
$25\% \times 1.5 = 37.5\%$ annual 150%-declining-balance rate

Year 1: $48,000 × 37.5% = $18,000 depreciation
          $48,000 − $18,000 = $30,000 book value

Year 2: $30,000 × 37.5% = $11,250 depreciation

Computing Depreciation with the Sum-of-the-Years-Digits Method

Learning Objective 4

Compute depreciation using the sum-of-the-years-digits method.

The **sum-of-the-years-digits (SYD) method** also is used to compute a greater depreciation amount in the earlier years of an asset's life. The book value decreases more slowly than under the declining-balance method. This method's name comes from the calculation done in Step 1.

18.2 Most companies use basic computer programs to compute monthly and annual depreciation figures, thus simplifying application of the sum-of-the-years-digits method to assets with many years of use.

STEPS to Compute Depreciation Using the SYD method

1. Compute the sum of all the years digits in the estimated life of the asset. Use this shortcut formula:

   $$\frac{(n + 1) \times n}{2},$$

   where n = number of years in the estimated life.

2. Determine the current year's depreciation fraction by using this formula: Estimated years of life remaining at the beginning of the current year ÷ Sum of all digits from Step 1.

3. Multiply the total depreciation amount (Cost − SV) of the asset by the depreciation fraction from Step 2 to determine depreciation for the current year.

Note that each year a new depreciation fraction from Step 2 is determined and Step 3 is repeated. The sum of all digits in Step 1 and the total depreciation amount in Step 3 are the same every year.

EXAMPLE G

Under the sum-of-the-years-digits method, the office computer in example A would be depreciated as follows.

STEP 1 $\quad \dfrac{(5 + 1) \times 5}{2} = 15$ (or $1 + 2 + 3 + 4 + 5 = 15$)

Year	STEP 2 Fraction		Depreciation Total Amount		STEP 3 Depreciation
1	$\dfrac{5}{15}$	×	$11,600	=	$ 3,866.67
2	$\dfrac{4}{15}$	×	11,600	=	3,093.33
3	$\dfrac{3}{15}$	×	11,600	=	2,320.00
4	$\dfrac{2}{15}$	×	11,600	=	1,546.67
5	$\dfrac{1}{15}$	×	11,600	=	$ 773.33
				Total depreciation	$11,600.00

✓ CONCEPT CHECK 18.3

On January 1, Oakdale Appliances bought a new delivery truck for $48,000. Oakdale's accountant estimated a truck life of 4 years and a scrap value of $4,000. Compute the depreciation for the first 2 years using the sum-of-the-years-digits method.

$48,000$ cost $- \$4,000$ SV $= \$44,000$ to be depreciated

$$\frac{(4 + 1) \times 4}{2} = 10 \text{ (or } 1 + 2 + 3 + 4 = 10)$$

Year 1: $\frac{4}{10} \times \$44,000 = \$17,600$ depreciation

Year 2: $\frac{3}{10} \times \$44,000 = \$13,200$ depreciation

Computing Depreciation with the Modified Accelerated Cost Recovery System

Businesses use the depreciation methods previously described for financial reporting. However, federal tax laws regulate how depreciation must be taken for income tax purposes. The IRS requires that the **Modified Accelerated Cost Recovery System (MACRS)** be used for depreciation of property purchased and put into service after 1986. MACRS "recovers" the entire cost of depreciable property over the allowable period. No scrap value is permitted.

For common business assets, MACRS provides depreciation periods of 3, 5, 7, 10, 15, and 20 years. Examples of assets from each of these categories are as follows:

- 3 years: Property with a life of 4 years or less—some types of equipment used for research and development, some machine tools, some tractors, and racehorses more than 2 years old when placed in service.
- 5 years: Property with a life of 4 to 10 years—computers, automobiles and taxis, office machines, certain telephone equipment, and trucks and buses.
- 7 years: Property with a life of 10 to 15 years—office furniture and fixtures, some agricultural and horticultural structures, and commercial airplanes.
- 10 years: Property with a life of 16 to 19 years—tugboats, vessels, and barges.
- 15 years: Property with a life of 20 to 24 years—this category usually contains certain municipal, public utility, and telephone distribution plants.
- 20 years: Property with a life of 25 or more years—farm buildings and certain municipal infrastructure items such as sewers.

Figure 18-1 shows IRS annual percentages used to compute depreciation by MACRS.

Learning Objective

Compute depreciation for income tax purposes using the Modified Accelerated Cost Recovery System (MACRS).

18.3 Some businesses compute depreciation for their assets twice: once for business purposes, using one of the first three methods previously explained, and once for tax purposes, using MACRS. Doing so is perfectly legal so long as the business always reports annual depreciation for IRS tax purposes on the basis of MACRS.

18.4 Current-year MACRS depreciation tables are published annually by the IRS and will usually be mailed in response to a phone request to a local IRS office. The IRS provides a number of tables for MACRS, such as the one shown in Figure 18-1. The computations provided in the examples illustrate the correct mathematical procedures. Be sure to consult current tax law for appropriate tables, rates for other assets, and additional rules, such as maximum annual depreciation amounts for luxury automobiles.

Figure 18-1 | MACRS Depreciation Schedule

Appropriate Percentage

Year	3-Year Class	5-Year Class	7-Year Class	10-Year Class	15-Year Class	20-Year Class
1	33.33	20.00	14.29	10.00	5.00	3.750
2	44.45	32.00	24.49	18.00	9.50	7.219
3	14.81	19.20	17.49	14.40	8.55	6.677
4	7.41	11.52	12.49	11.52	7.70	6.177
5		11.52	8.93	9.22	6.93	5.713
6		5.76	8.92	7.37	6.23	5.285
7			8.93	6.55	5.90	4.888
8			4.46	6.55	5.90	4.522
9				6.56	5.91	4.462
10				6.55	5.90	4.461
11				3.28	5.91	4.462
12					5.90	4.461
13					5.91	4.462
14					5.90	4.461
15					5.91	4.462
16					2.95	4.461
17						4.462
18						4.461
19						4.462
20						4.461
21						2.231

Note: The MACRS percentage for the first year is applicable to a partial or full year.

EXAMPLE H

Use the MACRS Depreciation Schedule shown in Figure 18-1 to depreciate the office computer in example A for tax purposes.

Year	Rate (%)	Cost	Depreciation (Rounded)	Beginning Book Value	Current Depreciation	Ending Book Value
1	20.00	× $12,500 =	$2,500	$12,500	− $2,500	= $10,000
2	32.00	× 12,500 =	4,000	10,000	− 4,000	= 6,000
3	19.20	× 12,500 =	2,400	6,000	− 2,400	= 3,600
4	11.52	× 12,500 =	1,440	3,600	− 1,440	= 2,160
5	11.52	× 12,500 =	1,440	2,160	− 1,440	= 720
6	5.76	× 12,500 =	720	720	− 720	= 0

CONCEPT CHECK 18.4

On April 10, Oakdale Appliances bought a new delivery truck for $48,000. Compute the depreciation for the first year and for the second year using the MACRS table (5-year class).

MACRS depreciation first year: $48,000 \times 20.00\% = \$9,600$ (effectively allows for $\frac{1}{2}$ year's depreciation)

MACRS depreciation second year: $48,000 \times 32.00\% = \$15,360$ (full year's depreciation)

Computing Partial-Year Depreciation

Frequently, businesses are faced with the need to compute depreciation for only part of the year. Partial-year depreciation can be computed with any of the methods described in this chapter.

With the straight-line method, compute the depreciation amount for a partial year by dividing the annual depreciation amount by 12 and then multiplying that result by the number of months of use.

With the units-of-production method, simply multiply the number of units (miles or hours) used by the per-unit amount.

With the declining-balance method, find the current year's annual depreciation and then divide by 12; multiply that result by the number of months of use.

With the sum-of-the-years-digits method, first consider the overlapping years. To find the annual depreciation for the first partial year, divide by 12 and multiply the result by the number of months of use. From then on, every year will include the remaining fraction of the prior year's depreciation and the partial-year depreciation for the remainder of the current year.

MACRS tables automatically consider partial-year depreciation for the first and last years regardless of the date the item was placed in service.

Learning Objective 6

Compute partial-year depreciation using the five primary different depreciation methods covered.

EXAMPLE 1

Office furniture costing $18,000 and put in use on May 1 is expected to have a useful life of 10 years. Its estimated resale value is $1,500. Using each of the four methods, compute the depreciation expense for May 1 through December 31 of the first tax year and all 12 months of the second year.

Method	Year	Calculation (rounded to the nearest dollar)
SL	1st	$(\$18,000 - \$1,500) \div 10 \times \frac{8}{12} = \$1,100$
	2nd	$(\$18,000 - \$1,500) \div 10 = \$1,650$
DB (200%)	1st	$\left(\frac{100\%}{10}\right) \times 2 \times \$18,000 \times \frac{8}{12} = \$2,400$
	2nd	$(\$18,000 - \$2,400) \times 20\% = \$3,120$

Method	Year	Calculation (rounded to the nearest dollar)
SYD	1st	$\dfrac{(10+1) \times 10}{2} = 55$
		$(\$18{,}000 - \$1{,}500) \times \dfrac{10}{55} \times \dfrac{8}{12} = \$2{,}000$
	2nd	$(\$18{,}000 - \$1{,}500) \times \dfrac{10}{55} \times \dfrac{4}{12} = \$1{,}000$
		$(\$18{,}000 - \$1{,}500) \times \dfrac{9}{55} \times \dfrac{8}{12} = \$1{,}800$
MACRS (7-year class)	1st	$\$18{,}000 \times 14.29\% = \$2{,}572.20$
	2nd	$\$18{,}000 \times 24.49\% = \$4{,}408.20$

$\$2{,}800$

✓ CONCEPT CHECK 18.5

In October, Oakdale Appliances bought a new mid-size van for $34,000. It had an estimated scrap value of $4,000 and useful life of 5 years. Compute the depreciation expense for the 3 months of the first year and for the full second year, using the 150%-declining-balance and the sum-of-the-years-digits methods.

$34,000 cost − $4,000 scrap value = $30,000 to be depreciated

Declining Balance

$100\% \div 5 \text{ years} \times 1.5 = 30\%$

$30\% \times \$34{,}000 = \$10{,}200$

Year 1: $\$10{,}200 \times \dfrac{3}{12} = \$2{,}550$ (3 months)

Year 2: $(\$34{,}000 - \$2{,}550) \times 30\% = \$9{,}435$ (full year)

Sum of the Years Digits

$\dfrac{(5+1) \times 5}{2} = 15$ (or $1 + 2 + 3 + 4 + 5 = 15$)

$\dfrac{5}{15} \times \$30{,}000 = \$10{,}000$

Year 1: $\$10{,}000 \times \dfrac{3}{12} = \$2{,}500$ (3 months)

Year 2: $\$10{,}000 \times \dfrac{9}{12} = \$7{,}500$ (9 months)

$\dfrac{4}{15} \times \$30{,}000 = \$8{,}000$

$\$8{,}000 \times \dfrac{3}{12} = \$2{,}000$ (3 months)

$\$7{,}500 + \$2{,}000 = \$9{,}500$ in year 2

COMPLETE ASSIGNMENTS 18.1 AND 18.2.

Chapter Terms for Review

accumulated depreciation

basic depreciation rate

book value

declining-balance (DB) method

declining-balance depreciation rate

depreciation

double-declining-balance

estimated service life

Modified Accelerated Cost Recovery System (MACRS)

obsolescence

150%-declining-balance

original cost

scrap value (SV)

straight-line (SL) method

sum-of-the-years-digits (SYD) method

units-of-production method

THE BOTTOM LINE

Review of chapter learning objectives:

Learning Objective	Example
18.1 Compute depreciation, using the straight-line method	1. On January 1, 2000, the local Pepsi-Cola bottling franchise purchased a bottling machine for $320,000. Freight was added for $12,000. The cost of installation was $68,000. It was estimated that the machine could be used for 80,000 hours, after which there would be no resale value. The machine was used 4,600 hours the first year, 4,300 hours the second year, and 5,200 hours the third year. Determine the straight-line depreciation per year based on the hours of use and the book value at the end of each year.
18.2 Compute depreciation using the units-of-production method	2. The Yellow Cab Company bought a new taxi for $42,000 and estimated its useful life to be 200,000 miles, after which it would have a scrap value of $2,000. Compute the depreciation for the first 7 months if the vehicle had been driven 37,600 miles.
18.3 Compute depreciation, using the declining-balance method	3. For $56,000, a Gap clothing store bought display racks with an estimated life of 20 years and a scrap value of $4,000. After 3 years, this store closed and sold the display racks for $32,000. If the racks were depreciated by the declining-balance method (150% annual rate), how much less than the book value did the company receive? Round to the nearest dollar.
18.4 Compute depreciation, using the sum-of-the-years-digits method	4. A local Ford dealership purchased, for $60,000, a hydraulic lift unit with an estimated life of 7 years and a scrap value of $4,000. Compute the depreciation for each of the first 2 years using the sum-of-the-digits method. Round to the nearest dollar.
18.5 Compute depreciation for income tax purposes, using the Modified Accelerated Recovery Systems (MACRS)	5. Bank One bought new calculators in July for $12,000. Using the MACRS method (5-year class), show the rate, depreciation, and ending book value for the first 2 years.
18.6 Compute partial-year depreciation using the four different depreciation methods covered	6. On October 1, 2001, Corner Grocery bought and installed a new cash register for $1,400. It has an estimated service life of 6 years and an estimated scrap value of $200. The company decided to use the straight-line method of depreciation. What was the depreciation for 2001? What was it for 2002?

Answers: 1. 2000: $23,000/$377,000; 2001: $21,500/$355,500; 2002: $26,000/$329,500 2. $7,520 3. $12,321 4. First year: $14,000; Second year: $12,000 5. First year: 20.00% rate, $2,400 depreciation, $9,600 EBV; Second year: 32.00% rate, $3,840 depreciation, $5,760 EBV 6. 2001: $50; 2002: $200

SELF-CHECK

Review Problems for Chapter 18

1 Determine the annual declining-balance depreciation rate to be used for each of the following:

  a. 150% declining balance, 12-year life _____

  b. 200% declining balance, 8-year life _____

  c. 125% declining balance, 5-year life _____

  d. 200% declining balance, 5-year life _____

2 What fraction is to be used each year for sum-of-the-years-digits depreciation for an asset with a useful life of 4 years? _____

3 For which depreciation method(s) is salvage value *not* subtracted to calculate depreciation? _____

4 Lopez Construction Company purchased construction equipment for $116,000 at the beginning of the year. It is estimated that the equipment will have a useful life of 12 years and will have a scrap value of $8,000.

  a. Calculate the annual depreciation if Lopez uses straight-line depreciation. _____

  b. Calculate the book value of the equipment at the end of 5 years, assuming that Lopez uses straight-line depreciation. _____

  c. Compute the depreciation for the first year ending December 31 if Lopez purchased the equipment September 1. _____

  d. Determine the depreciation per hour if Lopez uses the straight-line method based on 120,000 hours of useful life and an $8,000 scrap value. _____

  e. Using the rate determined in (d), what is the depreciation for the year if the equipment is used for 2,360 hours? _____

5 Jurgenson Manufacturing uses the double-declining-balance method of depreciation. A piece of equipment costing $37,500 has an estimated useful life of 5 years and an estimated scrap value of $2,700.

  a. Compute the amount of depreciation taken in the second year. _____

  b. What is the book value at the end of the second year? _____

6 Young Manufacturing uses the sum-of-the-years-digits method of depreciation. Equipment costing $37,500 has an estimated life of 5 years and an estimated scrap value of $2,700.

  a. Compute the amount of depreciation expense for the second year. _____

  b. What is the book value at the end of the second year? _____

7 Calculate the depreciation expense for tax purposes using MACRS for each asset. Use Figure 18-1 on page 370 to determine the proper life and rate for each asset. (Round to the nearest dollar.)

  a. Computer equipment purchased this year for $5,200. _____

  b. Office furniture purchased 2 years ago for $8,500. (This is the third year.) _____

Notes

Assignment 18.1: Business Depreciation

Name

Date   Score

Learning Objectives 1 2 3 4

A (30 points) Solve the following depreciation problems. (points for correct answers as marked)

1. A pharmaceutical company has testing machines on which it estimates depreciation by the straight-line method. The following table shows cost, estimated life, years used, and scrap value for each machine. Find the annual depreciation, total depreciation, and book value after the indicated number of years of use. ($\frac{1}{2}$ point for each correct answer)

	Original Cost	Estimated Life (years)	Years Used	Scrap Value	Annual Depreciation	Total Depreciation to Date	Book Value
a.	$30,000	10	4	$3,000	$2,700	$10,800	$19,200
b.	48,000	7	4	$5,300	$6,100	$24,400	$23,600
c.	84,000	8	2	none	$10,500	$21,000	$63,000
d.	34,600	6	2	$1,000	$5,600	$11,200	$23,400

Illustrative solution:

a. ($30,000 cost − $3,000 SV) ÷ 10 years = $2,700 annual depreciation

$2,700 × 4 years = $10,800 total depreciation to date

$30,000 − $10,800 = $19,200 book value

2. Ace Delivery Service bought two new trucks. The following table shows the cost, scrap value, estimated life (in miles), and mileage for the first year. Using the straight-line method based on mileage driven, compute the first year's depreciation and the book value at the end of the first year for each truck. (2 points for each correct depreciation amount and 1 point for each correct book value)

	Original Cost	Scrap Value	Estimated Life (miles)	Mileage for First Year	Depreciation for First Year	Book Value after 1 Year
a.	$49,500	$1,500	150,000	21,700	$6,944	$42,556
b.	$23,000	$ 600	80,000	9,500	$2,660	$20,340

Illustrative solution:

a. $49,500 cost − $1,500 scrap value = $48,000 to be depreciated; $48,000 ÷ 150,000 miles = $0.32 depreciation per mile; 21,700 miles × $0.32 = $6,944 depreciation first year; $49,500 original cost − $6,944 = $42,556 book value after 1 year

3. Dole Fruit Company's equipment cost $214,000. Its useful life is estimated to be 15 years, and its scrap value is $4,000. The company uses straight-line depreciation. (2 points for each correct answer)

a. What is the annual depreciation? $14,000

$214,000 − $4,000 = $210,000 total depreciation

$210,000 ÷ 15 years = $14,000 annual depreciation

b. What is the book value of the equipment at the end of 14 years? $18,000

$14,000 × 14 years = $196,000

$214,000 − $196,000 = $18,000 book value after 14 years

Chapter 18 Depreciation

Assignment 18.1 Continued

4. Carlucci and sons purchased a machine for $13,645 at the beginning of the year. Additional costs included $250 freight and $175 for installation. It was estimated that the machine could be operated for 30,000 hours, after which its resale value would be $570. Determine the straight-line depreciation based on hours of operation and the book value at the end of each of the first 7 years. (1 point for each correct answer)

Year	Hours of Operation	Depreciation	Book Value
1	2,300	$1,035.00	$13,035.00
2	2,750	$1,237.50	$11,797.50
3	2,500	$1,125.00	$10,672.50
4	2,480	$1,116.00	$9,556.50
5	2,800	$1,260.00	$8,296.50
6	3,100	$1,395.00	$6,901.50
7	2,950	$1,327.50	$5,574.00

Illustrative solution:
Year 1: $13,645 + $250 + $175 = $14,070 full cost
$14,070 − $570 = $13,500 total depreciation
$13,500 ÷ 30,000 = $0.45 depreciation per hour
2,300 × $0.45 = $1,035 depreciation first year
$14,070 − $1,035 = $13,035 book value end of first year
Year 2: 2,750 × $0.45 = $1,237.50 depreciation second year
$13,035 − $1,237.50 = $11,797.75 book value end of second year

Score for A (30)

B (56 points) Solve the following depreciation problems. Round dollar amounts to two decimal places. (points for correct answers as marked)

5. Anderson Tool and Die Company owns a group of machines, the details of which are shown in the following table. Anderson uses the double-declining-balance method of calculating depreciation. Compute the depreciation for the specific years indicated. (2 points for each correct answer)

	Original Cost	Estimated Life (years)	Scrap Value	Year	Depreciation	Year	Depreciation
a.	$32,000	16	$1,200	1	$4,000.00	3	$3,062.50
b.	$25,800	5	$3,000	3	$3,715.20	5	$343.68
c.	$ 8,000	4	$ 300	2	$2,000.00	3	$1,000.00
d.	$15,000	10	—	3	$1,920.00	5	$1,228.80
e.	$12,600	8	$1,200	2	$2,362.50	4	$1,328.91
f.	$95,000	20	—	3	$7,695.00	5	$6,232.95

Illustrative solutions:
a. 100% ÷ 16 years = 6.25% × 2 = 12.5% annual rate
$32,000 × 12.5% = $4,000 depreciation 1st year
$32,000 − $4,000 = $28,000 new balance
$28,000 × 12.5% = $3,500 depreciation 2nd year
$28,000 − $3,500 = $24,500 new balance
$24,500 × 12.5% = $3,062.50 depreciation 3rd year

b. 100% ÷ 5 yrs = 20% × 2 = 40% annual rate
$25,800 × 40% = $10,320 depreciation 1st year
$25,800 − $10,320 = $15,480 new balance
$15,480 × 40% = $6,192 depreciation 2nd year
$15,480 − $6,192 = $9,288 new balance
$9,288 × 40% = $3,715.20 depreciation 3rd year
$9,288 − $3,715.20 = $5,572.80 new balance
$5,572.80 × 40% = $2,229.12 depreciation 4th year
$5,572.80 − $2,229.12 = $3,343.68 new balance
$3,343.68 − $3,000 scrap value = $343.68 depreciation 5th year
(40% in year 5 would decrease book value below scrap value)

Assignment 18.1 Continued

6. Machinery purchased from Telecom, Inc., by Blazedales cost $69,800. Depreciation was determined by the double-declining-balance method for an estimated life of 16 years. Compute the following:

   a. Book value after 4 years (8 points): $40,915.47

   100% ÷ 16 = 6.25%; 6.25% × 2 = 12.5% annual rate
   $69,800 × 0.125 = $8,725 depreciation first year
   $69,800 − $8,725 = $61,075 new balance
   $61,075 × 0.125 = $7,634.38 depreciation second year
   $61,075.00 − $7,634.38 = $53,440.62 new balance
   $53,440.62 × 0.125 = $6,680.08 depreciation third year
   $53,440.62 − $6,680.08 = $46,760.54 new balance
   $46,760.54 × 0.125 = $5,845.07 depreciation fourth year
   $46,760.54 − $5,845.07 = $40,915.47 book value after 4 years

   b. Total depreciation after 6 years (4 points): $38,474.09

   $40,915.47 × 0.125 = $5,114.43 depreciation fifth year
   $40,915.47 − $5,114.43 = $35,801.04 new balance
   $35,801.04 × 0.125 = $4,475.13 depreciation sixth year
   $35,801.04 − $4,475.13 = $31,325.91 new balance
   $69,800 − $31,325.91 = $38,474.09 total depreciation after 6 years

7. The Dugan Manufacturing Company bought an engine for $31,500. The engine had an estimated life of 20 years and a scrap value of $5,250. After 6 years, the company went out of business and sold the engine for $15,200. If the machine was depreciated by the double-declining-balance method, how much did the company lose on the sale (the difference between the book value and the selling price)? (20 points) $1,540.39

   100% ÷ 20 = 5%; 5% × 2 = 10% annual rate
   $31,500 × 0.10 = $3,150 depreciation first year
   $31,500 − $3,150 = $28,350 new balance
   $28,350 × 0.10 = $2,835 depreciation second year
   $28,350 − $2,835 = $25,515 new balance
   $25,515 × 0.10 = $2,551.50 depreciation third year
   $25,515 − $2,551.50 = $22,963.50 new balance
   $22,963.50 × 0.10 = $2,296.35 depreciation fourth year
   $22,963.50 − $2,296.35 = $20,667.15 new balance
   $20,667.15 × 0.10 = $2,066.72 depreciation fifth year
   $20,667.15 − $2,066.72 = $18,600.43 new balance
   $18,600.43 × 0.10 = $1,860.04 depreciation sixth year
   $18,600.43 − $1,860.04 = $16,740.39 book value at end of sixth year
   $16,740.39 − $15,200 = $1,540.39 loss

Score for B (56)

Assignment 18.1 Continued

C **(14 points) Solve the following depreciation problems. (1 point for each correct answer)**

8. The Western Salvage Service bought three trucks. The following table shows the cost, estimated life, and resale estimate for each truck. Use the sum-of-the-years-digits method to find each truck's depreciation for the first and second years of use. Round answers to the nearest dollar.

	Original Cost	Estimated Life	Resale Estimate	Depreciation for First Year	Depreciation for Second Year
a.	$36,000	6 yr	$6,000	$8,571	$7,143
b.	$48,000	5 yr	8,000	$13,333	$10,667
c.	$60,000	7 yr	12,000	$12,000	$10,286

Illustrative solution:

a. $\dfrac{(6+1)6}{2} = 21$, or $1 + 2 + 3 + 4 + 5 + 6 = 21$

$36,000 - $6,000 = $30,000 total depreciation

First year: $\dfrac{6}{21} \times $30,000 = $8,571$ Second year: $\dfrac{5}{21} \times $30,000 = $7,143$

9. Use the information in problem 8b to compute the amount of depreciation for years 3–5.

Year 3: $8,000
Year 4: $5,333
Year 5: $2,667

Year 3: $\dfrac{3}{15} \times $40,000 = $8,000$

Year 4: $\dfrac{2}{15} \times $40,000 = $5,333$

Year 5: $\dfrac{1}{15} \times $40,000 = $2,667$

10. Use the information in problem 8 to compute the amount of depreciation for each vehicle for the first 2 years using the straight-line method. Round to the nearest dollar.

a. $10,000   b. $16,000   c. $13,714

a. ($36,000 − $6,000) ÷ 6 = $5,000; $5,000 × 2 = $10,000
b. ($48,000 − $8,000) ÷ 5 = $8,000; $8,000 × 2 = $16,000
c. ($60,000 − $12,000) ÷ 7 = $6,857; $6,857 × 2 = $13,714

11. Which method of depreciation would give the smaller amount of write-off, and how much less would it be for the three vehicles for the first 2 years? straight-line, $22,286

SYD: ($8,571 + $7,143) + ($13,333 + $10,667) + ($12,000 + $10,286) = $62,000

SL: $10,000 + $16,000 + $13,714 = $39,714

Difference: $62,000 − $39,714 = $22,286

Score for C (14)

Assignment 18.2: Business Depreciation

Name

Date   Score

Learning Objectives **1 4 5 6**

A (43 points) Solve the following depreciation problems. Round dollar amounts to two decimal places. (points for correct answers as marked)

1. An architect bought drafting equipment for $7,500. Its estimated life was 6 years, and its scrap value was $300. At the end of 4 years, the equipment wears out and is sold for scrap for $225. (4 points for each correct answer)

    a. By the straight-line method, how much difference is there between the book value and the cash value of the equipment on the date of the sale? __$2,475__
    $7,500 − $300 = $7,200 total depreciation
    $7,200 ÷ 6 years = $1,200 annual depreciation
    $1,200 × 4 years = $4,800
    $7,500 − $4,800 = $2,700 book value
    $2,700 − $225 = $2,475

    b. In April 2000, a computer and software costing $18,000 are purchased. Its estimated life is 5 years. What is the book value of the new computing equipment on December 31, 2001? Use MACRS. __$8,640__
    20% + 32% = 52% recovery for 2000 and 2001
    $18,000 × 0.52 = $9,360 depreciation
    $18,000 − $9,360 = $8,640 book value

2. E, F, and G were partners in a small textile company. They spent $54,000 for equipment that they agreed would last 8 years and have a resale value of 5% of cost. The three partners couldn't agree on the depreciation method to use. E was in favor of using the double-declining-balance system, F insisted on the 150%-declining-balance method, and G was sure that the sum-of-the-years-digits method would be better. Show the depreciation for the first 4 years for each method in the following table. At the end of 4 years, what would be the book value under each of the three methods? (2 points for each correct depreciation amount, 1 point for each correct total, and 1 point for the correct book value)

Year	Double-DB	150%-DB	SYD
1	$13,500.00	$10,125.00	$11,400
2	$10,125.00	$8,226.56	$9,975
3	$7,593.75	$6,684.08	$8,550
4	$5,695.31	$5,430.82	$7,125
Total	$36,914.06	$30,466.46	$37,050
Book value	$17,085.94	$23,533.54	$16,950

3. Baxter Company owned assets that cost $100,000. Depreciation was figured at a straight-line rate of 4% per year. After 12 years, the company sold the assets for $60,000. How much greater was the selling price than the book value at the time of the sale? (5 points) __$8,000__
    $100,000 × 0.04 = $4,000 depreciation per year
    $4,000 × 12 years = $48,000 depreciation for 12 years
    $100,000 − $48,000 = $52,000 book value after 12 years
    $60,000 − $52,000 = $8,000

Score for A (43)

Assignment 18.2 Continued

B (57 points) Solve the following depreciation problems. (points for correct answers as marked)

4. On March 1, Jarvis Realty spent $16,000 for a new company car with an estimated life of 4 years and an estimated scrap value of $4,000. Jarvis Realty elected to use the straight-line method for depreciation. On the same date, Carter Realty bought an identical car at the same price and also estimated the car's life and scrap value to be 4 years and $4,000, respectively. Carter Realty, however, chose the sum-of-the-years-digits method for depreciation.

   a. At the end of the first year (10 months of use) and second year, how much depreciation did each company calculate? (3 points for each correct answer)

   Jarvis: Year 1 $2,500   Carter: Year 1 $4,000
            Year 2 $3,000   Year 2 $3,800

   Jarvis (SL)   Year 1: $($16,000 - $4,000) \times \frac{1}{4} \times \frac{10}{12} = $2,500$

                 Year 2: $$12,000 \times \frac{1}{4} = $3,000$

   Carter (SYD) Year 1: $($16,000 - $4,000) \times \frac{4}{10} \times \frac{10}{12} = $4,000$

                 Year 2: $\left($12,000 \times \frac{4}{10} \times \frac{2}{12}\right) + \left($12,000 \times \frac{3}{10} \times \frac{10}{12}\right) = $3,800$

   b. At the end of the second year, which company had more recorded accumulated depreciation, and what was the difference in the amounts? (5 points) Carter Realty, by $2,300

   ($4,000 + $3,800) − ($2,500 + $3,000) = $2,300

   c. True or false: At the end of the fourth year, Carter Realty will have recorded more accumulated depreciation than Jarvis Realty. Explain your answer. (4 points) false
   At the end of the fourth year, the estimated life of both cars will be over. Each company will have recorded the same amount of accumulated depreciation ($16,000 cost − $4,000 SV = $12,000)

5. In May 2001, Jian & Ming bought a light-duty truck for $20,800. One year later, they bought an additional truck for $21,800. In June 2003, a third truck was purchased for $23,500. Use MACRS (5-year class) to determine the total allowable cost recovery for 2003. (12 points) $15,670

   2001 truck: $20,800 × 0.1920 = $ 3,993.60   or $ 3,994
   2002 truck: $21,800 × 0.32 =   6,976.00   or   6,976
   2003 truck: $23,500 × 0.20 =   4,700.00   or   4,700
   Total cost recovery for 2003: $15,669.60   or $15,670

6. David Marcus purchased new office furniture July 15, 2000, for $28,100. Use MACRS (7-year class) to show the rate, depreciation, and beginning and ending book values for 2000, 2001, and 2002. Round to the nearest dollar. (2 points for each correct answer)

Year	Rate		Cost		Depreciation	Beginning Book Value	Ending Book Value
2000	14.29%	×	$28,100	=	$4,015	$28,100	$24,085
2001	24.49%	×	$28,100	=	$6,882	$24,085	$17,203
2002	17.49%	×	$28,100	=	$4,915	$17,203	$12,288

Score for B (57)

Financial Statements

19

Learning Objectives
By studying this chapter and completing all assignments you will be able to:

Learning Objective 1 — Analyze balance sheets, comparing items and periods.

Learning Objective 2 — Analyze income statements, comparing items and periods.

Learning Objective 3 — Compute commonly used business operating ratios.

19.1 On the balance sheet, *current assets* are cash and those readily salable securities, short-term receivables, and inventory that are expected to be converted into cash or used up within 1 year of the balance sheet date. *Current liabilities* are those payment obligations that are expected to be paid off within 1 year of the balance sheet date.

Financial statements provide information that allows owners, managers, and others interested in a business to evaluate its current condition and past operating results. Two important financial statements are the balance sheet and income statement. The **balance sheet** shows the current condition of a business at a definite point in time. It lists what a business owns (**assets**), how much it owes (**liabilities**), and the difference between the two (**net worth**), usually referred to as owners' or shareholders' equity. The **income statement** shows the past operating results for a given period of time. It lists the revenues, the expenses, and the net income or loss for the period.

Financial statement data are typically analyzed three ways. The first, called *horizontal analysis*, is a comparison of data from year to year. This analysis shows the dollar amount of change and the percent of change for each item on the statement from one year to the next. The second, called *vertical analysis*, compares all other data on a statement with one figure for that same year. On the balance sheet, for example, each asset, liability, and equity amount is calculated as a percent of total assets (or total liabilities and owners' equity). The third type of analysis compares selected related data for the year such as current assets to current liabilities. These analyses are used by managers, owners, investors, creditors, and others to help them analyze and simplify the complex data and make decisions concerning the business.

Analyzing Balance Sheets

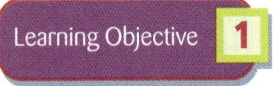

Analyze balance sheets, comparing items and periods.

On a balance sheet, total assets must always equal total liabilities plus owners' or shareholders' equity. Balance sheets are analyzed to compare individual items with other items and with the same item on different dates, usually 1 year apart. Many businesses use the form of balance sheet illustrated in Figure 19-1.

Figure 19-1 Balance Sheet

THE SKI CHALET
Balance Sheet as of December 31, 2004 and 2003

	2004 Amount	2004 Percent	2003 Amount	2003 Percent	Increase/Decrease Amount	Increase/Decrease Percent
ASSETS						
Current assets:						
Cash	$ 90,000	12.64%	$ 86,000	13.15%	$ 4,000	4.65%
Accounts receivable	134,000	18.82%	98,000	14.98%	36,000	36.73%
Notes receivable	28,000	3.93%	32,000	4.89%	(4,000)	−12.50%
Merchandise inventory	180,000	25.28%	148,000	22.63%	32,000	21.62%
Total current assets	$432,000	60.67%	$364,000	55.66%	$ 68,000	18.68%
Fixed assets:						
Equipment	$220,000	30.90%	$190,000	29.05%	$ 30,000	15.79%
Less depreciation	(60,000)	−8.43%	(50,000)	−7.65%	(10,000)	20.00%
Equipment net	$160,000	22.47%	$140,000	21.41%	$ 20,000	14.29%
Buildings	300,000	42.13%	300,000	45.87%	—	0.00%
Less depreciation	(180,000)	−25.28%	(150,000)	−22.94%	(30,000)	−20.00%
Buildings net	$120,000	16.85%	$150,000	22.94%	$(30,000)	−20.00%
Total fixed assets	$280,000	39.33%	$290,000	44.34%	$(10,000)	−3.45%
TOTAL ASSETS	$712,000	100.00%	$654,000	100.00%	$ 58,000	8.87%

384 Part 5 Business Applications

Figure 19-1 Balance Sheet (continued)

	2004 Amount	2004 Percent	2003 Amount	2003 Percent	Increase/Decrease Amount	Increase/Decrease Percent
LIABILITIES AND SHAREHOLDERS' EQUITY						
Current liabilities:						
Accounts payable	$ 18,000	2.53%	$ 24,000	3.67%	$ (6,000)	−25.00%
Accrued payroll	38,000	5.34%	30,000	4.59%	8,000	26.67%
Payroll taxes payable	6,000	0.84%	4,000	0.61%	2,000	50.00%
Notes payable	42,000	5.90%	48,000	7.34%	(6,000)	−12.50%
Total current liabilities	$104,000	14.61%	$106,000	16.21%	$ (2,000)	−1.89%
Long-term liabilities:						
Mortgage payable	$ 90,000	12.64%	$120,000	18.35%	$(30,000)	−25.00%
Notes payable (over 1 year)	36,000	5.06%	30,000	4.59%	6,000	20.00%
Total long-term liabilities	$126,000	17.70%	$150,000	22.94%	$(24,000)	−16.00%
Total liabilities	$230,000	32.30%	$256,000	39.14%	$(26,000)	−10.16%
Shareholders' equity:						
Common stock	$359,000	50.42%	$359,000	54.89%	—	0.00%
Retained earnings	123,000	17.28%	39,000	5.96%	84,000	215.38%
Total shareholders' equity	$482,000	67.70%	$398,000	60.86%	$ 84,000	21.11%
TOTAL LIABILITIES AND SHAREHOLDERS' EQUITY	$712,000	100.00%	$654,000	100.00%	$ 58,000	8.87%

In Figure 19-1, the amounts for various items such as cash and accounts payable are compared to total assets and total liabilities and shareholders' equity. Also, the amounts for 2004 are compared to the corresponding amounts for 2003, and the amounts and percents of increase or decrease are shown. When two statements are compared, the earlier period, usually the prior year, is *always* used as the base. The changes in balance sheet items between two periods measure the growth or decline of the business.

The first step in analyzing a balance sheet is to compute the percent each item is of the total assets or of the total liabilities and shareholders' equity (net worth). For example, the percent of cash for 2004 is calculated by dividing the amount of cash for 2004 by the total assets for 2004 and then converting the resulting decimal answer to a percent ($90,000 ÷ $712,000 = 0.1264 = 12.64%).

The second step is to compute the amount and percent of change between the two dates being compared. The amount of change in cash from 2003 to 2004 is calculated by subtracting the cash amounts for the two years ($90,000 − $86,000 = $4,000 increase). Increases are shown as positive numbers. Decreases, negative changes, are shown in parentheses. The percent of change in cash is calculated by dividing the amount of change by the prior year's amount ($4,000 ÷ $86,000 = 0.0465 = 4.65%).

Note three facts:

1. The totals for assets equal the totals for liabilities and shareholders' equity.
2. The percent listed for each item under assets is of the total assets; the percent listed for each item under liabilities and shareholders' equity is of the total liabilities and shareholders' equity.
3. The percent of increase or decrease between the two years is based on 2003, the *earlier* year.

CONCEPT CHECK 19.1

In its next year, 2005, The Ski Chalet had total assets of $720,000, total liabilities of $245,000, cash of $123,000, and mortgage payable of $60,000. Determine the following amounts and percents.

a. What was its total shareholders' equity in 2005?
$720,000 − $245,000 = $475,000

b. What was its balance sheet percent of cash?
$123,000 ÷ $720,000 = 17.08%

c. What was its balance sheet percent of mortgage payable?
$60,000 ÷ $720,000 = 8.33%

d. What was its percent of increase in cash?
($123,000 − $90,000) ÷ $90,000 = 36.67%

e. What was its percent of decrease in mortgage payable?
($90,000 − $60,000) ÷ $90,000 = 33.33%

Analyzing Income Statements

Learning Objective 2

Analyze income statements, comparing items and periods.

19.2 The income statement is commonly referred to as the *profit and loss statement* or the *P&L*.

19.3 Often, companies in the same trade or industry (or trade associations) compile and distribute industry-average percentage figures, which allow individual companies to compare their own percents with the industry averages. Almost all companies that have many branches derive percentage figures and compare each branch against all others.

The income statement shows revenue, expenses, and the difference between the two, or net income. Income statements are analyzed by comparing all other statement items with the **net revenue**, which is total revenue less any returns and allowances. Net revenue (frequently called net sales) is always 100%. All other items on the income statement are reported as a percent of net revenue/sales. The resulting percents are extremely important for all businesspeople. They are compared to budgeted amounts, to percents for competing businesses, and to percents for past periods.

Figure 19-2 shows a typical income statement for 1 year, in which dollar amounts are converted to percents based on net sales. Percents are rounded to two decimal places, and dollar amounts are rounded to the nearest whole dollar. Cents are seldom used in reporting annual figures.

Figure 19-2 | Income Statement

THE SKI CHALET
Income Statement for Year Ended December 31, 2004

	2004 Amount	2004 Percent
Revenue from sales:		
Sales	$ 988,900	101.43%
Less returns	13,900	1.43%
NET SALES	$ 975,000	100.00%

386 Part 5 Business Applications

Figure 19-2 | Income Statement (continued)

	2004 Amount	2004 Percent
Cost of goods sold:		
Inventory, January 1	$ 148,000	15.18%
Purchases	440,000	45.13%
Available for sale	$ 588,000	60.31%
Inventory, December 31	180,000	18.46%
Cost of goods sold	$ 408,000	41.85%
Gross profit on sales	$ 567,000	58.15%
Operating expenses:		
Salary and benefits	$ 290,000	29.74%
Rent and utilities	62,000	6.36%
Advertising	32,400	3.32%
Depreciation	40,000	4.10%
Equipment and supplies	15,800	1.62%
Administrative	12,500	1.28%
Total operating expense	$ 452,700	46.43%
Income before tax	$ 114,300	11.72%
Income tax	30,300	3.11%
NET INCOME	$ 84,000	8.62%

Most businesses want to compare the operations of the current year with those of the preceding year. The statement shown in Figure 19-3 has information for both the current and the preceding year. It also shows the amount and percent of increase or decrease from the preceding year.

Figure 19-3 | Comparative Income Statement

THE SKI CHALET
Income Statement for the Years Ended
December 31, 2004 and 2003

	2004 Amount	2004 Percent	2003 Amount	2003 Percent	Difference Amount	Difference Percent
Revenue from sales:						
Sales	$988,900	101.43%	$850,000	104.81%	$138,900	16.34%
Less returns	13,900	1.43%	39,000	4.81%	(25,100)	−64.36%
NET SALES	$975,000	100.00%	$811,000	100.00%	$164,000	20.22%
Cost of goods sold:						
Inventory, January 1	$148,000	15.18%	$152,000	18.74%	$ (4,000)	−2.63%
Purchases	440,000	45.13%	379,000	46.73%	61,000	16.09%
Available for sale	$588,000	60.31%	$531,000	65.47%	$ 57,000	10.73%
Inventory, December 31	180,000	18.46%	148,000	18.25%	32,000	21.62%
Cost of goods sold	$408,000	41.85%	$383,000	47.23%	$ 25,000	6.53%
Gross profit on sales	$567,000	58.15%	$428,000	52.77%	$139,000	32.48%

Figure 19-3 Comparative Income Statement (continued)

	2004 Amount	2004 Percent	2003 Amount	2003 Percent	Difference Amount	Difference Percent
Operating expenses:						
Salary and benefits	$290,000	29.74%	$242,000	29.84%	$ 48,000	19.83%
Rent and utilities	62,000	6.36%	61,400	7.57%	600	0.98%
Advertising	32,400	3.32%	25,700	3.17%	6,700	26.07%
Depreciation	40,000	4.10%	32,000	3.95%	8,000	25.00%
Equipment and supplies	15,800	1.62%	10,300	1.27%	5,500	53.40%
Administrative	12,500	1.28%	14,200	1.75%	(1,700)	−11.97%
Total operating expense	$452,700	46.43%	$385,600	47.55%	$ 67,100	17.40%
Income before tax	$114,300	11.72%	$42,400	5.23%	$ 71,900	169.58%
Income tax	30,300	3.11%	24,400	3.01%	5,900	24.18%
NET INCOME	$ 84,000	8.62%	$18,000	2.22%	$ 66,000	366.67%

19.4 Charitable, nonprofit, and government agencies also use budgets for planning and monitoring. Many of these budgets closely follow the sections and sequences of the business income statement.

Another analysis carried out by many businesses is a comparison between actual results and budgeted figures. Owners and managers note differences between budgeted and actual amounts and make adjustments where necessary. Most businesses and virtually all government entities use monthly and annual budgets to guide and monitor their operations. Figure 19-4 illustrates a monthly and year-to-date budget comparison at the end of June, the sixth month of the year.

To find the percent change, the budgeted amount is subtracted from the actual amount and the difference is divided by the *budgeted* amount.

Figure 19-4 Monthly/Year-to-Date Budget Comparison

THE SKI CHALET
Income Statement for the Month and the
Six-Month Period Ended June 30, 2004

	June 2004 Budget	June 2004 Actual	June 2004 Amount Difference	June 2004 Percent Difference	Six Months Year-to-Date Budget	Six Months Year-to-Date Actual	Six Months Year-to-Date Amount Difference	Six Months Year-to-Date Percent Difference
Revenue from sales:								
Sales	$85,000	$86,500	$ 1,500	1.76%	$510,000	$480,000	$(30,000)	−5.88%
Sales returns	5,000	3,500	$(1,500)	−30.00%	10,000	6,000	$ (4,000)	−40.00%
NET SALES	$80,000	$83,000	$ 3,000	3.75%	$500,000	$474,000	$(26,000)	−5.20%
Cost of goods sold	35,000	38,000	$ 3,000	8.57%	225,000	230,000	$ 5,000	2.22%
Gross profit	$45,000	$45,000	$ —	0.00%	$275,000	$244,000	$(31,000)	−11.27%
Operating expenses	31,000	39,000	$ 8,000	25.81%	185,000	196,000	$ 11,000	5.95%
Income before tax	$14,000	$ 6,000	$(8,000)	−57.14%	$ 90,000	$ 48,000	$(42,000)	−46.67%
Income tax	6,000	1,000	$(5,000)	−83.33%	40,000	16,000	$(24,000)	−60.00%
NET INCOME	$ 8,000	$ 5,000	$(3,000)	−37.50%	$ 50,000	$ 32,000	$(18,000)	−36.00%

CONCEPT CHECK 19.2

In its next year, 2005, The Ski Chalet had total sales of $1,480,000, net sales of $1,320,000, gross profit of $710,000, and advertising expense of $45,000. In 2005, the company budgeted gross profit of $800,000. Determine the following amounts and percents.

a. Amount of sales returns in 2005
   $1,480,000 − $1,320,000 = $160,000

b. Amount of cost of goods sold in 2005
   $1,320,000 − $710,000 = $610,000

c. Percent of net sales increase from 2004 to 2005
   ($1,320,000 − $975,000) ÷ $975,000 = 35.38%

d. Percent of advertising expense in 2005
   $45,000 ÷ $1,320,000 = 3.41%

e. Difference between percent gross profit and 2005 budgeted amount
   ($710,000 − $800,000) ÷ $800,000 = −11.25%

Computing Business Operating Ratios

In addition to comparing dollar amounts and percents on financial statements, business managers and owners frequently want to study relationships between various items on their income statements and balance sheets. These relationships generally are expressed by ratios. A **ratio** is the relation of one amount to another. Thus the ratio of one dollar to one quarter, or $1 to $0.25, is a ratio of 4 to 1, or 4:1, showing that a dollar is 4 times the value of a quarter.

In analyzing financial statements, six important financial analysis ratios are commonly used: the working capital ratio, the acid test ratio, the ratio of accounts receivable to net sales, the inventory turnover rate, the relation of net income to net sales, and the rate of return on investment (equity).

Learning Objective 3

Compute commonly used business operating ratios.

WORKING CAPITAL RATIO

Working capital and the working capital ratio come from the balance sheet. **Working capital** is the amount of current assets less current liabilities. It tells the amount of current assets that would remain if all the company's current liabilities were paid immediately. The **working capital ratio** shows the relationship between current assets and current liabilities. It calculates the amount of current assets per dollar of current liabilities. The working capital ratio helps the reader of the balance sheet understand how well the company is able to pay its current debts.

Working capital ratio = Total current assets ÷ Total current liabilities

EXAMPLE A

The working capital ratio for The Ski Chalet for 2004 from Figure 19-1 is

$432,000 ÷ $104,000 = 4.2 = 4.2 : 1

The ratio 4.2 to 1, or 4.2:1, means that the business has $4.20 in current assets to pay for each $1 in current liabilities.

ACID TEST RATIO

The **acid test ratio** is used to determine the relationship between assets that can be quickly turned into cash and current liabilities. Usually, these assets are cash and accounts receivable. **Accounts receivable** are amounts owed to a business for services performed or goods delivered.

Acid test ratio = (Total of cash + Accounts receivable) ÷ Total current liabilities

EXAMPLE B

The acid test ratio for The Ski Chalet for 2004 from Figure 19-1 is computed as follows:

Cash	$ 90,000
Accounts receivable	134,000
Total cash and receivables	$224,000

$224,000 ÷ $104,000 = 2.2 = 2.2:1

RATIO OF ACCOUNTS RECEIVABLE TO NET SALES

When businesses sell on credit, they need to be alert to the amount and quality of their accounts receivable. They need to compare the amount of their current receivables to the amounts for prior years and compare the extent of their receivables to those of similar companies. By computing the **ratio of accounts receivable to net sales** every year, management and investors can keep track of the percent of sales that have not yet been paid for by customers. An increasing ratio over the years can indicate problems with collecting payments and should be investigated.

Ratio of accounts receivable to net sales = Accounts receivable ÷ Net sales

EXAMPLE C

The Ski Chalet ratio for 2004 is

Figure 19-1   **Figure 19-3**
$134,000   ÷ $975,000 = 0.137 = 0.14:1

INVENTORY TURNOVER

In retail stores, the cost of inventory often is very high. One way to control inventory costs and increase profit is to maintain a high level of inventory turnover. Recall from Chapter 17 that *inventory turnover* lets management and others know the number of times average inventory is sold during the year. The higher the turnover number, the better is the movement of inventory. Recall also that *average inventory,* found by averaging monthly, quarterly, or yearly inventory amounts, must be computed first. Inventory turnover is given as the number of times instead of as a ratio to 1.

Average inventory = (Beginning inventory + Ending inventory) ÷ 2 (annual)
Inventory turnover = Cost of goods sold ÷ Average inventory

EXAMPLE D

Based on the information given in Figures 19-3 and 19-1, the 2004 inventory turnover for The Ski Chalet is found as follows:

01-Jan   31-Dec
($148,000 + $180,000) ÷ 2 = $328,000 ÷ 2 = $164,000 average inventory
$408,000 cost of merchandise sold ÷ $164,000 = 2.5 times inventory turnover rate

RELATIONSHIP OF NET INCOME TO NET SALES

An increase in total sales volume doesn't necessarily mean that a business is improving because expenses may be increasing at an equal or greater rate than revenues. Thus looking at the **relationship of net income to net sales** is important. The relationship is given as a percentage.

Relationship of net income to net sales = Net income ÷ Net sales

EXAMPLE E

Based on information from Figure 19-3, The Ski Chalet's 2004 relationship of net income to net sales is $84,000 ÷ $975,000 = 8.6%. Comparison with the relationship for 2003 of 2.2% ($18,000 ÷ $811,000) indicates an improvement.

RATE OF RETURN ON INVESTMENT

Shareholders and owners want a reasonable return on their investment (equity). A ratio that measures the **rate of return on investment** is the ratio of net income to shareholders'/owners' equity. The rate is given as a percentage.

Rate of return on investment = Net income ÷ Shareholders'/owners' equity

EXAMPLE F

Based on Figures 19-3 and 19-1, the rate of return on the shareholders' investment for The Ski Chalet for 2004 is

$84,000 ÷ $482,000 = 0.1742 = 17.4% rate of return

CONCEPT CHECK 19.3

Boswell Designs' financial statements showed the following:

Cash	$ 85,000	Current liabilities	$320,000	Net sales	$950,000
Accounts receivable	260,000	Total liabilities	560,000	Inventory 1/1/2004	240,000
Total current assets	580,000	Net income	80,000	Inventory 12/31/2004	200,000
Total assets	990,000	Shareholders' equity	430,000	Purchases for 2004	630,000

Using the above numbers, compute the following ratios:
a. Working capital ratio $580,000 ÷ $320,000 = 1.81 : 1
b. Acid test ratio ($85,000 + $260,000) ÷ $320,000 = 1.08 : 1
c. Average inventory ($240,000 + $200,000) ÷ 2 = $220,000
d. Inventory turnover $240,000 + $630,000 − $200,000 = $670,000
   $670,000 ÷ $220,000 = 3.05 turnovers
e. Net income to net sales ratio $80,000 ÷ $950,000 = 0.084, or 8.4%
f. Rate of return on investment $80,000 ÷ $430,000 = 0.186, or 18.6%

COMPLETE ASSIGNMENTS 19.1, 19.2, AND 19.3.

Chapter Terms for Review

- accounts receivable
- acid test ratio
- assets
- balance sheet
- financial statements
- income statement
- liabilities
- net revenue
- net worth
- rate of return on investment
- ratio
- ratio of accounts receivable to net sales
- relationship of net income to net sales
- working capital
- working capital ratio

THE BOTTOM LINE

Summary of chapter learning objectives:

Learning Objective	Example
19.1 Analyze balance sheets, comparing items and periods	1. A modified balance sheet for The Ski Chalet for December 2004 and 2003 is shown. Compute the percents for 2004 and the percents of increase/decrease between 2004 and 2003.

THE SKI CHALET
Balance Sheet as of
December 31, 2004 and 2003

	2004 Amount	2004 Percent	2003 Amount	2003 Percent	Increase/Decrease Amount	Increase/Decrease Percent
ASSETS						
Current assets:						
Cash	$ 90,000		$ 86,000	15.03%	$ 4,000	
Accounts receivable	134,000		98,000	17.13%	36,000	
Merchandise inventory	180,000		148,000	25.87%	32,000	
Total current assets	$404,000		$332,000	58.04%	$ 72,000	
Fixed assets:						
Equipment	$220,000		$190,000	33.22%	$ 30,000	
Less depreciation	(60,000)		(50,000)	−8.74%	(10,000)	
Equipment net	$160,000		$140,000	24.48%	$ 20,000	
Buildings	100,000		100,000	17.48%	—	
Total fixed assets	$260,000		$240,000	41.96%	$ 20,000	
TOTAL ASSETS	$664,000		$572,000	100.00%	$ 92,000	
LIABILITIES AND SHAREHOLDERS' EQUITY						
Current liabilities:						
Accounts payable	$ 18,000		$ 24,000	4.20%	$ (6,000)	
Accrued payroll	38,000		30,000	5.24%	8,000	
Payroll taxes payable	6,000		4,000	0.70%	2,000	
Total current liabilities	$ 62,000		$ 58,000	10.14%	$ 4,000	
Long-term liabilities:						
Mortgage payable	90,000		120,000	20.98%	(30,000)	
Total liabilities	$152,000		$178,000	31.12%	$(26,000)	
Shareholders' equity:						
Common stock	$359,000		$359,000	62.76%	—	
Retained earnings	153,000		35,000	6.12%	18,000	
Total shareholders' equity	$512,000		$394,000	68.88%	$118,000	
TOTAL LIABILITIES AND SHAREHOLDERS' EQUITY	$664,000		$572,000	100.00%	$ 92,000	

Answers: 1. 2004 percent: 13.55%; 20.18%; 27.11%; 60.84%; 33.13%; −9.04%; 24.10%; 15.06%; 39.16%; 100.00%; 2.71%; 5.72%; 0.90%; 9.34%; 13.55%; 22.89%; 54.07%; 7.98%; 77.11%; 100.00%. Increase/decrease percent: 4.65%; 36.73%; 21.62%; 21.69%; 15.79%; 20.00%; 14.29%; 0%; 8.33%; 16.08%; −25.00%; 26.67%; 50.00%; 6.90%; −25.00%; −14.61%; 0%; 51.43%; 29.95%; 16.08%.

THE BOTTOM LINE

Summary of chapter learning objectives:

Learning Objective	Example
19.2 Analyze income statements, comparing items and periods	2. A modified income statement for The Ski Chalet for the years 2004 and 2003 is shown. Compute the percents for 2004 and the percents of difference between 2004 and 2003.

THE SKI CHALET
Income Statement for the Years Ended
December 31, 2004 and 2003

	2004 Amount	2004 Percent	2003 Amount	2003 Percent	Difference Amount	Difference Percent
Revenue from sales:						
Sales	$988,900	_____	$850,000	104.81%	$138,900	_____
Less returns	13,900	_____	39,000	4.81%	(25,100)	_____
NET SALES	$975,000	_____	$811,000	100.00%	$164,000	_____
Cost of goods sold:						
Inventory, January 1	$148,000	_____	$152,000	18.74%	($4,000)	_____
Purchases	440,000	_____	379,000	46.73%	61,000	_____
Available for sale	$588,000	_____	$531,000	65.47%	$57,000	_____
Inventory, December 31	180,000	_____	148,000	18.25%	32,000	_____
Cost of goods sold	$408,000	_____	$383,000	47.23%	$25,000	_____
Gross profit on sales	($408,000)	_____	($383,000)	52.77%	($25,000)	_____
Operating expenses:						
Salary and benefits	$221,000	_____	$225,000	27.74%	$(4,000)	_____
Rent and utilities	62,000	_____	61,400	7.57%	600	_____
Advertising	32,400	_____	25,700	3.17%	6,700	_____
Depreciation	40,000	_____	32,000	3.95%	8,000	_____
Equipment and supplies	15,800	_____	10,300	1.27%	5,500	_____
Administrative	12,500	_____	14,200	1.75%	(1,700)	_____
Total operating expense	$383,700	_____	$368,600	45.45%	$15,100	_____
Income before tax	$183,300	_____	$59,400	7.32%	$123,900	_____
Income tax	30,300	_____	24,400	3.01%	5,900	_____
NET INCOME	$153,000	_____	$35,000	4.32%	$118,000	_____

19.3 Compute commonly used business operating ratios	Using the Balance Sheet and Income Statement for 2004 from The Bottom Line problems 1 and 2, compute the following ratios: 3. Acid test 4. Average inventory 5. Net income to net sales 6. Rate of return on investment

Answers: 2. 2004 percent: 101.43%; 1.43%; 100%; 15.18%; 45.13%; 60.31%; 18.46%; 41.85%; 58.15%; 22.67%; 6.36%; 3.32%; 4.10%; 1.62%; 1.28%; 39.35%; 18.80%; 3.11%; 15.69%. Difference percent: 16.34%; −64.36%; 20.22%; −2.63%; 16.09%; 10.73%; 21.62%; 6.53%; 32.48%; −1.78%; 0.98%; 26.07%; 25.00%; 53.40%; −11.97%; 4.10%; 208.59%; 24.18%; 337.14% 3. 3.61 4. $218,500 5. 15.7% 6. 29.9%

SELF-CHECK

Review Problems for Chapter 19

1 Quality Construction Company, Inc., had total assets of $620,000 and total liabilities of $335,000 on December 31, 2004. On December 31, 2005, Quality Construction has total assets of $712,000 and total liabilities of $330,000.

   a. What was the amount of the owners' equity as of December 31, 2004?

   b. What is the amount of the owners' equity as of December 31, 2005?

   c. Calculate the percent of increase or decrease in total assets, total liabilities, and owners' equity. (Round to one decimal place.)

2 Quality Construction Company, Inc., had net sales of $460,250 and cost of merchandise sold of $320,600. Compute the gross profit amount and the percent of gross profit based on net sales.

3 The comparative income statement of Benson Electronics, Inc., showed sales of $425,000 in 2003 and $494,450 in 2004. Compute the percent of change in sales. (Round answer to one decimal place.)

4 Calculate the percent of increase or decrease for each of the following balance sheet items. If any percent cannot be calculated, give a brief explanation. (Answers correct to two decimal places.)

Item	2005	2004	Percent of Increase/Decrease
a. Cash	$35,000	$30,000	_____
b. Supplies	1,200	1,600	_____
c. Notes Receivable	2,000	-0-	_____
d. Merchandise Inventory	16,500	16,500	_____
e. Accounts Receivable	-0-	1,500	_____

5 Selected figures from the Balance Sheet and the Income Statement of Multimedia, Inc., follow. Use the data to calculate the ratios listed. (Give answers accurate to two decimal places.)

From the Balance Sheet		From the Income Statement	
Cash	$210,734	Net Sales	$244,750
Accounts Receivable	$138,126	Cost of Merchandise Sold	$190,000
Merchandise Inventory:		Net Income	$26,406
End of this year	$184,500		
End of last year	$178,300		
Total Current Assets	$533,360		
Total Current Liabilities	$324,152		
Total Stockholders' Equity	$149,000		

   a. Working capital ratio _____

   b. Acid test ratio _____

   c. Inventory turnover _____

   d. Rate of return on investment _____

   e. Net income as a percent of sales _____

   f. Ratio of accounts receivable to net sales _____

Notes

Assignment 19.1: Balance Sheet Analysis

Name

Date   Score

Learning Objective 1

A **(50 points) Solve the following balance sheet problems. (points for correct answers as marked)**

1. In the following balance sheet, find the percent for each 2004 and 2003 item. Then find the amount and percent of change. Round percents to two decimal places. (1/2 point for each correct answer)

Blair Merchandising Company
Balance Sheet
As of December 31, 2004 and 2003

	2004 Amount	2004 Percent	2003 Amount	2003 Percent	Increase/Decrease Amount	Increase/Decrease Percent
ASSETS						
Current assets:						
Cash	$ 230,000	13.98%	$ 212,000	15.40%	$ 18,000	8.49%
Accounts receivable	250,000	15.20%	175,000	12.71%	$ 75,000	42.86%
Inventory	420,000	25.53%	350,000	25.42%	$ 70,000	20.00%
Total current assets	$ 900,000	54.71%	$ 737,000	53.52%	$ 163,000	22.12%
Fixed assets:						
Machinery	$ 280,000	17.02%	$ 280,000	20.33%	—	0.00%
Less depreciation	120,000	7.29%	100,000	7.26%	$ 20,000	20.00%
Machinery net	$ 160,000	9.73%	$ 180,000	13.07%	(20,000)	−11.11%
Building	350,000	21.28%	270,000	19.61%	$ 80,000	29.63%
Land parcel holdings	235,000	14.29%	190,000	13.80%	$ 45,000	23.68%
Total fixed assets	$ 745,000	45.29%	$ 640,000	46.48%	$ 105,000	16.41%
TOTAL ASSETS	$1,645,000	100.00%	$1,377,000	100.00%	$ 268,000	19.46 %
LIABILITIES						
Current liabilities:						
Accounts payable	$ 96,000	5.84%	$ 62,000	4.50%	$ 34,000	54.84%
Accrued payroll	45,000	2.74%	35,000	2.54%	$ 10,000	28.57%
Payroll taxes payable	15,000	0.91%	20,000	1.45%	$ (5,000)	−25.00%
Total current liabilities	$ 156,000	9.48%	$ 117,000	8.50%	$ 39,000	33.33%
Long-term liabilities:						
Mortgages payable	$ 309,000	18.78%	$ 320,000	23.24%	$ (11,000)	−3.44%
Note payable—long-term	180,000	10.94%	210,000	15.25%	$ (30,000)	−14.29%
Total long-term liabilities	$ 489,000	29.73%	$ 530,000	38.49%	$ (41,000)	−7.74%
Total liabilities	$ 645,000	39.21%	647,000	46.99%	$ (2,000)	−0.31%
Shareholders' equity:						
Common stock	$ 520,000	31.61%	$ 467,000	33.91%	$ 53,000	11.35%
Preferred stock	330,000	20.06%	220,000	15.98%	$ 110,000	50.00%
Retained earnings	150,000	9.12%	43,000	3.12%	$ 107,000	248.84%
Total shareholders' equity	$1,000,000	60.79%	$ 730,000	53.01%	$ 270,000	36.99%
TOTAL LIABILITIES AND SHAREHOLDERS' EQUITY	$1,645,000	100.00%	$1,377,000	100.00%	$ 268,000	19.46%

2. Blair Merchandising's bookkeeper overlooked the fact that $15,000 cash had been paid to employees but not deducted from the cash account. Assume that the balance sheet in problem 1 was adjusted to reflect the correction. (1 point for each correct answer)
   a. What would be the adjusted amount for 2004 cash? $215,000
   b. What would be the adjusted amount for 2004 accrued payroll? $30,000

Score for A (50)

Assignment 19.1 Continued

B **(50 points) Solve the following balance sheet problems. (points for correct answers as marked)**

3. In the following balance sheet, find the percent for each 2004 and 2003 item. Then find the amount and percent of change. Round percents to one decimal place. Note that totals will sometimes be different from individual amounts because of rounding. (1/2 point for each correct answer)

Cozy Coffee Company
Balance Sheet
As of December 31, 2004 and 2003

	2004 Amount	2004 Percent	2003 Amount	2003 Percent	Increase/Decrease Amount	Increase/Decrease Percent
ASSETS						
Current assets:						
Cash	$ 52,500	10.0%	$ 37,900	8.10%	$ 14,600	38.5%
Accounts receivable	37,800	7.2%	29,790	6.37%	$ 8,010	26.9%
Inventory	62,000	11.8%	55,500	11.87%	$ 6,500	11.7%
Total current assets	$152,300	28.9%	$123,190	26.34%	$ 29,110	23.6%
Fixed assets:						
Equipment	$ 84,200	16.0%	$ 72,000	15.39%	$ 12,200	16.9%
Less depreciation	15,300	2.9%	12,500	2.67%	$ 2,600	22.4%
Machinery net	$ 68,900	13.1%	$ 59,500	12.72%	$ 9,400	15.8%
Building	235,000	44.7%	235,000	50.25%	—	0.0%
Land parcel holdings	70,000	13.3%	50,000	10.69%	$ 20,000	40.0%
Total fixed assets	$373,900	71.1%	$344,500	73.66%	$ 29,400	8.5%
TOTAL ASSETS	$526,200	100.0%	$467,690	100.00%	$ 58,510	12.5%
LIABILITIES						
Current liabilities:						
Accounts payable	$ 13,950	2.7%	$ 14,200	3.04%	$ (250)	−1.8%
Accrued payroll	8,200	1.6%	7,400	1.58%	$ 800	10.8%
Payroll taxes payable	1,200	0.2%	980	0.21%	$ 220	22.4%
Total current liabilities	$ 23,350	4.4%	$ 22,580	4.83%	$ 770	3.4%
Long-term liabilities:						
Mortgages payable	$ 81,500	15.5%	$ 83,700	17.90%	$ (2,200)	−2.6%
Note payable—long-term	25,000	4.8%	21,000	4.49%	$ 4,000	19.0%
Total long-term liabilities	$106,500	20.2%	$104,700	22.39%	$ 1,800	1.7%
Total liabilities	$129,850	24.7%	$127,280	27.21%	$ 2,570	2.0%
Shareholders' equity:						
Common stock	$195,000	37.1%	$180,000	38.49%	$15,000	8.3%
Preferred stock	82,000	15.6%	82,000	17.53%	—	0.0%
Retained earnings	119,350	22.7%	78,410	16.77%	$40,940	52.2%
Total shareholders' equity	$396,350	75.3%	$340,410	72.79%	$55,940	16.4%
TOTAL LIABILITIES AND SHAREHOLDERS' EQUITY:	$526,200	100.0%	$467,690	100.00%	$58,510	12.5%

4. Show what changes would have been made in the cash and preferred stock amount in 2004 if Cozy Coffee Company had sold an additional $6,000 in preferred stock. (1/2 point for each correct answer)

	Amount	Percent
Cash	$58,500	11.0%
Preferred stock	$88,000	16.5%

Score for B (50)

Assignment 19.2: Income Statement Analysis

Name

Date   Score

Learning Objective 2

A (50 points) Solve the following income statement problems. (points for correct answers as marked)

1. In the following income statement, find the percent for each 2004 and 2003 item. Then find the amount and percent of change. Round percents to two decimal places. (1/2 point for each correct answer)

Georgia Textiles
Income Statement
For the Years Ended December 31, 2004 and 2003

	2004 Amount	2004 Percent	2003 Amount	2003 Percent	Increase/Decrease Amount	Increase/Decrease Percent
Revenue from sales:						
Sales	$920,000	103.95%	$827,000	103.76%	$ 93,000	11.25%
Less returns	35,000	3.95%	30,000	3.76%	$ 5,000	16.67%
NET SALES	$885,000	100.00%	$797,000	100.00%	$ 88,000	11.04%
Cost of goods sold:						
Inventory, January 1	$210,000	23.73%	$197,000	24.72%	$ 13,000	6.60%
Purchases	460,000	51.98%	395,000	49.56%	$ 65,000	16.46%
Available for sale	$670,000	75.71%	$592,000	74.28%	$ 78,000	13.18%
Inventory, December 31	240,000	27.12%	210,000	26.35%	$ 30,000	14.29%
Cost of goods sold	$430,000	48.59%	$382,000	47.93%	$ 48,000	12.57%
Gross profit	$455,000	51.41%	$415,000	52.07%	$ 40,000	9.64%
Operating expenses:						
Wages	$132,600	14.98%	$120,000	15.06%	$ 12,600	10.50%
Rent	84,000	9.49%	80,000	10.04%	$ 4,000	5.00%
Advertising	18,000	2.03%	20,000	2.51%	$ (2,000)	−10.00%
Insurance	4,500	0.51%	4,200	0.53%	$ 300	7.14%
Depreciation	3,600	0.41%	3,100	0.39%	$ 500	16.13%
Equipment rental	1,200	0.14%	1,400	0.18%	$ (200)	−14.29%
Administrative	7,000	0.79%	5,200	0.65%	$ 1,800	34.62%
Miscellaneous	3,200	0.36%	2,100	0.26%	$ 1,100	52.38%
Total operating expenses	$254,100	28.71%	$236,000	29.61%	$ 18,100	7.67%
Income before tax	$200,900	22.70%	$179,000	22.46%	$ 21,900	12.23%
Income tax	32,000	3.62%	28,000	3.51%	$ 4,000	14.29%
NET INCOME	$168,900	19.08%	$151,000	18.95%	$ 17,900	11.85%

2. Assume that the ending inventory was $220,000 in 2004. Compute the following items. (2 points for each correct answer)

2004 Gross profit amount $ 435,000   2004 Gross profit percent   49.15%
2004 NET INCOME amount $ 148,900   2004 NET INCOME percent   16.82%

Score for A (50)

Assignment 19.2 Continued

B **(100 points) Solve the following income statement problems. (points for correct answers as marked)**

3. In the following income statement, find the percent for each 2004 and 2003 item, then find the amount and percent of change. Round percents (no decimal places). (84 points, 1 point for each correct answer)

Baldwin Field Enterprises
Income Statement
For the Years Ended December 31, 2004 and 2003

	2004 Amount	2004 Percent	2003 Amount	2003 Percent	Difference Amount	Difference Percent
Revenue from sales:						
Sales	$ 87,000	102%	$ 74,800	102%	$ 12,200	16%
Less returns	2,000	2%	1,800	2%	$ 200	11%
NET SALES	$ 85,000	100%	$ 73,000	100%	$ 12,000	16%
Cost of goods sold:						
Inventory, January 1	$ 22,000	26%	17,500	24%	$ 4,500	26%
Purchases	38,000	45%	35,000	48%	$ 3,000	9%
Available for sale	$ 60,000	71%	52,500	72%	$ 7,500	14%
Inventory, December 31	24,100	28%	22,000	30%	$ 2,100	10%
Cost of goods sold	$ 35,900	42%	30,500	42%	$ 5,400	18%
Gross profit	$ 49,100	58%	$ 42,500	58%	$ 6,600	16%
Operating expenses:						
Salary	$ 11,200	13%	10,900	15%	$ 300	3%
Rent	7,500	9%	6,000	8%	$ 1,500	25%
Advertising	1,400	2%	1,200	2%	$ 200	17%
Delivery	450	1%	380	1%	$ 70	18%
Depreciation	650	1%	600	1%	$ 50	8%
Equipment rental	350	0%	420	1%	$ (70)	−17%
Administrative	1,900	2%	1,700	2%	$ 200	12%
Miscellaneous	190	0%	220	0%	$ (30)	−14%
Total operating expenses	$ 23,640	28%	$ 21,420	29%	$ 2,220	10%
Income before tax	$ 25,460	30%	21,080	29%	$ 4,380	21%
Income tax	2,200	3%	2,000	3%	$ 200	10%
NET INCOME	$ 23,260	27%	$ 19,080	26%	$ 4,180	22%

4. Assume that the beginning inventory was $18,000 in 2003 and $20,500 in 2004 and that the rent was $6,400 in 2003 and $8,800 in 2004. Compute the following amounts and percents to reflect the revised beginning inventory and rent numbers. (8 points for each correct row)

	2004 Amount	2004 Percent	2003 Amount	2003 Percent	Difference Amount	Difference Percent
Gross profit	$50,600	60%	$40,500	55%	$10,100	25%
NET INCOME	$23,460	28%	$16,680	23%	$ 6,780	41%

Score for B (100)

Assignment 19.3: Financial Statement Ratios

Name

Date   Score

Learning Objectives 1 2 3

A (26 points) Solve the following financial statement ratio problems. (1/2 point for each correct answer)

1. Alice Anderson was considering investing in a business. She used the following statement in analyzing the Dover Clock Shop. Compute the net changes in the balance sheet and income statement. Round to one decimal place.

Dover Clock Shop
Comparative Balance Sheet
As of December 31, 2004 and 2003

	2004 Amount	2003 Amount	Increase/Decrease Amount	Increase/Decrease Percent
ASSETS				
Current assets:				
Cash	$110,000	$104,600	$ 5,400	5.2%
Accounts receivable	135,000	115,900	19,100	16.5%
Merchandise inventory	185,000	145,000	40,000	27.6%
Total current assets	$430,000	$365,500	$ 64,500	17.6%
Fixed assets:				
Building improvements	$ 45,000	$ 48,500	$ (3,500)	−7.2%
Equipment	145,000	132,000	13,000	9.8%
Total fixed assets	$190,000	$180,500	$ 9,500	5.3%
TOTAL ASSETS	$620,000	$546,000	$ 74,000	13.6%
LIABILITIES				
Current liabilities:				
Salaries payable	$ 33,000	$ 28,200	$ 4,800	17.0%
Accounts payable	120,000	112,900	7,100	6.3%
Total current liabilities	$153,000	$141,100	$ 11,900	8.4%
Long-term liabilities:				
Note payable	$100,000	$120,000	$(20,000)	−16.7%
Total liabilities	$253,000	$261,100	$ (8,100)	−3.1%
Owner's equity:				
J. C. Dover, capital	367,000	284,900	82,100	28.8%
TOTAL LIABILITIES AND OWNER'S EQUITY	$620,000	$546,000	$ 74,000	13.6%

Assignment 19.3 Continued

Dover Clock Shop
Comparative Income Statement
For the Years Ended December 31, 2004 and 2003

	2004	2003	Difference Amount	Difference Percent
NET SALES	$780,000	$835,000	$(55,000)	−6.6%
Cost of goods sold:				
Merchandise inventory, January 1	$145,000	$138,000	$ 7,000	5.1%
Purchases	585,000	620,000	(35,000)	−5.6%
Merchandise available for sale	$730,000	$758,000	$(28,000)	−3.7%
Merchandise inventory, December 31	185,000	145,000	40,000	27.6%
Cost of goods sold	$545,000	$613,000	$(68,000)	−11.1%
Gross profit on sales	$235,000	$222,000	$ 13,000	5.9%
Expenses:				
Selling	$ 82,000	$ 78,600	$ 3,400	4.3%
Other	29,200	30,200	(1,000)	−3.3%
Total expenses	$111,200	$108,800	$ 2,400	2.2%
NET INCOME	$123,800	$113,200	$ 10,600	9.4%

Score for A (26)

B (24 points) Solve the following problems. (2 points for each correct answer)

2. Provide the following information for Alice Anderson's consideration. When the ratio is less than 1, give the ratio to three decimal places; otherwise, round to one decimal place.

	2004	2003
a. Working capital ratio	2.8:1	2.6:1

$430,000 ÷ $153,000 = 2.8; $365,500 ÷ $141,100 = 2.6

| b. Acid test ratio | 1.6:1 | 1.6:1 |

$245,000 ÷ $153,000 = 1.6; $220,500 ÷ $141,100 = 1.6

| c. Ratio of accounts receivable to net sales | 0.173:1 | 0.139:1 |

$135,000 ÷ $780,000 = 0.173; $115,900 ÷ $835,000 = 0.139

| d. Inventory turnover | 3.3 times | 4.3 times |

($145,000 + $185,000) ÷ 2 = $165,000; $545,000 ÷ $165,000 = 3.3
($138,000 + $145,000) ÷ 2 = $141,500; $613,000 ÷ $141,500 = 4.3

| e. Ratio of net income to net sales | 15.90% | 13.60% |

$123,800 ÷ $780,000 = 0.159 = 15.9%; $113,200 ÷ $835,000 = 0.136 = 13.6%

| f. Rate of return on investment | 33.70% | 39.70% |

$123,800 ÷ $367,000 = 0.337 = 33.7%; $113,200 ÷ $284,900 = 0.397 = 39.7%

Score for B (24)

C (26 points) Solve the following problems. (1/2 point for each correct answer)

3. Alice Anderson was offered a second business. She received the following statements for 2004 and 2003. Complete calculations for a comparative balance sheet and a comparative income statement for The Grandfather Clock Shop, showing the amount and percent of change.

Assignment 19.3 Continued

Grandfather Clock Shop
Comparative Balance Sheet
As of December 31, 2004 and 2003

	2004		2003		Increase/Decrease	
	Amount	Percent	Amount	Percent	Amount	Percent
ASSETS						
Current assets:						
Cash	$ 25,000	18.2%	$ 16,000	14.7%	$ 9,000	56.3%
Accounts receivable	12,000	8.8%	8,000	7.3%	4,000	50.0%
Merchandise inventory	46,000	33.6%	31,000	28.4%	15,000	48.4%
Total current assets	$ 83,000	60.6%	$ 55,000	50.5%	$28,000	50.9%
Fixed assets:						
Store fixtures	$ 39,000	28.5%	$ 43,000	39.4%	$ (4,000)	−9.3%
Office equipment	15,000	10.9%	11,000	10.1%	4,000	36.4%
Total fixed assets	$ 54,000	39.4%	$ 54,000	49.5%	$ 0	0.0%
TOTAL ASSETS	$137,000	100.0%	$109,000	100.0%	$28,000	25.7%
LIABILITIES						
Current liabilities:						
Sales tax payable	$ 4,500	3.3%	$ 5,500	5.0%	$ (1,000)	−18.2%
Accounts payable	9,500	6.9%	6,000	5.5%	3,500	58.3%
Total current liabilities	$ 14,000	10.2%	$ 11,500	10.6%	$ 2,500	21.7%
Long-term liabilities:						
Note payable	$ 30,000	21.9%	$ 38,000	34.9%	$ (8,000)	−21.1%
Total liabilities	$ 44,000	32.1%	$ 49,500	45.4%	$ (5,500)	−11.1%
Owner's equity						
R. A. Banner, capital	$ 93,000	67.9%	$ 59,500	54.6%	$33,500	56.3%
TOTAL LIABILITIES AND OWNER'S EQUITY	$137,000	100.0%	$109,000	100.0%	$28,000	25.7%

Assignment 19.3 Continued

Grandfather Clock Shop
Comparative Income Statement
For the Years Ended December 31, 2004 and 2003

	2004		2003		Difference	
	Amount	Percent	Amount	Percent	Amount	Percent
NET SALES	$205,000	100.0%	$120,000	100.0%	$85,000	70.8%
Cost of goods sold:						
Merchandise inventory, January 1	$ 31,000	15.1%	$ 27,500	22.9%	$ 3,500	12.7%
Purchases	154,000	75.1%	84,500	70.4%	69,500	82.2%
Merchandise available for sale	$185,000	90.2%	$112,000	93.3%	$73,000	65.2%
Merchandise inventory,						
December 31	46,000	22.4%	31,000	25.8%	15,000	48.4%
Cost of goods sold	$139,000	67.8%	$ 81,000	67.5%	$58,000	71.6%
Gross profit on sales	$ 66,000	32.2%	$ 39,000	32.5%	$27,000	69.2%
Expenses:						
Selling	$ 31,000	15.1%	$ 21,500	17.9%	$ 9,500	44.2%
Other	13,000	6.3%	7,250	6.0%	5,750	79.3%
Total expenses	$ 44,000	21.5%	$ 28,750	24.0%	$15,250	53.0%
NET INCOME	$ 22,000	10.7%	$ 10,250	8.5%	$11,750	114.6%

Any differences of 0.1% from individual items are due to rounding.

Score for C (26)

International Business

20

Learning Objectives
By studying this chapter and completing all assignments you will be able to:

Learning Objective 1 — Compute currency exchange rates.

Learning Objective 2 — Compute the effects of exchange rate changes.

Learning Objective 3 — Compute duties on imports.

Learning Objective 4 — Convert between U.S. weights and measures and metric weights and measures.

Businesses in the United States **import** goods made in other countries and **export** domestic goods made in the United States. International business transactions amount to billions of dollars annually and constitute an important part of the economies of most nations in the world.

International trade between U.S. companies and those in other countries is under the jurisdiction of the International Trade Administration (ITA), a branch of the Department of Commerce. All international trade is subject to a set of ITA rules and regulations known as the **Export Administration Regulations.** Any company in the United States planning to sell goods to companies in other countries must have an ITA export license for the transactions.

Computing Currency Exchange Rates

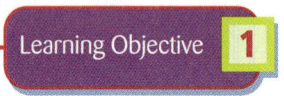

Compute currency exchange rates.

In order to conduct international trade, U.S. companies must exchange U.S. dollars for other currencies and vice versa. Figure 20-1 lists the names of the currency units used in the major countries, the U.S. dollars per unit, and the number of units per U.S. dollar.

20.1 In addition to being major players in the export/import business, Americans are among the world's most frequent international travelers. Today, international travel and communication constitute major components of international business and monetary exchange.

Figure 20-1 | Foreign Currency–U.S. Dollar Exchange Rates

CURRENCY EXCHANGE RATES
Quotes delayed at least 20 minutes.
All Currencies

Name	In US$	Per US$
Algerian Dinar	0.01310	76.350
Argentine Peso	0.33750	2.963
Australian Dollar	0.72530	1.379
Bahraini Dinar	2.6524	0.377
Bolivian Boliviano	0.12479	8.014
Brazilian Real	0.35026	2.855
British Pound	1.7983	0.556
Botswana Pula	0.21277	4.700
Canadian Dollar	0.79586	1.257
Chilean Peso	0.00164	609.8
Chinese Yuan	0.12068	8.287
Colombian Peso	0.00039	2,567
Cyprus Pound	2.1678	0.461
Czech Koruna	0.03978	25.139
Danish Krone	0.16799	5.953
Ecuador Sucre	0.00004	25,500
Euro	1.2495	0.800
Ghana Cedi	0.00011	9,102
Guatemalan Quetzal	7.9975	0.125
Hong Kong Dollar	0.12837	7.790
Hungarian Forint	0.00504	198.3

406 Part 5 Business Applications

Figure 20-1 Foreign Currence–U.S. Dollar Exchange Rates *(continued)*

Name	In US$	Per US$
Israeli New Shekel	0.22292	4.486
Indian Rupee	0.02181	45.845
Indonesian Rupiah	0.00011	9,112
Japanese Yen	0.00914	109.4
Jordanian Dinar	1.4094	0.710
Kenyan Shilling	0.01228	81.420
Korean Won	0.00088	1,143
Kuwaiti Dinar	3.3931	0.295
Moroccan Dirham	0.11307	8.844
Malaysian Ringgit	0.26312	3.801
Mexican Peso	0.08716	11.473
Namibian Dollar	0.15319	6.528
New Zealand Dollar	0.68520	1.459
Norwegian Krone	0.15192	6.583
Omani Rial	2.5963	0.385
Peruvian Nuevo Sol	0.30187	3.313
Pakistani Rupee	0.01662	60.160
Qatari Riyal	0.27467	3.641
Russian Ruble	0.03433	29.130
Saudi Arabian Riyal	0.26663	3.751
Singapore Dollar	0.59584	1.678
South African Rand	0.15686	6.375
Swedish Krona	0.13763	7.266
Swiss Franc	0.81162	1.232
Taiwanese Dollar	0.02961	33.775
Tanzanian Shilling	0.00094	1,067
Thai Baht	0.02411	41.470
Tunisian Dinar	0.79650	1.256
UAE Dirham	0.27224	3.673
Venezuelan Bolivar	0.00052	1,920
Vietnamese Dong	0.00006	15,760
Zimbabwe Dollar	0.00019	5,407

Four other governments call their currency the dollar—Australia, Canada, Hong Kong, and New Zealand. These dollars are not U.S. dollars; each is a separate currency. Several currencies share names, such as the franc, mark, peso, pound, and euro.

EXAMPLE A

A person planning a trip to Denmark wants to change $100 U.S. dollars to Danish kroner. How many kroner will the person get for the $100 U.S. dollars? (Round answer to nearest krone.)

$5.953 \times 100 = 595$ kroner

EXAMPLE B

A traveler from Argentina is planning a trip to the United States and wants to change 1,000 Argentine pesos to U.S. dollars. How many U.S. dollars will the traveler receive for the 1,000 pesos? (Round answer to nearest dollar.)

1,000 ÷ 2.963 = $337

EXAMPLE C

An American tourist shopping in a Canadian store purchased an item for 100 Canadian dollars. How much did his purchase cost him in U.S. dollars? (Round answer to nearest U.S. penny.)

100 Canadian dollars × 0.79586 U.S. dollars per Canadian dollar = $79.59

✓ CONCEPT CHECK 20.1

Using the "Per US$" column from Figure 20.1, compute the number of euros one would receive for $300 U.S. dollars. (Round answer to nearest euro.)

$300 × 0.80 = 240 euros

Using the "In US$" column from Figure 20.1, compute the number of U.S. dollars one would receive for 400 Japanese yen. (Round answer to nearest U.S. penny.)

400 × 0.00914 = $3.66

Computing the Effects of Exchange Rate Changes

Learning Objective 2

Compute the effects of exchange rate changes.

One hazard of foreign trade is the uncertainty of future exchange rates between currencies. The relationship between the values of the U.S. dollar and a foreign currency can change between the time a contract is signed and the time payment is received. If a U.S. exporter agrees to accept foreign currency, a devaluation in the foreign currency could cause the exporter to lose money on the transaction.

EXAMPLE D

Global Industries, a U.S. company, sold merchandise to Europa, a company in Hungary. Europa agreed to pay 500,000 Hungarian forint for the goods. On the date of the sale, the Hungarian forint was valued at 198.3 per U.S. dollar, as noted in Figure 20.1. Global Industries expected to receive $2,521.43. (500,000 Hungarian forint ÷ 198.3 per U.S. dollar = $2,521.43.)

Between the date the sale was made and the date the goods were shipped and paid for by Europa, the value of the forint changed to 204.7 per U.S. dollar. How much did Global Industries lose by accepting the forint as the medium of payment?

Value of merchandise at time of sale: (500,000 Hungarian forint ÷ 198.3 per U.S. dollar = $2,521.43. Value of merchandise at time shipped and paid for: (500,000 Hungarian forint ÷ 197.0 = $2,538.07.) (Value of 500,000 forint at time of sale $2,521.43 − value of 500,000 forint at time shipped and paid $2,442.60 = loss to Global Industries $78.83.)

EXAMPLE E

Global Industries investigated a purchase of raw materials from a company in England. The price of the materials was 150,000 British pounds. At the time, the value of the British pound was $1.652. Three months later, when Global actually made the purchase, the value of the British pound was as shown in Figure 20.1. How many more dollars did Global have to pay as a result of the change in the value of the British pound?

150,000 × $1.652 = $247,800 cost when investigated
150,000 × $1.7983 = $269,745 cost when purchase was made
$269,745 − $247,800 = $21,945 more dollars at time of purchase

✓ CONCEPT CHECK 20.2

Global Industries contracts to sell a printing press to a company in Denmark. The Danish company agreed to pay $300,000 U.S. dollars for the press.

On the date the agreement was made, the Danish krone was worth 0.16799 U.S. dollars. On the date payment was made, the krone had changed to 0.1592 U.S. dollars. How many more or less Danish kroner did the Danish company pay by stipulating a purchase price of $300,000 U.S. dollars?

$300,000 ÷ 0.16799 = 1,785,821 kroner at time of agreement
$300,000 ÷ 0.1592 = 1,884,422 kroner at time of payment
1,884,422 − 1,785,821 = 98,601 more kroner at time of payment

If the Danish company had agreed to pay 1,785,821 kroner instead of $300,000 for the purchase, how many U.S. dollars would it have saved between the time of agreement and the time of payment?

1,785,821 kroner to be paid × 0.1592 value of krone at payment = $284,302.70
$300,000 value of kroner at time of agreement − $284,302.70 = 15,697.30 saved

Computing Duties on Imports

All items imported into the United States must go through the U.S. Customs Agency. Many imported items have a **duty** (charge or tax) imposed by the Customs Agency to protect U.S. manufacturers against foreign competition in domestic markets. Duties vary widely from item to item. A duty may be a set amount—such as $0.50 per item—or an **ad valorem duty**, which is a percent of the value of the item.

Learning Objective 3

Compute duties on imports.

20.2 U.S. travelers returning from foreign countries must go through Customs and report any purchases made abroad in excess of a certain total amount. Any purchases beyond this limit are subject to immediate payment of the appropriate duty. This duty protects the business interests of American companies that have to pay duty when importing the same items for resale.

EXAMPLE F

Assume that a wristwatch in a leather case with a metal band has four duty rates imposed: $0.40 per wristwatch + 6% of the value of the case + 14% of the value of the metal band + 5.3% of the value of the battery. Anderson Jewelry Company imported four dozen wristwatches. The value of the case was $16; the metal band, $10; and the battery, $6. How much duty did the Anderson Jewelry Company pay for the four dozen wristwatches? (Round answer to nearest cent.)

Duty per wristwatch:	$0.40
Ad valorem duty on case: $16 × 0.06 =	0.96
Ad valorem duty on metal band: $10 × 0.14 =	1.40
Ad valorem duty on battery: $6 × 0.053 =	$0.318
Total duty per watch	$3.078

$3.078 per watch × 48 watches = $147.74 total duty paid

20.3 The flexibility of duties imposed on like items from different countries is a means of helping developing countries participate in international trade and to punish countries for imposing high tariffs on U.S. goods.

EXAMPLE G

A computer printer costs $150 whether purchased from country A or country B. However, it has an ad valorem duty rate of 3.5% if purchased from country A and an ad valorem duty rate of 28% if purchased from country B. How much more would it cost a company to purchase the printer from country B than from country A?

Country A: $150 × 0.035 = $5.25
                  $150 + $5.25 = $155.25 total cost

Country B: $150 × 0.28 = $42
                  $150 + $42 = $192 total cost

$192 country B − $155.25 country A = $36.75 more

    Foreign trade zones are domestic sites in the United States considered to be outside U.S. Customs territory. These foreign trade zones are used for import and export activities. No duty or federal excise taxes are charged on foreign goods moved into the zone until the goods or products made from them are moved into U.S. Customs territory. No duty is charged on imports that later are exported for sale, because they never entered U.S. Customs territory. Recently, there were more than 150 foreign trade zones in port communities in the United States. Operations in them include storage, repacking, inspection, exhibition, assembly, and manufacturing.

EXAMPLE H

A U.S. company located in a foreign trade zone imported $500,000 worth of goods. The duty rate on the goods is 5%. If 30% of the goods were moved into U.S. Customs territory for sale and 70% were exported for sale, how much money did the company save by being located in a foreign trade zone?

$500,000 × 5% duty = $25,000 duty if goods are sold in U.S. Customs territory
$25,000 × 70% exported = $17,500 saved

CONCEPT CHECK 20.3

a. Downtown Toy Store ordered from a foreign country 400 dolls on which an ad valorem duty of 4.5% is charged. Payment is to be made in U.S. dollars. The price of each doll is $23. What is the total cost to Downtown?

400 × $23 = $9,200 cost before duty
$9,200 × 0.045 duty = $414
$9,200 + $414 = $9,614 total cost to buyer

b. A company located in a foreign trade zone purchased $1 million worth of electronic equipment having an ad valorem duty of 4.1%. Forty percent of the products were moved into U.S. Customs territory for sale, and 60% were repackaged and exported. How many dollars did the company save by being located in a foreign trade zone?

$1,000,000 × 60% = $600,000
$600,000 × 4.1% = $24,600 saved

Converting Between U.S. Weights and Measures and Metric Weights and Measures

Some businesses, especially in the area of import–export activities, must convert U.S. customary units of weight and measure to the **metric system** of weights and measures used in most other countries. Figure 20-2 shows the conversion values for the U.S./metric units used most frequently in business.

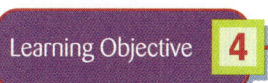

Convert between U.S. weights and measures and metric weights and measures.

Figure 20-2 U.S./Metric Unit Conversions

To Convert U.S. to	Metric	Multiply by Number of Metric in U.S.	To Convert Metric to U.S.	Multiply by Number of U.S. in Metric
Inches	Meters	0.0254	Meters	Inches	39.37
Feet	Meters	0.305	Meters	Feet	3.281
Yards	Meters	0.914	Meters	Yards	1.09
Miles	Kilometers	1.609	Kilometers	Miles	0.621
Ounces	Grams	28.35	Grams	Ounces	0.035
Pounds	Grams	454	Grams	Pounds	0.0022
Pounds	Kilograms	0.454	Kilograms	Pounds	2.2
Pints	Liters	0.473	Liters	Pints	2.113
Quarts	Liters	0.946	Liters	Quarts	1.057
Gallons	Liters	3.785	Liters	Gallons	0.264

20.4 At one time there was a worldwide campaign to convert all nations to the metric system. This movement has now subsided, but most European and Asian countries have adopted the metric system. When business is conducted in those countries, the metric system is the system of choice.

EXAMPLE I

Convert the following U.S. measures to metric measures.

a. Convert 30 inches to meters.
30 in. × 0.0254 = 0.7620 m

b. Convert 15 feet to meters.
15 ft × 0.305 = 4.5750 m

c. Convert 10 yards to meters.
10 yd × 0.914 = 9.14 m

d. Convert 20 miles to kilometers.
20 mi × 1.609 = 32.18 km

e. Convert 15 ounces to grams.
15 oz × 28.35 = 425.25 g

f. Convert 20 pounds to grams.
20 lb × 454 = 9,080 g

g. Convert 10 pounds to kilograms.
10 lb × 0.454 = 4.54 kg

h. Convert 20 pints to liters.
20 pt × 0.473 = 9.46 L

i. Convert 40 quarts to liters.
40 qt × 0.946 = 37.84 L

j. Convert 20 gallons to liters.
20 gal × 3.785 = 75.7 L

EXAMPLE J

Convert the following metric measures to U.S. measures.

a. Convert 20 meters to inches.
20 m × 39.37 = 787.4 in.

b. Convert 20 meters to feet.
20 m × 3.281 = 65.62 ft

c. Convert 30 meters to yards.
30 m × 1.09 = 32.7 yd

d. Convert 15 kilometers to miles.
15 km × 0.621 = 9.315 mi

e. Convert 20 grams to ounces.
20 g × 0.035 = 0.7 oz

f. Convert 20 grams to pounds.
20 g × 0.0022 = 0.044 lb

g. Convert 40 kilograms to pounds.
40 kg × 2.2 = 88 lb

h. Convert 30 liters to pints.
30 L × 2.113 = 63.39 pt

i. Convert 20 liters to quarts.
20 L × 1.057 = 21.14 qt

j. Convert 20 liters to gallons.
20 L × 0.264 = 5.28 gal

✓ CONCEPT CHECK 20.4

Using Figure 20-2, make the following conversions:

a. Convert 28 inches to meters.
28 × 0.0254 = 0.7112

b. Convert 17 feet to meters.
17 × 0.305 = 5.185

c. Convert 3 meters to inches.
3 × 39.37 = 118.11

d. Convert 18 meters to feet.
18 × 3.281 = 59.058

e. Convert 3 ounces to grams.
3 × 28.35 = 85.05

f. Convert 7 pounds to grams.
7 × 454 = 3,178

g. Convert 36 grams to pounds.
36 × 0.0022 = 0.0792

h. Convert 18 kilograms to pounds.
18 × 2.2 = 39.6

i. Convert 8 pints to liters.
8 × 0.473 = 3.784

j. Convert 2 quarts to liters.
2 × 0.946 = 1.892

COMPLETE ASSIGNMENT 20.2.

Chapter Terms for Review

ad valorem duty

duty

export

Export Administration Regulations

foreign trade zones

import

metric system

THE BOTTOM LINE

Summary of chapter learning objectives:

Learning Objective	Example
20.1 Compute currency exchange rates	1. Using the In U.S.$ column in Figure 20-1, compute the value in U.S. dollars of 5,000 units of each of the following foreign currencies. Round answers to the nearest cent. a. Czech Republic's koruna _____ b. Russian ruble _____ c. Korean won _____ d. Thailand's baht _____ 2. Using the Per U.S.$ column in Figure 20-1, compute the amount of U.S. dollars necessary to buy 5,000 units of each of the following foreign currencies. Round answers to the nearest cent. a. Sweden's krona _____ b. Hungary's forint _____ c. Zimbabwe dollar _____ d. Mexican peso _____
20.2 Compute the effects of exchange rate changes.	3. A U.S. company has contracted to sell certain goods to a company in Mexico. The Mexican company has contracted to pay 700,000 pesos for the goods. At the time the contract was signed, the In U.S. $ column in the newspaper showed that the Mexican peso was worth $0.0812. On the date payment was due, the peso changed to a value of $0.08716 U.S. How much did the U.S. company gain or lose by having agreed to accept payment in pesos instead of U.S. dollars?
20.3 Compute duties on imports	4. Broadway Department Store ordered from a foreign country 300 sets of dishes on which an ad valorem duty of 5.8% is charged. The price of each set of dishes is $72. Payment is to be made in U.S. currency. What is the total cost to Broadway?
20.4 Convert between U.S. weights and measures and metric weights and measures	5. Using Figure 20-2, make the following conversions. a. Convert 100 inches to meters. b. Convert 1,000 meters to feet. c. Convert 6 miles to kilometers. d. Convert 100 grams to ounces. e. Convert 3 gallons to liters. f. Convert 7 liters to quarts.

Answers: 1a. $198.90 b. $171.65 c. $4.40 d. $120.55; 2a. $688.14 b. $25.21 c. $92 d. $435.81 3. $56,840 expected; 61,012 received; 4,172 gain 4. $22,852.80 5a. 2.54 m b. 3,281 ft c. 9.654 km d. 3.5 oz e. 11.355 L f. 7.399 qt

SELF-CHECK

Review Problems for Chapter 20

(In all cases, round to the nearest U.S. penny.)

1. How many Thai baht can a person get for $15 U.S. dollars?

2. How many U.S. dollars can a person get for 15 Thai baht?

3. How many South African rand can a person get for $540 U.S. dollars?

4. How many U.S. dollars can a person get for 540 South African rand?

5. A U.S. exporter agrees to accept 300,000 South African rand in payment for goods. The South African rand is valued as shown in Figure 20-1. Compute the value in U.S. dollars that the U.S. exporter will receive.

6. In problem 5, suppose that the value of the South African rand changes to 7.185 per U.S. dollar. How much will the exporter gain or lose in this transaction?

7. Tonaka Manufacturing, Inc. contracted to sell goods to a company in Sweden for 630,000 Swedish kronor. Using the data in Figure 20-1, compute the U.S. dollar value that Tonaka expects to receive.

8. Assume that the value of the Swedish krona decreased by 20%; compute the U.S. dollar value that Tonaka would then expect to receive.

9. Princess Jewelry contracted to purchase 144 bracelets from a foreign manufacturer. The price of each bracelet is $40. An ad valorem duty of 17% is charged on each bracelet. Compute the duty Princess Jewelry will pay for the shipment.

10. ABC, Inc., plans to purchase 250 units of computer components. ABC can buy the components from country Y at a price of $60 each plus an ad valorem duty of 35% or from country YY at a price of $64 plus an ad valorem duty of 13%. Compute the amount ABC will save by purchasing from the lowest-cost source.

11. Convert 8 pints to liters.

12. The length of trip A is stated as 300 miles. The length of trip B is stated as 300 kilometers. In miles, how much farther is trip A than trip B?

Notes

Assignment 20.1: Trading with Other Countries

Name

Date   Score

Learning Objectives 1 2

A (44 points) Solve the following problems. (4 points for each correct answer)

1. Using the data in Figure 20-1, find the amount of U.S. dollars needed to buy 300 units of each foreign currency listed.

Foreign Currency	Price of 300 Units
a. Australian dollar	$217.59
b. Bahraini dinar	$795.72
c. Bolivian boliviano	$37.42
d. Brazilian real	$105.08
e. Canadian dollar	$238.76
f. Chinese yuan	$36.20
g. South African rand	$47.06

   Illustrative solution:
   a. $0.7253 × 300 = $217.59

2. Using the data in Figure 20-1, determine the value in U.S. dollars of 3,000 units of each foreign currency listed below. (Round answers to the nearest cent.)

Foreign Currency	Value of 3,000 units
a. Argentinean peso	$1,012.50
b. British pound	$5,394.90
c. Danish krone	$503.97
d. Indian rupee	$65.43

   Illustrative solution:
   a. $0.3375 × 3,000 = $1,012.50

   Score for A (44)

Chapter 20   International Business

Assignment 20.1 Continued

B (56 points) Solve the following problems. Round pennies to the nearest dollar. (8 points for each correct answer)

3. Hadley Enterprises has contracted to sell certain goods to a company in Britain. The price agreed on for the goods is 80,000 British pounds. On the date the contract was signed, the financial section of the local paper showed that the British pound was valued at $1.6554 U.S.

   a. How much in U.S. dollars does Hadley Enterprises expect to receive for the goods? $132,432

   80,000 pounds × $1.6554 U.S. per pound = $132,432

   b. If the value of the British pound fell from 1.6554 to 1.550 on the date of payment, how much would Hadley Enterprises lose by having contracted in British pounds instead of U.S. dollars? $8,432

   80,000 pounds × $1.550 U.S. per pound = $124,000
   $132,432 expected − $124,000 received = $8,432 lost

   c. If the British pound rose to 1.7500 on the date of payment, how much would Hadley Enterprises gain by having contracted in British pounds instead of U.S. dollars? $7,568

   80,000 pounds × $1.7500 U.S. per pound = $140,000
   $140,000 received − $132,432 = $7,568 gain

4. Miller Furniture Company imported 150 chairs from a Danish firm. Each chair is valued at 890 Danish kroner. What is the value of the chairs in U.S. dollars if the Danish krone is currently valued at 0.1694? $22,615

   150 chairs × 890 Danish kroner × $0.1694 U.S. per krone = $22,615

5. Oldtown Industries, Inc., is contracting to sell its product to a country whose currency is unstable and difficult to convert to U.S. currency. The value of the goods is $20,000 U.S. The currency of the country to which the goods will be shipped is currently valued at 0.0040 per U.S. dollar. Oldtown Industries is willing to accept the currency of a third country. The Singapore dollar is agreed on. The Singapore dollar is shown as 0.6428 on the date the contract is signed.

   a. How many Singapore dollars does Oldtown Industries expect to receive?
   (Round the answer to the nearest dollar.) 31,114

   $20,000 ÷ 0.6428 = 31,114

   b. If the Singapore dollar does not change before the date of payment, but the value of the currency of the receiving country falls from 0.0040 to 0.0003, how much did Oldtown Industries save by using the Singapore dollar? $18,500

   $20,000 U. S. ÷ 0.0040 currency of importing country = 5,000,000 units of currency expected
   5,000,000 × 0.0003 new value of currency = $1,500 U.S. on payment day
   $20,000 price − $1,500 on payment day = $18,500 U.S. saved

6. If the British pound is valued at 1.9000 per U.S. dollar and the Egyptian pound at 0.3700, how many more Egyptian pounds than British pounds could a U.S. citizen buy for $1,000 U.S.? (Round the answer to the nearest pound.) (10 points) $2,177

   $1,000 ÷ 1.9000 = 526 British pounds
   $1,000 ÷ .3700 = 2,703 Egyptian pounds
   2,703 − 526 = 2,177 more Egyptian pounds

Score for B (56)

Assignment 20.2: Duties and Metric Conversion

Name

Date　　　　　　　　Score

Learning Objectives **3** **4**

A **(56 points) Solve the following problems. (points for correct answers as marked)**

1. Benjamin's Department Store ordered from a foreign country 150 music boxes on which an ad valorem duty rate of 3.2% is charged. Payment is to be made in U.S. dollars. The price of each music box is $18. (2 points for each correct answer)

    a. What is the price of the 150 music boxes before duty is added? $2,700

    　　150 music boxes × $18 each = $2,700 cost before duty

    b. What is the amount of duty charged on the shipment? $86.40

    　　$2,700 × 0.032 duty = $86.40

    c. What is the total cost to Benjamin's? $2,786.40

    　　$2,700 + $86.40 = $2786.40 total cost

2. Gems International Company is purchasing from a foreign country one gross (144) of 20-inch gold necklaces at $75 each and six dozen 18-inch silver necklaces at $55 each. The ad valorem duty rate for gold and silver jewelry is 7%. What is the total cost of the shipment to the buyer? (8 points) $15,793.20

    144 gold necklaces × $75 each = $10,800 cost before duty

    72 (6 × 12) silver necklaces × $55 each = $3,960 cost before duty

    $10,800 + $3,960 = $14,760 total cost before duty

    $14,760 × 0.07 = $1,033.20 duty

    $14,760 + $1,033.20 = $15,793.20 total cost with duty

3. Sutter's Department Store is going to buy four gross (one gross = 144) of vases for the next Christmas season. It can buy porcelain vases or lead crystal vases for $45 each. The duty on porcelain vases is 9%. The duty on lead crystal vases is 4%. How much will Sutter's save in total cost by purchasing lead crystal instead of porcelain? (8 points) $1,296

    4 gross = 4 × 144 = 576 vases

    576 × $45 = $25,920 without duty

    9% duty on porcelain − 4% duty on lead crystal = 5% less on crystal

    $25,920 × 0.04 = $1,296 saved

4. Melody Piano Store can purchase pianos domestically for $1,360 each. It can purchase pianos from a foreign country for $1,300 plus 5.3% ad valorem duty.

    a. Melody Piano Store purchases the pianos with the lower total cost. Does it purchase from a domestic or a foreign manufacturer? (6 points) Domestic

    　　$1,300 × 0.053 = $68.90 duty import

    　　$1,300 + $68.90 = $1,368.90 total cost from foreign manufacturer

    b. How much does it save on each piano? (2 points) $8.90

    　　$1,368.90 foreign − $1,360 domestic = $8.90 saved per piano

Assignment 20.2 Continued

5. Broadway Office Equipment Company purchased the following equipment from a foreign country:

   72 automatic typewriters at $150 each + 2.2% duty
   24 addressing machines at $30 each + 4.2% duty
   144 pencil sharpeners at $12 each + 6% duty
   24 check-writing machines at $60 each, duty free
   80 calculators at $24 each + $3.9% duty

   a. What was the cost of the order before duty? (8 points) $16,608

   72 typewriters × $150 = $10,800
   24 addressing machines × $30 = $720
   144 pencil sharpeners × $12 = $1,728
   24 check-writing machines × $60 = $1,440
   80 calculators × $24 = $1,920
   $10,800 + $720 + $1,728 + $1,440 + $1,920 = $16,608

   b. What was the cost of the order after duty? (Round each calculation to the nearest cent.) (8 points) $17,042.30

   Typewriters: $10,800 × 0.022 = $237.60
   Addressing machines: $720 × 0.042 = $30.24
   Pencil sharpeners: $1,728 × 0.06 = $103.68
   Check-writing machines: No duty
   Calculators: $1,920 × 0.039 = $74.88
   $237.60 + $30.24 + $103.68 + $74.88 = $446.40
   $16,608.00 before duty + $446.40 duty = $17,054.40

   c. If the 144 pencil sharpeners had been purchased at $12 each from a country with which trade was discouraged and the ad valorem duty rate was 50%, how much would the pencil sharpeners have cost? (4 points) $2,592

   $1,728 × 0.5 = $864 duty
   $1,728 + $864 = $2,592 total cost

   d. How much more duty would a buyer pay on the pencil sharpeners at the ad valorem rate of 50% than at an ad valorem duty rate of 6%? (2 points) $760.32

   0.50 ad valorem rate − 0.06 ad valorem rate = 0.44 difference
   $1,728 × 0.44 difference = $760.32 more

6. Adams Industries could purchase $30,000 worth of textiles from country A with an ad valorem duty rate of 2.5% or from country B with an ad valorem duty rate of 1.2%.

   a. How much would the shipment cost if purchased from country A? (2 points) $30,750

   $30,000 × 0.025 = $750 duty
   $30,000 + $750 = $30,750 country A

   b. How much would Adams Industries save by purchasing from country B? (2 points) $390

   $30,000 × 0.012 = $360 duty
   $30,000 + $360 = $30,360 country B
   $30,750 − $30,360 = $390 savings

Score for A (56)

Assignment 20.2 Continued

B **(24 points) Solve the following problems. (points for correct answers as marked)**

7. The Allied Computer Company imports some computer components and manufactures other components and then assembles computers for sale within the United States or for export to foreign countries. The company is located in a district that has been designated by the International Trade Administration as a foreign trade zone. The company imported $250,000 worth of monitors having an ad valorem duty rate of 3.7%, $300,000 worth of power supplies having an ad valorem duty rate of 3.0%, and $500,000 worth of printers having an ad valorem duty rate of 3.7%. All products were finished and sold 1 year later.

   a. If all products were sold within U.S. Customs territories, how much duty—in U.S. dollars—did the company pay at the end of the year? (2 points) $36,750

   $250,000 × 0.037 = $9,250 duty on monitors

   $300,000 × 0.03 = $9,000 duty on power supplies

   $500,000 × 0.037 = $18,500 duty on printers

   $9,250 + $9,000 + $18,500 = $36,750 total duty

   b. If 40% of the finished products were moved into U.S. Customs territories for sale and 60% were exported for sale in foreign countries, how many dollars of duty did the company pay at the end of the year? (8 points) $14,700

   $36,750 total duty × 0.40 subject to duty = $14,700 duty paid

   c. If all products were exported for sale, how much duty did the company pay at the end of the year? (2 points) 0

   No duty is paid on exports.

8. The Allied Computer Company imported $260,000 worth of portable computers having an ad valorem duty rate of 3.9% and kept 20% of them for exhibition and company use on the premises.

   a. If the company repackaged and sold the remaining portable computers in U.S. Customs territories, how many dollars of duty did the company pay on the portable computers? (4 points) $8,112

   100% − 20% = 80%, or 0.8 subject to duty

   $260,000 × 0.8 = $208,000 subject to duty

   $208,000 × 0.039 duty rate = $8,112 duty paid

   b. If the company repackaged and exported 50% of the portable computers for sale in foreign countries and moved the remaining 30% into U.S. Customs territories for sale, how many dollars did the company pay in duty on the portable computers? (4 points) $3,042

   $260,000 × 0.3 = $78,000 subject to duty

   $78,000 × 0.039 duty rate = $3,042 duty paid

9. A company imported $5 million worth of laptop computers having an ad valorem duty rate of 3.9%. The company repackaged and exported all the computers for resale. How many dollars did the company save by being located in a foreign trade zone? (4 points) $195,000

   $5,000,000 × 0.039 duty rate = $195,000 saved

Score for B (24)

Assignment 20.2 Continued

C (20 points) Solve the following problems using Figure 20-2. (1 point for each correct answer)

10. Make the following conversions from U.S. measures to metric:
    a. Convert 15 inches to meters: 0.381 m   15 × 0.0254
    b. Convert 15 feet to meters: 4.575 m   15 × 0.305
    c. Convert 15 yards to meters: 13.71 m   15 × 0.914
    d. Convert 15 miles to kilometers: 24.135 km   15 × 1.609
    e. Convert 25 ounces to grams: 708.75 g   25 × 28.35
    f. Convert 25 pounds to grams: 11,350 g   25 × 454
    g. Convert 25 pounds to kilograms: 11.35 kg   25 × 0.454
    h. Convert 30 pints to liters: 14.19 L   30 × 0.473
    i. Convert 30 quarts to liters: 28.38 L   30 × 0.946
    j. Convert 30 gallons to liters: 113.55 L   30 × 3.785

11. Make the following conversions from metric to U.S. measures:
    a. Convert 15 meters to inches: 590.55 in.   15 × 39.37
    b. Convert 15 meters to feet: 49.215 ft   15 × 3.281
    c. Convert 15 meters to yards: 16.35 yd   15 × 1.09
    d. Convert 15 kilometers to miles: 9.315 mi   15 × 0.621
    e. Convert 25 grams to ounces: .875 oz   25 × 0.035
    f. Convert 25 grams to pounds: .055 lb   25 × 0.0022
    g. Convert 25 kilograms to pounds: 55 lb   25 × 2.2
    h. Convert 30 liters to pints: 63.39 pt   30 × 2.113
    i. Convert 30 liters to quarts: 31.71 qt   30 × 1.057
    j. Convert 30 liters to gallons: 7.92 gal   30 × 0.264

Score for C (20)

Notes

Part 6

Corporate and Special Applications

21 Corporate Stocks
22 Corporate and Government Bonds
23 Annuities
24 Business Statistics

Corporate Stocks

21

Learning Objectives
By studying this chapter and completing all assignments you will be able to:

Learning Objective 1 — Compute the costs and proceeds of stock buy-and-sell transactions.

Learning Objective 2 — Compute the costs and proceeds of round and odd lots.

Learning Objective 3 — Compute rates of yield and gains or losses on the purchase and sale of stocks.

Learning Objective 4 — Compute comparative earning potential of the major classes of corporate stocks.

21.1 Students are generally familiar with the New York Stock Exchange (NYSE), and it's proper to point out that *all* companies listed on the New York and other major stock exchanges are corporations with an issue of some form of corporate stock.

Many companies operate as corporations. A **corporation** is a body that is granted a charter by a state legally recognizing it as a separate entity, having its own rights, privileges, and liabilities distinct from those of its owners. A corporation acquires assets, enters into contracts, sues or is sued, and pays taxes in its own name. Two primary reasons for forming a corporation are to limit liability and facilitate broadening the ownership base. A corporation raises capital by selling shares of ownership, which increases its assets without increasing its debt.

The general term applied to the shares of a corporation is **capital stock**. Each share of capital stock is a share of the ownership of the company's net assets (assets minus liabilities). The number of shares that a corporation is authorized to *issue,* or offer for sale, is set forth in its **charter,** the basic approval document issued by the state, under which the corporation operates. Ownership of stock is evidenced by a **stock certificate**.

Frequently, the shares of capital stock are assigned a value known as **par**, which is stated on the stock certificate. For example, a company incorporated with capital stock of $1,000,000 and 100,000 shares has a par value of $10 per share. Stock issued without par value is known as **no-par** *stock*. The par value may differ from the market price. In the marketplace, stock may be sold for any amount agreed upon by the buyer and seller.

Computing the Costs and Proceeds of Stock Transactions

Learning Objective 1

Compute the costs and proceeds of stock buy-and-sell transactions.

21.2 The NYSE has been in continuous operation since 1792, or more than 200 years. It traces its beginning to the Revolutionary War, when huge amounts of money had to be mobilized.

21.3 Many students have never seen a copy of *The Wall Street Journal*, so a copy of a recent edition should be obtained and shown to the class.

21.4 Today, the distinction between "full-service" and "discount" brokers is becoming less meaningful as more institutions such as banks are becoming eligible to conduct stock exchange transactions. The number of discount brokers has increased dramatically in the past decade.

After purchasing stock, a buyer may sell that stock at any price on the open market, regardless of the par value. Stocks are usually bought and sold on **stock exchanges**, the formal marketplaces set up for the purpose of trading stocks. Major exchanges in the United States are the New York Stock Exchange (NYSE), the American Stock Exchange (AMEX), and the National Association of Securities Dealers Automated Quotations (NASDAQ). A **stockbroker** usually handles **stock transactions**—the purchase and sale of stocks for clients. Today, many people also trade via the Internet.

The trading of shares of stock is published daily in newspapers. Figure 21-1 shows a sample stock market report, in which stocks are quoted in the traditional manner—dollars and fractions of a dollar. The NYSE, NASDAQ, and AMEX quote prices in hundredths. Consequently, the smallest increase or decrease in a stock price that will be reported is .01.

Both the buyer and the seller of stock pay commissions to the stockbroker. The total amount paid by a buyer to purchase a stock includes the market price of the stock and the stockbroker's commission (charge). The **total cost** paid by the purchaser is equal to the purchase price plus a broker's commission. The **proceeds** received by the seller are equal to the selling price minus the commission.

Broker commissions may be a flat rate per transaction, a percent of the value of the stock, an amount per share traded, or an amount negotiated between the client and the broker. Generally, commissions for brokers are less than 1% of the value of the stock, ranging from $0.02 to $0.50 per share bought or sold. A number of discount brokerages operating on the Internet now charge $7.00 to $22.99 per transaction, normally for up to 5,000 shares. Figure 21-2 shows a broker's confirmation report of a stock purchase with a commission rate of $50 and a transaction fee of $3.

We use a transaction charge of $0.20 per share or a flat fee of $19.95 per transaction in computing the cost of commissions in this chapter.

426 Part 6 Corporate and Special Applications

Figure 21-1 Daily Stock Report from the NYSE

52 weeks High [1]	52 weeks Low [2]	Stock [3]	Sym [4]	Div [5]	Yld % [6]	PE [7]	Vol 100s [8]	Hi [9]	Low [10]	Close [11]	Chg. [12]
60.45	50.45	WalMart	WMT	1.12	2.2	21	4672	51.7	51.12	51.45	+ .14
25.80	45.95	C Timber	CRT	2.11	6.42	19	242	44.35	43.80	44.29	− .5
45.59	37.70	Kellogg	K	1.01	2.36	19	68146	43.61	42.60	42.81	+ .62
8.3	3.36	SixFlags	PKS			dd*	9621	4.52	4.32	4.35	− .12
99.96	68.50	Caterpillar	CAT	1.56	1.7	17	44329	97.87	95.53	96.40	+ 1.05
58.94	38.04	Boeing	BA	0.77	1.75	25	36988	57.19	56.57	57.16	+ .2
53.50	38.30	CocaCola	KO	1.00	2.7	21	156186	41.52	41.35	41.46	− .18
69.8	51.21	Deere Co	DE	1.06	1.02	12	25415	68.49	67.49	68.37	− .4
59.39	31.21	Sears	S	0.92	1.62	37	53205	57.43	56.88	56.95	+ .13

*dd = Loss in the most recent four quarters.

[1] The highest price per share in the previous 52 weeks.
[2] The lowest price per share in the previous 52 weeks.
[3] Company names, often abbreviated to fit in stock tables, are listed alphabetically.
[4] The symbol is a stock's designation on databases and quote machines.
[5] The dividend shown usually is the annual rate based on the company's last payout.
[6] The dividend divided by the closing share price gives the stock's yield.
[7] One measure of a stock's value is its **price/earnings ratio (P/E)**. It is based on the per-share earnings as reported by the company for the four most recent quarters. The PE number is found by dividing the current price by those most recent four-quarter earnings.
[8] Volume is the number of shares traded that day, shown in hundreds of shares.
[9] The high for the day's trading range.
[10] The low for the day's trading range.
[11] The closing price on that day.
[12] The net change in price lets you calculate something that isn't in the stock table: the previous day's closing price.

Figure 21-2 Confirmation Report of a Stock Purchase

A.G. Edwards & Sons, Inc.
INVESTMENTS SINCE 1887
ONE NORTH JEFFERSON   ST. LOUIS, MISSOURI 63103   (314) 955-3000

WE CONFIRM THE FOLLOWING TRANSACTION SUBJECT TO THE AGREEMENT ON THE REVERSE SIDE

YOU	QUANTITY	PRICE	SECURITY DESCRIPTION	CUSIP NUMBER
BOUGHT	50	35.47	GENERAL ELECTRIC CO	

ACCOUNT NUMBER	IB	T	TRF	MKT	OFFICE PHONE NUMBER	SYMBOL
	47	1	4	3		GE

WHEN COMMUNICATING WITH US PLEASE REFER TO YOUR ACCOUNT NUMBER

TRADE DATE			SETTLEMENT DATE		
12	31	04	01	06	05

PLEASE PAY OR DELIVER BY THIS DATE

PRINCIPAL	STATE TAX	ACCRUED INTEREST	COMMISSION	SEC FEE	TRANSACTION CHARGE	AMOUNT
1,773.50			50.00		3.00	1,720.50

EXAMPLE A

Jennifer Low bought 200 shares of Sears stock at 50. What was her cost, including commission of $0.20 per share?

200 shares × $50 price	= $10,000	purchase price
200 shares × $0.20 commission	= + 40	commission
	$10,040	total cost

EXAMPLE B

Ken Yeager sold 800 shares of Applebee's International at 22.16, less commission of $0.20 per share. What were the proceeds of the sale?

800 shares × $22.16	= $17,728	selling price
800 shares × $0.20 commission	= − 160	commission
	$17,568	proceeds

EXAMPLE C

Juan Hernandez bought 500 shares of PepsiCo stock at 45.38. What was his cost, including a flat fee of $19.95?

500 shares × $45.38 price	= $22,690.00	purchase price
commission	= + 19.95	flat fee
	$22,709.95	total cost

✓ CONCEPT CHECK 21.1

David Cooper purchased 300 shares of Safeway at 19.02. He later sold the stock at 21.5. What was his gain/loss on the purchase and sale, after counting commissions of $0.20 per share on the purchase and the sale?

Purchase: 300 shares × $19.02 price = $5,706 purchase price
300 shares × $0.20 commission = +   60 commission
$5,766 total cost
Sale: 300 shares × $21.50 price = $6,450 selling price
300 shares × $0.20 commission = −   60 commission
$6,390 proceeds

$6,390 proceeds − $5,766 cost = $624 gain

Computing the Costs and Proceeds of Round and Odd Lots

Stocks are sold in round lots, odd lots, or a combination of the two. A **round lot** usually is 100 shares. An **odd lot** consists of any number of shares less than 100 (1 to 99 shares is an odd lot for a stock with a 100-share round lot). When odd lots are purchased, a small extra charge, or **odd-lot differential,** is commonly added to the round-lot price. The differential is added to the price for a purchaser and deducted from the price for the seller. In this book, we use a differential of 12.5 cents as the odd-lot rate.

Learning Objective 2

Compute the costs and proceeds of round and odd lots.

EXAMPLE D

Carson Grant bought 160 shares of U.S. Steel at 43. What was his cost?

Odd-lot purchase price = $43 + $0.125 = $43.125 per odd-lot share

100 shares × $43.00 round-lot price =	$4,300.00	round-lot total cost
60 shares × $43.125 odd-lot price =	2,587.50	odd-lot total cost
160 shares × $0.20 commission =	+ 32.00	commission
	$6,919.50	total cost

EXAMPLE E

Carson sold 160 shares of U.S. Steel at 43. What was the amount of his net proceeds?

Odd-lot selling price = $43 − $0.125 = $42.875

100 shares × $43.00 round-lot price =	$4,300.00	round-lot price
60 shares × $42.875 odd-lot price =	2,572.50	odd-lot price
160 shares × $0.20 commission =	− 32.00	commission
	$6,840.50	net proceeds

✓ CONCEPT CHECK 21.2

James O'Brien bought 160 shares of PG&E at 25.5. What was his total cost?

Odd-lot purchase price = $25.50 + $0.125 = $25.625
100 shares × $25.50 round-lot price = $2,550.00
60 shares × $25.625 odd-lot price = 1,537.50
160 shares × $0.20 commission = + 32.00
Total cost $4,119.50

Sarah Loeb sold 220 shares of Aetna at 153.25. What was the amount of her net proceeds?

Odd-lot selling price = $153.25 − $0.125 = $153.125
200 shares × $153.25 round-lot price = $30,650.00
20 shares × $153.125 odd-lot price = + 3,062.50
220 shares × $0.20 commission = − 44.00
Net proceeds $33,668.50

Computing the Rate of Yield and Gains or Losses

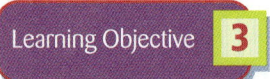

Learning Objective **3**

Compute rates of yield and gains or losses on the purchase and sale of stocks.

THE RATE OF YIELD

The **board of directors** is a group of people elected by shareholders to oversee the operations of the corporation. The board has sole authority to distribute earnings to shareholders. When such action is taken, the directors are said to **declare a dividend.** The rate of dividend is either a certain percent of the par value of the stock or a flat amount of money per share. Thus a dividend of 8% on a stock with a par value of $100 would be $8.00 per share. Most large corporations pay dividends quarterly.

The **rate of yield** from an investment in stock is the ratio of the dividend to the total cost of the stock.

EXAMPLE F

Aaron Ramos bought 300 shares of Wells Fargo stock at 32 and paid a $19.95 commission. A dividend of $2.15 per share was paid this year. What was the rate of yield?

$300 \times \$32 = \$9,600.00$ purchase price
$ + \ \ 19.95$ commission
$ \ \$9,619.95$ total cost

$300 \times \$2.15 \ \ = \$645$ dividend for first year
$\$645 \div \$9,619.95 = 6.7\%$ rate of yield

21.5 For a realistic business assignment, ask each student to select one NYSE company appearing in the business section of the local newspaper or the *Wall Street Journal*. The student can then use the current market quotation to calculate the gain/loss that would result from a sale today if the share had been purchased at the current year's low and high.

GAIN OR LOSS ON SALE OF STOCK

For income tax and accounting purposes, the amount of gain or loss on a sale of stock is determined by comparing the sale proceeds to the total cost.

EXAMPLE G

Refer back to example F. If Aaron sold his stock after 3 years at 36.5, less $19.95 commission, what were the amount and the percent of gain or loss?

$300 \times \$36.50 = \$10,950.00$ selling price
$ - \ \ \ 19.95$ commission
$ \ \$10,930.05$ proceeds

$\$10,930.05$ proceeds $- \$9,619.95$ cost (example F) $= \$1,310.10$ net gain
$\$1,310.10 \div \$9,619.95 = 13.6\%$ gain on sale

EXAMPLE H

Suppose that Aaron held his stock for 3 years and received a $645 dividend each year. Then to determine the total change in value (example G) he would need to add to his proceeds the $1,935 in dividends received.

Proceeds	Total Dividends	Total Cost		
($10,930.05	+ $1,935)	− $9,619.95	= $3,245.10	total gain in value

$\$3,245.10$ total gain $\div \$9,619.95$ initial cost $= 33.7\%$ gain in value

430 Part 6 Corporate and Special Applications

CONCEPT CHECK 21.3

a. Maria Sanchez owns 700 shares of stock with a par value of $100. If she receives a dividend of 5%, how much will her total dividend be?

$100 par value × 5% per share = $5.00 per-share dividend
700 shares × $5.00 per share = $3,500 total dividend

b. Maria also owns 300 shares of a stock without a stated par value. If she receives a dividend of $2.00 per share, what will her total dividend be?

300 shares × $2.00 per share = $600 total dividend

c. Magdalena Kaur bought 200 shares of Clorox at 32.25. A dividend of $0.45 per share was paid this year. What was the rate of yield?

200 shares × $32.25 = $6,450   purchase price
200 shares × $0.20 = + 40   commission
                      $6,490   total cost

200 shares × $0.45 dividend = $90 for first year
$90 dividend ÷ $6,490 total cost = 1.39% rate of yield

d. After 4 years, Magdalena sold the Clorox stock for 32.50. What were the amount and percent of gain or loss on the sale?

200 shares × $32.50 selling price = $6,500   selling price
200 shares × $0.20 commission   = − 40   commission
                                    $6,460   proceeds

$6,460 proceeds − $6,490 total cost = $(30) loss
$(30) loss ÷ $6,490 total cost = 0.46% loss

e. If Magdalena held the Clorox stock for 4 years, receiving the same $90 dividend each year, what was the total change in the value over the 4 years?

Proceeds	+	Total Dividends (4 years)	−	Total Cost	=	Gain in Value
($6,460	+	$360)	−	$6,490	=	$330

$330 gain in value ÷ $6,490 total cost = 5.08% gain

Computing Comparative Earning Potential

Common stock is the usual type of stock issued by a corporation. Another type frequently issued, **preferred stock,** gives holders a right to share in earnings and liquidation before common shareholders do. For example, a company that has a 7% preferred stock must first pay dividends of 7% of the par value to the holders of preferred stock before anything is paid to the holders of common stock. Preferred stock may be designated as **cumulative**—that is, if the corporation doesn't pay the specified percentage, the unpaid amount, called a **dividend in arrears**, carries over to the following year or years. If dividends aren't paid on noncumulative preferred stock during one year, the unpaid amount doesn't carry over to the next year.

Learning Objective 4

Compute comparative earning potential of the major classes of corporate stocks.

21.6 Preferred stock becomes most important when a company is having financial difficulty. In case of liquidation, funds are used first to pay creditors and then to pay shareholders. In the event that there are insufficient funds to pay all shareholders, those holding preferred stock are paid first.

EXAMPLE I

The ABC Company earned $48,000 last year. The capital stock of the company consists of 10,000 shares of 7% preferred stock, with a par value of $40 per share, and 50,000 shares of no-par common stock. If the board of directors declared a dividend of the entire earnings, what amount would be paid in total to the preferred and common shareholders and how much would each common shareholder receive?

Preferred: 10,000 shares × $40 par value = $400,000 total value
$400,000 value × 0.07 = $28,000 paid to preferred

Common: $48,000 total earnings − $28,000 paid to preferred = $20,000
$20,000 ÷ 50,000 shares = $0.40 paid per share to common

EXAMPLE J

Assume in example I that the preferred stock is cumulative and that for the preceding year the company had declared a dividend of only $16,000, or enough to pay a 4% dividend on preferred stock. The earnings of $48,000 for this year would be divided as follows:

Unpaid dividend from preceding year: 7% − 4% = 3%
$400,000 preferred par value × 0.03   = $12,000 cumulative (dividend in arrears)
$400,000 × 0.07 dividend for current year = $28,000
Total paid on preferred stock   = $40,000
$48,000 total earnings − $40,000 paid to preferred = $ 8,000
$8,000 ÷ 50,000 common shares   = $0.16 dividend per common share

Another feature that sometimes makes preferred stock an attractive investment is the possibility of converting the preferred stock into common stock. **Convertible preferred stock** gives the owner the option of converting those preferred shares into a stated number of common shares. For example, a stated conversion of 1 to 3 means that 1 share of preferred stock could be changed into 3 shares of common stock. The conversion feature combines the safety of preferred stock with the possibility of growth through conversion to common stock.

EXAMPLE K

Joel Turner owned 200 shares of GM convertible preferred stock at $20 par value. He converted each share of preferred into 3 shares of common. How many shares of common stock did Joel receive when he converted?

200 × 3 = 600 shares of common stock

If common stock was selling at $22 per share on the date of conversion, how much was Joel's common stock worth?

$22 × 600 shares = $13,200 common stock value

If Joel paid $42 per share for his preferred stock, how much had his investment increased?

$42 × 200 preferred = $8,400 preferred stock value
$13,200 − $8,400 = $4,800 increase in value

If the convertible stock pays 7% annually and the common stock usually pays $0.60 per share, how much more dividend might Joel expect to receive annually?

$20 par value × 200 shares = $4,000
$4,000 × 0.07 = $280 preferred stock dividend
600 shares × $0.60 = $360 common stock dividend
$360 − $280 = $80 more dividend annually

CONCEPT CHECK 21.4

a. The XYZ Corporation had a net profit of $120,000 in the fiscal year just ended. The capital stock consists of 8,000 shares of 8% convertible preferred stock with a par value of $50 per share and 20,000 shares of no-par common stock. If the board of directors declared a dividend of the entire earnings, what amount would be paid to preferred and common shareholders?

Preferred: 8,000 shares × $50 per share = $400,000 total par value
$400,000 par value × 8% = $32,000 paid to preferred shareholders
Common: $120,000 total earnings − $32,000 paid to preferred = $88,000 to be paid to common shareholders

b. Seth Ames owns 1,000 shares of convertible preferred stock in the XYZ Corporation, with a current market price of $52.00 per share. The preferred stock is convertible to common stock at the rate of 2 shares of common for each share of preferred. After the end of the year in part (a), common stock was selling for $32 per share. What would be the current market value of his stock before and after a conversion?

Preferred: 1,000 shares × $52 per share = $52,000 current value
Common: 1,000 shares preferred × 2 = 2,000 shares common
2,000 shares × $32 per share = $64,000 current value

COMPLETE ASSIGNMENTS 21.1 AND 21.2.

Chapter Terms for Review

board of directors	par
capital stock	price/earnings ratio (P/E)
charter	preferred stock
common stock	proceeds (from sale of stock)
convertible preferred stock	rate of yield
corporation	round lot
cumulative preferred stock	stockbroker
declare a dividend	stock certificate
dividend in arrears	stock exchanges
no-par stock	stock transactions
odd lot	total cost (for purchaser of stock)
odd-lot differential	

Try Microsoft® Excel

Try working the following problems using the Microsoft Excel templates found on your Student CD. Solutions for the problems are also shown on the CD.

1. Insert formulas in the shaded cells that will calculate the column amounts for **Total Cost, Total Proceeds, Amount of Gain or Loss,** and **Percent of Gain or Loss**.

   Hint: In calculating the total percent of gain or loss, be sure to use the total from the Amount of Gain or Loss column divided by the total from the Total Cost column.

Number of Shares	Cost Per Share to Purchase	Total Cost	Proceeds Per Share When Sold	Total Proceeds	Amount of Gain or Loss	Percent of Gain or Loss
200	$48.18		$51.60			
150	21.75		18.20			
190	15.00		28.85			
120	87.50		90.22			
550	16.10		15.90			
Total						

2. Add formulas to the following spreadsheet to calculate the **PE** (price to earnings) **Ratio** and the **Dividend Yield** for each stock.

Market Price	Earnings Per Share	Quarterly Dividends Per Share	PE Ratio	Percent of Dividend Yield
$65.80	$4.82	$0.95		
21.00	1.75	0.15		
125.00	8.1	1.75		
12.75	0.55	0.12		
34.00	1.92	0.45		

THE BOTTOM LINE

Summary of chapter learning objectives:

Learning Objective	Example
21.1 Compute the costs and proceeds of stock buy-and-sell transactions	For calculations throughout, use $0.20 a share for commissions and $0.125 for the odd-lot differential. Round all percents to two places. 1. Ahmad Ansari bought 100 shares of Disney at 26.59. What was the total cost of the purchase of common stock? 2. Ahmad sold 200 shares of Hasbro at 16.5. What were the proceeds of the sale?
21.2 Compute the costs and proceeds of round and odd lots	Elaine Fisher purchased 1,000 shares of Sysco common stock at 46 and 340 shares of preferred stock at 92. 3. What was the total cost of the purchase of common stock? 4. What was the total cost of the purchase of preferred stock?
21.3 Compute rates of yield and gains or losses on the purchase and sale of stocks	5. Douglas Mason purchased 320 shares of MMM at 81 and sold them 1 year later at 92.35. What were his total cost, net proceeds, and amount of gain on these two transactions?
21.4 Compute comparative earning potential of the major classes of corporate stocks	The MB Leasing Corporation earned $350,000 last year. The capital stock of the company consists of 20,000 shares of 6% preferred stock, with a par value of $50 per share, and 40,000 shares of no-par common stock. The board of directors declared a dividend of $280,000. 6. What amount will be paid to the preferred shareholders? 7. What amount per share will be paid to the common shareholders? 8. Sam Sosa owned 250 shares of Dow Chemical convertible preferred stock with a $50 par value. He converted each share of preferred into 3 shares of common. How many shares of common stock did he receive? 9. If the Dow Chemical common stock was selling at 26.50 on the day of the conversion, how much was his common stock worth?

Answers: 1. $2,679 2. $3,260 3. $46,200 4. $31,348 5. Total cost, $25,986.50; net proceeds, $29,485.50; gain $3,499.00 6. $60,000 7. $5.50 per share 8. 750 shares 9. $19,875

SELF-CHECK

Review Problems for Chapter 21

1 Use the following stock quotes from the NYSE to answer questions (a) through (e) below.

52 Weeks		Stock	Div	% Yld	PE	Vol 100s	Hi	Low	Close	Chg.
High	Low									
58	41	Boeing	1.06	2.2	21	2880	48.25	46	48.22	+2.21
96	80	Chevron	2.60	4.5	18	3267	83	81	82.45	−1.16

  a. How many shares of Boeing were traded?
  b. What was the closing price per share of Chevron in dollars and cents?
  c. What was the previous day's closing price for each stock?
  d. By how much has the price of 1 share of Boeing stock changed over the last 52 weeks?
  e. Use the P/E ratio to calculate the earnings per share for the last four quarters for Chevron.

2 Determine the total cost or proceeds of each purchase or sale. Include regular commission of $0.20 per share and an odd-lot differential of $0.125 per share.

  a. Purchased 300 shares of Caterpillar at 89.85.
  b. Purchased 550 shares of Hershey at 32.
  c. Sold 200 shares of Avon at 27.50.

3 Jason purchased 500 shares of XYZ stock at 17.12. One year later he sold the 500 shares at 18. He paid a transaction fee of $19.95 for each transaction.

  a. What was the amount of gain or loss on the sale?
  b. What was the rate of gain or loss?

4 Jason from question 3 received dividends of $0.65 per share during the year that he owned the stock.

  a. What was the rate of dividend yield?
  b. What was the total rate of gain or loss including the dividend?

5 Audrey owned 400 shares of Znix convertible preferred stock with a $20 par value. She converted all 400 shares into common stock at the rate of 4 to 1 (4 shares of common stock for each share of preferred). How many shares of common stock did she receive?

6 The Znix preferred stock from question 5 paid an annual dividend of 8%. Znix paid annual dividends on its common stock of $0.60 per share. How much more will Audrey receive each year in dividends by converting her stock from preferred to common?

7 Alpha Company's capital consists of 8,000 shares of $50 par 7.5% preferred stock and 50,000 shares of no-par common stock. The board of directors declared a dividend of $85,000. What is the dividend per share for preferred and common stock?

8 Assume the preferred stock in question 7 is cumulative and no dividends were declared the year before. Determine the dividend to be paid for each share of preferred and common if the board declares a total dividend of $90,000 the current year.

Assignment 21.1: Buying and Selling Stock

Name

Date   Score

Learning Objectives **1 2**

A (41 points) For calculations, use $0.20 a share for commissions unless the problem gives a flat fee and $0.125 for the odd-lot differential. Round all percents to two places. (5 points for a correct answer to problem 3; 4 points for each other correct answer)

1. Gail Sanders purchased 2,000 shares of JMK common stock at 18 and 180 shares of preferred stock at 60.

   a. What was the total cost of the purchase of common stock? $36,400

   2,000 shares × $18 = $36,000   purchase price
   2,000 × $0.20   = + 400   commission
                      $36,400   total cost

   b. What was the total cost of the purchase of preferred stock? $10,846

   $60 × 100   = $ 6,000   round lot
   ($60 + $0.125) × 80   = + 4,810   odd lot
                           $10,810   purchase price
   $0.20 × 180   = +   36   commission
                           $10,846   total cost

2. Three months later, Gail sold her 2,000 shares of JMK common stock at $21 and her 180 shares of preferred stock at $58.50.

   a. What were the proceeds on the sale of common stock? $41,600

   2,000 shares × $21 = $42,000 selling price
   2,000 × $0.20 = $400 commission
   $42,000 − $400 = $41,600 proceeds

   b. What were the proceeds on the sale of preferred stock? $10,484

   $58.50 × 100 = $5,850 round lot
   ($58.50 − $0.125) × 80 = $4,670 odd lot
   $5,850 + $4,670 = $10,520 sale price
   $0.20 × 180 = $36 commission
   $10,520 − $36 = $10,484 proceeds

   c. How much did Gail gain or lose on the purchase and sale of all of her JMK stock? $4,838 gain

   $ 36,400   total cost of common stock
   + 10,846   total cost of preferred stock
   $ 47,246   total cost of JMK stock
   $ 41,600   proceeds of common stock
   + 10,484   proceeds of preferred stock
   $ 52,084   proceeds of sale

   $52,084 − $47,246 = $4,838 gain

3. Susan Lu purchased 200 shares of Telmart common stock at $88.50 and paid a $19.95 transaction fee. A dividend of $7.00 per share was paid the first year. What was the rate of yield? 7.9%

   $88.50 × 200 = $17,700 purchase price
   $17,700 + $19.95 transaction fee = $17,719.95 total cost
   $7.00 × 200 = $1,400 dividend for 1 year
   $1,400 ÷ $17,719.95 = 7.9% rate of yield

Chapter 21 Corporate Stocks **437**

Assignment 21.1   Continued

4. Sheri Jeffers purchased stock for a total cost of $12,600, including commission. She sold the stock a month later for $13,960, after commission.

   a. What was her net gain on the sale? $1,360

   $ 13,960   proceeds of sale
   − 12,600   cost of stock
   $ 1,360   net gain

   b. What was her percent of gain on the sale? 10.79%

   $1,360 ÷ $12,600 = 10.79% gain on sale

   c. If Sheri had held her stock another week and sold for $12,280 after commission, what would her percent of loss on the sale have been? 2.54%

   $ 12,600   cost of stock
   − 12,280   proceeds of sale
   $ 320   net loss
   $320 ÷ $12,600 = 2.54% loss on sale

5. If Sheri hadn't sold her stock for $12,280 but had waited another 3 months while the stock fell to a price where she could have realized net proceeds of $11,275, what would have been her percent of loss? 10.52%

   $ 12,600   cost of stock
   − 11,275   proceeds of sale
   $ 1,325   net loss
   $1,325 ÷ $12,600 = 10.52% loss on sale

   Score for A (41)

B (59 points) Solve the following problems. (points for correct answers as marked)

6. Peter Roncalio, Paul Stevens, and Mary Petrakas each invested $10,000 in different areas. Calculate the value of each $10,000 investment at the end of 2 years. (5 points for each correct answer)

   a. Peter put his $10,000 in a savings account that paid 6.2% interest annually. (Add interest on the savings account the first year to the principal before figuring interest for the second year.) $11,278.44

   $10,000 savings × 0.062 annual interest = $620 first-year interest
   $10,000 + $620 = $10,620 account balance end of first year
   $10,620 × 0.062 annual interest = $658.44 second-year interest
   $10,620 + $658.44 = $11,278.44 value of Peter's investment

   b. Paul bought 9%, $50 par value preferred stock at $62.50 a share, including commission. He received his full dividend at the end of each year. He sold his stock at the end of the second year. The sales proceeds, after commission, were $62.50 a share. $11,440

   $10,000 investment ÷ $62.50 per share = 160 shares purchased
   $50 par value per share × 0.09 dividend = $4.50 dividend per share
   160 shares × $4.50 dividend = $720 total annual dividend
   $720 × 2 years = $1,440 total dividend paid on investment
   $10,000 value of shares + $1,440 dividend = $11,440 value of Paul's investment

Assignment 21.1 Continued

    c. Mary bought common stock at $40 a share, including commission. Her stock paid quarterly dividends of 90 cents per share. In 2 years, the stock decreased to a value of $38.50 a share. $11,425

        $10,000 investment ÷ $40 per share = 250 shares
        250 shares × $0.90 per share = $225 dividend per quarter
        $225 × 8 quarters in 2 years = $1,800 total dividend paid on investment
        $38.50 value per share × 250 shares = $9,625 stock value
        $9,625 + $1,800 dividends = $11,425 value of Mary's investment

7. Find the amount of the dividend per share and the rate of yield per share for each of the following preferred stocks. The cost per share includes all commissions. (2 points for each correct answer)

    a. Cost per share $32; dividend declared $2.10.

        Amount of dividend $2.10
        Rate of yield 6.56%

        $2.10 ÷ $32 = 6.56% rate of yield

    b. Cost per share $80; par value $100; dividend declared 6%.

        Amount of dividend $6
        Rate of yield 7.5%

        $100 × 0.06 = $6.00 amount of dividend
        $6.00 ÷ $80 = 7.5% rate of yield

    c. Cost per share $44.50; dividend declared $2.00.

        Amount of dividend $2
        Rate of yield 4.49%

        $2.00 ÷ $44.50 = 4.49% rate of yield

    d. Cost per share $90; par value $100; dividend declared 5.5%.

        Amount of dividend $5.50
        Rate of yield 6.11%

        $100 × 0.055 = $5.50 amount of dividend
        $5.50 ÷ $90 = 6.11% rate of yield

    e. Cost per share $58; par value $50; dividend declared 6.5%.
        Amount of dividend $3.25
        Rate of yield 5.6%

        $50 × 0.065 = $3.25 amount of dividend
        $3.25 ÷ $58 = 5.6% rate of yield

Assignment 21.1 Continued

8. Determine the amount and percent of gain or loss for each of the following transactions. Show an amount of loss in parentheses ( ). The purchase costs and the sale proceeds include commissions. Round percents to two decimal places. (3 points for each correct answer)

	Number of Shares	Per-Share Purchase Cost	Per-Share Sale Proceeds	Amount of Gain or Loss	Percent of Gain or Loss
a.	100	$47.20	$52.85	$565	11.97%
b.	250	12.00	14.50	625	20.83%
c.	140	22.30	20.70	(224)	−7.17%
d.	640	17.00	12.75	(2,720)	−25.0%

a. 100 × $47.20 = $4,720 purchase cost
100 × $52.85 = $5,285 sale proceeds
$5,285 − $4,420 = $565 gain on sale
$565 ÷ $4,720 = 11.97% gain on sale

b. 250 × $12 = $3,000 purchase cost
250 × $14.50 = $3,625 sale proceeds
$3,625 − $3,000 = $625 gain on sale
$625 ÷ $3,000 = 20.83% gain on sale

c. 140 × $22.30 = $3,122 purchase cost
140 × $20.70 = $2,898 sale proceeds
$3,122 − $2,898 = $224 loss on sale
$224 ÷ $3,122 = 7.17% loss on sale

d. 640 × $17.00 = $10,880 purchase cost
640 × $12.75 = $8,160 sale proceeds
$10,880 − $8,160 = $2,720 loss on sale
$2,720 ÷ $10,880 = 25.0% loss on sale

Score for B (59)

Assignment 21.2: Capital Stock

Name

Date Score

Learning Objectives **3** **4**

A (34 points) The information in problem 1 also applies to problems 2 and 3. (2 points for each correct answer)

1. The Duval Company was incorporated with 7% preferred capital stock of $500,000 and common stock of $1,800,000. The par value of the preferred stock was $100, and the par value of the common stock was $20. How many shares of each kind of stock were there?

   Preferred stock 5,000 shares $500,000 ÷ $100 = 5,000 shares
   Common stock 90,000 shares $1,800,000 ÷ $20 = 90,000 shares

2. Last year, dividends were declared by the Duval Company, which had earnings totaling $359,000.

   a. What was the total amount of the preferred stock dividend? $35,000
      $500,000 × 0.07 = $35,000

   b. What amount would have been paid on each share of common stock if all the earnings had been distributed?
      $3.60
      $359,000 − $35,000 = $324,000
      $324,000 ÷ 90,000 = $3.60 per share

3. The directors of the Duval Company actually declared four quarterly dividends of $0.75 a share on the common stock and $\frac{1}{4}$ of the amount due annually on the preferred stock.

   a. What was the total amount paid by Duval to all common shareholders for each quarterly dividend? $67,500
      $0.75 × 90,000 shares = $67,500

   b. What was the total amount paid to preferred shareholders each quarter? $8,750
      $500,000 × 0.07 = $35,000 paid annually: $35,000 ÷ 4 = $8,750 paid quarterly

   c. What was the quarterly per-share payment to preferred shareholders? $1.75
      $8,750 ÷ 5,000 = $1.75 per share

   d. What was the year's total amount of the common stock dividends? $270,000
      $67,500 × 4 quarters = $270,000

   e. What was the total amount of all dividends paid by Duval during the year? $305,000
      $8,750 per quarter × 4 quarters = $35,000; $270,000 + $35,000 = $305,000

   f. How much more in dividends was paid to each share of preferred than to each share of common? $4.00
      Common: 4 × $0.75 = $3.00 per year
      Preferred: 4 × $1.75 = $7.00 per year
      $7.00 − $3.00 = $4.00 more

Chapter 21 Corporate Stocks

Assignment 21.2 Continued

4. The capital stock of the Shubert Company consists of 300,000 shares of preferred stock and 5,500,000 shares of common stock. Last year, a dividend of $3.60 a share was declared on preferred stock and four quarterly dividends of $0.35 a share on common stock. How much was the total dividend for the year on each class of stock?

   Preferred stock $1,080,000   300,000 shares × $3.60 = $1,080,000
   Common stock $7,700,000   5,500,000 shares × $0.35 × 4 quarters = $7,700,000

5. ComputerMart has 150,000 shares of 6.5% preferred stock at $1 par value and 1,500,000 shares of common stock. ComputerMart declared total dividends of $250,000 for the current year. How much was the total dividend for preferred stock and how much was the dividend per share on the common stock?

   Preferred stock $9,750   150,000 shares × $1 per share = $150,000 × 0.065 = $9,750 dividend
   Common stock $0.16 $250,000 − $9,750 = $240,250 dividend
   (per share) $240,250 ÷ 1,500,000 shares = $0.16 per share

6. Michael Wu bought 300 shares of XRT 8% preferred stock, $10 par value, when it was selling at $11 per share, including commission.

   a. What was Michael's stock worth at the time of purchase? $3,300
      $11 per share × 300 shares = $3,300

   b. What was the amount of Michael's quarterly dividend? $60
      $10 × 0.08 = $0.80 annual dividend per share
      $0.80 ÷ 4 = $0.20 quarterly dividend per share
      $0.20 × 300 shares = $60 quarterly dividend

   c. What was Michael's dividend yield? 7.27%
      $60 × 4 = $240 annual dividend
      $240 ÷ $3,300 = 7.27%

Score for A (34)

B (66 points) Do not consider commission in the following problems. (points for correct answers as marked)

7. Inland Sales, Inc., has issued 25,000 shares of 8%, $20 par, cumulative preferred stock and 50,000 shares of common stock. The board of directors declares 50% of net income each year as dividends. Inland Sales had net income of $76,000 for 2000, $112,000 for 2001, and $130,000 for 2002. Compute the annual dividends per share for preferred and common stock for each of the 3 years. (2 points for each correct answer)

Year	Preferred Dividends/Share	Common Dividends/Share
2000	$38,000 ÷ 25,000 = $1.52	-0-
2001	$42,000 ÷ 25,000 = $1.68	$14,000 ÷ 50,000 = $0.28
2002	$40,000 ÷ 25,000 = $1.60	$25,000 ÷ 50,000 = $0.50

25,000 shares × $20 par × 8% = $40,000/yr for preferred

Dividends declared: Dividends in arrears:

2000	$76,000 × 50% = $38,000	$40,000 − $38,000 = $2,000
2001	$112,000 × 50% = $56,000	
2002	$130,000 × 50% = $65,000	

Assignment 21.2 Continued

8. Dan Baxter owned 200 shares of Sony 6.5% convertible preferred stock, $50 par value, for which he paid $56 per share, including commission. Two years later, after receiving preferred dividends each year, he converted to 600 shares of Sony common stock, valued at $23.50 a share at the time of conversion. (4 points for each correct answer)

    a. What was the cost to Dan of the preferred stock? $11,200

    $56 × 200 shares = $11,200

    b. How much did Dan receive in dividends from the preferred stock? $1,300

    $50 × 200 = $10,000 par value
    $10,000 × 0.065 × 2 years = $1,300

    c. What was the value of the common stock that Dan received? $14,100

    $23.50 × 600 = $14,100 common stock value

    d. If he sells the 600 common shares immediately, how much gain will Dan realize, including his dividend? $4,200

    $14,100 + $1,300 = $15,400 total return
    $15,400 − $11,200 = $4,200 total gain

    e. What would be Dan's percent of gain? 37.5%

    $4,200 ÷ $11,200 = 37.5%

9. Texas Air Corporation issued 5,000,000 shares of 7% preferred stock at $100 par value and 10,000,000 shares of no-par common stock. Bob Thruston owned 100 shares of preferred. Barbara Beck owned 500 shares of common. In 2005, Texas Air paid $25,000,000 in dividends to its common shareholders. How much more than Bob did Barbara receive? (10 points) $550

    Bob: 100 shares × $100 × 0.07 = $700
    Barbara: $25,000,000 ÷ 10,000,000 shares = $2.50 per share
    $2.50 × 500 shares = $1,250
    $1,250 − $700 = $550 more for Barbara

10. Sonia Revas owned 700 shares of PIE 6% convertible stock, $50 par value, for which she paid $42 a share. She received a dividend for 1 year. She then converted the preferred stock to 400 shares of common stock valued at $98.50 a share. (4 points for each correct answer)

    a. What was the cost to Sonia for her preferred stock? $29,400

    $42 × 700 shares = $29,400 cost of preferred stock

    b. How much did Sonia receive as a dividend for her preferred stock? $2,100

    $50 × 700 shares = $35,000 par value
    $35,000 × 0.06 = $2,100 preferred dividend

    c. What was the value of her common stock at the time of conversion? $39,400

    $98.50 × 400 shares = $39,400 common stock value

    d. If the common stock paid an annual dividend of $6.00 a share, how much more dividend would she receive annually? $300

    $6.00 × 400 shares = $2,400 common dividend
    $2,400 − $2,100 = $300 more

    e. What was Sonia's percent of increase in annual return as a result of conversion to common stock? 14.29%

    $300 ÷ $2,100 = 14.29%

Assignment 21.2 Continued

11. Determine the price/earnings ratio (P/E) of each of the following stocks: (2 points for each correct answer)

    a. JBC common stock has a current market price of $49 and has had earnings per share of $0.72 each quarter for the last four quarters. <u>17</u>

    $49 ÷ ($0.72 × 4) = 17.01 rounded to 17

    b. The current market price of Cannon common stock is $72.88. Cannon has paid dividends of $1.20 per quarter for each of the last four quarters. <u>15</u>

    $72.88 ÷ ($1.20 × 4) = 15.18 rounded to 15

Score for B (66)

Corporate and Government Bonds

22

Learning Objectives
By studying this chapter and completing all assignments you will be able to:

Learning Objective 1 — Compute gains and losses on convertible and callable corporate bond transactions.

Learning Objective 2 — Compute annual interest on bonds.

Learning Objective 3 — Compute accrued interest on bond transactions made between interest payment dates.

Learning Objective 4 — Compute annual yield on bonds selling at a premium or a discount.

Learning Objective 5 — Compute a rate of yield to maturity.

When a corporation or government entity needs cash for a long period of time, usually 10 years or more, it often will issue long-term notes known as **bonds**. Bonds are bought and sold on the open market, much like stocks.

Two main types of **government bonds** are treasury bonds and municipal bonds. **Treasury bonds** are issued by the United States government. These bonds are fully guaranteed by the full faith and credit of the United States government. Bondholders are protected against default unless the federal government becomes insolvent. **Municipal bonds** are issued by states, cities, school districts, and other public entities. Unlike treasury bonds, municipal bonds pose a risk that the issuer might fail to repay the principal. Interest paid on municipal bonds generally is exempt from federal and state income taxes.

There are many kinds of **corporate bonds**, two of which are convertible bonds and callable bonds. **Convertible bonds** have a provision that they may be converted to a designated number of shares or a designated value of the corporation's stock. **Callable bonds** have a provision that the issuer can repurchase, or call in the bonds, at specified dates if the board of directors authorizes the retirement (payoff) of the bonds before their maturity date. Such action by the board of directors would be appropriate if interest rates fell significantly below the interest rate of the callable bond.

22.1 It should be noted that callable bonds are usually called during periods when interest rates fall significantly. Suppose XYZ callable bonds were issued at 12%. If interest rates fall to 8%, XYZ Corporation can secure a bank loan or issue new bonds at 8% and call the 12% bonds, thus saving 4% on borrowing costs.

Computing Gains and Losses on Corporate Bonds

Learning Objective 1

Compute gains and losses on convertible and callable corporate bond transactions.

EXAMPLE A

Steve Bando bought one ABC Corporation convertible bond for $1,000. The bond was convertible to 100 shares of stock. At the time of the purchase, the stock was selling for $10 per share. At the end of 1 year, the stock was selling for $15 per share. Steve converted his bond. Assuming that the market value of the bond hadn't changed, how much profit did Steve realize by converting?

100 shares of stock $\times$ $15 per share = $1,500
$1,500 stock value − $1,000 bond value = $500 profit

EXAMPLE B

XYZ Corporation issued $1,000,000 worth of callable bonds paying 8% interest. The maturity date for the bonds was in 10 years. Two years later, interest rates fell to 6%. The bonds were called, and new bonds were sold at the 6% rate. How much did XYZ Corporation save by calling the bonds?

10 years to maturity at issue − 2 years = 8 years remaining to maturity
8% − 6% = 2% savings per year
$1,000,000 $\times$ 2% = $20,000 interest saved per year
$20,000 $\times$ 8 years = $160,000 saved

CONCEPT CHECK 22.1

a. What would be the "stock" value of a bond that was convertible to 40 shares of stock if the stock was priced at 37.62?

40 shares × $37.62 = $1,504.80

b. If a company issued a callable bond at $7\frac{1}{2}$% interest, would it be likely to call the bond if the current rate of interest was 8%?

No, because it could invest the cash at an extra $\frac{1}{2}$% interest.

Computing Annual Interest on Corporate and Government Bonds

When first issued, bonds are sold either through brokerage houses or directly to investors at or near the price of $1,000, called face value. **Face value** represents the amount that will be paid to the holder when the bonds are redeemed at maturity. If the market value becomes less than the face value, the bond sells at a **discount**. If the market value becomes more than the face value, the bond sells at a **premium**. (The discount or premium amount is the difference between the market value and the face value.)

Bonds are rated. By checking a bond's rating, buyers can have some indication of how safe their bond investment is. **Bond ratings** are information based on experience and research; they are not a guarantee. One major firm rating bonds is Standard & Poor's.

In Standard & Poor's system, the ratings include AAA (the highest rating), AA, A, BBB, BB, B, CCC, CC, C, and D. A bond with a low rating is a higher-risk bond and sometimes is known as a **junk bond**. The lower a bond's rating, the higher are its yield and its risk.

Learning Objective 2

Compute annual interest on bonds.

22.2 Bond rating has become very important for private and public organizations. Today, many corporate and government bonds are purchased by pension funds, college reserve funds, and insurance companies. These organizations frequently work with a "safety net" set by a board of trustees or a management group. The safety net limits purchases to bonds at or above a certain rating, such as AAA or AA.

EXAMPLE C

Kiley Moore purchased a $1,000 bond with a rating of B, paying 14% per year. Mary Baker purchased a $1,000 bond with a rating of AAA, paying 5% per year. Jean Carlson purchased a $1,000 junk bond, paying 25% per year. Each bond was to mature in 10 years.

Kiley's B-rated bond paid faithfully for 4 years. Then the company filed for bankruptcy and paid 60 cents on the dollar. Mary's AAA-rated bond paid interest during its entire 10-year life and paid face value on maturity. Jean's junk bond paid interest for 3 years. Then the company filed for bankruptcy and paid 30 cents on the dollar.

Compute how much each investor received for her $1,000 investment.

Kiley: $1,000 × 14% = $140 annual interest
$140 × 4 years = $560 interest
$560 interest + (0.60 × $1,000) redemption = $1,160 total

Mary: $1,000 × 5% = $50 annual interest
$50 × 10 years = $500 interest
$500 interest + $1,000 redemption = $1,500 total

Jean: $1,000 × 25% = $250 annual interest
$250 × 3 years = $750 interest
$750 interest + (0.30 × $1,000) redemption = $1,050 total

How much would Kiley and Jean have received on their investments if the bonds had paid full interest for the 10-year period and face value on maturity?

Kiley: $1,000 × 14% × 10 years = $1,400
$1,400 + $1,000 = $2,400

Jean: $1,000 × 25% × 10 years = $2,500
$2,500 + $1,000 = $3,500

22.3 The business section of most metropolitan newspapers provides current real-world information. Examples taken directly from a current paper should be used to illustrate the difference between interest rates for government bonds with greater safety and those for corporate bonds with greater risk.

NEWSPAPER INFORMATION ON BONDS

Information about the market value and sale of bonds on the major exchanges is reported daily in financial newspapers. Figure 22-1 shows information usually included in a bond report.

Figure 22-1 Bond Market Report

Bonds	Current Yield	Volume	Close	Net Change
ATT 7½s09	7.2	10	104	+1
Aetna 6⅜s12	6.6	25	96.80	...
ClrkOil 9½s06	9.1	33	104.25	+.25
Hertz 7s12	7.0	13	99.70	+.70
IBM 7s25	7.4	102	94.50	+.80
RJR Nb 8s10	7.9	15	101.50	...

Prices of bonds are quoted in percents of face value. For example, a $1,000 bond quoted at 104 would sell at a premium price of $1,040 ($1,000 × 104%). If quoted at 87, the bond would sell at a discounted price of $870 ($1,000 × 87%).

Rule: Prices over 100 (100%) include a premium. Those under 100 (100%) include a discount.

The two main factors that influence the market price are the interest rate and the bond rating. For example, if a bond pays 8% interest and the current market rate of interest is greater than 8% for similarly rated bonds, the bonds will sell at a discount sufficient to make up for the difference in interest rates over the term of the bond.

Printed bond reports generally give a letter abbreviation for the company, the interest rate, a small s to designate *semiannual* (every 6 months) interest payments, and the maturity date, followed by the current yield, the number of bonds sold that day, the closing price of the bond, and the net change in price from the prior day.

The first line of the bond market report in Figure 22-1 would be interpreted as ATT (designating American Telephone and Telegraph), a $7\frac{1}{2}$ interest rate based on the face value of the bond, and interest paid semiannually. The bond matures in 2009. The current yield (average annual interest rate based on the current price of the bond) is 7.2%. The day's volume of bonds sold was 10. The closing price was 104, up 1 from the prior day.

EXAMPLE D

Calculate the amount of the semiannual interest check for a $1,000 bond reported in a financial paper as R&S Corp $7\frac{1}{2}$s21.

$1,000 face value $\times$ $7\frac{1}{2}$% = $75 $75 $\div$ 2 = $37.50 semiannual interest payment

COMMISSIONS FOR BUYING AND SELLING BONDS

The charge for buying and selling bonds varies among brokers, but there is no standard commission. Commissions are very small and thus comprise only a negligible part of the bond transaction. We do not use commission costs for problems in this textbook.

22.4 Some government bonds, such as treasury bonds, can be purchased directly from the government without commission; however, there is usually a minimum purchase amount of $5,000 or more.

✓ CONCEPT CHECK 22.2

If James Kun purchased 27 triple-A bonds that pay 7.1% and mature in 8 years, what amount of interest income could he expect annually?

$1,000 $\times$ 0.071 $\times$ 27 = $1,917

If James holds the bonds until maturity, how much will he receive on redemption of the bonds?

$1,000 $\times$ 27 = $27,000 total face value

Computing Accrued Interest on Bond Transactions

Most bonds specify that interest is payable quarterly, semiannually, or annually. The interest payment dates—such as January 1 (for interest through December 31) and July 1 (for interest through June 30)—are stated on the bond. When a bond is purchased between these dates, it is customary to add the **accrued interest** (interest earned from the last payment date to the purchase date). This interest is calculated by finding the number of days from the day on which interest was last paid through the day before the purchase and dividing this number by 360.

The buyer pays the seller for the interest accumulated or accrued on the bond since the last interest payment date. On the next regular interest payment date, the new owner receives the interest for the full interest period. This procedure allocates the interest correctly between the buyer and the seller for the split interest period because the corporation that issued the bond will pay the entire amount to whoever owns the bond as of each interest date.

Learning Objective 3

Compute accrued interest on bond transactions made between interest payment dates.

22.5 In computing the days of accrual for interest, interest is paid through December 31, which places the income in the year ending December 31. Interest for the next 6-month period will then begin on January 1, and that day is therefore counted in determining the number of days of accrual.

Chapter 22 Corporate and Government Bonds **449**

EXAMPLE E

A $1,000 bond, with interest at 8% payable semiannually on January 1 and July 1, was purchased on October 8 at 104 plus accrued interest. What is the number of days for which the accrued interest is paid?

Purchase date: October 8

Days of accrued interest: (July) 31 + (August) 31 + (September) 30 + (October) 7 = 99

What is the purchase payment for the bond?

$1,000 × 104% = $1,040 market value
$1,000 × 0.08 interest × $\frac{99}{360}$ accrued days = $22 accrued interest
$1,040 + $22 = $1,062 purchase payment for bond

In example E, although the accrued interest is an additional payment by the buyer, the buyer will get it back in the $40 ($1,000 × 8% × $\frac{1}{2}$) interest payment on January 1.

✓ CONCEPT CHECK 22.3

Ann Ahn purchased two Hertz 7s08 bonds at 95.6 on March 15. What amount did she pay her broker?

$2,000 × 0.956 = $1,912.00
Purchase date: March 15
(January) 31 + (February) 28 + (March) 14 = 73 days
$2,000 × 0.07 × $\frac{73}{360}$ = $28.39 accrued interest
$1,912.00 + $28.39 = $1,940.39 paid to her broker

Computing the Rate of Yield for Bonds

Learning Objective 4

Compute annual yield on bonds selling at a premium or a discount.

Interest on bonds provides income to bondholders. This income is referred to as **yield**. Newspapers and bond brokers refer to the annual yield of a bond as its **current yield**. Many newspaper bond reports include a column showing current yield. To calculate the current yield from an investment in bonds, use the following formula:

Annual interest ÷ Current purchase price = Current yield

When a bond is purchased at a discount, the current yield is greater than the face rate. For example, a $1,000 bond, purchased at 90, pays 7% interest and matures in 10 years. Interest of $70 ($1,000 × 7%) is paid annually, but as the bond was purchased for $900 ($1,000 × 90%), the effective rate, or yield, as a percent of cost is 7.8% ($70 ÷ $900).

When a bond is purchased at a premium, the current yield is less than the face rate. The reason is that the interest paid is calculated on the face value, and the yield is based on the higher market price.

● **EXAMPLE F**

Five $1,000 Levi Straus $9\frac{1}{2}$s19 bonds were purchased at 80. What was the current yield on the bonds?

$1,000 × 5 = $5,000 face value
$5,000 × 80% = $4,000 purchase price
$5,000 × 0.095 = $475 annual interest
$475 ÷ $4,000 = 0.11875 = 11.9 % current yield
or 9.5 ÷ 0.80 = 11.875 = 11.9% current yield

22.6 When bonds are sold at *discount,* the yield rate will always be *higher* than the stated (face) rate. Likewise, when bonds are sold at a *premium,* the yield rate will always be *lower* than the stated (face) rate.

In example F, the bonds sold at a discount of $1,000 ($5,000 − $4,000) because the investor paid that much less for them than the maturity (face) value. Therefore, the current yield of 11.8% is more than the stated interest rate of $9\frac{1}{2}$%.

✓ **CONCEPT CHECK 22.4**

The RJR Nb bonds listed in Figure 22-1 recently rose to a price of 109. Zelda Morantz purchased four at 109. What will be her annual current yield?

$4,000 × 109% = $4,360 purchase price
$4,000 × 0.08 = $320 annual interest
$320 ÷ $4,360 = 0.0734, or 7.34%,
or 0.08 ÷ 1.09 = 0.0734, or 7.34%

Computing the Rate of Yield to Maturity

Careful investors calculate the **rate of yield to maturity**, or the rate of interest they will earn if they hold the bond to its maturity date. The yield to maturity calculation involves use of the true annual interest by adding a part of the discount or subtracting a part of the premium and basing the rate on the average principal invested (the average of the investor's purchase price and the bond's maturity value).

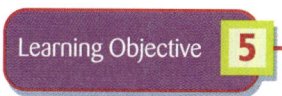

Learning Objective 5

Compute a rate of yield to maturity.

STEPS **to Compute the Rate of Yield to Maturity**

1. Compute the annual interest: multiply the face value by the stated (face) rate.
2. Determine the **annual discount** (or **premium**) **amortization**: Divide the discount (or premium) by the number of years from purchase to maturity.
3. Determine the **average principal invested**: Add the maturity value and the cost price and then divide by 2.
4. The following formula computes the rate:

$$\frac{\text{Annual interest} + \text{Annual discount amortization}}{\text{Average principal invested}}$$
(or − Annual premium amortization)

Again, because brokerage charges are such a small part of the cost, they usually are omitted from the calculations of yield to maturity.

EXAMPLE G

Assume that the Levi Straus bonds in example F matured 20 years after the purchase date.

STEP 1 $5,000 × 0.095 = $475 annual interest

STEP 2 $1,000 ÷ 20 years = $50 annual discount amortization

STEP 3   ($5,000 + $4,000) ÷ 2 = $4,500 average principal invested

STEP 4   ($475 + $50) ÷ $4,500 = 0.1167 = 11.67% yield to maturity

This rate is somewhat less than the 11.9% current yield, but it is more accurate with respect to actual income if the bond is held to maturity.

EXAMPLE H

To calculate the yield to maturity on bonds sold at a premium, assume that five IntTT $9\frac{1}{2}$s20 bonds were bought at a premium price of 124 and that the bonds will mature in 15 years. The market value of the five bonds is $6,200 ($5,000 × 124%).

STEP 1 $5,000 × 0.095 = $475 annual interest

STEP 2   ($6,200 − $5,000) ÷ 15 years = $80 annual premium amortization

STEP 3   ($5,000 + $6,200) ÷ 2 = $5,600 average principal invested

STEP 4   ($475 − $80) ÷ $5,600 = 0.0705 = 7.05% yield to maturity

This rate is less than the stated rate of $9\frac{1}{2}$% on the premium bonds.

✓ CONCEPT CHECK 22.5

If the four RJR Nb 8s10 bonds Zelda Morantz purchased at 109 (Concept Check 22.4) had 5 years to maturity, what would be her rate of yield to maturity?

$4,000 × 0.08 = $320 annual interest
$360 premium ÷ 5 years = $72 annual premium amortization
($4,000 + $4,360) ÷ 2 = $4,180 average principal invested
($320 − $72) ÷ $4,180 = 0.0593 = 5.93% yield to maturity

COMPLETE ASSIGNMENTS 22.1 and 22.2.

Chapter Terms for Review

accrued interest	discount
annual discount amortization	face value
annual premium amortization	government bonds
average principal invested	junk bond
bond ratings	municipal bonds
bonds	premium (bond)
callable bonds	rate of yield to maturity
convertible bonds	treasury bonds
corporate bonds	yield
current yield	

Try Microsoft® Excel

Try working the following problems using the Microsoft Excel templates found on your Student CD. Solutions for the problems are also found on the CD.

1. Complete the following Excel worksheet by entering formulas in the shaded cells to calculate the **Total Cost** and **Premium or (Discount)** for each bond purchase.
   Hint: Remember that each bond has a face value of $1,000.

Number Purchased	Price Paid	Total Cost	Premium (Discount)
5	92		
12	108		
8	112		
2	88		
16	92		

2. Complete the following Excel worksheet by entering formulas in the shaded cells to calculate the **Annual Interest, Current Purchase Price,** and **Current Yield** for each bond.
   Hint: Calculations are for one bond (face value $1,000). Current yield should be shown as a percent.

Bond	Price	Annual Interest	Current Purchase Price	Current Yield
IBM 7s12	90			
SBC 9s08	107			
CXL 6.2s09	86.5			

3. Complete the following Excel worksheet by entering formulas in the shaded cells to calculate the Yield to Maturity for six InTT 8.2s18 bonds purchased at a premium price of 120. The bonds will mature in 12 years.

   Hint: Use parentheses to do addition or subtraction before multiplication or division. Yield to maturity should be shown as a percent.

Market Value of Bonds	
Annual Interest	
Annual Premium Amortization	
Average Principal Invested	
Yield to Maturity	

THE BOTTOM LINE

Summary of chapter learning objectives:

Learning Objective	Example
22.1 Compute gains and losses on convertible and callable corporate bond transactions	1. John Jacobs bought five DVC bonds at $1,000 per bond. Each bond was convertible after 3 years to 50 shares of stock. At the end of 3 years, shares of DVC stock were selling at $32. The bond price had risen to 120. Should Mark exercise his option to convert? 2. Colton Mfg. Corp. issued $2,000,000 worth of callable bonds paying 9% interest. The maturity date for the bonds was in 20 years. Four years later, interest rates fell to $7\frac{1}{2}\%$. The bonds were called, and new bonds sold at the $7\frac{1}{2}\%$ rate. How much did Colton Mfg. Corp. save by calling the bonds?
22.2 Compute annual interest on bonds	3. Amy Coles purchased three 12-year, $1,000 bonds: one Boeing at 7%, one U.S. Treasury at 4.5%, and one Water World Sports at 12%. If the Water World Sports bond defaulted after 5 years and paid holders 60%, which bond produced the most income in the 5-year period, assuming that the $400 loss on the WWS bond was considered to be a reduction in income? How much did it produce?
22.3 Compute accrued interest on bond transactions made between interest payment dates	4. One BLM 9s18 bond was purchased at 102 on February 12. What was the amount of accrued interest if interest is paid January 1 and July 1?
22.4 Compute annual yield on bonds selling at a premium or a discount	5. Six Khol 7.4s25 bonds were purchased at 92. What was the current yield?
22.5 Compute a rate of yield to maturity	6. Three NYR 8s20 bonds were purchased at 120. The bonds will mature in 14 years. What is the rate of yield to maturity?

Answers: 1. The stock has $2,000 greater value; yes, he should convert 2. $480,000 3. Boeing; $350 4. $10.50 5. 8.04% 6. 5.97%

SELF-CHECK

Review Problems for Chapter 22

1 Alfred Tennyson purchased 15 IBM 7½s18 bonds at 104.
  a. What was the cost of the bonds?
  b. How often will interest be paid?
  c. How much interest will Alfred receive each interest period?
  d. Assuming the bonds pay interest on April 1 and October 1, calculate the accrued interest if the bonds were purchased June 6.
  e. What is the total amount Alfred paid for the bonds including accrued interest?
  f. Were the bonds purchased at a premium or a discount?
  g. What was the amount of the premium or discount?
  h. When do the bonds mature?
  i. What is the current yield on the bonds?
  j. Assume the bonds mature in 12 years. Calculate the yield to maturity.

2 Marta Samuals purchased six Xerox $1,000 convertible bonds at 95. Each bond was convertible into 30 shares of common stock. After 5 years, when the stock was selling at 42, Marta converted all six bonds.
  a. How many shares of stock did she receive?
  b. What was the value of the stock upon conversion?
  c. What was Marta's gain upon conversion of the bonds?
  d. Should Marta convert her bonds into stock if the stock's current market price is $45 per share? Why or why not?

3 Avis, Inc., issued $50,000,000 of 9½%, 20-year, callable bonds. After 6 years, the interest rate fell to 8%. How much interest would Avis save by calling the bonds and reissuing bonds at the lower rate?

4 Ron Nelson is considering purchasing one of the following bonds:
  MCD 7s15 at a market price of 90
  AOC 8s15 at a market price of 100
  JBC 9s15 at a market price of 110

Calculate the annual yield and yield to maturity for each bond assuming there are 10 years to maturity for each bond. Which bond would you recommend Ron purchase based on your computations?

Assignment 22.1: Corporate and Government Bonds

Name

Date   Score

Learning Objectives **1 2 3**

A (38 points) Solve the following problems. (points for correct answers as marked)

1. Jean Francis purchased seven IBM $1,000 convertible bonds at 105. Each bond was convertible to 25 shares of IBM stock in 5 years. At the end of 5 years, IBM stock was selling at 52. If Jean converted, what would be her 5-year capital gain? (4 points) $1,750

    7 × $1,000 × 1.05 = $7,350 purchase price
    175 × $52 = $9,100 stock value on conversion
    $9,100 − $7,350 = $1,750 capital gain

2. Return to problem 1 and assume that the stock price after 5 years was 35. How much more money would Jean get by cashing in the bonds rather than converting to stock? (4 points) $875

    7 × $1,000 = $7,000 cash from bonds
    175 × $35 = $6,125 stock value on conversion date
    $7,000 − $6,125 = $875 more money from cashing in bonds

3. The city of Jamestown, Virginia, issued $27,000,000 worth of callable bonds at 9% on January 1, 2000. The bonds were due in 2015. If interest rates were to fall to 6.5% on January 1, 2007, how much could Jamestown save by reissuing the bonds at the 6.5% rate on January 1, 2007? (4 points) $5,400,000

    $27,000,000 × 0.09 = $2,430,000 annual interest at 9%
    $27,000,000 × 0.065 = $1,755,000 annual interest at 6.5%
    $2,430,000 − $1,755,000 = $675,000 annual savings
    $675,000 × 8 years = $5,400,000 savings

4. Assume that an investor had purchased $500,000 worth of the Jamestown bonds referred to in problem 3. How much interest would he lose from having the bonds called if he reinvested in the new bond issue? (4 points) $100,000

    $500,000 × 0.09 = $45,000 annual interest at 9%
    $500,000 × 0.065 = $32,500 annual interest at 6.5%
    ($45,000 − $32,500) × 8 years = $100,000
    or $500,000 × 0.025 × 8 = $100,000

5. Devi Sharma purchased 22 corporate bonds, as shown. What was her total cost, and how much interest income would she realize annually? (1 point for each correct answer)

Bond	Number Purchased	Price	Total Cost	Annual Interest
a. Apex $7\frac{1}{2}$s09	4	100	$ 4,000.00	$ 300.00
b. DukeP $7\frac{7}{8}$s02	3	98	2,940.00	236.25
c. PGE $10\frac{1}{8}$s12	9	86	7,740.00	911.25
d. IBM $9\frac{3}{8}$s08	6	109	6,540.00	562.50
Total	22		$21,220.00	2,010.00

   a. 4 × $1,000.00 = $ 4,000.00 cost $4,000 × 0.075 = $ 300.00
   b. 3 × $ 980.00 = $ 2,940.00 cost $3,000 × 0.07875 = $ 236.25
   c. 9 × $ 860.00 = $ 7,740.00 cost $9,000 × 0.10125 = $ 911.25
   d. 6 × $1,090.00 = $ 6,540.00 cost $6,000 × 0.09375 = $ 562.50
                              $21,220.00 $2,010.00

Assignment 22.1 Continued

6. What is the dollar amount of interest per year and the maturity date for each of the following $1,000 bonds? (1 point for each correct answer)

Bond	Interest	Maturity date	Bond	Interest	Maturity date
a. PGE 6s08	$60	2008	d. Fldcst $12\frac{1}{2}$ s12	$125	2012
b. Avnet 8s13	$80	2013	e. OwCor 12s10	$120	2010
c. CPoWV 9s15	$90	2015	f. Cisco $7\frac{1}{2}$ s09	$75	2009

a. $1,000 × 0.06 = $60
b. $1,000 × 0.08 = $80
c. $1,000 × 0.09 = $90
d. $1,000 × 0.125 = $125
e. $1,000 × 0.12 = $120
f. $1,000 × 0.075 = $75

Score for A (38)

B **(50 points) Solve the following problems. (points for correct answers as marked)**

7. In each of the following problems, determine the number of days for which accrued interest is paid and the total purchase payment made for the bonds. (5 points for each correct answer)

  a. On September 12, Tracy Dean bought, at 103 plus accrued interest, two IBM 9s10 bonds with interest paid on January 1 and July 1.
  Number of days accrued interest: 73   Total payment: $2,096.50

  b. On October 9, Ben Blue bought, at 93 plus accrued interest, three IBM $7\frac{1}{2}$ s09 bonds with interest paid on January 1 and July 1.
  Number of days accrued interest: 100   Total payment: $2,852.50

  a. Purchase date: September 12
  (July) 31 + (Aug.) 31 + (Sept.) 11 = 73 days
  $2,000 × 1.03 = $2,060 market value
  $2,000 × 0.09 × $\frac{73}{360}$ = $36.50 accrued interest
  $2,060 + $36.50 = $2,093.50 total purchase payment

  b. Purchase date: October 9
  (July) 31 + (Aug.) 31 + (Sept.) 30 + (Oct.) 8 = 100 days
  $3,000 × 0.93 = $2,790 market value
  $3,000 × 0.075 × $\frac{100}{360}$ = $62.50 accrued interest
  $2,790 + $62.50 = $2,852.50 total purchase payment

8. Jack Mueller purchased a $1,000 corporate bond with a rating of AAA, paying 8% per year. Tom Bronkowski purchased a $1,000 junk bond paying 20%. Each bond was to mature in 10 years. Jack's bond paid interest for the 10-year period and face value at maturity. Tom's junk bond paid interest for 3 years before the company filed for bankruptcy and paid 45 cents on the dollar to its bondholders. How much more did Jack receive from his investment than Tom received from his? (10 points)

Jack: $1,000 × 0.08 × 10 years = $800 interest; $800 + $1,000 face value = $1,800 received
Tom: $1,000 × 0.20 × 3 years = $600 interest; $600 + $450 = $1,050 received
$1,800 − $1,050 = $750 more

9. Compute the current yield for the following bonds. (5 points for each correct answer)

Bond	Price	Current yield
a. PepsiCo 9s08	108	8.33%
b. IBM $7\frac{3}{8}$s08	93.5	7.89%
c. Avitar 10s12	112	8.93%
d. ABM 6s08	82	7.32%

a. ($1,000 × 0.09) ÷ $1,080 = 0.0833, or 8.33%   or   a. 9 ÷ 108 = 0.0833, or 8.33%
b. ($1,000 × 0.07375) ÷ $935 = 0.078877, or 7.89%   or   b. 7.375 ÷ 93.5 = 7.89%
c. ($1,000 × 0.10) ÷ $1,120 = 0.08929, or 8.93%   or   c. 10 ÷ 112 = 8.93%
d. ($1,000 × 0.06) ÷ $820.00 = 0.07317, or 7.32%   or   d. 6 ÷ 0.82 = 7.32%

Score for B (50)

Assignment 22.2: Bond Rate of Yield

Name

Date   Score

Learning Objectives **4** **5**

A   **(52 points) Solve the following problems. (points for correct answers as marked)**

1. An investor bought a 7.4% bond at 90. The bond would mature in 8 years. Round answers to two decimal places. (4 points for each correct answer)
   a. What was the average annual yield? <u>8.22%</u>   b. What was the rate of yield to maturity? <u>9.11%</u>

   $1,000 × 0.9 = $900 purchase price   ($1,000 − $900) ÷ 8 = $12.50 annual discount
   $1,000 × 0.074 = $74 annual interest $\frac{\$1,000 + \$900}{2}$ = $950 average principal invested
   $74 ÷ $900 = 8.22%
   or 7.4 ÷ 0.90 = 8.22%   ($74 + $12.50) ÷ $950 = 9.11%

2. In 2002, Jim Ayers bought six LTV 5s17 bonds for which he paid 82. Three years later, he sold the bonds at 84 and bought six Southern Electric $9\frac{1}{2}$ s24 bonds at 93. Did he increase or decrease the original rate of yield to maturity, and, if so, by how much? Round yields to one decimal place. (14 points) <u>3.4% increase</u>

   LTV   **Southern Electric**

   2017 − 2002 = 15 years to run   2024 − 2005 = 20 years to run
   $6,000 × 0.05 = $300 annual interest $6,000 × 0.095 = $570 annual interest
   $6,000 × 0.82 = $4,920 market value $6,000 × 0.93 = $5,580 market value
   ($6,000 − $4,920) ÷ 15 years = $72 annual discount amortization   ($6,000 − $5,580) ÷ 20 years = $21 annual discount amortization
   $\frac{\$6,000 + \$4,920}{2}$ = $5,460 average principal invested $\frac{\$6,000 + \$5,580}{2}$ = $5,790 average principal invested
   ($300 + $72) ÷ $5,460 = 0.0681 = 6.8% yield to maturity   ($570 + $21) ÷ $5,790 = 0.1021 = 10.2% yield to maturity

   10.2% − 6.8% = 3.4% increase

3. On July 29, Ann McCoy purchased four GMC $8\frac{1}{2}$ s09 bonds at 88. Interest was payable March 1 and September 1. Included in Ann's cost was accrued interest for 150 days. (4 points for each correct answer)
   a. What was the total purchase cost? <u>$3,661.67</u>   b. What was the average annual yield? Do not consider accrued interest when calculating this rate of yield.
   $4,000 × 0.88 = $3,520 market value   <u>9.66%</u>
   $4,000 × 0.085 × $\frac{150}{360}$ = $141.67 interest
   $3,520 + $141.67 = $3,661.67 $4,000 × 0.085 = $340 interest
                                                   $340 ÷ $3,520 = 9.66%, or 8.5 ÷ 0.88 = 9.66%

4. In 2005, Benito Cooper planned to purchase 20 $1,000 bonds and hold them to maturity. He had two choices: The first was EM&E $8\frac{1}{2}$ s18 at 106.50. The second was Standard of California 6s15 at 80. Benito purchased the issue that provided the higher rate of yield to maturity.
   a. Which issue did Benito purchase? (12 points) <u>Standard of California</u>

   EM&E   **Standard of California**

   For one bond: $1,000 × 1.065 = $1,065 market value   For one bond: $1,000 × 0.8 = $800 market value
   $1,000 × 0.085 = $85 annual interest per bond $1,000 × 0.06 = $60 annual interest paid per bond
   ($1,065 − $1,000) ÷ 13 years = $5 annual premium amortization per bond   ($1,000 − $800) ÷ 10 years = $20 annual discount amortization per bond
   $\frac{\$1,000 + \$1,065}{2}$ = $1,032.50 average principal invested $\frac{\$1,000 + \$800}{2}$ = $900 average principal invested
   ($85 − $5) ÷ $1,032.50 = 0.0775 = 7.7% yield to maturity   ($60 + $20) ÷ $900 = 0.0889 = 8.89% yield to maturity

Assignment 22.2 Continued

b. How much income would Benito have earned monthly if Standard of California had been purchased? (3 points)
  $133.33
  $20,000 face value × 0.06 = $1,200 annual yield ÷ 12 = $100 monthly
  $1,000 − $800 = $200 × 20 = $4,000 ÷ 10 = $400 annually ÷ 12 = $33.33 monthly amortization
  $100 + $33.33 = $133.33 monthly

c. If, in 2008, Benito had purchased EM&E $8\frac{1}{2}$ s18 bonds at a price of 97.5, what would have been the yield to maturity? (6 points) 8.9%
  $1,000 × 0.975 = $975 market value
  $1,000 × 0.085 = $85 annual interest per bond
  ($1,000 − $975) ÷ 10 years = $2.50 annual discount amortization per bond
  $$\frac{\$1,000 + \$975}{2} = \$987.50 \text{ average principal invested}$$
  ($85 + $2.50) ÷ $987.50 = 0.886 = 8.9% yield to maturity

d. Which company's bonds would be the better buy: EM&E at 97.5 or Standard of California? (1 point)
  Either, because their maturity levels are the same.

  8.9% yield to maturity for EM&E
  8.9% yield to maturity for Standard of California; 0.0889 is slightly better than 0.0886

Score for A (52)

B (48 points) Complete the following table. Show yield to maturity to one decimal place. (2 points for each correct answer)

	Number Purchased	Price Paid	Discount or Premium	Years to Maturity	Interest Rate	Annual Interest	+Discount −Premium Amortization	Average Principal Invested	Yield to Maturity
a.	8	105	$−400	5	8%	$ 640.00	$ −80.00	$8,200.00	6.83%
b.	10	97	+300	10	6%	$ 600.00	+30.00	9,850.00	6.40%
c.	12	86	+1,680	8	7.50%	$ 900.00	+210.00	11,160.00	9.95%
d.	5	112	−600	3	10.20%	$ 510.00	−200.00	5,300.00	5.85%
e.	1	90	+100	5	7%	$ 70.00	+20.00	950.00	9.47%
f.	20	102.5	−500	8	9.75%	$1,950.00	−62.50	20,250.00	9.32%

a. $8,000 × 0.08 = $640 annual interest
  $400 ÷ 5 = $80 premium amortization
  $8,000 × 1.05 = $8,400 market value
  ($8,400 + $8,000) ÷ 2 = $8,200 average principal
  ($640 − 80) ÷ $8,200 = 0.6829, or 6.83% yield
b. $10,000 × 0.06 = $600 annual interest
  $300 ÷ 10 = $30 discount amortization
  $10,000 × 0.97 = $9,700 market value
  ($9,700 + $10,000) ÷ 2 = $9,850 average principal
  ($600 + $30) ÷ $9,850 = 0.063959, or 6.40%
c. $12,000 × 0.075 = $900 annual interest
  $1,680 ÷ 8 = $210 discount amortization
  $12,000 × 0.86 = $10,320 market price
  ($10,320 + $12,000) ÷ 2 = $11,160 average principal
  ($900 +$210) ÷ $11,160 = 0.099462 = 9.95%

d. $5,000 × 0.102 = $510 annual interest
  $600 ÷ 3 = $200 premium amortization
  $5,000 × 1.12 = $5,600 market value
  ($5,600 + $5,000) ÷ 2 = $5,300 average principal
  ($510 − $200) ÷ $5,300 = 0.058491, or 5.85%
e. $1,000 × 0.07 = $70 annual interest
  $100 ÷ 5 = $20 discount amortization
  $1,000 × 0.90 = $900 market value
  ($900 + $1,000) ÷ 2 = $950 average principal
  ($70 +$20) ÷ $950 = 0.094737, or 9.47%
f. $20,000 × 0.0975 = $1,950 annual interest
  $500 ÷ 8 = $62.50 premium amortization
  $20,000 × 1.025 = $20,500 market price
  ($20,500 + $20,000) ÷ 2 = $20,250 average principal
  ($1,950 − $62.50) ÷ $20,250 = 0.09321, or 9.32%

Score for B (48)

Annuities

23

Learning Objectives
By studying this chapter and completing all assignments you will be able to:

Learning Objective 1 — Compute the future value of an annuity.

Learning Objective 2 — Compute the regular payments of an annuity from the future value.

Learning Objective 3 — Compute the present value of an annuity.

Learning Objective 4 — Compute the regular payments of an annuity from the present value.

Learning Objective 5 — Compute the loan payment required to amortize a loan.

Learning Objective 6 — Create a loan amortization schedule.

23.1 Chapter 16 must be covered before Chapter 23.

23.2 If consumers purchase on credit, they must pay the interest. If they save in advance for their purchases, they will receive the interest.

John and Joan Popplewell just won their state's lottery and the prize was listed as $5,000,000. When they purchased the winning ticket, they had a choice of taking the prize over 20 years or taking one cash payment now. The $5,000,000 represents the sum of 20 annual payments of $250,000 each. The series of equal payments is called an **annuity**. Because they chose the single cash payment, they do not actually receive $5,000,000 in cash. The amount that they receive is the **present value of an annuity**.

In Chapter 22, we discussed corporate and government bonds. When a corporation issues $10,000,000 worth of 8%, 20-year bonds, the corporation is simply borrowing money from the public for 20 years. Each $1,000 bond pays 8% (or $80) each year. The $80 is paid out in two $40 payments every 6 months for 20 years. The series of $40 interest payments is an annuity. The amount that someone pays for the bond is the present value of the annuity. Some investors may worry that the corporation won't have $10,000,000 available in 20 years to repay the bonds. Therefore, the corporation may decide to make 20 equal annual payments into a separate account managed by a neutral third party. At the end of the 20 years, the deposits plus accumulated interest will be worth the $10,000,000. This fund of deposits is called a **sinking fund**. Equal deposits into a sinking fund form an annuity. The total amount is the **future value of an annuity**.

Computing the Future Value of an Annuity

Learning Objective 1

Compute the future value of an annuity.

An annuity is made up of a series of equal payments that occur at regular time intervals. The payments go into—or come out of—an interest-bearing account or investment. The constant interest rate is compounded at the same time the payments are made. (Perhaps obviously, the number of periods in an annuity is the same as the number of payments.)

We can illustrate an annuity by drawing a straight line, called a **time line**. On the time line, we insert equal marks and the payment dates and write in the payment amount.

23.3 In a very technical course, you would emphasize that, for the basic tables and formulas to be used, (1) the payments must be equal; (2) the payments must be regular; (3) the interest rate must stay the same; and (4) the compounding must coincide with the payments. In this book, all of these details are assumed so that students can concentrate on the fundamental concept.

● **EXAMPLE A**

An annuity has four annual payments of $1,000, always on December 31. The date of the first $1,000 payment is December 31, 2005. Draw a time line showing the four years—2005, 2006, 2007, and 2008—and the four payments.

The annuity illustrated in Figure 23-1, with the payments occurring at the end of each period, is called an **ordinary annuity**. In this book, every annuity will have its payments at the end of each period. The date December 31, 2004, is the *beginning of the annuity*, and the date December 31, 2008, is the *end of the annuity*.

23.4 Another common type of annuity is the *annuity due*, in which the payments come at the *beginning* of the periods. Change Figure 23-1 so that the first payment is on 12/31/04 but the time line still stops at 12/31/08. An example of annuity due is a lease or rental contract; you don't move into an apartment or office unless you pay at least a month's rent in advance of the first month. We don't cover annuities due because the calculations aren't quite as straightforward.

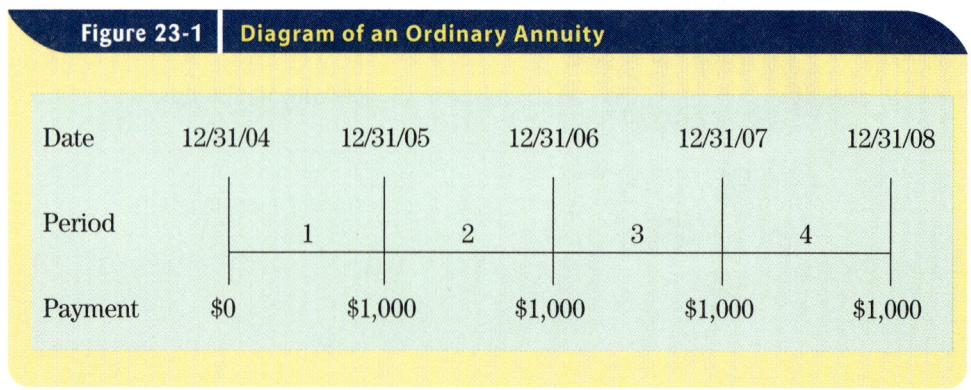

Figure 23-1 | Diagram of an Ordinary Annuity

Date	12/31/04	12/31/05	12/31/06	12/31/07	12/31/08
Period		1	2	3	4
Payment	$0	$1,000	$1,000	$1,000	$1,000

Again, the value of the annuity at the end of the annuity is called the *future value of the annuity*. In example A, it is the total value of all payments plus the compound

462 Part 6 Corporate and Special Applications

interest from the date of each payment until December 31, 2008. When a business or individual decides to deposit the same amount of money every year (or month or quarter) into an interest-bearing account for a specified amount of time, the future value of the annuity is the amount that will be in the account when the last deposit is made.

EXAMPLE B

In December, 2004, Mary Currie accepted a job with a manufacturing company. Mary decided to save $1,000 at the end of each year for 4 years. The company credit union allowed Mary to open a savings account on December 31, 2004, but Mary will not make any deposit until December 31, 2005. She also will make deposits on December 31 of 2006, 2007, and 2008. The credit union pays interest of 10% compounded annually. How much will be in the account after the last deposit? (*Hint:* Make a time line diagram and compute the future value of each of the four deposits.)

To find the future value of the annuity on December 31, 2008, first use Table 16-1 (see Chapter 16) to determine the future value of each of the four payments as of December 31, 2008. Then compute the total.

Amount of Payment	Date of Payment	Years of Interest	Future Value on 12/31/08
$1,000	12/31/05	3	$1,000 × 1.33100 = $1,331
$1,000	12/31/06	2	$1,000 × 1.21000 = $1,210
$1,000	12/31/07	1	$1,000 × 1.10000 = $1,100
$1,000	12/31/08	0	$1,000 × 1.00000 = $1,000
			Total = $4,641

23.5 Normally, a savings account would not be opened without the person making a deposit. But an annuity has the same number of periods as payments, so there must be a year either before the first payment or after the last one. Because we want the payments to come at the end of each period, the first year precedes the first payment. Also, example B would be more realistic if the deposits were made monthly instead of annually. See problem 19 of Assignment 23.1.

23.6 A savings interest rate of 10% per year is not realistic in our current economy. However, for teachers who want to illustrate all of the arithmetic, 10% permits the easier calculations.

Figure 23-2 illustrates how each of the four payments moves *forward* in time to December 31, 2008.

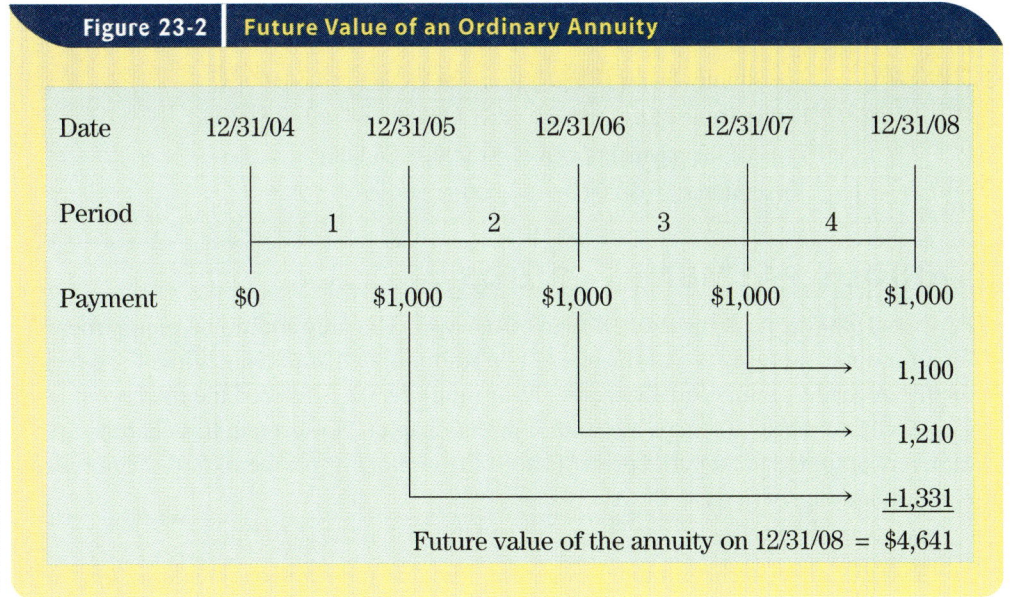

Figure 23-2 Future Value of an Ordinary Annuity

ANNUITY TABLES

Annuity calculations can be time-consuming, even with just four payments. With 20 or 30 payments, the calculations could be tiresome. Computers, financial calculators, and tables eliminate tedious computations. Table 23-1 on pages 490–491 is an abbreviated sample of a table of **future value of annuity factors** (**FVAF**). It is used the same way as Table 16-1. As in Chapter 16, the columns indicate the periodic interest rate and the rows indicate the number of periods.

> **STEPS** to Use Table 23-1 to Compute Future Value and Total Interest Earned
>
> 1. Locate the annuity factor (FVAF) in the correct row and column of Table 23-1, on pages 490–491.
> 2. Multiply the payment amount by the annuity factor (FVAF). The product is the future value of the annuity (FVA).
> 3. Multiply the payment amount by the number of payments. The product is the total of all payments.
> 4. Subtract the total of all payments from the future value of the annuity. The difference is the total interest earned.

FUTURE VALUE OF AN ANNUITY FORMULA

If you prefer, Step 2 above may be summarized as a formula, in words or in symbols:

Future value of an annuity = Periodic payment × Future value of annuity factor (Table 23-1)   or $FVA = Pmt \times FVAF$

EXAMPLE C

Find the future value of an annuity of four annual payments of $1,000. Each payment is made at the end of the year, and 10% interest is compounded each year. Also find the total interest earned over the 4 years.

STEP 1   The annuity factor (FVAF) from Table 23-1 is 4.64100.
STEP 2   Future value of the annuity = $1,000 × 4.64100 = $4,641
STEP 3   Total of the payments = 4 × $1,000 = $4,000
STEP 4   Total interest = $4,641 − $4,000 = $641

VARIOUS PAYMENT PERIODS

Payments may be made more often than once a year. The only additional requirement for an ordinary annuity is that the interest be compounded at the same time the payments are made—semiannually, quarterly, or monthly. We described the method in Chapter 16, and also use Steps i, ii, and iii in this chapter. However, in this chapter, the number computed in Step iii represents both the number of payments and the number of compounding periods.

Just as in Chapter 16, we use Steps i, ii, and iii in Chapter 23 to find

STEP i m = the number of compounding periods (and payments) in one year;
STEP ii i = periodic interest rate = *annual rate* ÷ m; and
STEP iii n = number of periods (payments) in the entire annuity = $m \times$ *number of years*.

These three steps are required whether we use Table 23-1 or a calculator to find the FVAF.

EXAMPLE D

Find the future value of an annuity in which $200 is deposited at the end of each quarter for 5 years. Interest is 6% compounded quarterly.

STEP i   There are $m = 4$ compounding periods in 1 year. $[m = 4]$

STEP ii   Periodic interest rate $i = 6\% \div 4 = 1.5\%$ per period $\left[i = \dfrac{0.06}{4} = 0.015\right]$

STEP iii   Number of payments $n = 4 \times 5$ years $= 20$ payments $[n = 4 \times 5 = 20]$

STEP 1   Use Table 23-1, 1.5% column and row 20: annuity factor $= 23.12367$

STEP 2   Future value $= \$200 \times 23.12367 = \$4,624.734$, or $4,624.73

✓ CONCEPT CHECK 23.1

Assume that $2,000 is invested every 6 months for 5 years in an account that pays 6% compounded semiannually. Compute the future value of the investment. Then compute the total interest earned by the investment.

Semiannual means $m = 2$ periods per year. $[m = 2]$
Periodic rate $= 6\% \div 2 = 3\%$ per period $\left[i = \dfrac{0.06}{4} = 0.03\right]$
Number of payments 2×5 years $= 10$ payments
The future value annuity factor from row 10 of the 3.00% column in Table 23-1 is 11.46388. $[n = 2 \times 5 = 10]$
Future value of the annuity $= \$2,000 \times 11.46388 = \$22,927.76$

Total of all payments $= \$2,000 \times 10$ payments $= \$20,000$
Total interest earned $=$ Future value $-$ Total payments $= \$22,927.76 - \$20,000.00 = \$2,927.76$

USING A CALCULATOR TO COMPUTE ANNUITY FACTORS (OPTIONAL)

Recall from Chapter 16 on Compound Interest that Tables 16-1 and 16-2 had the "future value factors" (FVF) and the "present value factors" (PVF), respectively. Recall also that you could use a calculator to find the FVF and PVF with these simple formulas: **FVF** $= (1 + i)^n$ and **PVF** $= 1 \div (1 + i)^n$ (or **PVF** $= (1 + i)^{-n}$), where i is the *periodic* interest rate and n is the total number of *periods*. To find the future value of $5,000 invested at 8% compounded quarterly for 3 years, you used either Table 16.1 or a calculator to find **FVF** $= 1.268$. The future value is FV $=$ PV $\times$ FVF $= \$5,000 \times 1.26824 = \$6,341.20$.

Earlier, we learned that the terms in Table 23-1 are "future value of an annuity factors" (**FVAF**s). Just as there was a calculator formula for **FVF**, there is a formula for **FVAF**. It is

$$\text{FVAF} = \frac{(1 + i)^n - 1}{i}$$

where i is the periodic interest rate *written as a decimal* (as in Chapter 16), and
         n is the total number of payments (or the number of periods)

© ROSE ALCORN/THOMSON

Applying the formula to example C where $n = 4$ years and $i = 10\%$ compounded annually, we find the same FVAF = 4.46100 as in row 4, column 10%, of Table 23-1:

$$\text{FVAF} = \frac{(1+i)^n - 1}{i} = \frac{(1+0.10)^4}{0.10} = \frac{1.46410000 - 1}{0.10}$$

$$= \frac{0.46410000}{0.10} = 4.6410000$$

Depending on your calculator, one set of calculator keystrokes to calculate this **FVAF** is

$$1\,[+]\,.1\,[=]\,[y^x]\,4\,[=]\,[-]\,1\,[=]\,[\div]\,.1\,[=]$$

To compute the future value of an annuity with a calculator, the formula is

$$\text{FVA} = Pmt \times \text{FVAF} \quad \text{or} \quad \text{FVA} = Pmt \times \left[\frac{(1+i)^n - 1}{i}\right]$$

In example C, FVA = $Pmt \times \text{FVAF}$ = $1,000 \times 4.64100 = \$4,641$.

In example D, Steps i, ii, iii give $m = 4$, $i = 6\% \div 4 = 1.5\%$ or 0.015, and $n = 4 \times 5$ years = 20. Using the formula and a calculator, we get

$$\text{FVA} = Pmt \times \text{FVAF} = Pmt \times \left[\frac{(1+i)^n - 1}{i}\right] = \$200 \times \left[\frac{(1+0.015)^{20} - 1}{0.015}\right]$$

$$= \$200 \times 23.1236671 = \$4,624.73$$

After first calculating $i = 0.015$ and $n = 20$, one typical set of calculator keystrokes to find the future value is

$$1\,[+]\,.015\,[=]\,[y^x]\,20\,[=]\,[-]\,1\,[=]\,[\div]\,.015\,[=]\,[\times]\,200\,[=]$$

Calculators differ. If your calculator has parentheses, you could use one or more pairs of parentheses to make an expression that you think is simpler. Use the keystrokes that seem simplest to you.

Computing Regular Payments of an Annuity from the Future Value

Compute the regular payments of an annuity from the future value.

In examples A–D, the amounts of the payments were known and the future values were unknown. If, however, the future value is known, then you can compute the amount of each payment. The procedure is identical whether you use Table 23-1 or a calculator to find the FVAF.

STEPS to Find the Size of the Payment in an Annuity, Given Its Future Value

1. Determine the annuity factor (FVAF) using Table 23-1 or a calculator.
2. Divide the future value by the annuity factor. The quotient is the amount of each payment in the annuity.

As a formula, Step 2 could be written as $Pmt = \text{FVA} \div \text{FVAF}$.

EXAMPLE E

Nate and Nan Roth want to have $35,000 in their credit union account when their son Danny starts college. They will make equal payments every month for 4 years. The credit union will pay 6% compounded monthly. What should their payment amount be?

The value of the annuity at the end, or the future value of the annuity, is $35,000. Use Table 23-1.

STEP i — There are $m = 12$ compounding periods in 1 year.

STEP ii — Periodic interest rate = 6% ÷ 12 = 0.5% per period

STEP iii — Number of deposits = 12 × 4 years = 48 deposits

STEP 1 — Use Table 23-1, 0.5% column and row 48: annuity factor = 54.09783

STEP 2 — Future value of the annuity = $35,000
Payment amount = $35,000 ÷ 54.09783 = $646.976, or $646.48

$$[m = 12]$$
$$\left[i = \frac{0.06}{12} = 0.005 \right]$$
$$[n = 12 \times 4 = 48]$$
$$\left[\text{FVAF} = \frac{(1 + 0.005)^{48} - 1}{0.005} = 54.09783222 \right]$$

SINKING FUNDS

At the beginning of this chapter, we mentioned that a $10,000,000 corporate bond issue may include a sinking fund feature. Sometimes a sinking fund means that the corporation will set aside an equal amount of money each year so that by the end of the 20 years, the corporation will have accumulated the $10,000,000. At other times, perhaps, a sinking fund may be used by the corporation to buy back $500,000 worth of the bonds each year.

Although the term *sinking fund* may be most often associated with the repayment of a bond issue, its use isn't restricted to bonds. A corporation may set up a sinking fund to save money for an expensive piece of equipment that it knows it must replace in the future. The college fund set up by Nate and Nan Roth in example E was essentially a sinking fund.

EXAMPLE F

Micromedia Corporation is preparing a $10,000,000 bond issue. The company wants to make 25 equal payments into a sinking fund so that it will have a total of $10,000,000 available in 25 years to repay the bonds. What size should each of the payments be if the company can earn 5% per year on the payments?

STEP i — There is $m = 1$ compounding period in 1 year.

STEP ii — Periodic interest rate = 5% ÷ 1 = 5% per period

STEP iii — Number of deposits = 1 × 25 years = 25 deposits

STEP 1 — Use Table 23-1, 5% column and row 25: annuity factor = 47.72710

STEP 2 — Future value of the annuity = $10,000,000
Payment amount = $10,000,000 ÷ 47.72710 = $209,524.57

$$[m = 1]$$
$$\left[i = \frac{0.05}{1} = 0.05 \right]$$
$$[n = 1 \times 25 = 25]$$
$$\left[\text{FVAF} = \frac{(1 + 0.05)^{25} - 1}{0.05} = 47.72709882 \right]$$

CONCEPT CHECK 23.2

Assume that an equal amount is invested every quarter for 7 years. After the last payment, the future value is $75,000. If the interest rate is 8% compounded quarterly, compute the size of each regular quarterly payment.

Quarterly means $m = 4$ periods per year.
Periodic rate = 8% ÷ 4 = 2% per period
Number of payments is 4 × 7 years = 28 payments
The future value annuity factor from row 28 of the 2% column in Table 23-1 is 37.05121.

Regular quarterly payment = $75,000 ÷ 37.05121 = $2,024.2254, or $2,024.23

COMPLETE ASSIGNMENT 23.1.

$$\left[m = 4 \right]$$
$$\left[i = \frac{0.08}{4} = 0.02 \right]$$
$$\left[n = 4 \times 7 = 28 \right]$$
$$\left[\text{FVAF} = \frac{(1 + 0.02)^{28} - 1}{0.02} = 37.05121031 \right]$$

Computing the Present Value of an Annuity

Learning Objective 3

Compute the present value of an annuity.

The annuity shown in Figure 23-3 begins December 31, 2004. Again, the value of the annuity on this date is called the **present value of the annuity.** For example, when a person deposits a large amount in a bank account and then makes a series of equal withdrawals from the account until it is empty, the series of withdrawals (the equal payments) is the annuity and the amount deposited is the present value. The interest earned equals the difference between the total amount withdrawn and the amount deposited.

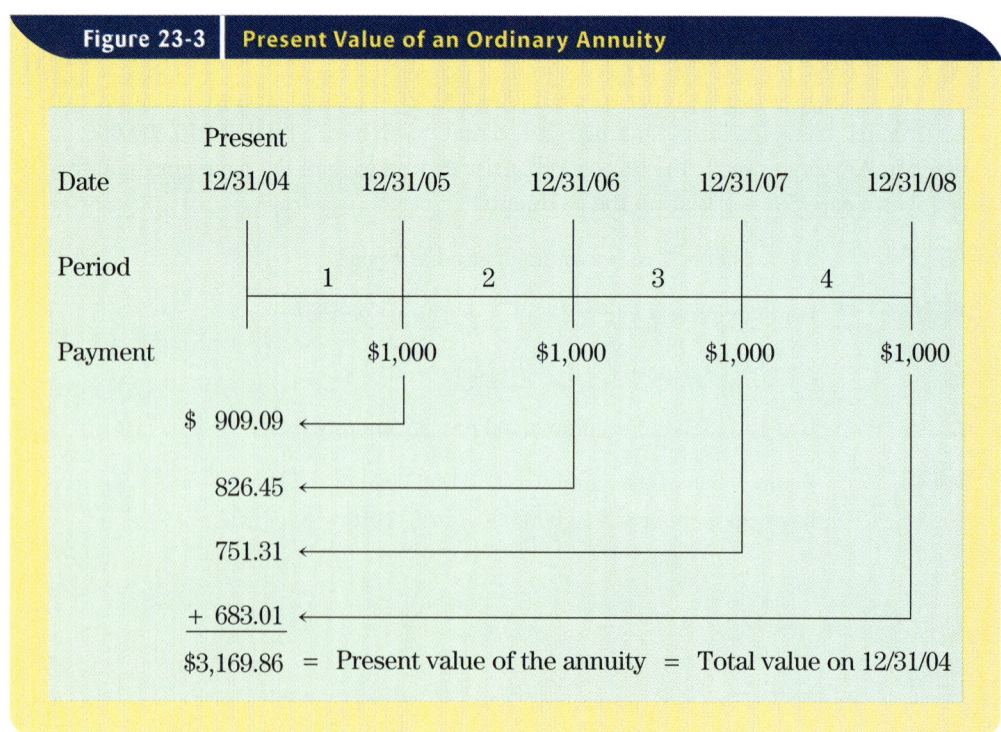

Figure 23-3 | Present Value of an Ordinary Annuity

EXAMPLE G

In November 2004, Ashley Hamilton inherited some money. She planned to donate part of the money immediately to the American Cancer Society and then to make four equal donations of $1,000 each on December 31 of 2005, 2006, 2007, and 2008. To prepare for the four future payments, Ashley went to her bank on December 31, 2004 and deposited money into a new account. The account paid 10% compounded annually. Ashley would withdraw $1,000 each year; after the last withdrawal on December 31, 2008, the account would be empty.

How much must Ashley deposit on December 31, 2004? (*Hint:* Make a time line diagram, and compute the present value of each of the four withdrawals.)

To find the present value of the annuity on December 31, 2004, first use Table 16-2 to find the present value of each of the four payments on December 31, 2004. Then compute the total.

Amount of Payment	Date of Payment	Years of Interest	Present Value on 12/31/04		
$1,000	12/31/05	1	$1,000 × 0.90909	=	$ 909.09
$1,000	12/31/06	2	$1,000 × 0.82645	=	826.45
$1,000	12/31/07	3	$1,000 × 0.75131	=	751.31
$1,000	12/31/08	4	$1,000 × 0.68301	=	$ 683.01
			Present value of the annuity on 12/31/04	=	$3,169.86

Figure 23-3 illustrates example G. The time line shows the equal withdrawals as each payment is moved from the future backward to the present (to December 31, 2004). Compare Figure 23-3 with Figure 23-2 where each payment was projected forward into the future.

The method shown in Figure 23-3 aids instruction but is too time-consuming to be practical. To get the same solution quickly, use Table 23-2 on pages 492–93.

23.7 Point out that, for income tax planning purposes, taxpayers may make charitable contributions on the last day of the year. In some years, December 31 is a Sunday, so a withdrawal for a cash contribution would have to be on December 29 or 30. However you still might write the check on December 31.

Video

Annuities: Future Value/Present Value

STEPS to Use Table 23-2 to Compute Present Value and Total Interest Earned

1. Locate the **present value of annuity factor (PVAF)** in the correct row and column of Table 23-2 on pages 492–93.
2. Multiply the payment amount by the annuity factor (PVAF). The product is the present value of the annuity.
3. Multiply the payment amount by the number of payments. The product is the total of all payments.
4. Subtract the present value of the annuity from the total of all payments. The difference is the total interest earned.

PRESENT VALUE OF AN ANNUITY FORMULA

If you prefer, Step 2 may be summarized as a formula, in words or in symbols:
Present value of an annuity = Periodic payment × Present value of annuity factor (Table 23-2), or $PVA = Pmt \times PVAF$

For example G, the factor in the 10.00% column and row 4 of Table 23-2 is 3.16987 (Step 1), and $1,000 × 3.16987 = $3,169.87 (Step 2).

The application in example H may not sound complicated, but even it would be tedious to do without Table 23-2. As the payments and compounding are quarterly, use Steps i, ii, and iii to find the periodic rate and the number of periods.

EXAMPLE H

Nanda Cerrado just won first prize in a fund-raising raffle. Nanda has a choice: She can receive quarterly payments of $750 each for 6 years, with the first payment 3 months (one quarter) from now, or she can receive 1 lump sum today. Assuming an interest rate of 6% compounded quarterly, what lump sum today equals the future payments? (*Hint:* The series of $750 payments is an annuity, and the lump sum is the present value of the annuity.)

$$\left[m = 4 \right]$$
$$\left[i = \frac{0.06}{4} = 0.015 \right]$$
$$\left[n = 4 \times 6 = 24 \right]$$

STEP i There are $m = 4$ compounding periods in 1 year.

STEP ii Periodic interest rate = 6% ÷ 4 = 1.5% per period

STEP iii Number of payments 4 × 6 years = 24

STEP 1 Using Table 23-2, 1.50% column and row 24: the PVAF = 20.03041

STEP 2 Present value = $750 × 20.03041 = $15,022.8075, or $15,022.81

✓ CONCEPT CHECK 23.3

What present value (principal) must be invested today in an account to provide for 7 equal annual withdrawals (an annuity) of $5,000 each? The interest rate is 8% compounded annually.

> Annual means $m = 1$ period per year.
> Periodic rate = 8% ÷ 1 = 8% per year
> Number of payments = 1 × 7 years = 7 payments
> From row 7 of the 8.00% column of Table 23-2, the PVAF = 5.20637.
>
> $\left[m = 1 \right]$
> $\left[i = \dfrac{0.08}{1} = 0.08 \right]$
> $\left[n = 1 \times 7 = 7 \right]$
>
> Present value of the annuity = $5,000 × 5.20637 = $26,031.85

USING A CALCULATOR TO COMPUTE THE PRESENT VALUE OF AN ANNUITY (OPTIONAL)

Just as there is a calculator formula to compute the future value of an annuity factor (FVAF), there is also a calculator formula to compute the present value of an annuity factor (PVAF). The formula can be written several ways. Use whichever one you think is easier to understand.

$$\text{PVAF} = \frac{1 - (1 + i)^{-n}}{i} \quad \text{or} \quad \text{PVAF} = \frac{1 - (1 \div (1 + i)^n)}{i}$$

$$\text{or} \quad \text{PVAF} = \frac{1 - \dfrac{1}{(1 + i)^n}}{i}$$

where i is the periodic interest rate *written as a decimal* (as in Chapter 16),
    n is the number of payments (or the number of periods)

To compute the present value of an annuity (**PVA**) with a calculator, the formula is

$$PVA = Pmt \times PVAF \quad \text{or} \quad PVA = Pmt \times \left[\frac{1 - (1 + i)^{-n}}{i}\right]$$

where Pmt is the periodic payment
    i is the periodic interest rate written as a decimal
    n is the number of payments (or the number of periods)
    PVA is the present value of the annuity

Return to example H and use the formulas for PVA and PVAF to compute the present value of the annuity in example H: quarterly payments of $750 each for 6 years at an interest rate of 6% compounded quarterly.

$Pmt = \$750$
$m = 4$ compounding periods in 1 year
$i = 6\% \div 4 = 1.5\%$, or 0.015, is the periodic interest rate
$n = 4 \times 6$ years $= 24$ is the number of compounding periods

$$PVA = Pmt \times \left[\frac{1 - (1 + i)^{-n}}{i}\right] = \$750 \times \left[\frac{1 - (1 + 0.015)^{-24}}{0.015}\right]$$

$$= \$750 \times \left[\frac{1 - 0.69954392}{0.015}\right] = \$750 \times \left[\frac{0.30045608}{0.015}\right]$$

$$= \$750 \times (20.03040537) = \$15,022.80402, \text{ or } \$15,022.80$$

> 23.8 Remind students to use the [+/−] key on their calculators to make the negative exponent, not the [−] key.

After first calculating $i = 0.015$ and $n = 24$, one typical set of calculator keystrokes to find the present value is 1[+] .015 [=] [y^x] 24 [+/−] [=] [+/−] [+] 1 [=] [÷] .015 [=] [×] 750 [=]

And remember: Your calculator may be different. You may have to use different keystrokes and you may be able to find a more efficient sequence of keystrokes.

Computing Regular Payments of an Annuity from the Present Value

In examples G and H, the amounts of the payments were known and the present values were unknown. If, however, the present value is known, then you can compute the amount of the payments. The procedure is identical whether you use Table 23-1 or a calculator to find the **PVAF**.

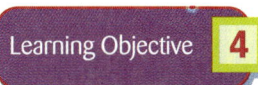

Compute the regular payments of an annuity from the present value.

> **STEPS** to Find the Size of the Payment in an Annuity, Given the Present Value
>
> 1. Determine the annuity factor (PVAF) using Table 23-1 or a calculator.
> 2. Divide the present value by the annuity factor (PVAF). The quotient is the amount of the payments in the annuity.

Chapter 23 Annuities **471**

● **EXAMPLE I**

Jim Schremp received a $25,000 bonus from his employer. Rather than spend it all at once, he decided to deposit it in a bank account that pays 9% compounded monthly. He will make equal monthly withdrawals for 4 years. After the last withdrawal, the account will be empty. How much will he withdraw each month?

The value of the annuity in the beginning (present value of the annuity) is $25,000. Use Table 23-2.

$[m = 12]$

$\left[i = \dfrac{0.09}{12} = 0.0075\right]$

$[n = 12 \times 4 = 48]$

$\left[\text{PVAF} = \dfrac{1 - (1 + 0.0075)^{-48}}{0.0075}\right.$

$\left. = 40.18478189\right]$

STEP i   There are $m = 12$ compounding periods in 1 year.

STEP ii   Periodic interest rate = 9% ÷ 12 = 0.75% per period

STEP iii   Number of withdrawals = 12 × 4 years = 48 withdrawals

STEP 1   Using Table 23-2, 0.75% column and row 48: the PVAF = 40.18478

STEP 2   Each withdrawal = $25,000 ÷ 40.18478 = $622.126, or $622.13

✓ CONCEPT CHECK 23.4

Assume that $50,000 is deposited today (the present value) to provide for 44 equal quarterly withdrawals (an annuity) over the next 11 years. If the interest rate is 8% compounded quarterly, what is the size of each regular quarterly payment? What is the total interest earned during the term of the annuity?

Quarterly means $m = 4$ periods per year.
Periodic rate = 8% ÷ 4 = 2% per quarter
Number of payments = 4 × 11 years = 44 payments
From row 44 of the 2.00% column in Table 23-2, the PVAF = 29.07996.
Regular quarterly payment = $50,000 ÷ 29.07996 = $1,719.3971, or $1,719.40
Total of all payments = $1,719.40 × 44 payments = $75,653.60
Total interest earned = Total payments − Present value
                          = $75,653.60 − $50,000.00 = $25,653.60

$[m = 4]$

$\left[i = \dfrac{0.08}{4} = 0.02\right]$

$[n = 4 \times 11 = 44]$

$\left[\text{PVAF} = \dfrac{1 - (1 + 0.02)^{-44}}{0.02}\right.$

$\left. = 29.07996307\right]$

Computing the Payment to Amortize a Loan

Recall from your study of loan amortization in Chapter 14 that the borrower repays the loan by making equal monthly payments and that the interest is computed on the unpaid balance each month. Loan amortization creates an annuity because there is a series of equal periodic payments. Computing the interest each month makes it compound interest. The amount of the loan is the present value of the annuity.

Stated another way, in amortization, when the amount of the loan is known, the present value of the annuity is known. As in example J, you can use Table 23-2 to compute the amount of the monthly payments.

Learning Objective 5

Compute the loan payment required to amortize a loan.

STEPS to Find the Size of the Payment to Amortize a Loan

1. Determine the annuity factor (PVAF) using Table 23-2 or a calculator.
2. Divide the loan amount by the annuity factor (PVAF). The quotient is the amount of the monthly loan payments.

EXAMPLE J

Barbara Luzardi wants to buy a new piano. Barbara pays $3,000 and also trades in her old piano. The balance is $2,400, and the piano dealer will amortize the $2,400 over 4 months at 12%. Find the size of the required monthly payments.

STEP i	There are 12 compounding periods in 1 year.	$[m = 12]$
STEP ii	Periodic interest rate = 12% ÷ 12 = 1% per period	$\left[i = \dfrac{0.12}{12} = 0.01\right]$
STEP iii	Number of monthly payments = 4	$[n = 4]$
STEP 1	Because the borrowing occurs at the *beginning* of the annuity, this is a present value problem and $2,400 is the present value of the annuity; use Table 23-2. In the 1.00% column and row 4, the PVAF = 3.90197.	$\left[\text{PVAF} = \dfrac{1 - (1 + 0.01)^{-4}}{0.01}\right.$ $\left. = 3.90196555\right]$
STEP 2	Size of each payment = $2,400 ÷ 3.90197 = $615.07392, or $615.07	

✓ CONCEPT CHECK 23.5

A bank loans $40,000 at an interest rate of 9% compounded monthly. Find the loan payment necessary to amortize the loan with monthly payments over 3 years.

Loan amortization involves an annuity. The amount borrowed is the present value of the annuity, and the monthly loan payment is the regular annuity payment.

Monthly means 12 periods per year.
Periodic rate = 9% ÷ 12 = 0.75% per period
Number of payments = 12 × 3 years = 36 payments
From row 36 of the 0.75% column of Table 23-2, the PVAF = 31.44681.

Loan payment = $40,000 ÷ 31.44681 = $1,271.98911, or $1,271.99

$[m = 12]$
$\left[i = \dfrac{0.09}{12} = 0.0075\right]$
$[n = 12 \times 3 = 36]$
$\left[\text{PVAF} = \dfrac{1 - (1 + 0.0075)^{-36}}{0.0075}\right.$
$\left. = 31.44680525\right]$

Creating a Loan Amortization Schedule

Recall from Chapter 14 that the following procedure is used to create an amortization schedule.

Create a loan amortization schedule.

STEPS to Create an Amortization Schedule

For each row except the last:
1. Interest payment = Unpaid balance × Monthly interest rate
2. Principal payment = Monthly payment − Interest payment
3. New unpaid balance = Old unpaid balance − Principal payment

For the last row:
1. Interest payment = Unpaid balance × Monthly interest rate
2. (Then ADD) Monthly payment = Unpaid balance + Interest payment
3. Principal payment = Unpaid balance

EXAMPLE K

Create an amortization schedule for the loan in example J: $2,400 to be amortized over 4 months with interest of 12% compounded monthly charged on the unpaid balance.

Month	Beginning Unpaid Balance	STEP 1 (1%) Interest Payment	Principal Payment	STEP 2 Total Payment	STEP 3 New Balance
1	$2,400.00	$24.00	$591.07	$615.07	$1,808.93
2	1,808.93	18.09	596.98	615.07	1,211.95
3	1,211.95	12.12	602.95	615.07	609.00
4	609.00	6.09	609.00	615.09	0

In example J, note that each month's beginning unpaid balance is multiplied by the monthly interest rate (1%) and rounded to the nearest cent.

474 Part 6 Corporate and Special Applications

CONCEPT CHECK 23.6

Amortize a $1,500 purchase over 3 months at an annual rate of 12%. First, use Table 23-2 or a calculator to calculate the first two monthly payments. Then show the calculations to construct a 3-month amortization schedule.

The periodic interest rate is 12% ÷ 12 = 1%, and the number of periods is 3. The present value annuity factor from row 3 of the 1.00% column of Table 23-2 is 2.94099. The loan payment is $1,500 ÷ 2.94099 = $510.03.

$$[m = 12]$$
$$\left[i = \frac{0.12}{12} = 0.01 \right]$$
$$[n = 3]$$
$$\left[\text{PVAF} = \frac{1 - (1 + 0.01)^{-3}}{0.01} \right.$$
$$\left. = 2.94098521 \right]$$

	Month 1		Month 2		Month 3	
Unpaid balance:	Purchase price	$1,500.00	From month 1	$1,004.97	From month 2	$504.99
Interest payment:	$1,500.00 × 0.01 =	$15.00	$1,004.97 × 0.01 =	$10.05	$504.99 × 0.01 =	$5.05
Total payment:	From above	$510.03	From above	$510.03	$504.99 + $5.05 =	$510.04
Principal payment:	$510.03 − $15.00 =	$495.03	$510.03 − $10.05 =	$499.98	(Unpaid balance)	$504.99
New balance:	$1,500 − $495.03 =	$1,004.97	$1,004.97 − $499.98 =	$504.99	$504.99 − $504.99 =	$0.00

COMPLETE ASSIGNMENT 23.2.

USING THE TEXAS INSTRUMENTS BA II PLUS BUSINESS CALCULATOR FOR ANNUITY CALCULATIONS (OPTIONAL)

Several inexpensive calculators are available to perform annuity calculations, as well as many other business and financial functions. A typical calculator is the Texas Instruments BA II Plus, shown in the photograph. Although we do not endorse this calculator above any others, it has the typical features for annuities. If you have a business or financial calculator, you should study the manual that came with it. We will give some very brief instruction on how to use the BA II Plus to do the revisited examples D, E, H and I beginning on the next page.

© ROSE ALCORN/THOMSON

The Basic Annuity Keys

Examine the picture of the BA II Plus. Notice especially the third row of keys labeled [N], [I/Y], [PV], [PMT], and [FV], and the [CPT] key in the upper left corner. These are the primary keys that are used to perform various annuity calculations. [N] is for the number of payments. [IY]* is for the periodic interest rate written as a *percent*, NOT a decimal. [PV] is the present value of the annuity. [PMT] is Pmt, the size of the equal payment each period. [FV] is the future value of the annuity. [CPT] is the "compute" key, which makes the final calculation.

*Business calculators permit you to make different entries for the annual interest rate, the number of compounding periods in 1 year, and the number of payments in 1 year. For example, you could make monthly payments into an account that paid 6% compounded daily. That computation is well beyond the capability of what we can do with only Tables 23.1 and 23.2. Therefore, in this explanation, we will simply assume that [N] = the number of payments, and [I/Y] = the periodic interest rate.

Additional Annuity Keys

Examine the notation above the line of annuity keys in row 3. You see second function keys that are used with the yellow [**2nd**] key. The most important second functions for annuities are [**xP/Y**], [**P/Y**], [**BGN**] and [**CLR TVM**].

[**P/Y**] stands for "Payments per Year." [**P/Y**] represents the same thing as m in Step i of example D. When the BA II Plus comes from the factory, [**P/Y**] is preset at 12 for monthly compounding. The calculator could then determine the monthly interest rate and the number of months. Since we will also want to use annual, semiannual, or quarterly compounding, we are going to change [**P/Y**] to 1 and leave it at 1. This will allow us to always determine for ourselves the periodic interest rate and number of periods without having to reset [**P/Y**] for every new problem.

First, to change [**P/Y**] to 1 (assuming that [**P/Y**] is preset to 12):

Instructions:	Calculator Display	
Press [**2nd**] [**P/Y**].	P/Y=	12
Press 1	P/Y	1
Press [**ENTER**]	P/Y=	1
Press [**2nd**] [**QUIT**]		0

Second, the [**BGN**] key is used to change between an annuity where the payments are at the "end" of each period and an annuity where the payments are at the "beginning" of each period. [**BGN**] is preset to "END," which is what we want for Chapter 23. To check,

Instructions:	Calculator Display	
Press [**2nd**] [**BGN**].	END	
Press [**2nd**] [**QUIT**]		0

NOTE: The [**BGN**] key may be set on "BGN" and, if so, it must be changed.

Instructions:	Calculator Display	
Press [**2nd**] [**BGN**].	BGN	
Press [**2nd**] [**SET**]	END	
Press [**2nd**] [**QUIT**]		0

Finally, the annuity memories are usually called the "Time Value of Money" memories and they can be cleared using [**2nd**] [**CLR TVM**]. Now, we are ready to revisit examples D, E, H and I.

EXAMPLE D

Payments of $200 each are invested at the end of each quarter for 5 years at 6% compounded quarterly. Find the future value of the annuity.

In example D, Steps i, ii, iii give $m = 4$, $i = 6\% \div 4 = 1.5\%$, and $n = 4 \times 5$ years = 20 payments. Using the Texas Instruments BA II Plus calculator,

Instructions:	Calculator Display	
Press [**2nd**] [**CLR TVM**]		
Press 20 [**N**]	N =	20
Press 1.5 [**I/Y**]	I/Y =	1.5
Press 0 [**PV**]	PV =	0
Press 200 [**PMT**]	PMT =	200
Press [**CPT**] [**FV**]	FV =	−4,624.733421

The future value of the annuity is $4,624.73. *Note*: The calculator shows a "negative" answer because of a normal calculator convention: If the amounts going INTO the bank account are "positive" $200 each quarter, then the amount of money that can come OUT OF the bank account is opposite in sign. If PMT had been set at -200, then FV would have been a positive 4,624.733421.

EXAMPLE E

What size monthly payment is required to reach a total future value of $35,000 after 4 years if the interest rate is 6% compounded monthly?

In example E, Steps i, ii, iii give $m = 12$, $i = 6\% \div 12 = 0.5\%$, and $n = 12 \times 4$ years $= 48$ payments. Using the Texas Instruments BA II Plus calculator,

Instructions:	Calculator Display	
Press [2nd] [CLR TVM]		
Press 48 [N]	N =	48
Press .5 [I/Y]	I/Y =	0.5
Press 0 [PV]	PV =	0
Press 35000 [FV]	FV =	35,000
Press [CPT] [PMT]	PMT =	−646.9760167

The amount of each monthly payment is $646.98.

EXAMPLE H

Payments of $750 each are paid at the end of each quarter for 6 years at 6% compounded quarterly. Find the present value of the annuity.

In example H, Steps i, ii, iii give $m = 4$, $i = 6\% \div 4 = 1.5\%$, and $n = 4 \times 6$ years $= 24$ payments.

Instructions:	Calculator Display	
Press [2nd] [CLR TVM]		
Press 24 [N]	N =	24
Press 1.5 [I/Y]	I/Y =	1.5
Press 750 [PMT]	PMT =	750
Press 0 [FV]	FV =	0
Press [CPT] [PV]	PV =	−15,022.80402

The present value of the annuity is $15,022.80.

EXAMPLE I

$25,000 is invested into an account that pays 9% compounded monthly. What equal amount can be withdrawn at the end of each month for 4 years? The account will be empty at the end of the 48th withdrawal. In example I, Steps i, ii, iii give $m = 12$, $i = 9\% \div 12 = 0.75\%$, and $n = 12 \times 4$ years $= 48$ payments.

Instructions:	Calculator Display
Press [2nd] [CLR TVM]	
Press 48 [N]	N = 48
Press .75 [I/Y]	I/Y = 0.75
Press 25000 [PV]	PV = 25,000
Press 0 [FV]	FV = 0
Press [CPT] [PMT]	PMT = −622.1260593

The amount of each monthly withdrawal is $622.13.

Chapter Terms for Review

- annuity
- future value of an annuity
- ordinary annuity
- future value of an annuity factor (FVAF)
- present value of an annuity factor (PVAF)
- present value of an annuity
- sinking fund
- time line

THE BOTTOM LINE

Summary of chapter learning objectives:

Learning Objective	Example
23.1 Compute the future value of an annuity	1. Compute the future value of $900 invested every month for 2.5 years, with interest at 6% compounded monthly.
23.2 Compute the regular payments of an annuity from the future value	2. Compute the regular annuity payment that is required to accumulate $6,000 after 17 quarterly payments at an interest rate of 8% compounded quarterly.
23.3 Compute the present value of an annuity	3. Compute the present value of $1,500 withdrawn every half-year for $7\frac{1}{2}$ years, with interest at 6% compounded semiannually.
23.4 Compute the regular payments of an annuity from the present value	4. An account starts at $4,000. Compute the regular annual withdrawal that is required to empty the account in 6 years if the interest is 10% compounded annually.
23.5 Compute the loan payment required to amortize a loan	5. Compute the loan payment that is required to amortize a $20,000 loan in 24 monthly payments, with an interest rate of 9% compounded monthly.
23.6 Create a loan amortization schedule	6. A $10,000 loan at a rate of 12% compounded monthly is amortized in 15 monthly payments of $721.24. Compute the entries for the first line of the amortization schedule.

Answers: 1. $29,052.01 2. $299.82 3. $17,906.90 4. $918.43 5. $913.69 6. $10,000; $100.00; $621.24; $721.24; $9,378.76

SELF-CHECK

Review Problems for Chapter 23

1 For each annuity, find either the future value (compound amount) or the payment, as indicated. Then compute the compound interest earned by the annuity. (Use Table 23-1)

Payment	Rate	Period	Time	Future Value	Interest
$3,500	6%	monthly	3.5 yr	a. _____	b. _____
$5,000	8%	semiannually	20 yr	c. _____	d. _____

Future Value	Rate	Period	Time	Payment	Interest
$30,000	5%	annually	18 yr	e. _____	f. _____
$28,000	6%	quarterly	6 yr	g. _____	h. _____

2 For each annuity, find either the present value or the payment, as indicated. Then compute the compound interest earned by the annuity. (Use Table 23-2)

Payment	Rate	Period	Time	Present Value	Interest
$1,750	12%	semiannually	2 yr	a. _____	b. _____
$2,875	6%	quarterly	11 yr	c. _____	d. _____

Present Value	Rate	Period	Time	Payment	Interest
$24,000	10%	annually	17 yr	e. _____	f. _____
$60,000	9%	monthly	2.5 yr	g. _____	h. _____

3 Sharon Wilder planned to save money for retirement. She put $750 every month in an investment that paid a return of 6% compounded monthly. How much would Sharon have in her account after 4 years?

4 Med-West wanted to set up a sinking fund to have $15,000,000 in 20 years. The company would make annual payments that would pay a return of 6% per year. What size should the payments be?

5 Nancy Duncan received a payment of $75,000 from a life insurance company. She put it in an account that would pay 6% compounded quarterly. Nancy wanted to make equal quarterly withdrawals from the account for 10 years, when the account would be empty. What size withdrawals can Nancy make?

6 Wayne Runn read about an investment opportunity on the Internet. The Website explained that Wayne would receive payments of $1,000 every 6 months for 14 years. If the returns are based on 8% compounded semiannually, what is the present value of this investment opportunity?

Assignment 23.1: Annuities—Future Value

Name

Date   Score

Learning Objectives **1** **2**

A (28 points) For each of the following annuities, find the future value or the amount of the periodic payment. Round answers to the nearest cent. (4 points for each correct answer)

	Payment Amount	Payment Periods	Interest Rate	Length of Annuity	Future Value
1.	$2,200	monthly	9% compounded monthly	2 years	$57,614.63

9% ÷ 12 = 0.75% $2,200 × 26.18847 = $57,614.63
12 × 2 = 24 payments

2.	$696.03	quarterly	6% compounded quarterly	7 years	$24,000

6% ÷ 4 = 1.5% $24,000 ÷ 34.48148 = $696.03
4 × 7 = 28 payments

3.	$3,500	semiannually	8% compounded semiannually	10 years	$104,223.28

8% ÷ 2 = 4% $3,500 × 29.77808 = $104,223.28
2 × 10 = 20 payments

4.	$4,634.23	annually	5% compounded annually	15 years	$100,000

5% ÷ 1 = 5% $100,000 ÷ 21.57856 = $4,634.23
1 × 15 = 15 payments

5.	$500	monthly	15% compounded monthly	3 years	$22,557.76

15% ÷ 12 = 1.25% $500 × 45.11551 = $22,557.76
12 × 3 = 36 payments

6.	$582.64	quarterly	5% compounded quarterly	10 years	$30,000

5% ÷ 4 = 1.25% $30,000 ÷ 51.48956 = $582.64
4 × 10 = 40 payments

7.	$1,000	semiannually	10% compounded semiannually	16 years	$75,298.83

10% ÷ 2 = 5% $1,000 × 75.29883 = $75,298.83
2 × 16 = 32 payments

Score for A (28)

Assignment 23.1 Continued

B (32 points) For each of the following annuities, find the future value, the amount of the periodic payment, or the total amount of interest paid. Round answers to the nearest cent. (4 points for each correct answer)

8. Calculate the future value of a 25-year annuity with payments of $3,000 each year and an interest rate of 5% compounded annually.

   $143,181.30

   5% ÷ 1 = 5%
   1 × 25 = 25 payments $3,000 × 47.72710 = $143,181.30

9. How much total interest is earned on an annuity with payments of $300 per month for 4 years and an interest rate of 6% compounded monthly?

   $1,829.35

   6% ÷ 12 = 0.5% $300 × 54.09783 = $16,229.35
   12 × 4 = 48 payments $16,229.35 − ($300 × 48) = $1,829.35

10. An annuity consists of quarterly payments of $1,600 each for 10 years at an interest rate of 6% compounded quarterly. Compute the future value of the annuity.

    $86,828.62

    6% ÷ 4 = 1.5%
    4 × 10 = 40 payments $1,600 × 54.26789 = $86,828.62

11. A 7-year annuity has semiannual payments of $8,000 each and an interest rate of 8% compounded semiannually. What will be the total amount of interest earned?

    $34,335.28

    8% ÷ 2 = 4% $8,000 × 18.29191 = $146,335.28
    2 × 7 = 14 payments $146,335.28 − ($8,000 × 14) = $34,335.28

12. A sinking fund has 7 annual payments, has an interest rate of 8% compounded annually, and has a future value of $15,000. Compute the amount of each annual payment.

    $1,681.09

    8% ÷ 1 = 8% $15,000 ÷ 8.92280 = $1,681.09
    1 × 7 = 7 payments

13. An 8-year annuity with quarterly payments and an interest rate of 5% compounded quarterly has a future value of $45,000. How much total interest does the annuity earn?

    $8,124.48

    5% ÷ 4 = 1.25% $45,000 ÷ 39.05044 = $1,152.36
    4 × 8 = 32 payments $45,000 − ($1,152.36 × 32) = $8,124.48

14. Calculate the amount of each monthly payment in a 1-year annuity that has a future value of $5,000 and an interest rate of 9% compounded monthly.

    $399.76

    9% ÷ 12 = 0.75%
    12 × 1 = 12 payments $5,000 ÷ 12.50759 = $399.76

Assignment 23.1 Continued

15. Determine the total interest earned by an annuity with semiannual payments for 18 years, an interest rate of 10% compounded semiannually, and a future value of $25,000.

$15,609.04

10% ÷ 2 = 5% $25,000 ÷ 95.83632 = $260.86
2 × 18 = 36 payments $25,000 − ($260.86 × 36) = $15,609.04

Score for B (32)

C **(40 points) In each of the following applications, find the future value of the annuity, the amount of the periodic payment, or the total amount of interest earned. Round answers to the nearest cent. (4 points for each correct answer)**

16. Jim Walter decides to make semiannual deposits in his credit union account because it is guaranteeing a rate of 8% compounded semiannually for the next 5 years. How much will Jim have after making equal semiannual deposits of $2,500 for 5 years?

$30,015.28

8% ÷ 2 = 4%
2 × 5 = 10 payments $2,500 × 12.00611 = $30,015.28

17. Calvin White is planning for his daughter's college education. An investment advisor recommends an investment whose prospectus claims it will return 9% compounded monthly. If the investment does return 9% compounded monthly, how much must Calvin invest each month for 4 years if he wants to have a total of $50,000 after the last deposit?

$869.25

9% ÷ 12 = 0.75%
12 × 4 = 48 payments $50,000 ÷ 57.52071 = $869.25

18. Maxfield International is raising $25,000,000 by selling bonds that will mature in 20 years. Maxfield plans to make equal annual payments in a sinking fund to repay the bonds. If Maxfield can earn 6% per year, what amount should it deposit each year in order to have $25,000,000 at the end of 20 years?

$679,613.95

6% ÷ 1 = 6%
1 × 20 = 20 payments $25,000,000 ÷ 36.78559 = $679,613.95

19. Ruben Mendoza is quite certain he will need to replace some construction equipment in 2 years. He decides to set up a sinking fund now to help buy the equipment. Ruben estimates that he can deposit $1,100 each month for 2 years in a sinking fund that will pay 9% compounded monthly. How much will Ruben's sinking fund be worth after the last deposit?

$28,807.32

9% ÷ 12 = 0.75%
12 × 2 = 24 payments $1,100 × 26.18847 = $28,807.32

20. Bill Starnes is planning that his twin daughters could get married in 6 years. He thinks that he should start saving now to try to accumulate $30,000 by the end of the 6 years. Assuming that Bill can find an investment that will pay 8% compounded quarterly, what amount must Bill deposit each quarter to have the necessary $30,000 at the end of the 6 years?

$986.13

8% ÷ 4 = 2%
4 × 6 = 24 payments $30,000 ÷ 30.42186 = $986.13

Assignment 23.1 Continued

21. Joseph Woo imports patio furniture from various countries. He prefers to have cash available when he goes on buying trips. Suppose that Joseph makes equal monthly deposits into a risky investment that promises to pay 12% compounded monthly. If he deposits enough each month to accumulate $60,000 by the end of 2 years, and if the investment pays as promised, how much of the $60,000 will the bank have paid in interest?

$6,614.16

12% ÷ 12 = 1% $60,000 ÷ 26.97346 = $2,224.41
12 × 2 = 24 payments $60,000 − ($2,224.41 × 24) = $6,614.16

22. Jeanne Knowles will graduate from high school in a few months. She has found a part-time job and is trying to determine how much money she can save in 6 years. Calculate the future value after 6 years if Jeanne makes semiannual deposits of $600 each in an investment account that promises a return of 8% compounded semiannually.

$9,015.49

8% ÷ 2 = 4%
2 × 6 = 12 payments $600 × 15.02581 = $9,015.49

23. Musical Instrument Manufacturing, Inc., (MIMI) just sold $40,000,000 in bonds. The bonds will mature in 20 years. MIMI will make equal semiannual payments into a sinking fund that will earn 10% compounded semiannually. If MIMI has the $40,000,000 after 20 years, what amount of the total was earned from the interest?

$26,754,941.60

10% ÷ 2 = 5% $40,000,000 ÷ 120.79977 = $331,126.46
2 × 20 = 40 payments $40,000,000 − ($331,126.46 × 40) = $26,754,941.60

24. Every three months, Katie Webb sends $750 to her granddaughter, Jenny. To encourage Jenny to save money, Jenny's father promises to give her interest of 12% compounded quarterly on everything that she saves. If Jenny always saves the entire $750 each quarter, and receives these payments every quarter for 7 years, how much money will Jenny's father pay her in interest?

$11,198.19

12% ÷ 4 = 3% $750 × 42.93092 = $32,198.19
4 × 7 = 28 payments $32,198.19 − ($750 × 28) = $11,198.19

25. This year, Doug McCombs charged all his family's Christmas gifts on a credit card, and the result was a minor financial disaster. Planning for next year, Doug decides to save money each month from January through November and put it into an account that will pay 9% compounded monthly. He plans to make 11 equal deposits, and he wants to have accumulated $2,000 once he makes the eleventh deposit. Calculate the size of each deposit.

$175.10

9% ÷ 12 = 0.75% $2,000 ÷ 11.42192 = $175.10
11 months = 11 payments

Score for C (40)

Assignment 23.2: Annuities—Present Value

Name

Date   Score

Learning Objectives **3** **4** **5** **6**

A (28 points) For each of the following annuities, find the present value or the amount of the periodic payment. Round answers to the nearest cent. (4 points for each correct answer)

	Payment Amount	Payment Periods	Interest Rate	Length of Annuity	Present Value
1.	$1,500	semiannually	8% compounded semiannually	9 years	$18,988.95

8% ÷ 2 = 4%
2 × 9 = 18 payments $1,500 × 12.65930 = $18,988.95

2.	$1,028.13	quarterly	6% compounded quarterly	12 years	$35,000

6% ÷ 4 = 1.5%
4 × 12 = 48 payments $35,000 ÷ 34.04255 = $1,028.13

3.	$800	monthly	6% compounded monthly	4 years	$34,064.26

6% ÷ 12 = 0.5%
12 × 4 = 48 payments $800 × 42.58032 = $34,064.26

4.	$4,772.24	annually	6% compounded annually	17 years	$50,000

6% ÷ 1 = 6%
1 × 17 = 17 payments $50,000 ÷ 10.47726 = $4,772.24

5.	$2,500	quarterly	8% compounded quarterly	8 years	$58,670.83

8% ÷ 4 = 2%
4 × 8 = 32 payments $2,500 × 23.46833 = $58,670.83

6.	$3,886.55	semiannually	6% compounded semiannually	25 years	$100,000

6% ÷ 2 = 3%
2 × 25 = 50 payments $100,000 ÷ 25.72976 = $3,886.55

7.	$750	monthly	9% compounded monthly	3 years	$23,585.11

9% ÷ 12 = 0.75%
12 × 3 = 36 payments $750 × 31.44681 = $23,585.11

Score for A (28)

Assignment 23.2 Continued

B (32 points) For each of the following annuities, find the present value, the amount of the periodic payment, or the total amount of interest paid. Round answers to the nearest cent. (4 points for each correct answer)

8. An annuity consists of quarterly payments of $1,200 each for 10 years at an interest rate of 5% compounded quarterly. Determine the present value of the annuity.

    $37,592.32

    5% ÷ 4 = 1.25%
    4 × 10 = 40 payments $1,200 × 31.32693 = $37,592.32

9. Compute the amount of each payment in an annuity that has a present value of $10,000 with 9 years of semiannual payments at an interest rate of 16% compounded semiannually.

    $1,067.02

    16% ÷ 2 = 8%
    2 × 9 = 18 payments $10,000 ÷ 9.37189 = $1,067.02

10. In a 20-year annuity, the annual payments are $5,000 each and the interest rate is 5% compounded annually. What is the present value of the annuity?

    $62,311.05

    5% ÷ 1 = 5%
    1 × 20 = 20 payments $5,000 × 12.46221 = $62,311.05

11. What is the total interest earned by an annuity that has a present value of $16,000 with monthly payments over a 2-year period at an interest rate of 9% compounded monthly?

    $1,543.04

    9% ÷ 12 = 0.75% $16,000 ÷ 21.88915 = $730.96
    12 × 2 = 24 payments   ($730.96 × 24) − $16,000 = $1,543.04

12. Calculate the size of the regular quarterly payments in a 10-year annuity that has a present value of $100,000 and an interest rate of 8% compounded quarterly.

    $3,655.57

    8% ÷ 4 = 2%
    4 × 10 = 40 payments $100,000 ÷ 27.35548 = $3,655.57

13. An annuity has a present value of $75,000. Compute the total interest earned by the annuity if there are annual payments over 10 years at an interest rate of 12% compounded annually.

    $57,738.20

    12% ÷ 1 = 12% $75,000 ÷ 5.65022 = $13,273.82
    1 × 10 = 10 payments   ($13,273.82 × 10) − $75,000 = $57,738.20

14. Find the present value of a 12-year annuity with semiannual payments of $6,000 each, which earns interest at a rate of 8% compounded semiannually.

    $91,481.76

    8% ÷ 2 = 4%
    2 × 12 years = 24 payments $6,000 × 15.24696 = $91,481.76

Assignment 23.2 Continued

15. Compute the amount of the regular monthly payments in a 1-year annuity that has a present value of $20,000 and an interest rate of 12% compounded monthly.

$1,776.98

12% ÷ 12 = 1%
1 × 12 = 12 payments $20,000 ÷ 11.25508 = $1,776.98

Score for B (32)

C **(28 points) In each of the following applications, find the present value of the annuity, the amount of the periodic payment, or the total amount of interest earned. Round answers to the nearest cent. (4 points for each correct answer)**

16. Walt Pierce is making a budget for the next 18 months. He estimates that his rent will be about $650 per month. For calculations, Walt considers his housing expense to be an annuity of 18 payments. If he uses an interest rate of 9% compounded monthly, what will be the present value of the annuity?

$10,906.47

9% ÷ 12 = 0.75%
18 payments $650 × 16.77918 = $10,906.47

17. After their children moved away from home, Barbara Cain and her husband sold their large house and bought a smaller condominium. Barbara invested $25,000 of their after-tax profit in an annuity that would give them equal quarterly payments for 10 years. The fund will pay a return of 8% compounded quarterly. At the end of the 10 years, their annuity will be finished. What amount will they receive each quarter?

$913.89

8% ÷ 4 = 2%
4 × 10 = 40 payments $25,000 ÷ 27.35548 = $913.89

18. Joe Littrell is considering an investment that is somewhat like a bond. The investment is an annuity that would pay Joe $800 every 6 months for 15 years. He is trying to determine how much the investment is worth today. If he uses an interest rate of 12% compounded semiannually, what is the present value of the annuity?

$11,011.86

12% ÷ 2 = 6%
2 × 15 = 30 payments $800 × 13.76483 = $11,011.86

19. Bonnie Bomar will receive a retirement bonus of $80,000. She has the option of either receiving the $80,000 now in one lump sum or having it invested and then receiving 15 equal annual annuity payments, the first payment arriving 1 year after retirement. If she selects payments over 15 years, the $80,000 is invested at a guaranteed rate of 8% compounded annually. Compute the amount of interest that Bonnie would earn by choosing the payments over 15 years instead of the lump sum.

$60,195.40

8% ÷ 1 = 8% $80,000 ÷ 8.55948 = $9,346.36
1 × 15 = 15 payments   ($9,346.36 × 15) − $80,000 = $60,195.40

20. Nellie Van Calcar inherited money from her grandfather. Nellie's daughter is in her second year of college, and Nellie wants to give her $1,600 every quarter for 3 years. Nellie can invest the money for her daughter at 6% compounded quarterly. How much should she invest now to provide for all the quarterly withdrawals and have an empty account after the last withdrawal?

$17,452.02

6% ÷ 4 = 1.5%
4 × 3 = 12 payments $1,600 × 10.90751 = $17,452.02

Assignment 23.2 Continued

21. Joyce Bodley plans to buy a pre-owned car. She can either finance the car through the dealer or borrow the money from the bank. Either way, the amount borrowed will be amortized in equal payments over 4 years. If the bank's 12% annual interest rate for pre-owned cars is compounded monthly, compute Joyce's monthly payments for a bank loan of $15,000.

$395.01

12% ÷ 12 = 1%
12 × 4 = 48 payments $15,000 ÷ 37.97396 = $395.01

22. Burton Hansen wanted to protect his home from fire and burglars, so he purchased a home security system. The total price including installation was $3,240. The alarm company convinced Burton to amortize the cost over 21 months at an interest rate of 1.25% per month (which is 15% compounded monthly). Determine the amount of each of the equal monthly payments.

$176.38

1.25% per month, 21 months $3,240 ÷ 18.36969 = $176.38

Score for C (28)

D (12 points) Gary Robinson purchased some new equipment and furniture for his office. Instead of charging it on a credit card, which had an 18% interest rate, Gary negotiated financing with the office supply dealer. The total purchase amount was $6,450 and it was amortized over 4 months. The interest rate was 6% per year, or 0.5% per month. The first three monthly payments were each $1,632.70. Complete the first three lines of the following amortization schedule. Round answers to the nearest cent. (1 point for each correct answer)

Month		Unpaid Balance	Monthly Interest	Principal Payment	Total Payment	New Balance
23.	1				$1,632.70	
24.	2				$1,632.70	
25.	3				$1,632.70	

Month 1:
$6,450.00 × 0.005 = $32.25
$1,632.70 − $32.25 = $1,600.45
$6,450.00 − $1,600.45 = $4,849.55

Month 2:
$4,849.55 × 0.005 = $24.25
$1,632.70 − $24.25 = $1,608.45
$4,849.55 − $1,608.45 = $3,241.10

Month 3:
$3,241.10 × 0.005 = $16.21
$1,632.70 − $16.21 = $1,616.49
$3,241.10 − $1,616.49 = $1,624.61

Solution

Month		Unpaid Balance	Monthly Interest	Principal Payment	Total Payment	New Balance
23.	1	6,450.00	32.25	1,600.45	1,632.70	4,849.55
24.	2	4,849.55	24.25	1,608.45	1,632.70	3,241.10
25.	3	3,241.10	16.21	1,616.49	1,632.70	1,624.61

Score for D (12)

Notes

Table 23-1 Future Value Annuity Factors

Period	0.50%	0.75%	1.00%	1.25%	1.50%	2.00%	3.00%	4.00%	5.00%	6.00%	8.00%	9.00%	10.00%	12.00%
1	1.00000	1.00000	1.00000	1.00000	1.00000	1.00000	1.00000	1.00000	1.00000	1.00000	1.00000	1.00000	1.00000	1.00000
2	2.00500	2.00750	2.01000	2.01250	2.01500	2.02000	2.03000	2.04000	2.05000	2.06000	2.08000	2.09000	2.10000	2.12000
3	3.01502	3.02256	3.03010	3.03766	3.04522	3.06040	3.09090	3.12160	3.15250	3.18360	3.24640	3.27810	3.31000	3.37440
4	4.03010	4.04523	4.06040	4.07563	4.09090	4.12161	4.18363	4.24646	4.31013	4.37462	4.50611	4.57313	4.64100	4.77933
5	5.05025	5.07556	5.10101	5.12657	5.15227	5.20404	5.30914	5.41632	5.52563	5.63709	5.86660	5.98471	6.10510	6.35285
6	6.07550	6.11363	6.15202	6.19065	6.22955	6.30812	6.46841	6.63298	6.80191	6.97532	7.33593	7.52333	7.71561	8.11519
7	7.10588	7.15948	7.21354	7.26804	7.32299	7.43428	7.66246	7.89829	8.14201	8.39384	8.92280	9.20043	9.48717	10.08901
8	8.14141	8.21318	8.28567	8.35889	8.43284	8.58297	8.89234	9.21423	9.54911	9.89747	10.63663	11.02847	11.43589	12.29969
9	9.18212	9.27478	9.36853	9.46337	9.55933	9.75463	10.15911	10.58280	11.02656	11.49132	12.48756	13.02104	13.57948	14.77566
10	10.22803	10.34434	10.46221	10.58167	10.70272	10.94972	11.46388	12.00611	12.57789	13.18079	14.48656	15.19293	15.93742	17.54874
11	11.27917	11.42192	11.56683	11.71394	11.86326	12.16872	12.80780	13.48635	14.20679	14.97164	16.64549	17.56029	18.53117	20.65458
12	12.33556	12.50759	12.68250	12.86036	13.04121	13.41209	14.19203	15.02581	15.91713	16.86994	18.97713	20.14072	21.38428	24.13313
13	13.39724	13.60139	13.80933	14.02112	14.23683	14.68033	15.61779	16.62684	17.71298	18.88214	21.49530	22.95338	24.52271	28.02911
14	14.46423	14.70340	14.94742	15.19638	15.45038	15.97394	17.08632	18.29191	19.59863	21.01507	24.21492	26.01919	27.97498	32.39260
15	15.53655	15.81368	16.09690	16.38633	16.68214	17.29342	18.59891	20.02359	21.57856	23.27597	27.15211	29.36092	31.77248	37.27971
16	16.61423	16.93228	17.25786	17.59116	17.93237	18.63929	20.15688	21.82453	23.65749	25.67253	30.32428	33.00340	35.94973	42.75328
17	17.69730	18.05927	18.43044	18.81105	19.20136	20.01207	21.76159	23.69751	25.84037	28.21288	33.75023	36.97370	40.54470	48.88367
18	18.78579	19.19472	19.61475	20.04619	20.48938	21.41231	23.41444	25.64541	28.13238	30.90565	37.45024	41.30134	45.59917	55.74971
19	19.87972	20.33868	20.81090	21.29677	21.79672	22.84056	25.11687	27.67123	30.53900	33.75999	41.44626	46.01846	51.15909	63.43968
20	20.97912	21.49122	22.01900	22.56298	23.12367	24.29737	26.87037	29.77808	33.06595	36.78559	45.76196	51.16012	57.27500	72.05244
21	22.08401	22.65240	23.23919	23.84502	24.47052	25.78332	28.67649	31.96920	35.71925	39.99273	50.42292	56.76453	64.00250	81.69874
22	23.19443	23.82230	24.47159	25.14308	25.83758	27.29898	30.53678	34.24797	38.50521	43.39229	55.45676	62.87334	71.40275	92.50258
23	24.31040	25.00096	25.71630	26.45737	27.22514	28.84496	32.45288	36.61789	41.43048	46.99583	60.89330	69.53194	79.54302	104.60289
24	25.43196	26.18847	26.97346	27.78808	28.63352	30.42186	34.42647	39.08260	44.50200	50.81558	66.76476	76.78981	88.49733	118.15524
25	26.55912	27.38488	28.24320	29.13544	30.06302	32.03030	36.45926	41.64591	47.72710	54.86451	73.10594	84.70090	98.34706	133.33387

Table 23-1 Future Value Annuity Factors *(continued)*

Period	0.50%	0.75%	1.00%	1.25%	1.50%	2.00%	3.00%	4.00%	5.00%	6.00%	8.00%	9.00%	10.00%	12.00%
26	27.69191	28.59027	29.52563	30.49963	31.51397	33.67091	38.55304	44.31174	51.11345	59.15638	79.95442	93.32398	109.18177	150.33393
27	28.83037	29.80470	30.82089	31.88087	32.98668	35.34432	40.70963	47.08421	54.66913	63.70577	87.35077	102.72313	121.09994	169.37401
28	29.97452	31.02823	32.12910	33.27938	34.48148	37.05121	42.93092	49.96758	58.40258	68.52811	95.33883	112.96822	134.20994	190.69889
29	31.12439	32.26094	33.45039	34.69538	35.99870	38.79223	45.21885	52.96629	62.32271	73.63980	103.96594	124.13536	148.63093	214.58275
30	32.28002	33.50290	34.78489	36.12907	37.53868	40.56808	47.57542	56.08494	66.43885	79.05819	113.28321	136.30754	164.49402	241.33268
31	33.44142	34.75417	36.13274	37.58068	39.10176	42.37944	50.00268	59.32834	70.76079	84.80168	123.34587	149.57522	181.94342	271.29261
32	34.60862	36.01483	37.49407	39.05044	40.68829	44.22703	52.50276	62.70147	75.29883	90.88978	134.21354	164.03699	201.13777	304.84772
33	35.78167	37.28494	38.86901	40.53857	42.29861	46.11157	55.07784	66.20953	80.06377	97.34316	145.95062	179.80032	222.25154	342.42945
34	36.96058	38.56458	40.25770	42.04530	43.93309	48.03380	57.73018	69.85791	85.06696	104.18375	158.62667	196.98234	245.47670	384.52098
35	38.14538	39.85381	41.66028	43.57087	45.59209	49.99448	60.46208	73.65222	90.32031	111.43478	172.31680	215.71075	271.02437	431.66350
36	39.33610	41.15272	43.07688	45.11551	47.27597	51.99437	63.27594	77.59831	95.83632	119.12087	187.10215	236.12472	299.12681	484.46312
37	40.53279	42.46136	44.50765	46.67945	48.98511	54.03425	66.17422	81.70225	101.62814	127.26812	203.07032	258.37595	330.03949	543.59869
38	41.73545	43.77982	45.95272	48.26294	50.71989	56.11494	69.15945	85.97034	107.70955	135.90421	220.31595	282.62978	364.04343	609.83053
39	42.94413	45.10817	47.41225	49.86623	52.48068	58.23724	72.23423	90.40915	114.09502	145.05846	238.94122	309.06646	401.44778	684.01020
40	44.15885	46.44648	48.88637	51.48956	54.26789	60.40198	75.40126	95.02552	120.79977	154.76197	259.05652	337.88245	442.59256	767.09142
41	45.37964	47.79483	50.37524	53.13318	56.08191	62.61002	78.66330	99.82654	127.83976	165.04768	280.78104	369.29187	487.85181	860.14239
42	46.60654	49.15329	51.87899	54.79734	57.92314	64.86222	82.02320	104.81960	135.23175	175.95054	304.24352	403.52813	537.63699	964.35948
43	47.83957	50.52194	53.39778	56.48231	59.79199	67.15947	85.48389	110.01238	142.99334	187.50758	329.58301	440.84566	592.40069	1081.08262
44	49.07877	51.90086	54.93176	58.18834	61.68887	69.50266	89.04841	115.41288	151.14301	199.75803	356.94965	481.52177	652.64076	1211.81253
45	50.32416	53.29011	56.48107	59.91569	63.61420	71.89271	92.71986	121.02939	159.70016	212.74351	386.50562	525.85873	718.90484	1358.23003
46	51.57578	54.68979	58.04589	61.66464	65.56841	74.33056	96.50146	126.87057	168.68516	226.50812	418.42607	574.18602	791.79532	1522.21764
47	52.83366	56.09996	59.62634	63.43545	67.55194	76.81718	100.39650	132.94539	178.11942	241.09861	452.90015	626.86276	871.97485	1705.88375
48	54.09783	57.52071	61.22261	65.22839	69.56522	79.35352	104.40840	139.26321	188.02539	256.56453	490.13216	684.28041	960.17234	1911.58980
49	55.36832	58.95212	62.83483	67.04374	71.60870	81.94059	108.54065	145.83373	198.42666	272.95840	530.34274	746.86565	1057.18957	2141.98058
50	56.64516	60.39426	64.46318	68.88179	73.68283	84.57940	112.79687	152.66708	209.34800	290.33590	573.77016	815.08356	1163.90853	2400.01825

Table 23-2 Present Value Annuity Factors

Period	0.50%	0.75%	1.00%	1.25%	1.50%	2.00%	3.00%	4.00%	5.00%	6.00%	8.00%	9.00%	10.00%	12.00%
1	0.99502	0.99256	0.99010	0.98765	0.98522	0.98039	0.97087	0.96154	0.95238	0.94340	0.92593	0.91743	0.90909	0.89286
2	1.98510	1.97772	1.97040	1.96312	1.95588	1.94156	1.91347	1.88609	1.85941	1.83339	1.78326	1.75911	1.73554	1.69005
3	2.97025	2.95556	2.94099	2.92653	2.91220	2.88388	2.82861	2.77509	2.72325	2.67301	2.57710	2.53129	2.48685	2.40183
4	3.95050	3.92611	3.90197	3.87806	3.85438	3.80773	3.71710	3.62990	3.54595	3.46511	3.31213	3.23972	3.16987	3.03735
5	4.92587	4.88944	4.85343	4.81784	4.78264	4.71346	4.57971	4.45182	4.32948	4.21236	3.99271	3.88965	3.79079	3.60478
6	5.89638	5.84560	5.79548	5.74601	5.69719	5.60143	5.41719	5.24214	5.07569	4.91732	4.62288	4.48592	4.35526	4.11141
7	6.86207	6.79464	6.72819	6.66273	6.59821	6.47199	6.23028	6.00205	5.78637	5.58238	5.20637	5.03295	4.86842	4.56376
8	7.82296	7.73661	7.65168	7.56812	7.48593	7.32548	7.01969	6.73274	6.46321	6.20979	5.74664	5.53482	5.33493	4.96764
9	8.77906	8.67158	8.56602	8.46234	8.36052	8.16224	7.78611	7.43533	7.10782	6.80169	6.24689	5.99525	5.75902	5.32825
10	9.73041	9.59958	9.47130	9.34553	9.22218	8.98259	8.53020	8.11090	7.72173	7.36009	6.71008	6.41766	6.14457	5.65022
11	10.67703	10.52067	10.36763	10.21780	10.07112	9.78685	9.25262	8.76048	8.30641	7.88687	7.13896	6.80519	6.49506	5.93770
12	11.61893	11.43491	11.25508	11.07931	10.90751	10.57534	9.95400	9.38507	8.86325	8.38384	7.53608	7.16073	6.81369	6.19437
13	12.55615	12.34235	12.13374	11.93018	11.73153	11.34837	10.63496	9.98565	9.39357	8.85268	7.90378	7.48690	7.10336	6.42355
14	13.48871	13.24302	13.00370	12.77055	12.54338	12.10625	11.29607	10.56312	9.89864	9.29498	8.24424	7.78615	7.36669	6.62817
15	14.41662	14.13699	13.86505	13.60055	13.34323	12.84926	11.93794	11.11839	10.37966	9.71225	8.55948	8.06069	7.60608	6.81086
16	15.33993	15.02431	14.71787	14.42029	14.13126	13.57771	12.56110	11.65230	10.83777	10.10590	8.85137	8.31256	7.82371	6.97399
17	16.25863	15.90502	15.56225	15.22992	14.90765	14.29187	13.16612	12.16567	11.27407	10.47726	9.12164	8.54363	8.02155	7.11963
18	17.17277	16.77918	16.39827	16.02955	15.67256	14.99203	13.75351	12.65930	11.68959	10.82760	9.37189	8.75563	8.20141	7.24967
19	18.08236	17.64683	17.22601	16.81931	16.42617	15.67846	14.32380	13.13394	12.08532	11.15812	9.60360	8.95011	8.36492	7.36578
20	18.98742	18.50802	18.04555	17.59932	17.16864	16.35143	14.87747	13.59033	12.46221	11.46992	9.81815	9.12855	8.51356	7.46944
21	19.88798	19.36280	18.85698	18.36969	17.90014	17.01121	15.41502	14.02916	12.82115	11.76408	10.01680	9.29224	8.64869	7.56200
22	20.78406	20.21121	19.66038	19.13056	18.62082	17.65805	15.93692	14.45112	13.16300	12.04158	10.20074	9.44243	8.77154	7.64465
23	21.67568	21.05331	20.45582	19.88204	19.33086	18.29220	16.44361	14.85684	13.48857	12.30338	10.37106	9.58021	8.88322	7.71843
24	22.56287	21.88915	21.24339	20.62423	20.03041	18.91393	16.93554	15.24696	13.79864	12.55036	10.52876	9.70661	8.98474	7.78432
25	23.44564	22.71876	22.02316	21.35727	20.71961	19.52346	17.41315	15.62208	14.09394	12.78336	10.67478	9.82258	9.07704	7.84314

Table 23-2 Present Value Annuity Factors *(continued)*

Period	0.50%	0.75%	1.00%	1.25%	1.50%	2.00%	3.00%	4.00%	5.00%	6.00%	8.00%	9.00%	10.00%	12.00%
26	24.32402	23.54219	22.79520	22.08125	21.39863	20.12104	17.87684	15.98277	14.37519	13.00317	10.80998	9.92897	9.16095	7.89566
27	25.19803	24.35949	23.55961	22.79630	22.06762	20.70690	18.32703	16.32959	14.64303	13.21053	10.93516	10.02658	9.23722	7.94255
28	26.06769	25.17071	24.31644	23.50252	22.72672	21.28127	18.76411	16.66306	14.89813	13.40616	11.05108	10.11613	9.30657	7.98442
29	26.93302	25.97589	25.06579	24.20002	23.37608	21.84438	19.18845	16.98371	15.14107	13.59072	11.15841	10.19828	9.36961	8.02181
30	27.79405	26.77508	25.80771	24.88891	24.01584	22.39646	19.60044	17.29203	15.37245	13.76483	11.25778	10.27365	9.42691	8.05518
31	28.65080	27.56832	26.54229	25.56929	24.64615	22.93770	20.00043	17.58849	15.59281	13.92909	11.34980	10.34280	9.47901	8.08499
32	29.50328	28.35565	27.26959	26.24127	25.26714	23.46833	20.38877	17.87355	15.80268	14.08404	11.43500	10.40624	9.52638	8.11159
33	30.35153	29.13712	27.98969	26.90496	25.87895	23.98856	20.76579	18.14765	16.00255	14.23023	11.51389	10.46444	9.56943	8.13535
34	31.19555	29.91278	28.70267	27.56046	26.48173	24.49859	21.13184	18.41120	16.19290	14.36814	11.58693	10.51784	9.60857	8.15656
35	32.03537	30.68266	29.40858	28.20786	27.07559	24.99862	21.48722	18.66461	16.37419	14.49825	11.65457	10.56682	9.64416	8.17550
36	32.87102	31.44681	30.10751	28.84727	27.66068	25.48884	21.83225	18.90828	16.54685	14.62099	11.71719	10.61176	9.67651	8.19241
37	33.70250	32.20527	30.79951	29.47878	28.23713	25.96945	22.16724	19.14258	16.71129	14.73678	11.77518	10.65299	9.70592	8.20751
38	34.52985	32.95808	31.48466	30.10250	28.80505	26.44064	22.49246	19.36786	16.86789	14.84602	11.82887	10.69082	9.73265	8.22099
39	35.35309	33.70529	32.16303	30.71852	29.36458	26.90259	22.80822	19.58448	17.01704	14.94907	11.87858	10.72552	9.75696	8.23303
40	36.17223	34.44694	32.83469	31.32693	29.91585	27.35548	23.11477	19.79277	17.15909	15.04630	11.92461	10.75736	9.77905	8.24378
41	36.98729	35.18307	33.49969	31.92784	30.45896	27.79949	23.41240	19.99305	17.29437	15.13802	11.96723	10.78657	9.79914	8.25337
42	37.79830	35.91371	34.15811	32.52132	30.99405	28.23479	23.70136	20.18563	17.42321	15.22454	12.00670	10.81337	9.81740	8.26194
43	38.60527	36.63892	34.81001	33.10748	31.52123	28.66156	23.98190	20.37079	17.54591	15.30617	12.04324	10.83795	9.83400	8.26959
44	39.40823	37.35873	35.45545	33.68640	32.04062	29.07996	24.25427	20.54884	17.66277	15.38318	12.07707	10.86051	9.84909	8.27642
45	40.20720	38.07318	36.09451	34.25817	32.55234	29.49016	24.51871	20.72004	17.77407	15.45583	12.10840	10.88120	9.86281	8.28252
46	41.00219	38.78231	36.72724	34.82288	33.05649	29.89231	24.77545	20.88465	17.88007	15.52437	12.13741	10.90018	9.87528	8.28796
47	41.79322	39.48617	37.35370	35.38062	33.55319	30.28658	25.02471	21.04294	17.98102	15.58903	12.16427	10.91760	9.88662	8.29282
48	42.58032	40.18478	37.97396	35.93148	34.04255	30.67312	25.26671	21.19513	18.07716	15.65003	12.18914	10.93358	9.89693	8.29716
49	43.36350	40.87820	38.58808	36.47554	34.52468	31.05208	25.50166	21.34147	18.16872	15.70757	12.21216	10.94823	9.90630	8.30104
50	44.14279	41.56645	39.19612	37.01288	34.99969	31.42361	25.72976	21.48218	18.25593	15.76186	12.23348	10.96168	9.91481	8.30450

Notes

Business Statistics

24

Learning Objectives
By studying this chapter and completing all assignments you will be able to:

- **Learning Objective 1** — Compute the mean.
- **Learning Objective 2** — Determine the median.
- **Learning Objective 3** — Determine the mode.
- **Learning Objective 4** — Construct frequency tables.
- **Learning Objective 5** — Construct histograms.
- **Learning Objective 6** — Construct bar graphs.
- **Learning Objective 7** — Construct line graphs.
- **Learning Objective 8** — Construct pie charts.

24.1 This chapter is independent of all other chapters except Chapter 8 where cost of goods sold, expenses, and profit were mentioned.

24.2 You might remind students that the word *data* is the plural of *datum*. It is worth mentioning because of the effect that it has on grammar.

Burger King has sold billions of hamburgers. Housing prices are higher in Boston than in Atlanta. The United States has a trade deficit, which means that the country has been importing more goods than it has been exporting. Families tend to spend more in retail stores during December than during any other single month of the year. These examples are based on collections of information about businesses. The information is called **business statistics**. The word **statistics** also refers to a field of study that includes the collection, organization, analysis, and presentation of data. Businesses use statistics for two primary purposes: (1) to summarize and report the performance of the business and (2) to analyze their options in making business decisions.

Individuals and groups who want information about the business performance of a company include the company's management, board of directors, investors, and government agencies like the IRS. Once statistics have been reported, individuals and groups use the statistics to make business decisions. For example, depending on the amount of profits, the board of directors decides how much dividend to pay the shareholders. Likewise, after hearing about current profits and projected profits, investors decide whether to purchase or sell shares of the company's stock. After studying sales figures for its products and those of competitors, management makes decisions about which markets to enter, what products to emphasize, and how to advertise.

If a Burger King analyst wants to report data on sales of hamburgers, she could list the number of hamburgers sold at every restaurant. But Burger King has so many restaurants that there would be too many numbers to be meaningful. To make the data meaningful, the analyst can make some summary calculations and/or organize the data in tables. To make her presentations of the data more meaningful and easier to interpret, she may draw charts, diagrams, and/or graphs.

Statistical Averages: Computing the Mean

Learning Objective 1

Compute the mean.

24.3 The word *average* is often used in an imprecise way as a synonym for the word *typical*, as in "What kind of calculator does the average student use?" The word *average* has many meanings, three of which are *mean*, *median*, and *mode*.

24.4 Bring in—or have students bring in—examples of news articles that discuss salaries or housing prices. Most often the article will mention the median rather than the mean.

The objective in reporting statistics is to summarize the data in a simple, yet meaningful manner. One way to simplify data is to compute an **average**. An average is a single number that is supposed to be "typical" or "representative" of the group. The most common way to find an average is to add all the data values and divide by the number of values. In statistics, this particular average is called the *mean*. When the mean isn't typical or representative of an entire group of data, another average might be more representative. We also discuss two other averages: the median and the mode.

The **mean** of a group of values is computed by dividing the sum of the group of values by the number of values in the group.

EXAMPLE A

Find the mean salary of five employees whose actual salaries are $51,500, $54,400, $57,600, $62,000, and $64,500.

Sum = $51,500 + $54,400 + $57,600 + $62,000 + $64,500 = $290,000

Mean = $290,000 ÷ 5 = $58,000

✓ CONCEPT CHECK 24.1

Find the mean for the following set of numbers: 14, 11, 12, 15, 10, 16, 15, 12, 13, 11, 15, 17, 13, 14, 15, 12, 18

> There are 17 numbers. The mean equals their sum divided by 17.
> Sum = 233
> Mean = 233 ÷ 17 = 13.706, or 13.7 rounded to one decimal place

Determining the Median

The **median** of a group of numbers is determined by arranging the numbers in numerical order and finding the middle number. The median is useful when one value in the group is much larger or much smaller than the rest of the numbers.

Learning Objective 2

Determine the median.

● EXAMPLE B

Find the median salary of five employees whose salaries are $51,500, $54,400, $57,600, $62,000, and $254,500.

The salaries are already in numerical order; the median is $57,600 because it is the middle number of the five numbers arranged in order.

In example B, the mean is $480,000 ÷ 5 = $96,000, but $96,000 is not representative of the salaries of the five employees. The mean is large because one employee (perhaps the owner) has a very large salary compared to the rest of the group. The median salary, $57,600, is more typical of the group.

If the number of values is even, the median will be halfway between the two middle values. (It is the mean of the middle two values.)

● EXAMPLE C

Find the median salary of six employees whose salaries are $57,600, $64,500, $51,500, $254,500, $62,000, and $54,400.

Rearranged in numerical order, the salaries are $51,500, $54,400, $57,600, $62,000, $64,500, and $254,500.

The median is halfway between the middle two numbers, $57,600 and $62,000. It is ($57,600 + $62,000) ÷ 2, or $119,600 ÷ 2 = $59,800.

✓ CONCEPT CHECK 24.2

Find the median for the following set of numbers: 14, 11, 12, 15, 10, 16, 15, 12, 13, 11, 15, 17, 13, 14, 15, 12, 18

> The median is the middle number, after all the numbers have been arranged by order of size:
> 10, 11, 11, 12, 12, 12, 13, 13, 14, 14, 15, 15, 15, 15, 16, 17, 18
> The median is the ninth number, or 14.

Determining the Mode

Learning Objective 3
Determine the mode.

The **mode** of a group of numbers is the number that occurs most often. None of examples A, B, and C has a mode because each number occurs only once. The mode is useful when the word *average* implies "most typical" or "happening most often." Retail businesses keep track of the items that sell most frequently so that they can avoid shortages of those items.

EXAMPLE D

Find the mode shoe size of 12 pairs of ASICS running shoes, sizes 6, 6, $7\frac{1}{2}$, $7\frac{1}{2}$, 8, $8\frac{1}{2}$, 9, 9, 9, 9, 9, and $9\frac{1}{2}$.

The mode is size 9, because 9 occurs most frequently.

Note that in example D neither the mean nor the median makes any sense. The mean is $98 \div 12 = 8.17$, or $8\frac{1}{6}$. The median is halfway between sizes $8\frac{1}{2}$ and 9, which would be 8.75, or $8\frac{3}{4}$. The store owner could not buy any shoes in either size $8\frac{1}{6}$ or size $8\frac{3}{4}$ because those sizes don't exist. However, the store owner does want to stock enough shoes in size 9.

✓ CONCEPT CHECK 24.3

Find the mode for the following set of numbers: 14, 11, 12, 15, 10, 16, 15, 12, 13, 11, 15, 17, 13, 14, 15, 12, 18

The mode is the number that occurs most often. It is easier to find if you arrange the numbers by size first:

10, 11, 11, 12, 12, 12, 13, 13, 14, 14, 15, 15, 15, 15, 16, 17, 18

There are four 15s, so the mode is 15.

Constructing Frequency Tables

Learning Objective 4
Construct frequency tables.

The data in examples A–D are sometimes called **ungrouped data** because the numbers are listed individually. Business applications, such as sales results for all Burger King restaurants, often involve hundreds or thousands of numbers. Interpreting data that are literally pages of raw numbers is impossible. To make sense of such data, we organize the individual values into groups called **classes of data** or *data classes*. Adjacent classes "touch each other," but cannot overlap, not even by one cent. Also, classes are normally the same width. In example E, the width of each class is $5,000. The number of values in each class, called the **frequency** of the class, is summarized in a table called a **frequency table**.

> **STEPS** to Develop a Frequency Table
>
> 1. Determine the classes of data, and list the classes in one column.
> 2. Tally the data by making one mark for each data item in the column next to the appropriate class.
> 3. Count the tally marks for each class and write the number in the column next to the tally marks.

EXAMPLE E

Listed are the salaries of 25 full-time employees of a large advertising agency. Make a frequency table with five classes: $40,000 up to *but not including* $45,000, $45,000 up to *but not including* $50,000, and so on.

$42,500	$41,300	$53,500	$62,400	$47,500
45,400	54,600	41,000	44,400	59,100
48,000	52,000	57,500	62,500	44,000
53,600	46,200	53,500	51,800	56,400
55,500	46,000	45,200	46,000	60,800

The frequency table for these salaries appears in Figure 24-1.

Figure 24-1 Frequency Table

Class	Tally	Frequency (F)
$40,000 up to $45,000	‖‖‖	5
$45,000 up to $50,000	‖‖‖ ‖‖	7
$50,000 up to $55,000	‖‖‖ ‖	6
$55,000 up to $60,000	‖‖‖‖	4
$60,000 up to $65,000	‖‖‖	3
Total		25

24.5 Continue to emphasize the exact words of example E: "$40,000 up to $45,000" means "from $40,000 up to *but not including $45,000*." The class could also be defined as "$40,000.00 up to *and including $44,999.99*." This, however, is somewhat more awkward and appears more complicated. We will try to avoid difficulties by not including a number, like $45,000, which is exactly on the boundary.

COMPUTING THE MEAN OF LARGE DATA SETS

When a data set contains many numbers, as in example E, a computer spreadsheet can be used to compute the mean. If you use a calculator, be sure to check your work. One way to do so is to add all the numbers twice; one way to add them twice, but in different order, is the following.

> **STEPS** to Compute the Mean for a Large Data Set
>
> 1. Add all the numbers in each column.
> 2. Add all the numbers in each row.
> 3. Compute the grand total by adding all the column totals.
> 4. Check the grand total by adding all the row totals.
> 5. Divide the grand total by the number of values to get the mean.

24.6 In statistics textbooks, authors often distinguish between *class limits* and *class boundaries*. We do not mention either. Class limits are the smallest and largest possible values in a class. In the first class, the lower class limit is $40,000.00; the upper class limit is $44,999.99. The class boundary is the number that separates two consecutive classes. The class boundary between the first two classes is $44,999.995, which is halfway between the upper class limit of $44,999.99 and the adjacent lower class limit of $45,000.00.

EXAMPLE F

Compute the mean of the 25 salaries in example E.

$ 42,500	$ 41,300	$ 53,500	$ 62,400	$ 47,500	$ 247,200
45,400	54,600	41,000	44,400	59,100	244,500
48,000	52,000	57,500	62,500	44,000	264,000
53,600	46,200	53,500	51,800	56,400	261,500
55,500	46,000	45,200	46,000	60,800	253,500
$245,000	$240,100	$250,700	$267,100	$267,800	$1,270,700

The sum of the row totals and the sum of the column totals are both $1,270,700.

Mean = $1,270,700 ÷ 25 = $50,828

✓ CONCEPT CHECK 24.4

Make a frequency table for the following set of data. Use the classes 1,500 up to 2,000, 2,000 up to 2,500, and so on.

2,550	3,275	3,410	2,650	3,140
3,480	3,400	2,860	3,810	3,480
1,660	3,280	2,940	2,480	3,325
1,975	4,270	3,520	2,440	2,325
4,110	3,300	2,290	4,140	3,990
2,570	2,150	2,840	4,325	2,720

Class	Tally	Frequency
1,500 up to 2,000	II	2
2,000 up to 2,500	IIII	5
2,500 up to 3,000	IIII II	7
3,000 up to 3,500	IIII IIII	9
3,500 up to 4,000	III	3
4,000 up to 4,500	IIII	4
Total		30

COMPLETE ASSIGNMENT 24.1.

Charts and Graphs: Constructing Histograms

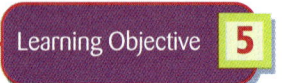

Construct histograms.

In business, statistical information is first summarized clearly in tables. For presentation, the results are then often displayed in charts or graphs. Popular graphs include the histogram, the bar graph, and the pie chart (circle graph). Histograms, bar graphs, and line graphs all have a rectangular shape. Labels are placed at the left (the vertical axis) and bottom (the horizontal axis).

A **histogram** is a diagram that presents the **grouped data** from a frequency table. The classes are positioned adjacent to each other along the horizontal axis, and the frequencies are written along the vertical axis. Figure 24-2 shows the histogram for the frequency table in Figure 24-1. The numbers on the horizontal axis increase from left to right. The numbers on the vertical axis increase from bottom to top.

24-7 Remind students of the old adage "A picture is worth a thousand words." There are over 100,000 references to it on Google. Many students will claim they are not "math-oriented." Explain to them that graphs are the artistic method of conveying the numeric information from the tables. Some people who are not "number-oriented" may be "graph-oriented."

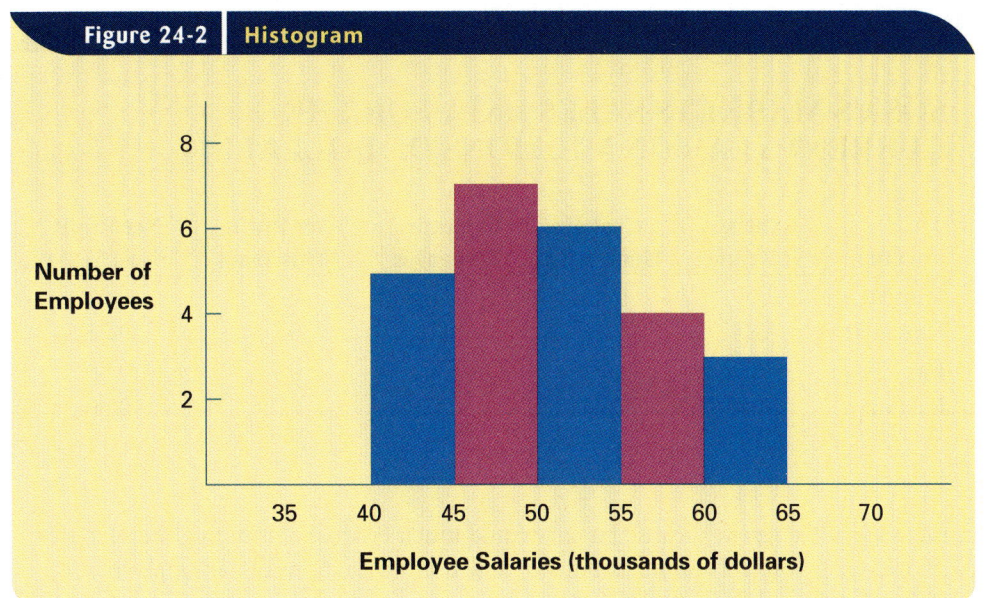

Figure 24-2 Histogram

24.8 Remind students that in a histogram, adjacent classes "touch," but do not overlap. In a very strict sense, the actual number that divides the first two rectangles in Figure 24-2 is $44,999.995. The actual dividing point should be a value that cannot go into either class. This dividing point would put $44,999.99 into the first rectangle and would put $45,000.00 into the second rectangle. You could not have a dividing point of exactly $45,000.00 because that value must go into one of the two classes.

✓ CONCEPT CHECK 24.5

Construct a histogram from the following frequency table.

Class	Frequency
1,500 up to 2,000	2
2,000 up to 2,500	5
2,500 up to 3,000	7
3,000 up to 3,500	9
3,500 up to 4,000	3
4,000 up to 4,500	_4_
Total	30

Constructing Bar Graphs

A **bar graph**, or bar chart, resembles the histogram except that there may not be a numeric scale on the horizontal axis and the bars normally do not touch each other. Sosa's Markets has grocery stores in four different towns: Warren, Hubbard, Bay City, and Easton, although the Warren store just opened last year in July. The table in Figure 24-3 shows the annual sales revenue, cost of goods sold, operation expenses, and net profits for the current year and the net profits for last year. The bar graph in Figure 24-4 illustrates the data from the current year. Data from the table in Figure 24-3 are used throughout the remainder of this chapter.

Note: It does not make sense to have the vertical bars "touch each other" as in a histogram. The four stores are distinct objects. If the horizontal axis were "time," like consecutive months of the year, then you could make a bar graph. But it would also make

Construct bar graphs.

24.9 Bar graphs can be drawn horizontally. Vertical bars are more common, especially in comparative or component bar graphs.

Figure 24-3 Revenues, Expenses and Net Profit (in millions of dollars)

SOSA'S MARKETS SALES DATA FOR THE CURRENT YEAR (IN MILLIONS OF DOLLARS)

Location	Sales Revenue	Cost of Good Sold	Operating Expenses	Net Profit (This Year)	Net Profit (Last Year)
Warren	1.50	0.75	0.50	0.25	0.15
Hubbard	3.25	1.75	1.00	0.50	0.75
Bay City	2.00	1.00	0.75	0.25	0.50
Easton	4.00	2.00	1.25	0.75	0.50

sense to use a histogram because January could touch February at midnight on January 31. However, as you will see, we can make some useful variations of bar graphs that we really cannot do with histograms.

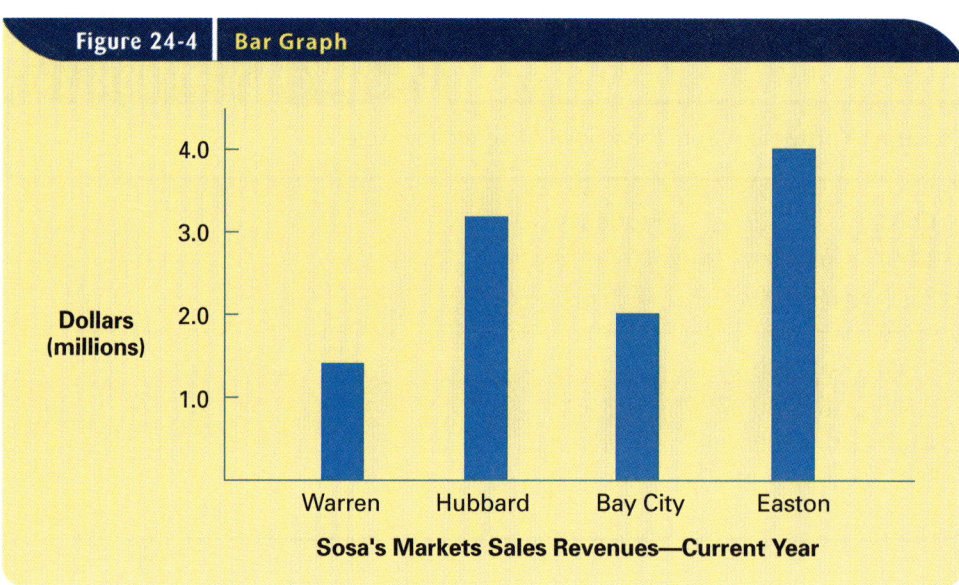

Figure 24-4 Bar Graph

Sosa's Markets Sales Revenues—Current Year

COMPARATIVE BAR GRAPH

Two bar graphs can be combined on one grid to make a **comparative bar graph**. This permits the statistician to make a graph that will compare two different sets of comparable data. The graph for Sosa's Markets in Figure 24-5 compares each store's net profit this year with its net profit last year. Each store has one pair of bars and the bars need to be colored or shaded differently to help the reader distinguish the two years.

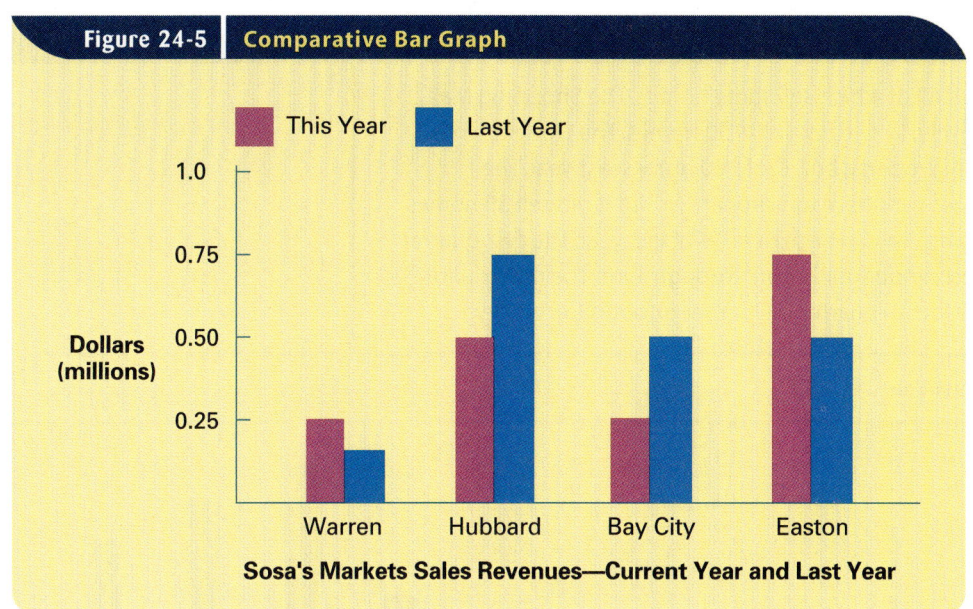

Figure 24-5 Comparative Bar Graph

Sosa's Markets Sales Revenues—Current Year and Last Year

COMPONENT BAR GRAPH

A bar graph constructed to show how certain data are composed of various parts is a **component bar graph**. Figure 24-6 shows how the current sales revenue is composed of cost of goods sold, operating expenses, and net profit. As in the comparative bar graph, the component parts are colored or shaded differently to permit easier reading.

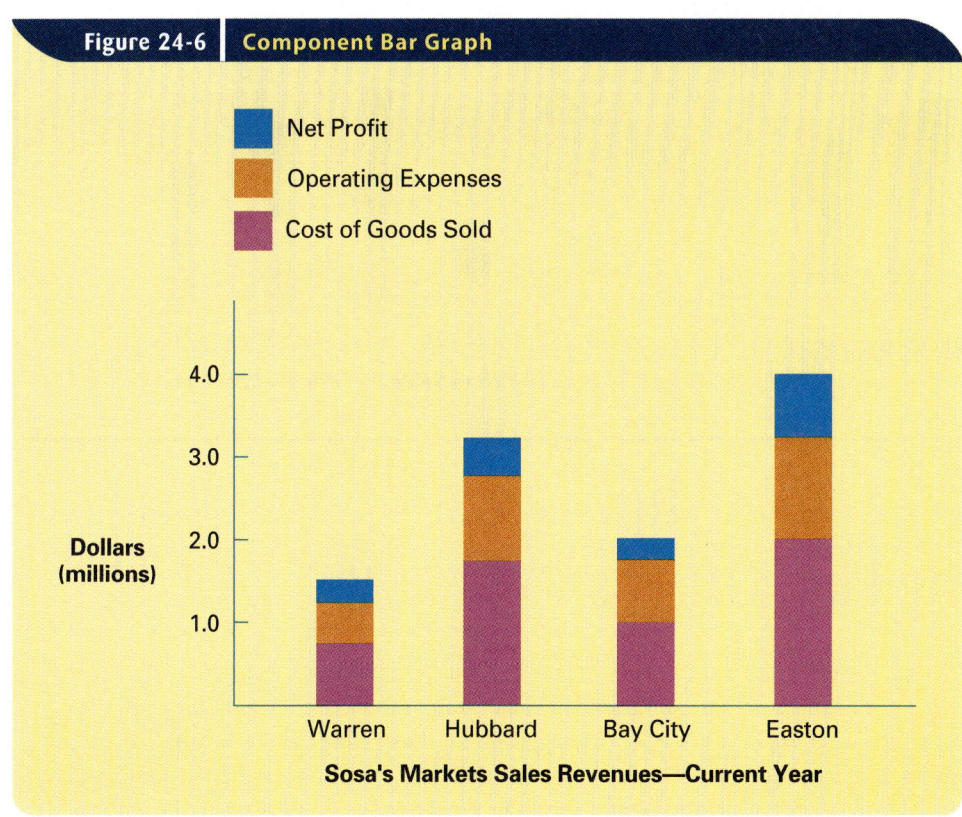

Figure 24-6 Component Bar Graph

Sosa's Markets Sales Revenues—Current Year

24.10 Discuss with students how difficult, misleading, or confusing it would be to try to use a histogram to illustrate the same information as either a comparative bar graph or a component bar graph.

CONCEPT CHECK 24.6

A real estate firm has three offices, all of which sell some homes. The Shopping Mall Office sells homes almost exclusively; last year it sold 150 homes. The Downtown Office handles mostly commercial property, but it sold 60 homes last year. The Mountain Office primarily manages various resort properties, but it did sell 30 homes. Following are the numbers of homes sold in each quarter of last year. The first quarter is January through March; the second quarter is April through June; the third quarter is July through September; and the fourth quarter is October through December.

Home Sales Last Year

Quarter	1st	2nd	3rd	4th
Shopping Mall Office	20	60	40	30
Downtown Office	5	20	25	10
Mountain Office	10	3	5	12
Total sales last year	35	83	70	52
Total sales prior year	30	75	65	55

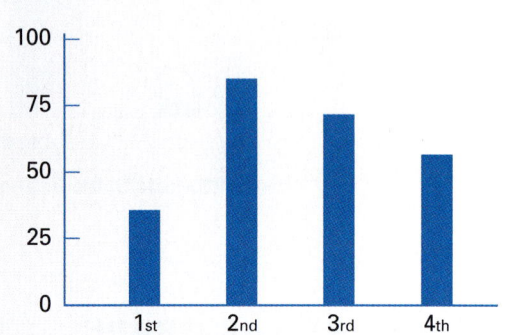
Home Sales by Quarter — Last Year

a. Construct a bar graph showing total home sales for each quarter last year. Make the vertical scale from 0 to 100, and mark the four quarters on the horizontal scale.

b. Construct a comparative bar graph showing quarterly home sales for last year and the prior year.

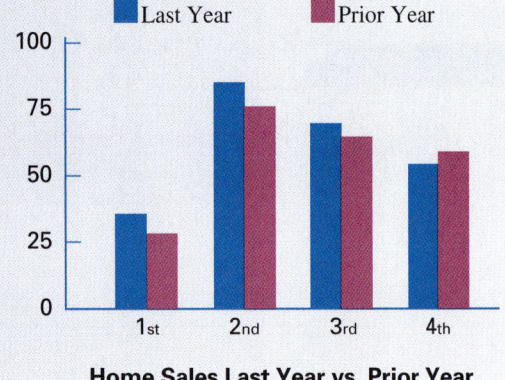
Home Sales Last Year vs. Prior Year

c. Construct a component bar graph showing quarterly home sales for each office last year.

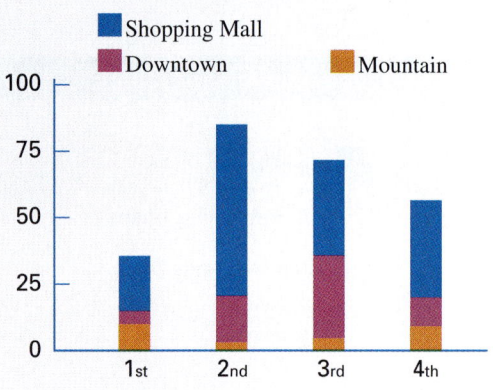
Home Sales by Office—Last Year

Constructing Line Graphs

Learning Objective 7
Construct line graphs.

Businesses very often want to view data over time, perhaps monthly or annually. As we mentioned earlier, both a histogram and a bar graph can be used when time is on the horizontal axis. However, another useful graph for illustrating data over time is the **line graph**. Plot the midpoint of each vertical bar and then connect consecutive points with straight line segments. Notice that it would not make sense to put time on the vertical axis.

Following are five months of expenses for materials for the residential and commercial divisions of New Age Metals, a custom metal fabricating business.

	Jan.	Feb.	Mar.	Apr.	May
Residential	12,000	15,000	13,000	18,000	16,000
Commercial	23,000	20,000	25,000	23,000	27,000

Figure 24-7 shows a comparative bar graph and Figure 24-8 shows a line graph with one line for the Residential Division and the other for the Commercial Division.

As we mentioned earlier, there is not a convenient, unconfusing method to make one histogram show all of the information. If you simply take the comparative bar graph, but draw the vertical bars all adjacent, the result is NOT a histogram. Histograms are simply not normally used for this kind of data. Histograms for the two divisions are shown in Figures 24-9 and 24-10, but their purpose is for you to see that the line graph and the comparative bar graph are better suited to illustrate the data.

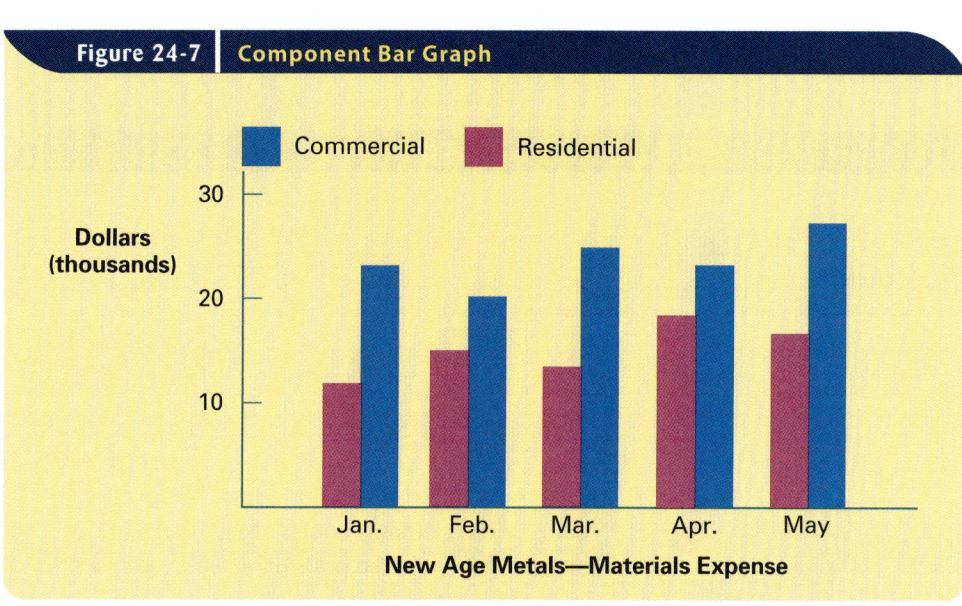

Figure 24-7 Component Bar Graph

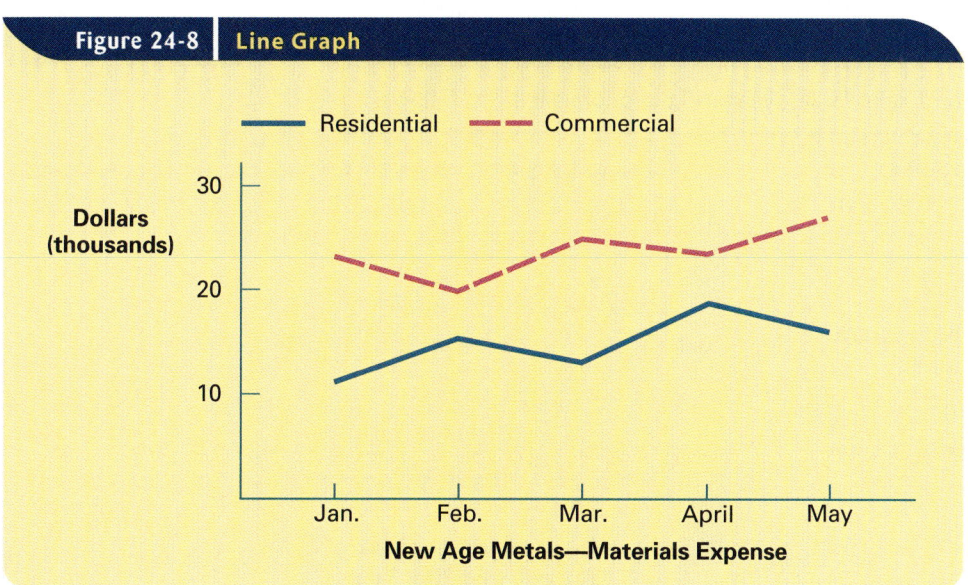

Figure 24-8 Line Graph

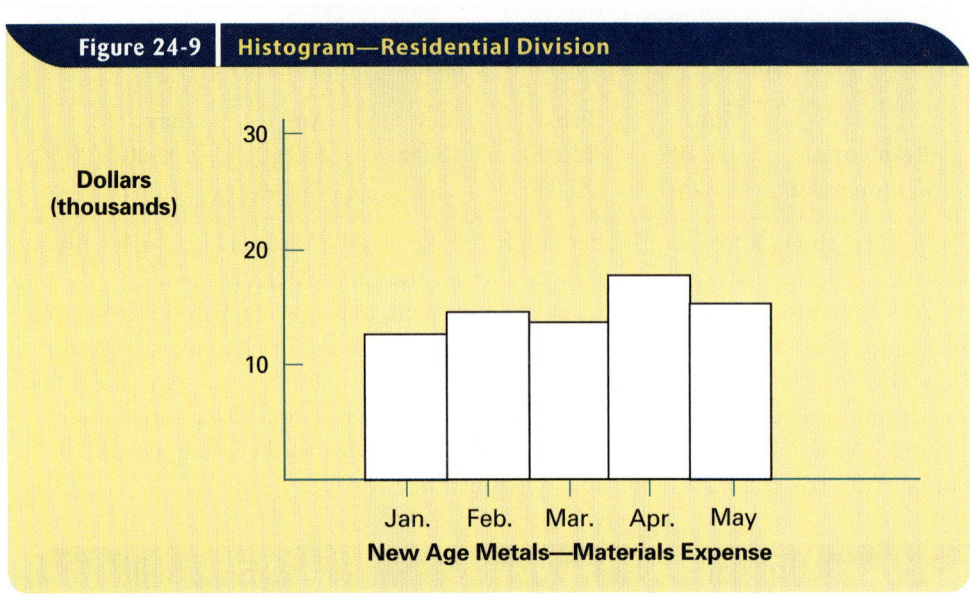

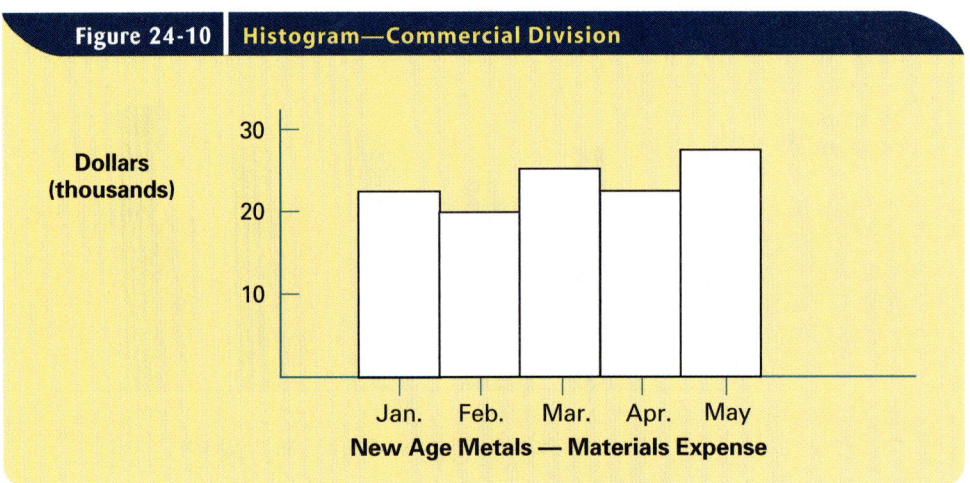

✓ CONCEPT CHECK 24.7

The real estate firm's quarterly home sales for last year and the prior year are as follows.

Quarterly Home Sales				
Quarter	1st	2nd	3rd	4th
Sales last year	35	85	75	62
Sales prior year	25	73	64	50

Construct two line graphs on the same grid showing quarterly home sales for last year and for the prior year. Make the vertical scale from 0 to 100, and mark the four quarters on the horizontal scale.

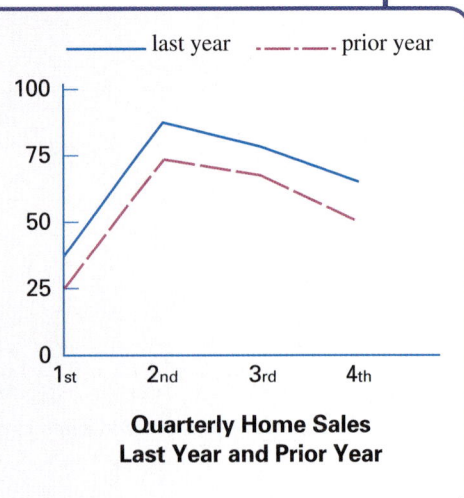

Constructing Pie Charts

Learning Objective 8
Construct pie charts.

A **pie chart**, sometimes called a circle graph, resembles a component bar graph because it shows how one quantity is composed of different parts. In a pie chart, however, the parts normally are written as percents. Figure 24-12 shows a version of the data from Bay City Market in Figure 24-11. The pie chart shown in Figure 24-12 indicates how sales revenue for March is composed of cost of goods sold, operating expenses, and net profit.

Before the graph is drawn, the data are changed into percents, as shown in Figure 24-11. The size of each part of the circle can be reasonably estimated by using the fractional equivalents of the percents. In Figure 24-12, cost of goods sold is 50%, or $\frac{1}{2}$, of the circle. Operating expenses make up 37.5%, or $\frac{3}{8}$, of the circle. The remaining $\frac{1}{8}$ represents net profit.

Figure 24-11 | Sales Revenue for Bay City Market

	Amount	Percent
Cost of Goods Sold	$1,000,000	50.0%
Operating Expenses	750,000	37.5%
Net Profit Last Year	250,000	12.5%
Sales Revenue	$2,000,000	100.0%

$1,000,000 \div $2,000,000 = 50.0\%$
$750,000 \div $2,000,000 = 37.5\%$
$250,000 \div $2,000,000 = 12.5\%$

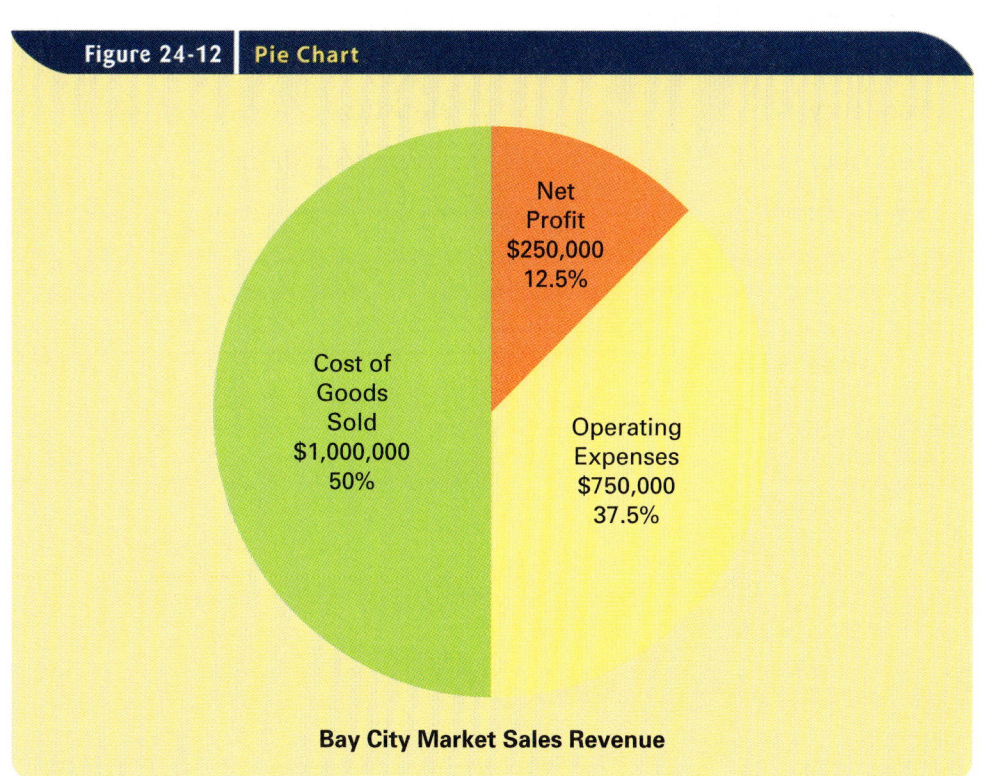

Figure 24-12 | Pie Chart

Bay City Market Sales Revenue

- Net Profit $250,000 — 12.5%
- Cost of Goods Sold $1,000,000 — 50%
- Operating Expenses $750,000 — 37.5%

Chapter 24 Business Statistics

CONCEPT CHECK 24.8

The total home sales by three real estate offices for the past year are shown. Calculate the percent of total sales for each office, and make a pie chart showing each office's share of the sales.

Office	Homes	Percent
Shopping Mall Office	150	150 ÷ 240 = 62.5%
Downtown Office	60	60 ÷ 240 = 25%
Mountain Office	30	30 ÷ 240 = 12.5%
Total	240	240 ÷ 240 = 100%

25% is $\frac{1}{4}$ of the circle; 12.5% is half of another quarter, or $\frac{1}{8}$; 62.5% is the remaining eighth plus the remaining half, or $\frac{5}{8}$.

COMPLETE ASSIGNMENT 24.2.

Home Sales Last Year

- Downtown: 60 homes, 25%
- Mountain: 30 homes, 12.5%
- Shopping Mall: 150 homes, 62.5%

Chapter Terms for Review

- average
- bar graph
- business statistics
- classes of data
- comparative bar graph
- component bar graph
- frequency
- frequency table
- grouped data
- histogram
- line graph
- mean
- median
- mode
- pie chart
- statistics
- ungrouped data

Part 6 Corporate and Special Applications

THE BOTTOM LINE

Summary of chapter learning objectives:

Learning Objective	Example
24.1 Compute the mean	1. Determine the mean (rounded to one decimal place) for these seven values: 34, 26, 17, 9, 21, 24, and 15.
24.2 Determine the median	2. Determine the median for these seven values: 15, 26, 17, 9, 21, 24, and 15.
24.3 Determine the mode	3. Determine the mode for these seven values: 24, 26, 17, 9, 21, 26, and 15.
24.4 Construct a frequency table	4. Use the following set of data to construct a frequency table. Use the classes 0 up to 100, 100 up to 200, and so on. 150 427 134 254 75 8 134 228 317 284 347 289 129 180 125 197 27 430 246 308 210 330 297 141 182
24.5 Construct histograms	5. Construct a histogram from the following frequency table. **Class** **Frequency** 0 up to 100 6 100 up to 200 4 200 up to 300 7 300 up to 400 5 400 up to 500 3 Total 25

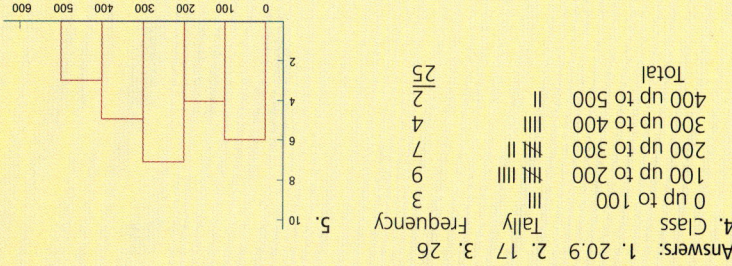

Answers: 1. 20.9 2. 17 3. 26

4.
Class	Tally	Frequency
0 up to 100	III	3
100 up to 200	IIII IIII	9
200 up to 300	IIII II	7
300 up to 400	IIII	4
400 up to 500	II	2
Total		25

5. [histogram]

THE BOTTOM LINE

Summary of chapter learning objectives:

Learning Objective	Example
24.6 Construct bar graphs	6. The monthly car sales for one car salesperson for April, May, June and July of this year are arranged as follows by type of vehicle. The total sales for these same 4 months of last year are also given. <table><tr><td>Vehicle Type</td><td>April</td><td>May</td><td>June</td><td>July</td></tr><tr><td>Two-door coupe</td><td>9</td><td>6</td><td>9</td><td>8</td></tr><tr><td>Four-door sedan</td><td>12</td><td>8</td><td>6</td><td>9</td></tr><tr><td>Sport utility vehicle</td><td>3</td><td>12</td><td>5</td><td>10</td></tr><tr><td>Totals this year</td><td>24</td><td>26</td><td>20</td><td>27</td></tr><tr><td>Totals last year</td><td>15</td><td>20</td><td>16</td><td>22</td></tr></table> Construct a bar graph showing the four monthly totals for this year. Make a vertical scale from 0 to 30, and mark the horizontal scale April, May, June, and July 7. Construct a comparative bar graph showing the four monthly totals for this year and last year. 8. Construct a component bar graph showing car sales by model for April through July of this year.

Answers: 6. 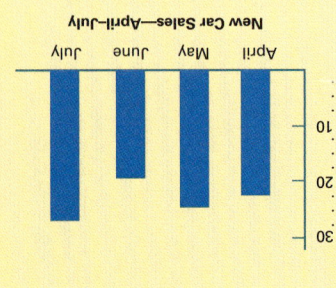 7. 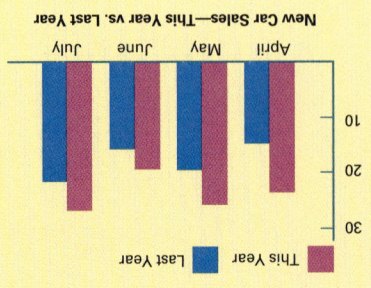 8.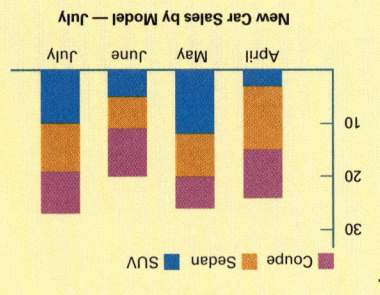

THE BOTTOM LINE

Summary of chapter learning objectives:

Learning Objective	Example
24.7 Construct line graphs	9. The monthly car sales for April, May, June, and July of this year and last year for one salesperson are as follows. | Period	April	May	June	July
---	---	---	---	---
This year	24	26	20	27
Last year	15	20	16	22	 On one grid, construct line graphs showing sales for these 4 months during this year and last year.
24.8 Construct pie charts	10. In April of this year, one car salesperson sold the following numbers of cars, arranged by type of vehicle. 	Vehicle Type	Sales	Percent
---	---	---
Two-door coupe	9	
Four-door sedan	12	
Sport utility vehicle	3	
Totals	24		 Calculate the percent for each model, and make a pie chart showing each model.

Answers: 9. 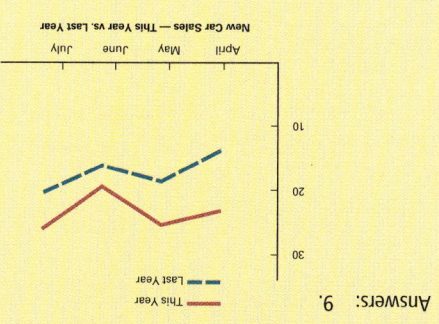 10. 50%, 37.5%, 12.5%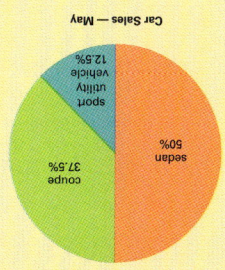

SELF-CHECK

Review Problems for Chapter 24

1 For the data 65, 53, 77, 88, 58, 82, 66, 52, 57, 62, 47, 68, 57, 78, 59, 45, and 57, find (a) the mean, (b) the median, and (c) the mode.

2 Use the data given to complete the following frequency distribution:

86, 67, 85, 57, 72
61, 77, 53, 85, 67
69, 83, 79, 68, 71
59, 62, 88, 64, 81

Class	Tally	Frequency
50 up to 60	_____	a. _____
60 up to 70	_____	b. _____
70 up to 80	_____	c. _____
80 up to 90	_____	d. _____

3 Use the frequency distributions from problem 2(a)–(d) to create the appropriate histogram. (Each vertical bar should represent one part of the problem.

4 Kevin and Al Bianchini own two markets, Bianchini's and Foodville. In a typical week, each store sells approximately 2,400lb of meat, fish, and poultry. Typical amounts are as follows:

Location	Meat	Fish	Poultry
Bianchini's	900	900	600
Foodville	1200	300	900

a. Draw a comparative bar graph showing the sales of the two markets (two vertical bars for each type of product).
b. Draw a component bar graph showing the sales of the two markets. Make one vertical bar for each store, and each bar should show the amount of each product sold in that store.
c. Draw a pie chart for the sales for Bianchini's market only.

Assignment 24.1: Statistical Averages

Name

Date   Score

Learning Objectives **1 2 3 4**

A (52 points) Solve the following problems. (points for correct answers as marked)

1. A department store has three local locations: Mason Plaza, Corbin Center, and Balbo Mall. The store gives every applicant for any type of managerial job a test of basic business skills. Listed here are the scores from the tests given to applicants at the three locations last week.

Mason Plaza	Corbin Center	Balbo Mall
59	46	65
88	60	44
62	89	53
47	55	66
68	46	58
88	74	43
78	64	77
59	89	82
45	46	66
59		62
87		
740	569	616

b. Combine all the scores into one frequency distribution with the classes as shown. (1 point for each correct answer)

Class	Tally	Frequency
40 up to 50	|||| ||	7
50 up to 60	||||		6
60 up to 70	|||| |||	8
70 up to 80	|||	3
80 up to 90	||||		6

a. Find the mean, median, and mode for each location. (3 points for each correct answer)

	Mason	Corbin	Balbo
Sum	740	569	616
Mean	67.3	63.2	61.6
Median	62	60	63.5
Mode	59	46	66

Mason Plaza: 88, 88, 87, 78, 68, 62, 59, 59, 59, 47, 45
Corbin Center: 89, 89, 74, 64, 60, 55, 46, 46, 46
Balbo Mall: 82, 77, 66, 66, 65, 62, 58, 53, 44, 43

Mason Plaza mean $= \dfrac{740}{11} = 67.3$

Corbin Center mean $= \dfrac{569}{9} = 63.2$

Balbo Mall mean $= \dfrac{616}{10} = 61.6$

Balbo Mall median $= \dfrac{65 + 62}{2} = \dfrac{127}{2} = 63.5$

2. Cirano Aguilar operates a popular coffee cart from which he also sells sandwiches. He has the opportunity to open another cart in the inner patio of a complex of office buildings, but he won't be allowed to sell sandwiches. Perform a statistical analysis on Cirano's sales receipts for nonsandwich items for the first 15 work days of April and October. (3 points for each correct answer)

April			October		
$430	$470	$450	$200	$320	$430
240	350	240	340	240	295
280	260	340	280	230	360
305	360	370	320	370	420
310	190	250	220	250	180

a. Find the mean for April. $323
b. Find the mean for October. $297
c. Find the median for April. $310
d. Find the median for October. $295
e. Find the combined mean for all 30 days. (*Hint:* Add the two sums and divide by 30.) $310

April sum = $4,845; April mean $= \dfrac{\$4{,}845}{15} = \$323$

April: 470, 450, 430, 370, 360, 350, 340, 310, 305, 280, 260, 250, 240, 240, 190

Chapter 24 Business Statistics **513**

Assignment 24.1 Continued

April median is $310

Oct. sum = $4,455; Oct. mean = $\frac{\$4,455}{15}$ = $297

Oct.: 430, 420, 370, 360, 340, 320, 320, 295, 280, 250, 240, 230, 220, 200, 180

Oct. median is $295

Combined mean = $\frac{\$4,845 + \$4,455}{30}$ = $\frac{\$9,300}{30}$ = $310

Score for A (52)

B **(48 points) Solve the following problems. (points for correct answers as marked)**

3. La Morra Bank & Trust Co. has several retail branches. Bank management wants to compare the ages of personal banking customers at two specific branches—the Financial District Branch, downtown, and the University Branch, located in a residential area between the local university and a retirement community. The bank's analyst randomly selects 30 personal banking customers from each bank and writes down their ages. The following two tables show the results.

Financial District Branch

43	30	43	51	60	227
68	32	72	52	27	251
28	73	43	19	64	227
70	35	56	55	31	247
63	24	47	44	34	212
52	61	66	57	58	294
324	255	327	278	274	1,458

University Branch

74	82	46	19	20	241
21	36	73	57	18	205
54	17	18	75	84	248
76	22	24	19	68	209
27	21	75	34	18	175
81	64	22	60	70	297
333	242	258	264	278	1,375

a. Compute the mean age of the group of customers from the Financial District Branch. (8 points) __48.6__

Mean = 1,458 ÷ 30 = 48.6

b. Compute the mean age of the group of customers from the University Branch. (8 points) __45.8__

Mean = 1,375 ÷ 30 = 45.8

c. Make two frequency tables of customer ages, one for the Financial District Branch and one for the University Branch. For each table, use frequency classes 10 up to 20, 20 up to 30, . . . , 80 up to 90. (2 points for each correct row in each table)

Financial District Branch

Class	Tally	Frequency
10 up to 20	I	1
20 up to 30	III	3
30 up to 40	IIII	5
40 up to 50	IIII	5
50 up to 60	IIII II	7
60 up to 70	IIII I	6
70 up to 80	III	3
80 up to 90		0
Total		30

University Branch

Class	Tally	Frequency
10 up to 20	IIII I	6
20 up to 30	IIII II	7
30 up to 40	II	2
40 up to 50	I	1
50 up to 60	II	2
60 up to 70	III	3
70 up to 80	IIII I	6
80 up to 90	III	3
Total		30

Score for B (48)

Assignment 24.2: Graphs and Charts

Name

Date   Score

Learning Objectives 5 6 7 8

A **(18 points) Complete the following problem as directed. (9 points for each correct graph)**

1. After doing the initial research in problem 3 of Assignment 24.1, the analyst from La Morra Bank randomly selected 100 customers from the Financial District Branch and 100 customers from the University Branch. She found the age of each customer and summarized the data in the following two frequency tables.

Financial District Branch			University Branch	
Class	Frequency		Class	Frequency
10 up to 20	6		10 up to 20	19
20 up to 30	10		20 up to 30	21
30 up to 40	16		30 up to 40	11
40 up to 50	21		40 up to 50	8
50 up to 60	19		50 up to 60	6
60 up to 70	15		60 up to 70	8
70 up to 80	11		70 up to 80	15
80 up to 90	2		80 up to 90	12
Total	100		Total	100

   a. Draw a histogram for the Financial District Branch. Label each axis, and write a title under the graph.

   b. Draw a histogram for the University Branch. Label each axis, and write a title under the graph.

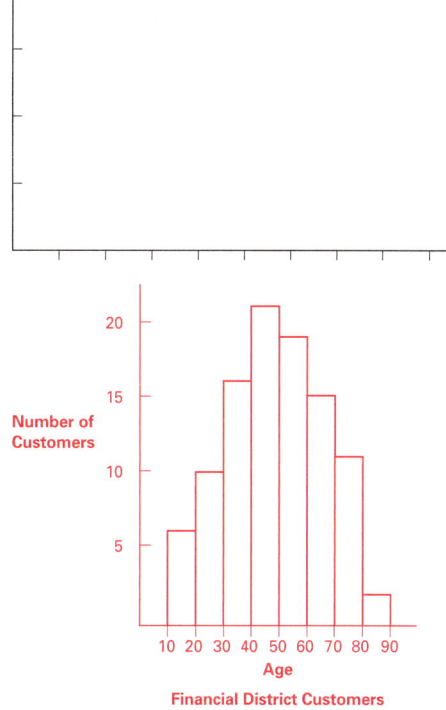

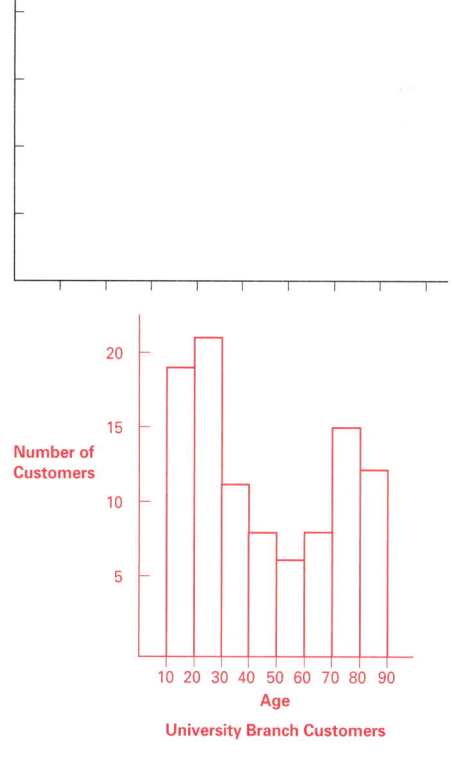

Score for A (18)

Assignment 24.2 Continued

B (54 points) Complete the following problems as directed. (18 points for each correct graph)

2. Carla Cortez owns two printing/copying businesses: Cortez Printing and Carla's Copies. Cortez Printing is near City Hall and does most of its work for corporations. Carla's Copies is in a residential district and does primarily printing and copies for individuals and small businesses. The following table shows sales revenues for the two shops for the last 4 months of the year.

Shop	September	October	November	December
Cortez Printing	$300,000	$225,000	$275,000	$200,000
Carla's Copies	125,000	150,000	100,000	175,000

a. Make a comparative bar graph showing the monthly sales revenue for each shop. Label each axis, and write a title under the graph. Shade the bars for each shop differently.

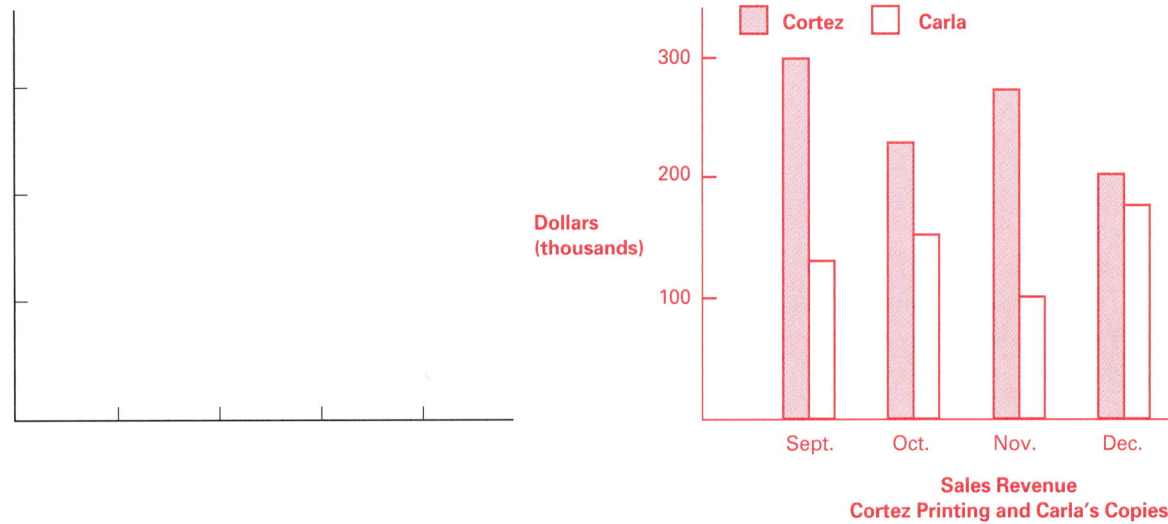

b. On the same grid, make line graphs showing the monthly sales revenue for each shop. Label each axis, and write a title under the graph. Use a solid line for Cortez Printing and a dashed line for Carla's Copies.

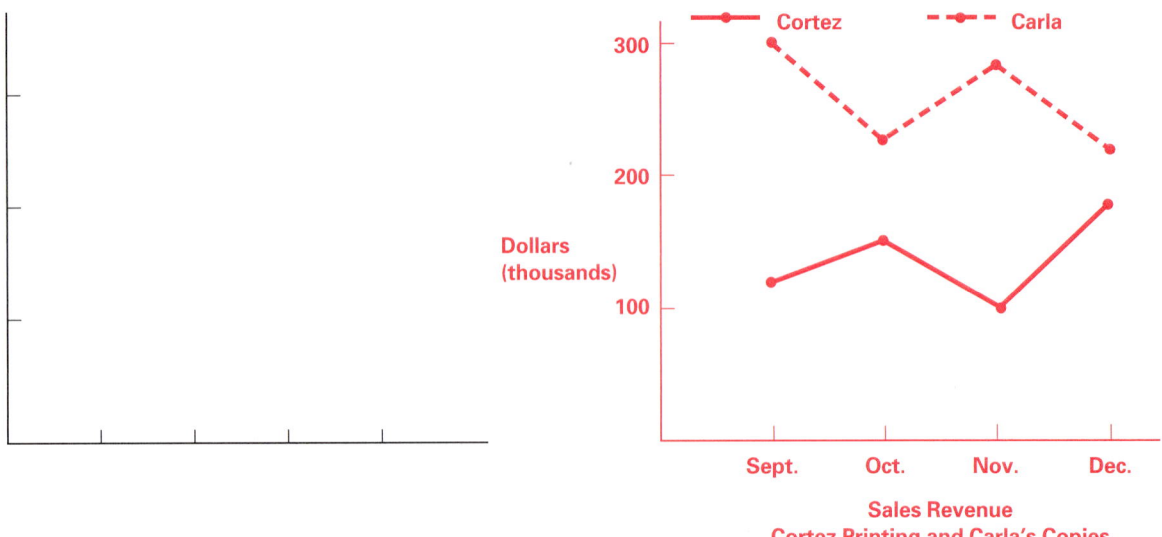

Part 6 Corporate and Special Applications

Assignment 24.2 Continued

3. New England Insurance Agency records the totals of residential (as opposed to commercial) insurance policy premiums billed each month. The results for the first 4 months of the year are shown classified by automobile insurance, homeowner's insurance, and life insurance. Construct a component bar graph showing the premiums for each insurance type each month. Label each axis, and write a title under the graph. Shade the three types of insurance differently.

Insurance Type	January	February	March	April
Auto	$200,000	$200,000	$160,000	$240,000
Home	360,000	320,000	440,000	360,000
Life	160,000	120,000	200,000	160,000
Total	$720,000	$640,000	$800,000	$760,000

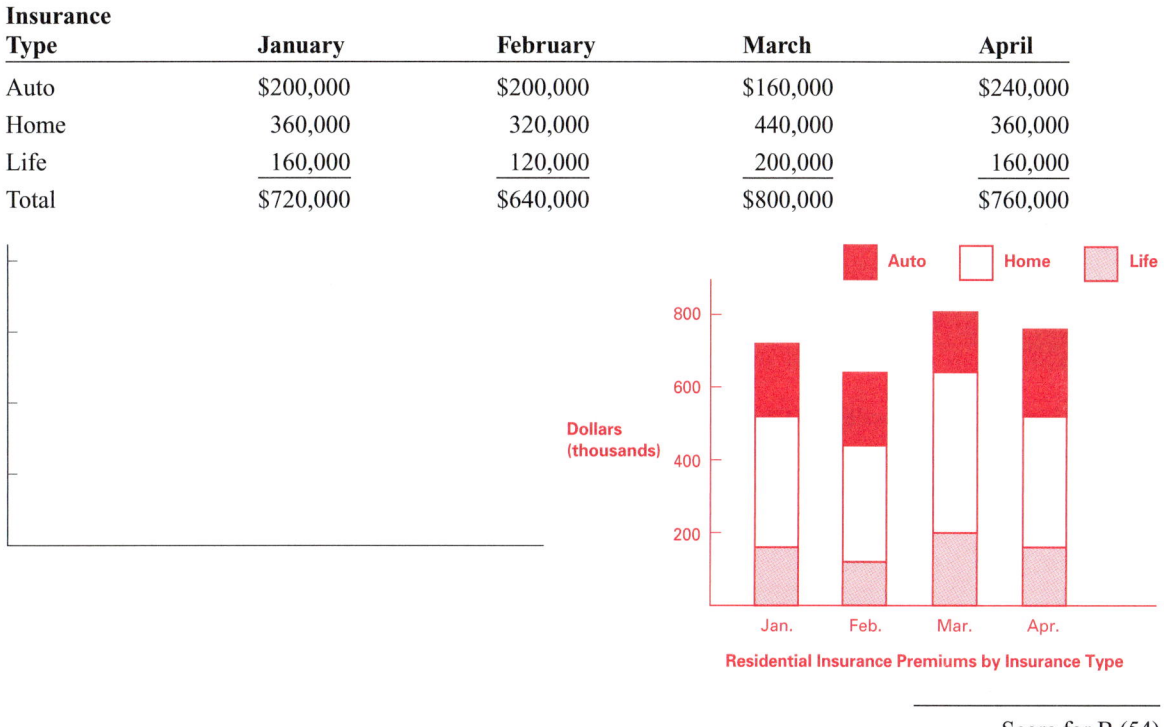

Score for B (54)

C (28 points) Complete the following problem. (points for correct answers as marked)

4. Mobile Media Warehouse is a large discount store selling audio and video products. For its internal analysis, the store classifies all music sales as Rock, Folk, Classical, or Jazz. Every music sale is included in one of these four categories. In November, the store recorded the sales shown.

   a. Compute the percent of the total and the fraction of the total represented by each category of music. (2 points for each correct percent, 1 point for each correct fraction)

Music Type	Sales	Percent	Fraction	
Rock	$276,000	50%	$\frac{1}{2}$	$276,000 \div 552,000 = 0.50$ or 50% or $\frac{1}{2}$
Folk	138,000	25%	$\frac{1}{4}$	$138,000 \div 552,000 = 0.25$ or 25% or $\frac{1}{4}$
Classical	69,000	12.5%	$\frac{1}{8}$	$69,000 \div 552,000 = 0.125$ or 12.5% or $\frac{1}{8}$
Jazz	69,000	12.5%	$\frac{1}{8}$	$69,000 \div 552,000 = 0.125$ or 12.5% or $\frac{1}{8}$
Total	$552,000	100.0%	$\frac{8}{8}$, or 1	

Assignment 24.2 Continued

b. Complete the pie chart to approximate the percent of total November sales for each category of music. Label each section with the category and percent and write a title under the graph. (8 points)

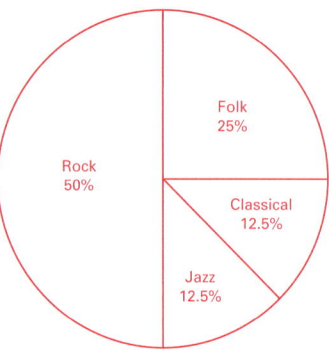

Mobile Media Warehouse
November Music Sales

c. The percents of music sales at Mobile Media Warehouse for October are shown. Complete the pie chart to approximate the percent of total October sales for each category of music. Label each section with the category and percent, and write a title under the graph. (*Hint:* 37.5% is $\frac{3}{8}$; 12.5% is $\frac{1}{8}$; 30% is somewhere between 25% and 37.5%; 20% is between 12.5% and 25%.) (8 points)

Music Type	Percent
Rock	37.5%
Folk	30.0%
Classical	12.5%
Jazz	20.0%
	100.0%

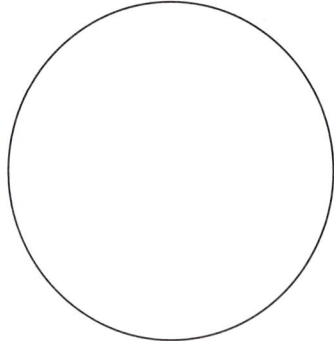

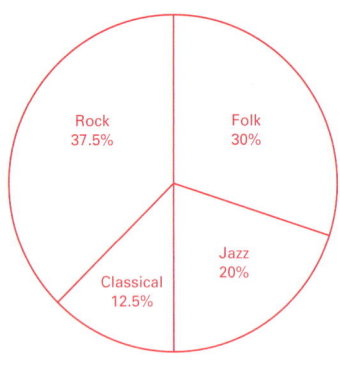

Mobile Media Warehouse
October Music Sales

Score for C (28)

Answers to Odd-Numbered Problems

Chapter 1

Assignment 1.1
1. 300
3. 377
5. 491
7. 639
9. 337
11. 1,215
13. 2,437
15. 1,626
17. 1,589
19. 2,362
21. 1,897.20
23. 1,286.33
25. 829.90
27. 1,904.78
29. 7,269.37
31. 175.93
33. 132.44
35. 265.86
37. 296.36
39. 224.25

Assignment 1.2
1. 61
3. 47
5. 36
7. 76
9. 60
11. 7
13. 59
15. 29
17. 14
19. 584
21. 103
23. 616
25. $73.98
27. $60.82
29. $38.61
31. $4,642.81
33. $8,216.01
35. $3,151.61
37. $6,983.78
39. $48.80
41. $1,790,906.69

Assignment 1.3
1. 24
3. 520
5. 90
7. 240
9. 72
11. 144
13. 48
15. 80
17. 36
19. 88
21. 28
23. 136
25. 72,576
27. 317,327,062
29. 1,080,000
31. 4,184,998
33. 548,784
35. 2,266,875
37. 184,200
39. 166,050
41. 37,500
43. 52,640
45. 9,800
47. 1,000
49. 585,514
51. 144.00
53. 366.08
55. 1,787.50
57. 2,352
59. 3,234
61. 26,400

Assignment 1.4
1. 12
3. 3
5. 42
7. 4
9. 18
11. 30
13. 13
15. 99
17. 52
19. 17
21. 5 (153)
23. 976
25. 390
27. 90 (5)
29. 7 (600)
31. 22 (16)
33. 612
35. 178 (28)
37. 184 (137)
39. 1,000 (7)
41. 20 (118)
43. 517 (597)
45. 1,111 (49)
47. $2.20
49. 1 (49)
51. 1,112 (36)
53. 260 (49)
55. 2,000,148 (24)
57. 45
59. 105 (9)

Assignment 1.5
1. 400,000
3. 2,400,000
5. 5,400,000
7. 30,000
9. 2,000,000
11. 640,000
13. 7,000,000
15. 1,000,000
17. 4,000
19. 4000
21. 270,000; 259,602
23. 10,000,000; 9,822,780
25. 160,000; 157,807
27. 60; 51
29. 200; 208

Chapter 2

Assignment 2.1
1. $2\frac{1}{6}$
3. 3
5. $1\frac{4}{7}$
7. $\frac{37}{10}$
9. $\frac{21}{8}$
11. $\frac{33}{5}$
13. $\frac{2}{5}$
15. $\frac{5}{6}$
17. $\frac{2}{3}$
19. $\frac{7}{10}$

21. $\dfrac{13}{18}$

23. $\dfrac{15}{24}$

25. $\dfrac{88}{48}$

27. $\dfrac{36}{45}$

29. $\dfrac{6}{10} = \dfrac{3}{5}$

31. $3\dfrac{17}{12} = 4\dfrac{5}{12}$

33. $7\dfrac{12}{6} = 9$

35. $6\dfrac{58}{45} = 7\dfrac{13}{45}$

37. $1\dfrac{6}{12} = 1\dfrac{1}{2}$

39. $1\dfrac{8}{12} = 1\dfrac{2}{3}$

41. $1\dfrac{17}{20}$

43. $2\dfrac{17}{30}$

45. $7\dfrac{17}{36}$ gallons

47. $\dfrac{7}{8}$ in.

Assignment 2.2
1. $\dfrac{4}{9}$
3. $\dfrac{5}{8}$
5. 7
7. $6\dfrac{3}{4}$
9. $1\dfrac{1}{6}$
11. $\dfrac{6}{7}$
13. $1\dfrac{3}{7}$
15. $4\dfrac{1}{6}$

17. $13\dfrac{1}{3}$ cu yd
19. $1\dfrac{1}{2}$ qt
21. $14\dfrac{2}{3}$ times

Chapter 3

Assignment 3.1
1. 0.0613
3. 0.64
5. 860.00098
7. twenty-six and eighty-five thousandths
9. four hundred ninety-two and three tenths
11. forty-two and four hundred eighty-one ten-thousandths
13. one thousand seven and four tenths
15. 48.8 mi
17. 374.3 lb
19. 6.4 oz
21. $0.10
23. $8.10
25. $51.38
27. 0.005 gal
29. 5.041 ft
31. 0.200 lb
33. $0.16
35. $2.10
37. $0.66
39. 22.2363
41. 104.4996
43. 29.281
45. 249.202
47. 0.364
49. 17.415
51. 7.63
53. 0.4095
55. 0.176
57. 1.677

Assignment 3.2
1. $1,072.00
3. $338.52
5. 79.3354
7. 79.9969128
9. $1.85
11. $45.25
13. 6.12
15. 62.5

17. 470
19. 0.632
21. $21,723.00
23. $280.00
25. $0.43
27. c. 0.04
29. c. 28
31. b. 0.048
33. c. 270
35. d. 120,000
37. a. 0.004
39. a. 0.14
41. a. 70

Assignment 3.3
1. 6.75 ft
3. 16.85 mi
5. $302.13
7. $285
9. $125
11. $0.08
13. 7.8 gal

Chapter 4

Assignment 4.1
1. 28
3. 2
5. 1
7. 60
9. 23
11. 15
13. 9
15. 114
17. 253
19. 1,000
21. $16.40
23. 2
25. 11
27. $400
29. $310
31. $28
33. $62.50
35. $114
37. 22
39. 385
41. 11
43. 7
45. 3
47. 16
49. 21
51. 50
53. 5
55. 15

57. a. 30, 25
    b. 36, 31
    c. 66, 60
59. a. 25, 5
    b. 9, 3
    c. 100, 20

Assignment 4.2
1. $0.72
3. $6.12
5. 103 lb
7. $1.50
9. 900 mi
11. $9.95
13. $79.92
15. $14.85
17. $23.70
19. $760
21. $801
23. $799.60
25. $240
27. $55.79
29. $89.40
31. 6 + 4 + 2 = 17 − 5
33. 9 − 3 − 1 = 2 + 3
35. 20 + 1 + 2 = 16 + 7
37. 12 + 3 − 3 = 7 + 5
39. 64 − 32 − 8 = 8 + 16

Chapter 5

Assignment 5.1
1. 0.31
3. 0.0333
5. 300%
7. 15%
9. 175%
11. 2.245
13. 52%
15. 8.25%
17. 400%
19. 0.001
21. 0.21
23. 11.17
25. 0.34
27. $0.29
29. $1.65
31. 16
33. 75
35. 0.96
37. 20%
39. 200%
41. $1.20
43. 150%
45. $48
47. $8,000
49. 56
51. 480
53. 40%
55. $21.00
57. 160%
59. 25

Assignment 5.2
1. 210
3. 30
5. $8,320
7. 544
9. $170
11. 16%
13. 25%
15. 20%
17. (25); (4.6%)
19. +230; +12.7%
21. (1,318); (8.9%)
23. (189); (17.4%)
25. +310; +17.2%
27. ($63.53); (9.4%)
29. +55.60; +14.9%
31. +22.74; +15%
33. +193.39; +4.0%
35. (216.61); (4.7%)

Assignment 5.3
1. 220
3. 6,500
5. 25%
7. 280,000
9. $720
11. $3,250
13. $52,942
15. 10%
17. $62,500
19. 100%

Assignment 5.4
1. a. 2,400; 32%; $5,120
   b. 1,800; 24%; 3,840
   c. 2,100; 28%; 4,480
   d. 1,200; 16%; 2,500
3. $6,400; $3,200; $4,800; $5,600
5. $8,840; $6,760; $4,940; $5,460

Chapter 6

Assignment 6.1
1. $3,600; $3,600
3. 2,100; 3,600
5. 3,840; 5,640
7. $3,040
9. 3,720
11. $4,900
13. $1,152; $36,995.75
15. $504; $7,612.00
17. $196; $5,207.00
19. $539; $5,634.00
21. $388; $5,456.00

Assignment 6.2
1. $5,340
3. $3,450
5. $3,680
7. $1,298.15
9. $952
11. $10,800

Chapter 7

Assignment 7.1
1. $441; $819
3. $2,120; $6,360
5. 60%; $2,250
7. $720; $420; —; $1,260
9. 70%; 85%; —; $1,071
11. 70%; 80%; 95%; 46.8%
13. $466

Assignment 7.2
1. June 1; June 21; $18.68; $603.88
3. Sept. 4; Oct. 4; $6.75; $443.25
5. Apr. 8; 98%; $570.85
7. $412.37; $251.90

Chapter 8

Assignment 8.1
1. $655.95
3. $455.48
5. $280.99
7. $340; $1,190
9. $1,050; $2,550
11. $480; $1,120
13. $2,250; $3,750
15. 160%; $775
17. 200%; $55
19. 135%; $440
21. 250%; $420
23. $1,575; $3,675
25. $1,116; 55%

Assignment 8.2
1. $149.49
3. $1,819
5. $37.49
7. $66; $54
9. $144; $216
11. $999; $999
13. $494.40; $329.60
15. 60%; $1,425
17. 55%; $260
19. 70%; $3,600
21. 65%; $820
23. $174; $174
25. $72.96; 60%

Chapter 9
Assignment 9.1
1. 585.00; 4,782.50; 3,262.50; 2,272.50; 2,207.50; 1,917.50; 5,762.75; 5,636.33; 4,671.33; 4,021.33
3. 1,190.85; 1,190.85; 1,190.85; 878.05
5. 877.76; 3,037.76; 3,037.76; 2,901.36
7. $1,669.35
9. 2,141; 70; 1,993; 50; 2,970; 30; 2,156; 30; $1,871; 13
11. 3,020; 10; 2,754; 38; 2,668; 68; 3,604; 30; $2,374; 16

Assignment 9.2
1. 802.50; 752.90; 678.71; 904.21; 791.89; 758.56; 746.56; 678.79; 466.79; 328.79; 422.79
3. a. $728.47 b. $1,630.27
   c. $951.41 d. $737.40
   e. $962.18

Assignment 9.3
1. Cogswell Cooling, Inc.
   Reconciliation of Bank Statement, November 30

Checkbook balance		$ 668.45
Minus unrecorded bank charges:		
Service charge		9.50
		$ 658.95
Plus bank interest credit		12.00
Adjusted checkbook balance		$ 670.95
Bank balance on statement		$1,050.82
Minus outstanding checks:		
No. 148	$ 13.90	
No. 156	235.10	
No. 161	96.35	
No. 165	$ 34.52	379.87
Adjusted bank balance		$ 670.95

3. Linberg Floors
   Reconciliation of Bank Statement, May 31

Checkbook balance		$19,512.54
Plus bank interest credited		35.20
		$19,547.74
Minus unrecorded bank charges:		
Service charge	$ 18.00	
Automatic transfer—insurance	1,765.00	
Returned check	920.00	2,703.00
Adjusted checkbook balance		$16,844.74
Bank balance on statement		$18,120.16
Plus deposit not recorded by bank		2,004.35
		$20,124.51
Minus outstanding checks:		
No. 730	$ 85.17	
No. 749	1,216.20	
No. 753	462.95	
No. 757	512.80	
No. 761	19.75	
No. 768	982.90	3,279.77
Adjusted bank balance		$16,844.74

Chapter 10
Assignment 10.1
1. $360.00; $108.00; $18.00; $486.00
   320.00; —; —; 320.00
   400.00; 120.00; 40.00; 560.00
   360.00; 67.50; —; 427.50
   352.00; —; —; 352.00
   280.00; —; —; 280.00
   320.00; 84.00; —; 404.00
   360.00; 13.50; —; 373.50
   352.00; 105.60; 17.60; 475.20
   352.00; —; 352.00
   380.00; 114.00; 38.00; 532.00
   400.00; 60.00; 460.00
   $4,235.00; $672.60; $113.60; $5,022.20
3. $2,808.38
5. $633.54
7. $11.21; $11.00; $0.21
9. $43.46; $43.00; $0.46

Assignment 10.2
1. $496.00; $496.00; $30.75; $7.19; $51.00; $103.04; $392.00
   400.00; 15.00; 45.00; 445.00; 27.59; 6.45; 17.00; 63.04; 381.90;
   432.00; 432.00; 26.78; 6.26; 51.00; 96.04; 335.96
   600.00; 600.00; 37.20; 8.70; 27.00; 90.90; 509.10
   368.00; 13.80; 110.40; 478.40; 29.66; 6.94; 20.00; 74.60; 403.80
   592.00; 22.20; 88.80; 680.80; 42.21; 9.87; 30.00; 100.08; 580.72
   384.00; 384.00; 23.81; 5.57; 34.00; 75.38; 308.62
   571.20; 21.42; 42.84; 614.04; 38.07; 8.90; 16.00; 74.97; 539.07
   500.00; 500.00; 31.00; 7.25; 52.00; 105.25; 394.75
   $4,343.20; $287.04; $4,630.24; $287.07; $67.13; $298.00; $784.20; $3,846.04
3. $27.94; $6.53; $17.74; $56.21; $394.39
   25.54; 5.97; 13.88; 49.39; 362.61
   25.54; 5.97; 13.88; 49.39; 362.61
   29.48; 6.89; 20.23; 60.60; 414.90
   25.74; 6.02; 14.20; 49.96; 365.24
   30.40; 7.11; 21.70; 63.21; 427.04
   26.51; 6.20; 15.43; 52.14; 375.36
   27.03; 6.32; 16.27; 53.62; 382.28
   31.62; 7.40; 23.68; 66.70; 443.30
   31.35; 7.33; 23.24; 65.92; 439.68
   31.99; 7.48; 24.28; 67.75; 448.25
   30.91; 7.23; 22.53; 64.67; 433.83
   33.22; 7.77; 26.26; 71.25; 464.55
   $377.27; $88.22; $253.32; $770.81; $5,314.04
5. a. $22,528.40
   b. $1,396.75
   c. $326.67
   d. $2,500.95
   e. $5,947.79
7. a. $19,500; $7,000
   b. $56
   c. $378
   d. $434

Chapter 11

Assignment 11.1
1. $0.43; $6.61; $3.39
   0.31; 4.71; 0.30
   0.90; 13.79; 6.21
   1.37; 20.93; 4.07
   0.41; 6.21; 3.79
   2.06; 31.47; 8.53
   1.30; 19.85; 0.15
   0.07; 1.05; 0.20
   0.98; 14.97; 0.03
   1.10; 16.79; 3.21
3. $96.55
5. a. Discount Carpets
   b. $312

Assignment 11.2
1. a. $625,000,000
   b. $732,997,500
   c. $361,760,000
3. $1.30
   $0.98
5. $2,565
7. $337.50
9. a. 1.7% (0.017)
      1.5% (0.015)
      1.35% (0.0135)
      2.0% (0.02)
   b. 17 mills
      15 mills
      13.5 mills
      20 mills
11. $1,392

Assignment 11.3
1. a. 20,750
   b. $32,900
   c. $8,000
   d. $7,392
   e. $14,888
3. a. $2,250
   b. $225
5. a. $38,050
   b. $4,993

Chapter 12

Assignment 12.1
1. a. $960
   b. $220
   c. $1,650
   d. $1,430
3. a. $3,600
   b. $2,400
   c. $279
   d. $3,600
5. a. $53,340
   b. $50,000
   c. $6,000
   d. $3,440
   e. $56,000

Assignment 12.2
1. a. $3,724
   b. $2,793
   c. $558.60
3. $200,000
5. a. $165,000
   b. $55,000
   c. $180,000
   d. $120,000
7. $360,000

Assignment 12.3
1. $19.30; $3,860.00
   $8.26; $2,643.20
   $27.04; $540.80
   $4.91; $2,356.80
   $16.83; $3,366.00
   $53.86; $4,578.10
3. $3,990
5. a. $9,050
   b. $9,500
   c. $6,545
7. a. $574
   b. $2,524

Chapter 13

Assignment 13.1
1. $30
3. $48
5. $187.50
7. $2,240
9. $130
11. $48.00; $47.34; $0.66
13. $480.00; $473.42; $6.58
15. $375.00; $369.86; $5.14
17. $6.38; $6.25; $0.13
19. $60.32; $60; $0.32
21. $4,800
23. 8%
25. 225 days
27. $33.73
29. 7.5%

Assignment 13.2
1. $12.75
   $862.75
3. $90
   $3,690
5. $1,600
   $76,600
7. $924.66
   $45,924.66
9. $67.81
   $5,067.81

Chapter 14

Assignment 14.1
1. a. 1.5%
   b. 1.25%
   c. 1.4%
   d. 0.6%
   e. 0.5%
   f. 1.6%
   g. 1.2%
   h. 0.7%
   i. 0.75%
   j. 0.8%
3. $29.34; $1,748.68
5. $45.15; $1,151.95
7. $23.63; $993.55
9. $1,098.40; $12.23;
   $1,783.02
11. $790.12; $9.15; $1,571.62

Assignment 14.2
1. $36.00; $1,636.00;
   $3,200.00
   3,200.00; 24.00; 1,624.00;
   1,600.00
   1,600.00; 12.00; 1,612.00
3. $36.00; $1,636.00;
   $3,200.00
   3,200.00; 36.00; 1,636.00;
   1,600.00
   1,600.00; 36.00; 1,636.00
5. a. $3,200
   b. $102
   c. 12.75%
7. a. $3,200
   b. $108
   c. 13.5%

Assignment 14.3
A. 1. $170.33143; $851.66
   3. $6.44301; $1,127.53
B. 5. $254.70501; $851.66
   7. 4,516.77; 33.88;
      1,494.35; 3,022.42
   9. 1,516.86; 11.38;
      1,528.24; 1,516.86

C. 11. 4,845.00; 36.34; 1,163.66; 3,681.34
13. 2,508.95; 18.82; 2,527.77; 2,508.95

Chapter 15

Assignment 15.1
1. 188
3. 122
5. 121
7. January 30, 2006
9. December 8, 2008
11. March 7, 2006
13. Jan. 9, 2007; $403; $26,403
15. Oct. 28, 2005; $583.92; $36,333.92
17. 125; $198.01; $11,998.01
19. Aug. 9, 2005; $2,115.62; $54,115.62

Assignment 15.2
1. $31.25
   $2,531.25
   May 15
   32
   $24.75
   $2,506.50
3. $0
   $4,500
   Jan. 23
   39
   $48.75
   $4,451.25
5. $71.01
   $3,671.01
   July 19
   44
   $57.53
   $3,613.48
7. $0
   $4,000
   Oct. 18
   45
   $49.32
   $3,950.68

Assignment 15.3
1. $250; $7,250; 10.34%
3. $825; $15,675; 12.63%
5. $27.18; $952.82; 7.71%
7. $100.00; 20; $26.85; $73.15
9. $525.00; 20; $74.41; $450.59

11. $92.00; 30; $71.87; $20.13
13. $650.00; 20; $118.07; $531.93

Chapter 16

Assignment 16.1
1. $7,622.94; $1,622.94
3. $37,690.80; $17,690.80
5. $5,719.80; $719.80
7. $5,713.00; $1,713.00
9. $4,381.50
11. $46,140.66
13. $1,626.84
15. $22,510.44
17. $7,590.85
19. $3,046.95
21. $31,622.58
23. $1,750.71
25. $308.99

Assignment 16.2
1. $3,266.17; $633.83
3. $22,561.35; $12,438.65
5. $5,512.60; $4,487.40
7. $2,285.35; $214.65
9. $1,060.20
11. $9,230.00
13. $4,407.62
15. $2,714.50
17. $2,218.97
19. $4,273.90
21. $18,561.75
23. $4,884.72
25. $42.88

Chapter 17

Assignment 17.1
1. $765.60
   $655.20
   $368.00
   $744.00
   $1,785.00
   $486.00
   $4,803.80
3. a. $22,950
   b. $22,200
   c. $21,700

Assignment 17.2
1. $120,000; $96,000 120,000; 72,000; 93,000 86,000

$75,000; 82,000; 87,000; $61,000
3. a. $46,300
   b. 6.41
5. a. $30,123; $\left(2\frac{1}{2} \text{ points}\right)$
   b. $50,205; $\left(2\frac{1}{2} \text{ points}\right)$
7. $1,555,829
9. a. $200,000; $4,000
   b. $182,000; $86,000
   c. $255,500; $188,500
   d. $275,591; $168,409
   e. $24,000; $13,500
   f. $160,000; $208,000
   g. $360,000; $60,000
   h. $313,043; $126,957
   i. $112,500; $12,500
   j. $100,000; $30,000

Chapter 18

Assignment 18.1
1. a. $2,700; $10,800; $19,200
   b. $6,100; $24,400; $23,600
   c. $10,500; $21,000; $63,000
   d. $5,600; $11,200; $23,400
3. a. $14,000
   b. $18,000
5. a. $4,000.00; $3,062.50
   b. $3,715.20; $343.68
   c. $2,000.00; $1,000.00
   d. $1,920.00; $1,228.80
   e. $2,362.50; $1,328.91
   f. $7,695.00; $6,232.95
7. $1,540.39
9. $8,000
   $5,333
   $2,667
11. straight-line, $22,286

Assignment 18.2
1. a. $2,475
   b. $8,640
3. $8,000
5. $15,670

Chapter 19

Assignment 19.1
1. 13.98%; 15.40%; $18,000; 8.49%

15.20%; 12.71%; $75,000; 42.86%
25.53%; 25.42%; $70,000; 20.00%
54.71%; 53.52%; $163,000; 22.12%
17.02%; 20.33%; —; 0.00%
7.29%; 7.26%; $20,000; 20.00%
9.73%; 13.07%; (20,000); −11.11%
21.28%; 19.61%; $80,000; 29.63%
14.29%; 13.80%; $45,000; 23.68%
45.29%; 46.48%; $105,000; 16.41%
100.00%; 100.00%; $268,000; 19.46 %
5.84%; 4.50%; $34,000; 54.84%
2.74%; 2.54%; $10,000; 28.57%
0.91%; 1.45%; $(5,000); −25.00%
9.48%; 8.50%; $39,000; 33.33%
18.78%; 23.24%; $(11,000); −3.44%
10.94%; 15.25%; $(30,000); −14.29%
29.73%; 38.49%; $(41,000); −7.74%
39.21%; 46.99%; $(2,000); −0.31%
31.61%; 33.91%; $53,000; 11.35%
20.06%; 15.98%; $110,000; 50.00%
9.12%; 3.12%; $107,000; 248.84%
60.79%; 53.01%; $270,000; 36.99%
100.00%; 100.00%; $268,000; 19.46%

3. 10.0%; 8.10%; $14,600; 38.5%
7.2%; 6.37%; $ 8,010; 26.9%
11.8%; 11.87%; $6,500; 11.7%
28.9%; 26.34%; $29,110; 23.6%
16.0%; 15.39%; $12,200; 16.9%
2.9%; 2.67%; $2,600; 22.4%
13.1%; 12.72%; $9,400; 15.8%
44.7%; 50.25%; —; 0.0%
13.3%; 10.69%; $20,000; 40.0%
71.1%; 73.66%; $29,400; 8.5%
100.0%; 100.00%; $58,510; 12.5%
2.7%; 3.04%; $(250); −1.8%
1.6%; 1.58%; $800; 10.8%
0.2%; 0.21%; $220; 22.4%
4.4%; 4.83%; $770; 3.4%
15.5%; 17.90%; $(2,200); −2.6%
4.8%; 4.49%; $ 4,000; 19.0%
20.2%; 22.39%; $1,800; 1.7%
24.7%; 27.21%; $2,570; 2.0%
37.1%; 38.49%; $15,000; 8.3%
15.6%; 17.53%; —; 0.0%
22.7%; 16.77%; $40,940; 52.2%
75.3%; 72.79%; $55,940; 16.4%
100.0%; 100.00%; $58,510; 12.5%

Assignment 19.2
1. 103.95%; 103.76%; $93,000; 11.25%
3.95%; 3.76%; $5,000; 16.67%
100.00%; 100.00%; $88,000; 11.04%
23.73%; 24.72%; $13,000; 6.60%
51.98%; 49.56%; $65,000; 16.46%
75.71%; 74.28%; $78,000; 13.18%
27.12%; 26.35%; $30,000; 14.29%
48.59%; 47.93%; $48,000; 12.57%
51.41%; 52.07%; $40,000; 9.64%
14.98%; 15.06%; $12,600; 10.50%
9.49%; 10.04%; $4,000; 5.00%
2.03%; 2.51%; $(2,000); −10.00%
0.51%; 0.53%; $300; 7.14%
0.41%; 0.39%; $500; 16.13%
0.14%; 0.18%; $(200); −14.29%
0.79%; 0.65%; $1,800; 34.62%
0.36%; 0.26%; $1,100; 52.38%
28.71%; 29.61%; $18,100; 7.67%
22.70%; 22.46%; $21,900; 12.23%
3.62%; 3.51%; $4,000; 14.29%
19.08%; 18.95%; $17,900; 11.85%

3. 102%; 102%; $12,200; 16%
2%; 2%; $200; 11%
100%; 100%; $12,000; 16%
26%; 24%; $4,500; 26%
45%; 48%; $3,000; 9%
71%; 72%; $7,500; 14%
28%; 30%; $2,100; 10%
42%; 42%; $5,400; 18%
58%; 58%; $6,600; 16%
13%; 15%; $300; 3%
9%; 8%; $1,500; 25%
2%; 2%; $200; 17%
1%; 1%; $70; 18%
1%; 1%; $50; 8%
0%; 1%; $(70); −17%
2%; 2%; $200; 12%
0%; 0%; $(30); −14%
28%; 29%; $2,220; 10%
30%; 29%; $4,380; 21%
3%; 3%; $200; 10%
27%; 26%; $4,180; 22%

Assignment 19.3
1. $5,400; 5.2%
19,100; 16.5%
40,000; 27.6%
$64,500; 17.6%
$(3,500); −7.2%
13,000; 9.8%
$9,500; 5.3%
$74,000; 13.6%
$4,800; 17.0%
7,100; 6.3%
$11,900; 8.4%
$(20,000); −16.7%
$(8,100); −3.1%

82,100; 28.8%
$74,000; 13.6%
$(55,000); −6.6%
$7,000; 5.1%
(35,000); −5.6%
$(28,000); −3.7%
40,000; 27.6%
$(68,000); −11.1%
$13,000; 5.9%
$3,400; 4.3%
(1,000); −3.3%
$2,400; 2.2%
$10,600; 9.4%

3. $9,000; 56.3%
4,000; 50.0%
15,000; 48.4%
$28,000; 50.9%
$(4,000); −9.3%
4,000; 36.4%
$0; 0.0%
$28,000; 25.7%
$(1,000); −18.2%
3,500; 58.3%
$2,500; 21.7%
$(8,000); −21.1%
$(5,500); −11.1%
$33,500; 56.3%

$28,000; 25.7%
$85,000; 70.8%
$3,500; 12.7%
69,500; 82.2%
$73,000; 65.2%
15,000; 48.4%
$58,000; 71.6%
$27,000; 69.2%
$9,500; 44.2%
5,750; 79.3%
$15,250; 53.0%
$11,750; 114.6%

Chapter 20

Assignment 20.1
1. a. $217.59
   b. $795.72
   c. $37.42
   d. $105.08
   e. $238.76
   f. $36.20
   g. $47.06
3. a. $132,432
   b. $8,432
   c. $7,568
5. a. 31,114
   b. $18,500

Assignment 20.2
1. a. $2,700
   b. $86.40
   c. $2,786.40
3. $1,296
5. a. $16,608
   b. $17,042.30
   c. $2,592
   d. $760.32
7. a. $36,750
   b. $14,700
   c. 0
9. $195,000
11. a. 590.55 in.
    b. 49.215 ft
    c. 16.35 yd
    d. 9.315 mi
    e. .875 oz
    f. .055 lb
    g. 55 lb
    h. 63.39 pt
    i. 31.71 qt
    j. 7.92 gal

Chapter 21

Assignment 21.1
1. a. $36,400
   b. $10,846
3. 7.9%
5. 10.52%
7. a. $2.10; 6.56%
   b. $6; 7.5%
   c. $2; 4.49%
   d. $5.50; 6.11%
   e. $3.25; 5.6%

Assignment 21.2
1. 5,000 shares; 90,000 shares
3. a. $67,500
   b. $8,750
   c. $1.75
   d. $270,000
   e. $305,000
   f. $4.00
5. $9,750; $0.16
7. $38,000 ÷ 25,000 = $1.52;
   -0-
   $42,000 ÷ 25,000 = $1.68;
   $14,000 ÷ 50,000 = $0.28
   $40,000 ÷ 25,000 = $1.60;
   $25,000 ÷ 50,000 = $0.50
9. $550
11. a. 17
    b. 15

Chapter 22

Assignment 22.1
1. $1,750
3. $5,400,000
5. a. $4,000.00; $300.00
   b. 2,940.00; 236.25
   c. 7,740.00; 911.25
   d. 6,540.00; 562.50
   $21,220.00; 2,010.00
7. a. 73; $2,096.50
   b. 100; $2,852.50
9. a. 8.33%
   b. 7.89%
   c. 8.93%
   d. 7.32%

Assignment 22.2
1. a. 8.22%
   b. 9.11%
3. a. $3,661.67
   b. 9.66%

Chapter 23

Assignment 23.1
1. $57,614.63
3. $104,223.28
5. $22,557.76
7. $75,298.83
9. $1,829.35
11. $34,335.28
13. $8,124.48
15. $15,609.04
17. $869.25
19. $28,807.32
21. $6,614.16
23. $26,754,941.60
25. $175.10

Assignment 23.2
1. $18,988.95
3. $34,064.26
5. $58,670.83
7. $23,585.11
9. $1,067.02
11. $1,543.04
13. $57,738.20
15. $1,776.98
17. $913.89
19. $60,195.40
21. $395.01
23. 6,450.00; 32.25; 1,600.45; 4,849.55
25. 3,241.10; 16.21; 1,616.49; 1,624.61

Chapter 24

Assignment 24.1

1. **a.** Mean: 67.3; 63.2; 61.6
      Median: 62; 60; 63.5
      Mode: 59; 46; 66
   b. 7; 6; 8; 3; 6
3. **a.** 48.6
   b. 45.8
   c. Financial District Branch
      10 up to 20: 1
      20 up to 30: 3
      30 up to 40: 5
      40 up to 50: 5
      50 up to 60: 7
      60 up to 70: 6
      70 up to 80: 3
      80 up to 90: 0
      Total: 30

   University Branch
      10 up to 20: 6
      20 up to 30: 7
      30 up to 40: 2
      40 up to 50: 1
      50 up to 60: 2
      60 up to 70: 3
      70 up to 80: 6
      80 up to 90: 3
      Total: 30

Assignment 24.2

1. **a.**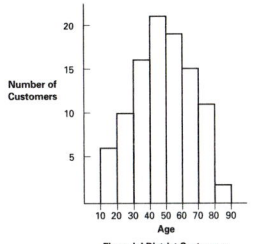

   b.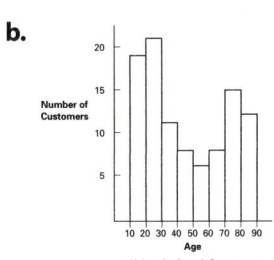

3.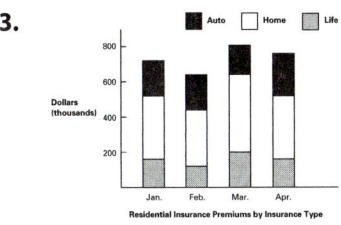

Answers to Self-Check Review Problems

Chapter 1
1. 38
2. 127; 67; 240; 204; 638
3. 2,696
4. 51 (3)
5. 7 (7)
6. 21 (33)
7. 4 (42)
8. 81 (2)
9. 32 R.12
10. 609,824
11. 5 (1)
12. 32
13. 10,000
14. 222; 313; 205; 740
15. 41,216
16. 705,408
17. 28 (4)
18. 640
19. 20,000 (6)
20. 110 (7)
21. 80 × 30 = 2400
22. 100 × 20 = 2,000
23. 400 × 200 = 80,000
24. 4000 × 100 = 400,000
25. 1,500 × 600 = 900,000
26. 400 ÷ 80 = 5
27. 900 ÷ 30 = 30
28. 10,000 ÷ 500 = 20
29. 3000 ÷ 60 = 50
30. 6000 ÷ 3000 = 2

Chapter 2
1. $\frac{17}{6}$
2. $7\frac{1}{2}$
3. $\frac{6}{7}$
4. $\frac{40}{56}$
5. $1\frac{17}{30}$
6. $1\frac{19}{24}$
7. $7\frac{11}{20}$
8. $\frac{7}{15}$
9. $1\frac{17}{18}$
10. $1\frac{34}{45}$
11. $\frac{3}{10}$
12. $1\frac{1}{20}$
13. $\frac{7}{3}$
14. $2\frac{1}{4}$
15. $2\frac{1}{10}$
16. $\frac{8}{9}$
17. $16\frac{1}{2}$
18. $24\frac{5}{16}$
19. $6\frac{1}{8}$
20. $9;\frac{7}{8}$ inches

Chapter 3
1. 116.0014
2. six thousand, four hundred thirty one and seven hundred nineteen thousandths
3. 3.5
4. $12.67
5. 743.64475
6. 20.807
7. 2.717
8. 178.4694
9. 1.797726
10. $259.51
11. 3.23
12. .74
13. 8649.3
14. 2.76235
15. d. 500
16. c. $0.80
17. $3,825.75
18. $148,235.96
19. 590.8 cubic feet
20. 21.88

Chapter 4
1. 30
2. 42
3. 96
4. 2
5. $31,256
6. $43,244
7. 427 miles
8. $400
9. $225.75
10. $250
11. 23 hours
12. 19 hours
13. 12 hours
14. 8
15. 156
16. 3
17. 3
18. 11
19. 20
20. $2.00

Chapter 5
1. .171
2. 62.5%
3. 1.5
4. $\frac{3}{4}$%
5. .0006
6. 40%
7. 7
8. 150
9. 180
10. 70
11. 87.5
12. 160
13. $120,000
14. $96,000
15. 100%
16. 50%
17. 1,625 rose bushes
18. 225%
19. $3,440
20. 64%

Chapter 6
1. a. $3,480,
   b. $6,480

2. a. $4,300
   b. $6,800
3. a. $2,601
   b. $7,101
4. a. $6,926
   b. $6,926
5. $6,000
6. $2,550
7. $7,750
8. $3,300
9. $1,400
10. $6,900
11. $4,250
12. $5,500
13. $8,550
14. $33,910
15. $3,210
16. $25,256

Chapter 7
1. a. $130
   b. $520
2. a. $360
   b. $168
   c. $672
3. a. 60%
   b. $525
4. a. 75%
   b. 90%
   c. $1,080
5. a. 60%
   b. 80%
   c. 90%
   d. 56.8%
6. a. Aug 4
   b. Aug 24
   c. $17.49
   d. $857.06
7. a. Jan. 2
   b. Feb. 11
   c. 97%
   d. $1,787.15
8. a. $10,204.08
   b. $6,335.92

Chapter 8
1. a. $43.35
   b. $207.83
   c. $1,570
   d. $572.63

2. a. $250
   b. $750
3. a. $23.40
   b. $59.40
4. a. 160%
   b. $360
5. a. 140%
   b. $231
6. a. 200%
   b. $420
7. a. 140%
   b. $70
8. a. $240
   b. 100%
9. a. $400
   b. 25%
10. a. $72
    b. $168
11. a. $36
    b. $108
12. a. 60%
    b. $744
13. a. 25%
    b. $132
14. a. 40%
    b. $2,400
15. a. 75%
    b. $48
16. a. $320
    b. 40%
17. a. $2,250
    b. 60%
18. a. $10
    b. 25%
    c. 20%

Chapter 9
1. a. B
   b. D
   c. A
   d. D
   e. C
   f. C
   g. D
   h. D
2. Bank Balance $10,961.65
   + Deposit
     in transit   1,850.15
                  12,811.80
   − O/S checks   342.90
   Adj. Bank
     Balance   12,468.90

Book Balance $12,583.40
+ Interest   52.50
+ Error   3.00
                12,638.90
− Svc Ch   200.00
+ 300 NSFV   150.00
Adj. Book
  Balance   12,468.90

Chapter 10
1. a. Gross pay = $712.50
   b. Social Security = $ 44.18
      Medicare = $ 10.33
   c. FIT withheld = $ 89.57
   d. Net pay = $568.42
2. a. Percentage
      method = $ 42.77
      Wage-bracket
      method = $ 44.00
   b. Percentage
      method = $ 55.36
      Wage-bracket
      method = $ 55.00
3. Jan. $1,260.35;
   Feb. $1,198.35;
   Mar. $888.35
4. Social Security, $7,688;
   Medicare $1,798, Federal
   income tax, $7,800;
   Total, $17,286
5. Social Security, 111.60;
   Medicare, $94.25;
   Total, $205.85
6. $614.08; $532.00; $464.40

Chapter 11
1. Choose A because the cost
   is less than B.
2. a. 1.5%
   b. $4,200, $2,322
3. a. $443.50
   b. 295.67
4. $27,300
5. $15,500
6. $11,600
7. $24,650
8. $9,725

Chapter 12
1. Jim's insurance pays $5,300,
   Jim's medical expenses.
   Joshua's insurance pays -0-.

2. $313.20
3. $2,695
4. $29.250
5. $30,000
6. $4,389
7. $3,255
8. $1,440

Chapter 13
1. a. $75.60
   b. $74.56
   c. $1.04
2. a. $140.00
   b. $138.08
   c. $1.92
3. a. $114.94
   b. $120.00
   c. $5.06
4. a. $58.98
   b. $60.00
   c. $1.02
5. $1,500
6. 5%
7. 219 days
8. $2,512.50
9. $289.97

Chapter 14
1. a. 9.0%
   b. 7.2%
   c. 14.4%
   d. 4.8%
2. a. 0.5%
   b. 1.25%
   c. 1.1%
   d. 0.8%
3. a. $26.72
   b. $2,387.35
4. a. $30.00
   b. $1,030.00
   c. $2,000.00
   d. $2,000.00
   e. $20.00
   f. $1,020.00
   g. $1,000.00
   h. $1,000.00
   i. $10.00
   j. $1,010.00
5. 12%
6. $1,158.77
7. a. $30.00
   b. $990.07

c. $2,009.93
d. $2,009.93
e. $20.10
f. $999.97
g. $1,009.96
h. $1,009.96
i. $10.10
j. $1,020.96
k. $1,020.66

Chapter 15
1. a. Feb. 7, 2007
   b. $3551.04
2. a. 151 days
   b. $4,510.73
3. a. Jan. 6, 2008
   b. $15,255.21
4. a. 123 days
   b. $3,045.27
5. a. $77.85
   b. $3,037.85
   c. September 12
   d. 59 days
   e. $73.66
   f. $2,964.19
6. a. $3,100
   b. February 8
   c. 60 days
   d. $61.15
   e. $3,038.85
7. a. $135.00
   b. $4,365
   c. 9.28%
8. a. $32.00
   b. 20 days
   c. $8.59
   d. $23.41

Chapter 16
1. a. $4,786.72
   b. $786.72
   c. $20,892.24
   d. $8,892.24
   e. $51,608.60
   f. $31,608.60
   g. $21,226.40
   h. $13,226.40
2. a. $21,320.40
   b. $8,679.60
   c. $2,340.72
   d. $3,659.28
   e. $10,479.15

f. $4,520.85
g. $29,698.80
h. $10,301.20
3. $7,927.74
4. $4,997.88
5. $6,691.12
6. $4,104.25

Chapter 17
1. 80
2. a. 86,371; 352,129
   b. 87,562.50; 350,937.50
   c. 83,125; 355,375
3. $346,000
4. a. $38,600
   b. $271,800
   c. 7.04 times

Chapter 18
1. a. 12.5%
   b. 25%
   c. 25%
   d. 40%
2. $\frac{4}{10}, \frac{3}{10}, \frac{2}{10}, \frac{1}{10}$
3. Declining Balance
4. a. $9,000
   b. $71,000
   c. $3,000
   d. $.90/hr
   e. $2,124
5. a. $9,000
   b. $13,500
6. a. $9,280
   b. $16,620
7. a. $1,040.00
   b. $1,487.00

Chapter 19
1. a. $285,000
   b. $382,000
   c. 14.84%; 1.49%; 34.04%
2. 30.34%; $139,650
3. 16.34%
4. a. 16.67% increase
   b. 25.00% decrease
   c. can't be calculated
   d. 0% no change
   e. can't be calculated
5. a. 1.65:1
   b. 1.08:1

c. 1.05 times
   d. 17.72%
   e. 10.79%
   f. 56.44%

Chapter 20
1. 622.05
2. $0.36
3. 3,442.50
4. $84.70
5. $47,058.82 or $47,058 if 300,000 × .15686 is used as calculation
6. $5,305.17 less
7. $86,706.90 ($86,705.20).
8. $69,365.52 ($69,364.16)
9. $979.20
10. $2,170.
11. 3.784 liters
12. 113.70 miles farther

Chapter 21
1. a. 2,880,000 shares
   b. $82.45
   c. Boeing $48.22 − $2.21 = $46.01
      Chevron $82.45 + $1.16 = $83.61
   d. $58 − $41 = $17
   e. $82.45 ÷ 18 = $4.58
2. a. $27,015
   b. $17,716.25
   c. $5,460
3. a. $400.10 gain
   b. $400.10 ÷ $8,579.95 = 4.7%
4. a. $0.65 ÷ $17.12 = 3.8%
   b. $325 + $400.10 = $725.10 ÷ $8,579.95 = 8.5%
5. 400 × 4 = 1,600 shares
6. 400 × $20 × 8% = $640 preferred dividend
   1,600 × $0.60 = $960 common dividend
   $960 − $640 = $320 more
7. 8,000 × 50 × 7.5% = $30,000 ÷ 8,000 = $3.75/share preferred;
   $85,000 − $30,000 = $55,000 ÷ 50,000 sh = $1.10/share common
8. $30,000 × 2 = $60,000 ÷ 8,000 = $7.50/share preferred; $90,000 − $60,000 = $30,000 ÷ 50,000 = $0.60/share common

Chapter 22
1. a. $15,600 ($15,000 × 1.04)
   b. Semiannually
   c. $562.50 $\left(\$15{,}000 \times 7.5\% \times \dfrac{1}{2}\right)$
   d. $206.25 ($15,000 × 7.5% × 66 days ÷ 360)
   e. $15,806.25 ($15,600 + $206.25)
   f. Premium (104 = 4% above face value)
   g. $600 ($15,000 × 4%) or ($15,600 − $15,000)
   h. 2018
   i. 7.21% ($1,125 annual interest ÷ $15,600)
   j. 7.03%; $600 premium ÷ 12 yrs = $50 amortization
      $1,125 − $50 = $1,075 annual interest adjusted for amortization
      ($15,000 + $15,600) ÷ 2 = $15,300 average principal invested
      $1,075 ÷ $15,300 = 7.03% yield to maturity
2. a. 180 (30 × 6)
   b. $7,560 (180 share × $42)
   c. $1,860 gain; $7,560 − $5,700 ($6,000 × 95%)
   d. $6,300 (180 share × $35)
3. $10,500,000
4. MCD 1/m

Chapter 23
1. a. $163,122.89
   b. $16,122.89
   c. $475,127.60
   d. $275,127.60
   e. $1,066.39
   f. $10,804.98
   g. $977.87
   h. $4,531.12
2. a. $6,063.94
   b $936.06
   c. $92,116.78
   d. $34,383.22
   e. $2,991.94
   f. $26,862.98
   g. $2,240.89
   h. $7,226.70
3. $40,573.37
4. $407,768.36
5. $2,507.03
6. $16,663.03

Chapter 24
1. a. 63
   b. 59
   c. 57
2. a. 3
   b. 7
   c. 4
   d. 6
3. a.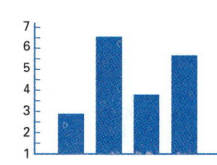
4. a.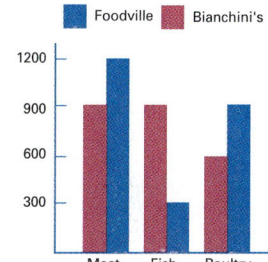
   b.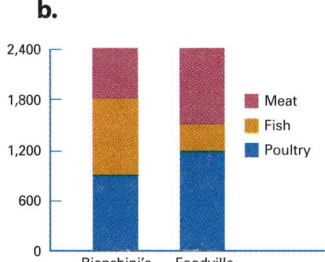
   c.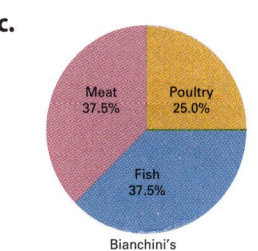

GLOSSARY

A

Account purchase. A detailed statement from the commission merchant to the principal.

Account sales. A detailed statement of the amount of the sales and the various deductions sent by the commission merchant to the consignor.

Accounts receivable. Amounts owed to a business for services performed or goods delivered.

Accrued interest. Interest earned from the last payment date to the purchase date.

Accumulated depreciation. The total of all the depreciation recognized up to a specified time.

Acid test ratio. Used to determine the amount of assets that can be quickly turned into cash to pay current liabilities; acid test ratio = total of cash plus receivables ÷ total current liabilities.

Ad valorem duty. A tax charged as a percent of the value of the item.

Addends. Any of a set of numbers to be added.

Additional death benefit (ADB). Benefits, available with some life insurance policies, that allow the insured to purchase, at a low rate per thousand dollars of coverage, additional insurance up to the full face value of the policy. In case of death of the insured by accident, both the full value of the policy and the ADB would be paid to the beneficiaries of the insured. If death occurs other than by accident, the full value of the policy is paid but no ADB is paid. Sometimes referred to as accidental death benefit.

Adjusted bank statement balance. The dollar amount obtained by adding to or subtracting from the bank statement balance checkbook activities not yet known to the bank. This amount should equal the adjusted checkbook balance.

Adjusted checkbook balance. The dollar amount obtained by adding to or subtracting from the checkbook balance those activities appearing on the bank statement that do not yet appear in the checkbook. This amount should equal the adjusted bank statement balance.

Adjusted gross income (AGI). Gross income minus certain income adjustments.

Amortization. The process by which a loan's monthly payments are always equal in dollar amount while the interest amount, which is calculated on the unpaid balance, always varies.

Amortization payment factor. A number which, when multiplied by the per $1,000 loan amount, calculates the amount of each loan payment.

Amortization schedule. A schedule of payments; the schedule shows the amount of interest and the amount of principal in each payment.

Amount credited. The total amount paid plus the amount of cash discount.

Amount of decrease. The rate of decrease times the base amount.

Amount of increase. The rate of increase times the base amount.

Annual discount amortization. Also known as the annual premium amortization, determined by dividing the discount (or premium) by the number of years from purchase to maturity.

Annual percentage rate (APR). The annual equivalent interest rate charged.

Annual premium amortization. Also known as the annual discount amortization, determined by dividing the premium (or discount) by the number of years from purchase to maturity.

Annuity. A sum of money paid out in a series of equal payments.

Annuity insurance. Life insurance that pays a certain sum of money to the insured every year after the insured reaches a specified age or until the insured's death.

Assessed valuation. A property value determined by a representative of the local or state government.

Assets. Things of value owned by a business or a person.

Auto collision insurance. Insurance that protects the vehicle of the insured against collision damage.

Auto comprehensive insurance. Insurance that protects the vehicle of the insured against fire, water, theft, vandalism, falling objects, and other damage not caused by collision.

Auto liability and property damage insurance. Insurance that protects the insured against claims resulting from personal injuries and property damage.

Automatic teller machine (ATM). A computerized electronic machine, many of which are located outside of banks and in numerous other locations, that allows customers to perform various banking functions, such as checking balances, making deposits, and withdrawing funds.

Average. A single number that is supposed to be "typical" or "representative" of the group, such as the mean, median, or mode.

Average cost method. A method of valuing inventory that is based on the assumption that the costs of all items on hand are averaged and shared evenly among all units.

Average daily balance. The sum of each day's balance divided by the number of days in the month. Payments are usually included; new purchases may or may not be included.

Average inventory. The inventory average calculated by summing each inventory valuation (determined by physical inventory) and divided by the number of physical inventories over a specified period of time; average annual inventory = (beginning inventory value + ending inventory value) ÷ 2.

Average principal. The average unpaid balance of a note or loan.

Average principal invested. Determined by adding the maturity value and the cost price and then dividing by 2.

Average unpaid balance. The sum of all of the unpaid monthly balances divided by the number of months.

B

Balance sheet. The financial statement of what is owned (assets), what is owed (liabilities), and the difference between the two (net worth) on a specific date.

Bank charge. A fee for services performed by the bank.

Bank discount. The decrease in value of a discounted note.

Bank statement. A formal accounting by a bank of the adding and subtracting activities that have occurred in one bank account over a stated period of time (usually a month).

Bar graph. Also known as a bar chart, a graphic presentation of statistical information resembling the histogram except that there may not be a numeric scale on the horizontal axis and the bars normally do not touch each other.

Base (B). The whole quantity, or 100%, of an amount.

Basic depreciation rate. A rate of depreciation determined by dividing 100% by the estimated total years of useful life of the item.

Bearer. The lender of a note.

Beginning inventory (BI). The cost of inventory on hand at the beginning of a time period.

Beneficiary. A person, a company, or an organization that benefits from an insurance policy.

Board of directors. A group of people elected by shareholders to oversee the operation of the corporation.

Bonds. Long-term notes that are bought and sold on the open market, much like stocks.

Bond ratings. Information on the presumed safety of a bond investment, provided by firms such as Standard & Poor's and based on experience and research.

Book value. The original cost of an asset minus accumulated depreciation.
Broker. A person who performs services of buying and/or selling for a commission.
Business statistics. Collections of information about businesses.

C

Callable bonds. Bonds that have a provision that the issuer can repurchase, or call in the bonds, at specified dates if the board of directors authorizes the retirement (payoff) of the bonds before their maturity date.
Cancel. "Divide out" common factors that occur in both the numerator and denominator.
Cancellation. Process of dividing out common factors.
Capital stock. The general term applied to the shares of a corporation.
Cash discount. A reduction in an invoice amount available to the buyer for paying all or part of the amount due within a stated period of time.
Cash surrender value. The amount of cash that a company will pay the insured on the surrender, or "cashing-in," of an insurance policy.
Charges. The commission and any other sales expenses, such as transportation, advertising, storage, and insurance.
Charter. A corporation's basic approval document, issued by the state, under which the corporation operates.
Check. A written order directing the bank to pay a certain sum to a designated party.
Checkbook. Checks and check stubs to record deposits, withdrawals, check numbers, dates of transactions, other additions or subtractions, and the account balance.
Check register. A place for recording important information about each transaction.
Child Tax Credit. Taxpayers with dependent children under age 17 can receive a credit of $1,000 per qualifying child. The credit phases out at higher income levels.
Classes of data. Individual values organized into groups, to more easily make sense of raw numbers.
Coinsurance clause. An insurance policy clause specifying that, if a property is not insured up to a certain percentage of its value, the owner is the bearer of part of the insurance and will not be covered for the full amount of damages.
Commercial paper. Documentation of a promise to repay a loan or pay for merchandise.
Commission. Payment to an employee or to an agent for performing or helping to perform a business transaction or service.
Commission merchant. A person who performs services of buying and/or selling for a commission.
Common denominator. A denominator that is shared by two or more fractions. The product of the denominators of two or more fractions is always a common denominator.
Common stock. The usual type of stock issued by a corporation, often with different rights compared to preferred stock.

Comparative bar graph. Two bar graphs combined on one grid, to compare two different sets of comparable data.
Complement method. A method for finding the net price.
Complement rate. A rate equal to 100% minus the discount rate; used with the complement method in determining trade or cash discounts.
Component bar graph. A bar graph constructed to show how certain data are composed of various parts.
Compound amount. Also known as the future value, the total value of an investment; equal to the principal plus all the compound interest.
Compound amount factors. Also known as future value factors, the numbers in a compound interest or future value table that are used to compute the total amount of compound interest.
Compound interest. Interest computed by performing the simple interest formula periodically during the term of the investment.
Compound interest tables. Tables of numbers, known as future value factors or compound amount factors, that can be used to compute future values (compound amounts) and compound interest.
Consignee. The party to whom a consignment shipment is sent.
Consignment. Goods from a producer to a commission merchant for sale at the best possible price.
Consignor. The party who sends a consignment.
Convertible bonds. Corporate bonds that have a provision that they may be converted to a designated number of shares or to a designated value of the corporation's stock.
Convertible preferred stock. Preferred stock that gives the owner the option of converting those preferred shares into a stated number of common shares.
Corporate bonds. Long-term notes, such as convertible bonds and callable bonds, issued by a corporation.
Corporation. A body that is granted a charter by a state legally recognizing it as a separate entity, with its own rights, privileges, and liabilities distinct from those of its owners.
Cost of goods sold. The seller's cost of items (goods) that have been sold during a certain time.
Credit. A deposit to a bank account.
Credit balance. A negative difference.
Credit card. Credit extended by a third party.
Cross-checking. Adding columns vertically and then adding these totals horizontally.
Cumulative preferred stock. Preferred stock that, if the corporation doesn't pay the specified percentage, has the unpaid amount (the dividend in arrears) carried over to the following year or years.
Current yield. The annual interest income of a bond, calculated by dividing the annual interest by the current purchase price.

D

Decimal equivalent. The presentation of a non-decimal number in decimal form.
Decimal places. The places for digits to the right of the decimal point, representing tenths, hundredths, thousandths, and so forth.
Decimal point. The period between two numerals.
Declare a dividend. A board of directors' distribution of earnings to shareholders.
Declining-balance depreciation rate. A multiple of the basic depreciation rate, such as two (double-declining-balance) or 1.5 (150%-declining-balance).
Deductible clause. An insurance policy clause that stipulates that the insured will pay the first portion of collision damage and that the insurance company will pay the remainder up to the value of the insured vehicle.
Denominator. In a fraction, the number below the line.
Dependency exemptions. Reductions to taxable income for each of one or more dependents.
Deposit slip. A written form that lists cash and checks being deposited in a bank account and cash received from the amount being deposited.
Depreciation. The decrease in the value of an asset through use.
Difference. The result of subtracting the subtrahend from the minuend.
Discount. A fee charged when someone buys the note before maturity. With regard to bonds, a bond sells at a discount if the market value becomes less than the face value.
Discount amount. The decrease in value of a discounted note.
Discount date. The last day on which a cash discount may be taken. The day on which a note is discounted (sold).
Discount period. A certain number of days after the invoice date, during which a buyer may receive a cash discount. The time between a note's discount date and its maturity date.
Discount rate. The percent used for calculating a trade or cash discount. The interest percent charged by the buyer of a discounted note.
Discounting a note. Selling a note before its maturity date.
Dividend. The number being divided.
Dividend in arrears. The unpaid amount carried over to the following year or years due to holders of cumulative preferred stock.
Divisor. The number used to divide another number.
Dollar markup. The total of operating expenses and net profit. Markup expressed as an amount rather than as a percent.
Double-declining-balance. A method that determines a depreciation amount for the first year that is approximately twice the straight-line rate.
Down payment. A partial payment made at the time of a purchase with the balance due at a later time.

Glossary 533

Due date. The final day by which time an invoice is to be paid. After that day the buyer may be charged interest. Also the date by which a loan is to be repaid.

Duty. A charge or tax often levied against imported items to protect the domestic market against foreign competition.

E

Effective interest rate. The actual annual rate of interest.

Electronic fund transfers (EFTs). Money that is transmitted electronically, primarily via computers and automatic teller machines.

Employee's earnings record. Summary by quarter of the employee's gross earnings, deductions, and net pay.

Employer's Quarterly Federal Tax Return. A tax report, filed on Form 941 every three months by all employers, that provides the IRS with details about the number of employees, total wages paid, income and FICA taxes withheld, and other figures that determine whether a tax balance is due from the company.

Ending inventory. The cost of the inventory on hand at the end of a time period.

Endowment insurance. Insurance payable upon the insured's death if it occurs within a specified period, and an endowment of the same amount as the policy, payable if the insured is alive at the end of that period.

Equation. A sentence consisting of numbers and/or letters that represent numbers, divided into two sections by an equals sign (=).

Equivalent single discount rate. A single trade discount rate that can be used in place of two or more trade discount rates to determine the same discount amount.

Estimated service life. The amount of usefulness that an owner expects to get from an item before it will need to be replaced owing to obsolescence.

Exact interest method. The calculation of interest based on the assumption that a year is 365 (or 366) days long.

Excise tax. A tax assessed on each unit, such as is levied on the sale of gasoline, cigarettes, and alcoholic beverages.

Exponent. A number written above and to the right of a number used to indicate raising to the power.

Export. The shipment of goods made in one country for sale in other countries.

Export Administration Regulations. In the U.S., the set of International Trade Administration/Department of Commerce rules and regulations that governs trade between domestic and foreign companies.

Extend credit. To give a buyer immediate possession or immediate service with payment due in the future.

Extension. When taking an inventory, the dollar amount derived by multiplying the quantity of an item by its unit price or average cost.

F

Face value. The dollar amount written on a note; it is the same as the amount borrowed, or the principal (P). With regard to corporate and government bonds, the amount that will be paid to the holder when a bond is redeemed at maturity.

Factors. Term used in multiplication to mean numbers.

Federal Insurance Contributions Act (FICA). Provides for a federal system of old-age, survivors, disability, and hospital insurance.

Federal Unemployment Tax Act (FUTA). Law that requires employers to pay the IRS an annual tax of 6.2% on the first $7,000 paid to each employee. The federal government uses the money to help fund State Employment Security Agencies, which administer unemployment insurance and job service programs.

Filing status. One of five conditions, including single, married, and married filing separate return, that a taxpayer qualifies for on Form 1040 that will determine such factors as tax rates and allowable deductions.

Finance charge. The fee that the seller charges for the privilege of buying on credit.

Financial statements. Statements presenting financial information about a company; two of these statements are the balance sheet and the income statement.

First-in, first-out (FIFO) costing method. A method of valuing inventory that assumes that costs for units used or sold are charged according to the order in which the units were manufactured or purchased.

Fixed interest rate. An interest rate that stays the same for the entire length of the loan.

Foreign trade zones. Domestic sites in the United States that are used for import and export activity and are considered to be outside U.S. Customs territory.

Form 1040. One of the basic income tax return forms filed by taxpayers.

Form W-4. The form used to inform the government of a person's marital status and to claim withholding allowances.

Fractions. Number expressions of one or more equal parts of whole units.

Frequency. The number of values in a class of data.

Frequency table. A table that summarizes the number of values in each class.

Future value. Also known as the compound amount, the total value of an investment; equal to the principal plus all the compound interest.

Future value factors. Also known as compound amount factors, the numbers in a compound interest or future value table that are used to compute the total amount of compound interest.

Future value of an annuity. The total value of a set of equal deposits into a sinking fund.

Future value of annuity factors (FVAF). Numbers used in annuity tables to compute total interest earned.

G

Government bonds. Long-term notes such as the treasury bonds issued by the federal government and the municipal bonds issued by states, cities, school districts, and other public entities.

Graduated commission rates. A system of rates by which graduated commissions increase as the level of sales increase.

Gross cost. The prime cost and all charges paid by the principal.

Gross proceeds. The price that a commission merchant gets for a consignment; also, the full sales price before any allowances, returns, or other adjustments are considered.

Gross profit method. A method of estimating inventory without a physical count or perpetual inventory system.

Group insurance. Health insurance coverage extended to a group of people. The cost for each person's coverage is less expensive than it would be under an individual policy.

Grouped data. Individual values that have been organized into data classes, as for use in a frequency table.

H

Health maintenance organization (HMO). Group health insurance coverage with limited options as a means of keeping health insurance costs lower than that of regular group policies.

Higher terms. A fraction in which both the numerator and denominator have been multiplied by the same number.

High-risk driver. A driver with a record of numerous citations or accidents.

Histogram. A diagram that presents the grouped data from a frequency table.

I

Import. Acquiring and selling goods made in a foreign country.

Improper fraction. One whole unit or more. The numerator is greater than or equal to the denominator.

Income statement. The financial statement that shows the revenues, the expenses, and the net income for a certain period of time.

Installments. Monthly payments, which for a credit sale typically include the purchase price plus credit charges.

Insured. For life insurance, the person whose life is being insured; for other types of insurance, the person who receives the benefit of the insurance.

Interest. A fee, usually charged for the use of money.

Interest-bearing note. A note that has a maturity value greater than its face value.

Interest dollars. The interest stated as an amount of money rather than as a percent.

Interest period. The period of time between the loan date and the repayment date.

Inventory sheet. A form used for recording information when taking a physical inventory.

Inventory turnover. The number of times the average inventory is converted into sales during the year.

Inventory turnover at cost. Cost of goods sold divided by average inventory for the same period computed at cost prices.

Inventory turnover at retail. Net sales divided by average inventory for the same period computed at retail prices.

Invoice. A document from a seller requesting payment from the buyer; the supplier's bill.

Invoice date. The date stated on an invoice; the beginning of the discount period.

Itemized deductions. Potential reductions to income allowed for certain payments made during the tax year.

J

Junk bond. A high-risk bond with a low rating.

L

Last-in, first-out (LIFO) costing method. A method of valuing inventory based on the assumption that the inventory on hand at the end of a period of time is composed of the units received first.

Least common denominator. The lowest shared multiple of two or more denominators.

Levy. A government charge or fee.

Liabilities. The sum total of all that a business owes at any point in time; debt.

Limited-payment life insurance. A certain premium to be paid every year for a certain number of years specified at the time of insuring, or until the death of the insured, should that occur during the specified period. The policy is payable on the death of the insured, although there may be some options available at the end of the payment period.

Line graph. A type of graph often used for illustrating data over time.

List price. The price amount listed in the catalog.

Loan value. The amount that an insured may borrow on a policy from the insurance company.

Long-term credit. Loans that are for longer than 1 year.

Lower of cost or market value (LCM). An inventory valuation method by which the lower amount of either the market value or the cost value is chosen.

Lower terms. A fraction that has been reduced by a common divisor.

Lowest terms. A fraction that cannot be reduced by any common divisor.

Low-risk driver. A driver with a long-standing, clear driving record.

M

Maker. With regard to a note, the borrower.

Market value. The dollar amount required to replace the inventory as of the inventory date.

Markup. The difference between price and a seller's cost of an item for sale. In dollars it is the amount added to the cost of the goods in order to have a gross profit high enough to cover operating expenses and to make a net profit.

Markup percent. A percent that is used to compute the amount of dollar markup by multiplication. It could be a percent that multiplies the cost to find the dollar markup; or, it could be a percent that multiplies the selling price to find that dollar markup.

Markup percent based on cost. The percent that is calculated by dividing the desired amount of dollar markup by the cost.

Markup percent based on selling price. The percent that is calculated by dividing the desired amount of dollar markup by the selling price.

Markup rate. Markup percent.

Maturity date. The final day of a note on which the borrower (the maker of the note) pays the face value and any interest due to the holder of the note. The due date.

Maturity value (MV). For an interest-bearing note, it is the sum of the face value (principal) and the interest dollars: $MV = P + I$.

Mean. An average of a group of values, computed by dividing the sum of the group of values by the number of values in the group.

Median. An average of a group of values, computed by arranging the numbers in numerical order and finding the middle number.

Metric system. The decimal system of weights (grams, kilograms, etc) and measures (meters, kilometers, etc.) used in most countries of the world, with the major exception of the U.S.

Mill. One tenth of one cent, or $0.001; a tax rate may be expressed in mills.

Minuend. Number from which subtraction is being made.

Mixed decimal. A number containing a decimal point and both a whole-number part and a decimal part.

Mixed number. A number that represents more than one whole unit by combining a whole number and a proper fraction.

Mode. An average of a group of values, computed by identifying the number that occurs most often.

Modified Accelerated Cost Recovery System (MACRS). The accelerated depreciation method required by the IRS.

Mortgage. A loan, usually amortized over 15 to 30 years, used to purchase a home.

Multiplicand. The factor that is multiplied.

Multiplier. The factor that indicates how many times to multiply.

Municipal bonds. Long-term notes issued by states, cities, school districts, and other public entities.

N

Negotiable promissory note. A promissory note that may be sold to a third party.

Net price. The price that a distributor will charge a customer after any trade discounts have been subtracted from the list price.

Net proceeds. The amount sent to the consignor as a result of consignment sales; gross proceeds minus charges.

Net purchase amount. The price of the merchandise actually purchased, including allowances for returns and excluding handling and other costs.

Net revenue. Total revenue less any returns and allowances; frequently called net sales.

Net sales. Total sales for the time period minus sales returned and adjustments made during the same time.

Net worth. The difference between what a business owns (its assets) and what it owes (its liabilities). Also known as owners' or stockholders' equity.

No-fault insurance. Insurance coverage under which the driver of each vehicle involved in an injury accident submits a claim to his or her own insurance company to cover medical costs for injuries to the driver and passengers in that person's own vehicle. The insurance does not cover damage to either vehicle involved in an accident.

No-par stock. Stock issued without par value.

Non-interest-bearing promissory note. A note having a maturity value equal to its face value.

Number of compounding periods (n). The number of compounding periods per year times the number of years of the loan.

Numerical sentence. A mathematical or logical statement, such as an equation, expressed in numbers and symbols.

Numerator. In a fraction, the number above the line.

O

Obsolescence. Becoming out-of-date.

Odd lot. Shares of stock for sale, consisting of any number of shares less than 100.

Odd-lot differential. A small extra charge, commonly added to the round-lot price, when odd lots are purchased.

Of. "Multiply," particularly when "of" is preceded by the Rate and followed by the Base.

150%-declining-balance. A method that determines a depreciation amount for the first year that is approximately one and one-half the straight-line rate.

Ordinary annuity. An annuity in which the payments occur at the end of each period.

Ordinary interest method. The calculation of interest based on the assumption that a year is 360 days long.

Original cost. The cost of building or buying an asset and getting it into use.

Outstanding check. One that has been written but hasn't yet cleared the bank and been charged to the customer's account.

Outstanding deposit. A credit that hasn't yet been recorded by the bank.

Glossary **535**

Overhead costs. General costs not directly related to sales merchandise.

P

Par. A value assigned the shares of capital stock and stated on the stock certificate.

Payee. Party to whom a check is written.

Payroll register. A summary of wages earned, payroll deductions, and final take-home pay.

Percentage (P). A portion of the Base.

Percentage method. One of two primary methods for calculating the amount of income tax to withhold from employee paychecks. After the total withholding allowance is subtracted from an employee's gross earnings, the amount to be withheld is determined by taking a percentage of the balance. The percentage to be used is specified by the IRS.

Period. The unit of time of the compounding.

Periodic interest rate (i). The rate of interest charged each period.

Perpetual inventory. A running count of all inventory units and unit costs based on a physical tracking of every item as it comes into and goes out of inventory.

Personal exemptions. Reductions to taxable income for the primary taxpayer and a spouse.

Physical inventory. An actual counting of the inventory.

Pie chart. Also known as a circle graph, a graphic presentation of statistics resembling a component bar graph because it shows how one quantity is composed of different parts.

Power. The number of times as indicated by an exponent that a number is multiplied by itself.

Preferred provider organization (PPO). Group health insurance coverage with benefits based on use of contracted providers as a means of keeping health insurance costs lower than that of regular group policies.

Preferred stock. A type of stock issued by corporations, which gives holders a right to share in earnings and liquidation before common shareholders do.

Premium. Fee for insurance coverage, usually paid every year by the insured person. The difference between a bond's par value and its market value when the market value is more. When bonds are sold at a premium, the yield rate will be lower than the stated (face) rate.

Present value. The amount needed to invest today to reach a stated future goal, given a certain rate of return.

Present value factors (PVF). The numbers in a present value factors table that are used to compute present value.

Present value of an annuity. The current value of a series of future payments.

Present value of annuity factor (PVAF). The numbers in a present value annuity factors table that are used to compute present value and total interest earned.

Price/earnings ratio (P/E). A measure of a stock's value, based on the per-share earnings as reported by the company for the four most recent quarters.

Prime cost. The price that commission merchants pay for the merchandise when they purchase goods for their principals.

Principal. The person (client) for whom a service is performed. Amount that is borrowed using credit.

Proceeds. The amount that a seller receives from the buyer of a note being discounted; the difference between the maturity value and the discount amount. In a stock transaction, the proceeds received by the seller are equal to the selling price minus the commission.

Product. The answer to a multiplication problem.

Promissory note. An agreement signed by the borrower that states the conditions of a loan.

Proper fraction. Smaller than one whole unit. The numerator is smaller than the denominator.

Property insurance. Insurance against loss of or damage to property.

Property tax. A tax on real estate or other property owned by the business or an individual.

Purchases (P). Those goods for sale that have been acquired during the current time period.

Pure decimal. A number with no whole-number part.

Q

Quotient. The answer to a division problem.

R

Rate (R). The stated or calculated percent of interest.

Rate (percent) of decrease. The negative change in two values stated as a percent.

Rate (percent) of increase. The positive change in two values stated as a percent.

Rate of return on investment. A rate that approximates the interest rate that owners are earning on their investment in a company; rate of return on investment = net income ÷ owner's equity.

Rate of yield. From an investment in stock, the ratio of the dividend to the total cost of the stock.

Rate of yield to maturity. The rate of interest investors will earn if they hold a bond to its maturity date.

Ratio. The relation of one amount to another.

Ratio of accounts receivable to net sales. Indicates the percentage of sales that have not yet bean paid for by customers; ratio of accounts receivable to net sales = accounts receivable ÷ net sales.

Reconciliation of the bank balance. Comparison of the check stubs or check register with the bank statement to determine the adjusted bank balance.

Recovery amount. The maximum amount that an insurance company will pay on a claim.

Relationship of net income to net sales. This ratio indicates the portion of sales that is income; relationship of net income to net sales = net income ÷ net sales.

Remainder. A part of a dividend that is left after even division is complete. The leftover part of division into which the divisor cannot go a whole number of times.

Remittance. Amount that a buyer actually pays after deducting a cash discount.

Round lot. A unit of stocks for sale, usually 100 shares.

Rounding off. Rounding up or down.

S

Sales tax. A government charge on retail sales of certain goods and services.

Scrap value (SV). The amount the owner of an asset expects to receive upon disposing of it at the end of its estimated service life.

Series of discounts. Two or more trade discount rates available to a buyer for different volume purchases.

Short rates. Insurance premium rates charged for less than a full term of insurance.

Short-term credit. Loans that are 1 year or less in length.

Simple interest. The fundamental interest calculation.

Sinking fund. A fund of deposits made by the issuer of a corporate or government bond and managed by a neutral third party in order to ultimately pay off a bond.

State Unemployment Tax Act (SUTA). Any of various laws passed by states that require the employer to pay a tax, such as 5.4% on the first $7,000 paid to each employee, used to help fund unemployment programs.

Statistics. A field of study that includes the collection, organization, analysis, and presentation of data.

Stock certificate. A paper document that establishes ownership of a stock.

Stock exchanges. Formal marketplaces, such as the New York Stock Exchange and the National Association of Securities Dealers Automated Quotations, that are set up for the purpose of trading stocks.

Stock transactions. The purchase and sale of stocks.

Stockbroker. An agent who handles stock transactions for clients.

Straight (or ordinary) life insurance. Insurance requiring a certain premium to be paid every year until the death of the insured person. The policy then becomes payable to the beneficiary.

Straight-line (SL) method. A depreciation method that distributes the depreciable cost of an item in equal amounts to designated units or periods covering its useful life; (original cost − scrap value) ÷ estimated total life in units or periods of time = depreciation amount for 1 unit or period.

Subtrahend. Number being subtracted.

Sum. The total of two or more addends.

Sum-of-the-years-digits (SYD) method. A depreciation method based on the assumption that greater use (and greater productivity) occurs in the earlier years of an asset's life; the rate of depreciation is greater than the straight-line method but less than the declining-balance method in the earlier years.

T

Tax rate. The percent used to calculate a tax.

Tax Rate Schedules. Tables formulated by the IRS to compute, depending upon filing status, the tax owed for various levels of taxable income.

Taxable income. The amount of income on which the income tax is determined.

Term insurance. Insurance protection issued for a limited time. A certain premium is paid every year during the specified time period, or term. The policy is payable only in case of death of the insured during the term. Otherwise, neither the insured nor the specified beneficiaries receive any payment, and the protection stops at the end of the term.

Term of the loan (or note). The period of time between the loan date and the repayment date.

Terms of payment. A statement on the invoice that informs the buyer of any available discount rate and discount date as well as the due date.

Time (T). Stated in terms of all or part of a year, the length of time used for calculating the interest dollars, the rate, or the principal.

Time line. A line representing time onto which marks are placed to indicate the occurrence of certain activities.

Total cost (for purchaser of stock). The purchase price of the stock plus a brokerage fee.

Trade discounts. Discounts given to buyers that generally are based on the quantity purchased.

Treasury bonds. Bonds issued by the United States government.

Truth in Lending Act. A federal law to assist consumers in knowing the total cost of credit.

U

Ungrouped data. Numbers listed individually.

Units-of-production method. A method for determining depreciation that distributes depreciation based on how much the asset is used.

V

Variable-rate loans. Loans that permit the lender to periodically adjust the interest rate depending on current financial market conditions.

W

Wage-bracket method. One of two primary methods for calculating the amount of income tax to withhold from employee paychecks. This method starts by granting a deduction for each withholding allowance claimed. The amount for each withholding allowance is provided by the IRS in a table. This method involves use of a series of wage-bracket tables published by the IRS.

Withholding allowance. An amount claimed on tax Form W-4 by an employee that determines how much income tax the employer will withhold from each paycheck. Each allowance claimed (as for a spouse or dependents) reduces the amount of income tax withheld.

Working capital. The amount of current assets less current liabilities.

Working capital ratio. The amount of current assets that would remain if all a company's current liabilities were paid immediately; total current assets ÷ total current liabilities.

Y

Yield. Income from an investment; generally stated as a percent, or rate.

INDEX

A

Accidental death benefit, 235
Account,
   purchase, 112
   sales, 111
Addends, 4
Adding, decimal numbers, 51–52
Addition,
   checking, 5–6
   of decimal numbers, 5–6
   equations, 74
   of fractions and mixed numbers, 30–33
   horizontal, 6
   number combinations, 4
   repeated digits, 5
   of two-digit numbers, 5
Additional death benefit, 235
Adjusted,
   bank balance, 164
   checkbook balance, 164
   gross income, 211
Adjustments to Income section, 211
Ad valorem duty, 409
Aggie Office Supply, 109
Amortization, *274*
   payment factor, 278–81
   schedule, 282
Amortizing a loan, 278–81
   computing a monthly payment, 278–79, 473
   loan payment schedule, 280–81, 474–75
   steps to create a schedule, 280–81
Amount credited, 129
Annual discount (or premium)
   amortization, 451
Annual percentage rate, 271–72
Annuity. *See also* Calculators
   computing the future value of an, 462
   computing the present value of an, 468–493t
      formula for present value, 469–70
      using a calculator for, 470
   computing regular payments of an, from the future value, 466–67
   computing regular payments of an, from the present value, 471–72
      using a calculator for, 472
   future value of annuity factors, 464, 490t–492t
   future value of an annuity formula, 464
   ordinary, 462–63
   present value of an, 462
   sinking funds, 467–68
   steps to use the table to compute future value and total interest earned, 464
   steps to use the table to compute present value and total interest earned, 469

   tables, 463
   using calculators to compute annuity factors, 465–66
   various payment periods, 464
Annuity insurance, 235
Asia-Pacific Tours, 112
Assessed valuation, 204–05
Assets, 384
Athlete's World, 140–146
Auto,
   comprehensive insurance, 230
   insurance, 230–35
   liability and property damage insurance, 230
Automated teller machine, 159
Average, 496
   daily balance, 270
   principal invested, 451
   unpaid balance, 275

B

Balance sheets, analyzing, 384–85
Bank,
   charge, 161
   discounting, *274*, 296, 303–04
   statements, 161
Bar chart. *See* Bar graph
Bar graph, 501–504
Base, finding, 90–91
Basic depreciation rate, 366
Bayside Coffee Shop, 92
Beneficiary, 235
Board of directors, 430
Bond ratings, 447
Bonds,
   accrued interest on, 449
   commissions for buying and selling, 449
   computing annual interest on, 447–48
   corporate, gains and losses on, 446–47
   definition and types of, 446
   interest rate, 448
   junk, 447
   newspaper information on, 448
   prices of, 448
   printed reports, 448–49
   rate of yield for, 450–52
   rating, 448
Book value, 365
"Borrow 1", *35*
Broadway Motors, 126–128
Broker, 108
Budget, monthly and year-to-date comparison, 388
Burger King, 122, 496
Business operating ratios,

   acid test ratio, 390
   inventory turnover, 391
   rate of return on investment, 391–92
   ratio of accounts receivable to net sales, 390
   relationship of net income to net sales, 391
   working capital ratio, 389
Business statistics, 496

C

Calculators,
   and exponents, 319
   use of in interest applications, 253
Calculators (*continued*)
   using, to compute annuity factors, 465–66
   using a, to compute the periodic payment in an annuity, 472
   using a, to compute the present value of an annuity, 470–71
   using the Texas Instruments BA II Plus for annuity
      calculations, 475
         additional annuity keys, 475–77
         basic annuity keys, 475
Callable bonds, 446
Capital stock, 426
Cash discounts, 126–130, 305–06
   for partially paid invoices,
      steps to compute the unpaid balance, 129
   for fully paid invoices,
      steps to compute, 126
Cash surrender value, 236–37
   of life insurance policy, 236–37
Charges, 111
Charter, 426
Check, 158
Checkbook, 160–161
Check register, 161
Child Tax Credit, 215
Circle graph. *See* Pie charts
Classes of data. *See* Data classes
Coinsurance,
   clause, 233
   computing it on property losses, 253
   to determine the owner's share of property loss under, 234
   for a fire insurance policy, 234
   on property, 233
Collision damage, 230
Commissions, 108
   calculating sales and purchases for principals, 108–109
   computing graduated sales, 109–111
   computing sales and purchases for principals, 111

computing when a sale involves
  returned goods, 109
  definition and terms, 108
  merchant, 111
Common,
  denominator, 33
  stock, 431
Complement method, 122–124
  to compute the remittance, 128
Complement rate, 122
Comparative bar graph, 502–03
Component bar graph, 503
Compound amount factors, 317
Compounding periods, 318–19
Compound interest, 316–17. *See also* Annuity
Computing,
  an employee's Federal and state unemployment
    tax liability, 189
  an employer's quarterly Federal tax return, 187–188
  auto insurance costs, 230
  the interest variables, 257
    finding the interest amount, principal, rate, or time, 258
  Social Security, Medicare, and other withholdings, 184–186
  special assessments, prorations, and exemptions, 207–08
Consignee, 111
Consignment, 111
Consignor, 111
Consumer Credit Protection Act of 1968, 271
Consumer Handbook to Credit Protection Laws, 271
Consumer Leasing Act of 1976, 271
Cost of goods sold, 140
Convertible,
  bonds, 446
  preferred stock, 432
Corporate bonds, 446
Corporation, 426
Cost,
  of goods sold, 351
Credit, 162
  card, 270
  offered for an interest charge, 270
  purchaser, 270
Cross-checking, 6
Cumulative stock, 431
Currency exchange rates,
  computing by country, *406*–408
  computing the effects of changes, 408
Current yield, of bonds, 450

D

Data classes, 498
Decimal,
  numbers,
    changing to percents, 88–89
    and electronic displays, 48–49
    equivalents to fractions, 56
    reading, 49–50
    reading long, 49
    shortcuts in multiplying and dividing, 58
    steps to add, 51
    steps to change a percent to a, 88–89
    steps to divide, 54
    steps to multiply, 53–54
    steps to round, 50, 58
    steps to subtract, 52–53
    using multipliers and divisors that end in zeroes, 57–58
    vs. fractions, 48
    and whole numbers, 51
  places, 49
  point, 49
Declare a dividend, 430
Declining-balance depreciation rate, 366–67
Deductible clause, 230
Deductions, 211
  tax, 211–13
Delta Marine Sales, 108
Denominator, 30
  canceling common factors in, 37
Dependency exemption, 209
Deposit slips, 158
Depreciation,
  accumulated, 365
  declining-balance method, 366
  definition, 364
  Modified Accelerated Cost Recovery System, 369–71
  partial-year, 371
  straight-line method of determining, 364
  sum-of-the-years-digits method of computing, 368–69
  units-of-production method, 365
Determining taxes due, using Standard Form 1040, 213
Discounts,
  date, 127
  on interest-bearing note, 296
  method, 122
  period, 127
  rate, 123
  when selling bonds, 447
Dividend, 11
  in arrears, 431
Divide,
  by 100, 13
  by 10, 12
  definition and terms, 11–14
  estimating, 14
  of decimal numbers, 54
  steps for fractions, mixed numbers, and whole numbers, 11–12, 38
  steps in long, 11
  when divisor and dividend end in zeroes, 13
Divisor, 11
Dollar markup, 140
  and cost,
    steps to compute from the markup percent, 144
Double-declining-balance, 366–67
Down payment, 252
Due-date, of promissory note, 296
Duties on imports, computing, 409–411

E

Eastern Restaurant Supply, 122–124
Effective interest rates, 275, 320
  daily compounding, 321
  increasing, 276
Electronic fund transfers, 159
Employee's earnings record, *186–187*
Employee's Withholding Allowance Certificate, 177
Employer's Quarterly Federal Tax Return, 187
Employer's Tax Guide, The, 179, *181–183*
Endowment insurance, 235
Equivalent single discount rate,
  steps to compute, 125
Equation, 74
Estimated service life, 364
Estimating,
  when dividing, 14
  when multiplying, 14
Excise tax, 203–04
  as an amount per unit, 203
Exemptions, on property taxes, 207–08
Exponent, 319
Export Administration Regulation, 406
Exports, 406

F

Face value,
  on bonds, 447
  of promissory note, 296
Factors, 8
Fair Labor Standards Act, 176
Federal government,
  income, from taxes, 208
  spending, 208
Federal income tax, 176
Federal Income Taxation, 209
Federal income tax withholding,
  amounts computations, 178–179
    using the percentage method, 179–184
    using the wage-bracket method, *181–183*
  steps to compute using the percentage method, 179
Federal Insurance Contributions Act (FICA), 176, 184
Federal taxes, 176
Federal Unemployment Tax Act, 189
Federal Wage and Hour Law, 176
FICA. *See* Federal Insurance Contributions Act
Filing status, 209
Finance charges, 270
Financial,
  sales taxes, 203
  statements, 384
Fixed interest rate, 281
Floyd's Appliance Store, 144
Foreign trade zones, 410
Form, 941, *187–188*
Form 941. *See* Employer's Quarterly Federal Tax Return
Form 1040, 209–14
  Line 42, 213
  remaining sections of, 213
  to determine taxable income, 209–13
  to determine taxes due, 213–16
Form 1040A, 209
Form 1040EZ, 209
Form W-4, *176–177*

Index **539**

Fractions,
  adding, 33
  bar, 30
  canceling, 32, 37
  changing to percents, 89
  decimal equivalents of, 89
  definition and vocabulary of, 30
  division of, 38
  improper to mixed numbers, 36–37
  multiplying, 36
  raising and reducing, 32
  steps to add two or more fractions and/or mixed fractions, 33
  steps to change an improper to a mixed number, 31
  steps to change a mixed number to an improper, 31
  steps to divide, 38
  steps to multiply fractions, mixed numbers, or whole numbers, 36
  steps to raise to a higher terms, 32
  steps to subtract one fraction or mixed number from another, 35
  subtracting, 34–36
  versus decimal numbers, 48
Freight charges, 127
Frequency tables, 498
  comparative bar graphs, 502
  component bar graph, 503–04
  computing the mean of large data sets, 499–500
  constructing bar graphs, 501–04
  constructing histograms, 500–01
  constructing line graphs, 504–06
  constructing pie charts, 507–08
  grouped data from, 500
  large data sets, 499–500
Function hierarchy, 72
FUTA. *See* Federal Unemployment Tax Act
Future value,
  computing present values from, 316
  factors, 317
  formula, 317–18
  steps to use the table, 317
  tables, 325–26, *338–39*

G
Gifts, inheritance, and bequests, 209
Government bonds, 446
Graduated commission rates, 109
Gross,
  cost, 112
  pay calculations, 176
  proceeds, 111
  profit method, 349
Group insurance, 237–39
  annual deductible, 238
Group medical,
  insurance, 185
  premiums, 238

H
Hart Furniture Co., 160, 163
Health maintenance organization (HMO), 237
High-risk driver, 231
Histogram, 500–02

Home Ownership and Equity Protection Act of 1994, 271
Horizontal analysis, 384

I
Imports, 406
Improper fraction, 30
Income statement, 384
  analyzing, 386–89
Income taxes,
  determining taxable income, 209
Installment purchases, 273–74
Insured, 232
Insurance,
  auto, 230–35
  life, 235–37
  medical contributions and reimbursements, 237–39
  no-fault, 230
  premium per $1,000, *236*
  property, 230, 233–35
  risk rates, 231
  short rates, 232
Interest,
  comparing ordinary and exact, 255
  computing exact, 254–55
  computing ordinary, 254–55
  computing simple, 252,256
  computing the variables, 257
  definition, 252
  dollars, 296
  estimating exact simple, 256–57
  rate, comparing discount to interest rate on a loan, 304
  rates, converting, 270
    combinations of time and interest that yield 1%, 256
    estimating exact, 256–57
    other rates and times, 256
  values per $1,000, *237*
International Trade Administration, 406
Inventory,
  average, 350
  average cost method of, 346
  estimator of value of, 349
  FIFO method of computing, 346
  LIFO method of computing, 347
  perpetual systems, 344
  sheets, 344
  turnover, 391
  turnover, computing, 350
  turnover, at retail, 351
  turnover, at cost, 351
  physical, 344
Invoice, 126
Itemized deductions, 213

J
Johnson Hardware, 165
Johnson and Johnson, 189
Joslin Realty, 92
Junk bonds, 447

L
Least common denominator, 33
Levy, 202

Liabilities, 384
Life insurance,
  computing premiums, 235–37
Limited-payment life insurance, 235
Line graphs, 504–06
List price, 122
Loan value, of a life insurance policy, 236–37
Lower of cost or market value, 347–48
Low-risk driver, 231

M
Macy's Department Store, 252
Market value, 204
Markup, 140
  computing based on cost, 141
  computing based on selling price, 144
  percent, 141–146
    to compute the cost from, 142, 145
    computing based on cost, 143
    steps to compute dollar markup and cost from, 144
    steps to compute from the selling price, 146
  rate, 141
  variables, 140
Maturity date, of promissory note, 296
Maturity value, 296
McDonald's, 122
  Quarter Pounder, 69
Mean, 496–97
  of large data sets, 499–500
Median, 497
Medical insurance contributions and reimbursements,
  computing, 237–39
Medicare, 176
  amounts, 187
  taxes, 184–185
    provides income for the Federal government, 208
Mental computations, 70
Merchandise returns, 127
Metric system, 411
Mills, 205–06
Minuend, 7
Mixed,
  decimal, 48
  number, 30
Mode, 498
Mortgage, 281–82
Multiplicand, 8
Multiplication,
  by 50, 10
  by 25, 10
  checking, 9
  of decimal numbers, 53–54
  definition and terms, 53–54
  estimating the answer, 14
  of fractions, 36
  of numbers ending in zero, 9
  of the product of two factors, 8, 10
  steps for fractions, mixed numbers, and whole numbers, 36
  when multiplier contains zero not at the end, 9
  See also Cancellation

Multiplier, 8
Municipal bonds, 446

N
National Automotive Supply, 126–128
Negotiable promissory note, 296
Net price, 122–123
   steps to compute with the discount method, 122
   steps to compute with the complement method, 123
Net,
   proceeds, 111
   purchase amount, *126*
   revenue, 386
   sales, 349, 386
   tax, 215
   worth, 384
No-fault insurance, 230
Nonprofit organizations,
   and exempt from property taxes, 207
No-par stock, 426
Numerator, 30
   canceling common factors in, 37
Numeric,
   equations,
     solving simple, 74–76
   sentence, 74

O
Obsolescence, 364
Odd lot, 429
Odd-lot differential, 429
Original cost, 364
Outstanding,
   check, 161
   deposit, 163
Overhead costs, 94
   steps to allocate based on total floor space, 95
Owner's share of property
   loss under coinsurance,
     steps to determine, 234

P
Par stock, 426
Payee, 158
Payroll,
   periods, 176
   register, 176–*178*, 185
Percentage, 90–91
   method, 179–184
Percents,
   in business, 92
   changing fractions and decimals to, 89
   changing to decimals, 88–89
   definition of, 88, 90
   and property taxes, 206
   sales tax as, 202–03
   using to allocate overhead expenses, 94–95
   using to measure increase and decrease, 92–94
   steps to change a fraction or a decimal to a, 89
Periodic interest rate, 318

Personal,
   exemptions, 209
   income taxes,
     provide income for the Federal government, 208
Pie charts, 507–08
Power, 319
Preferred provider organization, 237
Preferred stock, 431
Premiums, 230–33
   for property insurance, 233
   if the policy is cancelled, 232
   when selling bonds, 447
Present values,
   formula for computing, 322–23
   tables, 323–326, *340–41*
Prime cost, 112
Principal, 108, 252
Product, 8
   steps to approximate, 58
Promissory note,
   computing the interest period of, 297
   computing the maturity value of, 300
   determining due date of, 298–99
   discount amount, 301
   discount date on, 301
   discount period, 301
   discount rate, 301
   negotiable, 296, 300
   non-interest bearing, 302
   proceeds of, 301
   steps to compute the number of interest days between two dates, 297
Proper fraction, 30
Property,
   insurance, 233–35
   taxes,
     computing, 204–06
     definition, 204
     special assessments, prorations, and exemptions, 204, 207–08
Prorations, 207–08
Pure decimal, 48
P/Y, 476

Q
Quotient, 11
   steps to approximate, 59

R
Rates,
   percentage,
     finding, 90–91
     of increase or decrease, 92–94
   time, and distance problems, 72
Reconciliation, of bank balance, 161–164
Recovery amount, 234
Regal Meals, 122
Remainder, 11
Remittance,
   steps to compute, 126
   steps to compute when there are merchandise returns and/or freight charges, 127
   steps to compute with the complement method, 128
Retail sales taxes, 202
Rossi & Shanley Real Estate, 93
Rounding off, 50, 77

S
Sales,
   commissions,
     computing, 108
     steps to compute when a sale involves returned goods, 109
     steps to compute under a graduated rates plan, 109
   tax,
     as an amount per unit, 203
     as a percent of price, 202–03
     computing, 202
     definition of, 202–03
     excise taxes, 203
     financial sales taxes, 203
     goods and services exempt from, 202
     as percentage of price, 202
     social sales taxes, 203
     and total sales amount, steps to compute, 202
Selling price,
   computing cost from, 142, 144–145
   computing directly from cost, 141–142, 145
   computing from cost, 145
   steps to compute from the markup percent, 142
   steps to compute the markup percent from, 146
   steps to computing based on cost, 141–142
Series of discounts, 123
Short rates, 232–33
Simple interest, 252
   computing, 252
   formula for, 252
Sinking funds, 467–68
Social sales taxes, 203
Social Security, 176
   amounts, 187
   provides income for the Federal government, 208
   tax, 184–185
Space Savers, 130
Special assessments, for property, 207–08
Special payroll deductions, 185
Specialty Marketing Group, *112–113*
Standard deduction, 211–213
State,
   income taxes, 185
   taxes on cigarettes, 203
   Unemployment Tax Act, 189
Statistics, 496
Steps in long division, *11–12*
Straight (ordinary) life insurance coverage, 235–36
Subtraction,
   checking, 7
   of decimal numbers, 52–53
   of fractions, 34–36
   horizontal, 7
Subtrahend, 7
Sum, 4

SUTA. *See* State Unemployment Tax Act
Suzi's Muffins, 122–125

T

Tables for percentage method of withholding, *180*
Taxable income, 209–13
    computing, 213
    definition of, 210
    determining using Form 1040, 209
    what it does and does not include, 210
Tax,
    assessment bases, are expected to change, 209
    credits, 215–216
    rate, 202
        computing in percents and mills, 88–89, 205
            percents, 205
            mills, 205–06
        are expected to change, 209
Tax Rate Schedules, 213–15
Term insurance, 235
Terms of payment, 126
Trade discounts,
    computing, 122
    for 30-day payment series, calculating, 123–124
Taxes. *See* Income taxes, Property taxes, Sales taxes, Unemployment taxes
Truncating, *50*

U

Unemployment tax liability, 189
Ungrouped data, 498
Uniform Product Code, 203
United Food Services, 125
Unpaid balance, 129
User of calculators, in computing interest, 253

W

Wage-bracket method, *181–184*
Warner-Lambert Company, 189
Wells Fargo Bank, *158–159*, *162*
Willowbrook Farms, 111
Withholding allowance, 176–177
Word problems,
    percentage, 179–184
    rate, time, and distance, 72
    relationship problems, 76
    rounding, 77
    solving, 70–72, 74–76

Y

Yeager Manufacturing, *187*
Yield, of bonds, 450

Progress Report

Part	Chapter	Assignment	Title	Page	Date Assigned	Date Completed	Score/Grade
1	1	1.1	Addition	19			
		1.2	Subtraction	21			
		1.3	Multiplication	23			
		1.4	Division	25			
		1.5	Estimating	27			
	2	2.1	Addition and Subtraction of Fractions	43			
		2.2	Multiplication and Division of Fractions	45			
	3	3.1	Addition and Subtraction of Decimal Numbers	63			
		3.2	Multiplication and Division of Decimal Numbers	65			
		3.3	Decimal Numbers at Business	67			
	4	4.1	Word Problems, Equations, and Series	81			
		4.2	Word Problems, Formulas, and Equations	83			
2	5	5.1	Base, Rate, and Percentage	99			
		5.2	Rate of Increase and Rate of Decrease	101			
		5.3	Business Applications	103			
		5.4	Allocation of Overhead	105			
	6	6.1	Commission	117			
		6.2	Applications with Commission	119			
	7	7.1	Trade Discounts	135			
		7.2	Cash Discounts	137			
	8	8.1	Markup Based on Cost	151			
		8.2	Markup Based on Selling Price	153			
3	9	9.1	Check Register and Check Stubs	169			
		9.2	Check Register and Bank Statements	171			
		9.3	Bank Balance Reconciliation Statements	173			
	10	10.1	Payroll Problems	195			
		10.2	Payroll, Earnings Record, Payroll Tax Returns	197			
	11	11.1	Sales Tax	221			
		11.2	Property Taxes	223			
		11.3	Federal Income Tax	227			
	12	12.1	Auto Insurance	243			
		12.2	Property Insurance	245			
		12.3	Life and Medical Insurance	247			
4	13	13.1	Simple Interest	263			
		13.2	Simple Interest Applications	267			

Part	Chapter	Assignment	Title	Page	Date Assigned	Date Completed	Score/Grade
	14	14.1	Monthly Finance Charges	287			
		14.2	Installment Sales and Effective Rates	291			
		14.3	Amortization and Mortgages	293			
	15	15.1	Dates, Times, and Maturity Value	309			
		15.2	Discounting Promissory Notes	311			
	16	16.1	Future Value (Compound Amount)	329			
		16.2	Present Value	333			
5	17	17.1	Inventory Cost	357			
		17.2	Inventory Estimating and Turnover	359			
	18	18.1	Business Depreciation Part 1	377			
		18.2	Business Depreciation Part 2	381			
	19	19.1	Balance Sheet Analysis	397			
		19.2	Income Statement Analysis	399			
		19.3	Financial Statement Ratio	401			
	20	20.1	Trading with Other Countries	417			
		20.2	Duties and Metric Conversion	419			
6	21	21.1	Buying and Selling Stock	437			
		21.2	Capital Stock	441			
	22	22.1	Corporate and Government Bonds	457			
		22.2	Bond Rate of Yield	459			
	23	23.1	Annuities—Future Value	481			
		23.2	Annuities—Present Value	485			
	24	24.1	Statistical Averages	513			
		24.2	Graphs and Charts	515			